Lecture Notes in Computer Science   **8781**

Commenced Publication in 1973
Founding and Former Series Editors:
Gerhard Goos, Juris Hartmanis, and Jan van Leeuwen

More information about this series at http://www.springer.com/series/7410

Antoine Joux · Amr Youssef (Eds.)

Selected Areas
in Cryptography –
SAC 2014

21st International Conference
Montreal, QC, Canada, August 14–15, 2014
Revised Selected Papers

 Springer

Editors
Antoine Joux
Fondation Partenariale de l'UPMC
Paris Cedex
France

Amr Youssef
Concordia University
Montreal, QC
Canada

ISSN 0302-9743   ISSN 1611-3349   (electronic)
Lecture Notes in Computer Science
ISBN 978-3-319-13050-7   ISBN 978-3-319-13051-4   (eBook)
DOI 10.1007/978-3-319-13051-4

Library of Congress Control Number: 2014954580

LNCS Sublibrary: SL4 – Security and Cryptology

Springer Cham Heidelberg New York Dordrecht London
© Springer International Publishing Switzerland 2014

Printed on acid-free paper

[Springer International Publishing AG Switzerland] is part of Springer Science+Business Media
(www.springer.com)

Preface

This volume contains revised versions of the papers presented at the 21st conference on Selected Areas in Cryptography (SAC 2014), held during August 14–15, 2014 at Concordia University in Montreal, Canada. The conference Selected Areas in Cryptography (SAC) series was initiated in 1994, when SAC 1994 was held at Queen's University in Kingston, Ontario, Canada. At that time, it was called the Workshop on Selected Areas in Cryptography. Since then, SAC has been held annually in various Canadian cities, including Calgary, Kingston, Montreal, Ottawa, Sackville, St. John's, Toronto, Waterloo, and Windsor. SAC is currently the only cryptography conference series that is held annually in Canada. Information on previous SAC conferences can be found at the main SAC conferences website: http://sacconference.org/.

There are four areas covered at each SAC conference. The three permanent areas are:

- Design and analysis of symmetric key primitives and cryptosystems, including block and stream ciphers, hash function, MAC algorithms, cryptographic permutations, and authenticated encryption schemes.
- Efficient implementations of symmetric and public key algorithms.
- Mathematical and algorithmic aspects of applied cryptology.

This year, the fourth area for SAC 2014 is: Algorithms for cryptography, cryptanalysis, and their complexity analysis.

We greatly appreciate the hard work of the SAC 2014 Program Committee. We are also very grateful to the many others who participated in the review process. This year, we received a total of 103 submissions (co-authored by 260 authors from 30 countries), 22 of them were accepted for presentations at the conference. The 36 Technical Program Committee members were from 13 countries and involved 92 external reviewers. On average, each submitted paper was reviewed by about 3.8 TPC members.

The program also included three invited talks: Nigel Smart, from the University of Bristol, UK, presented a talk entitled "Practical Multi-party Computation." Pierrick Gaudry, from Université de Lorraine, France, presented a talk entitled "NFS: Similarities and Differences Between Integer Factorization and Discrete Logarithm." The Stafford Tavares Lecture was dedicated to the memories of Scott Vanstone and was given by Alfred Menezes from the University of Waterloo. The talk was entitled "Scott Vanstone and the Early Years of Elliptic Curve Cryptography."

SAC 2014 was generously supported by Microsoft Research. We would also like to thank Springer for publishing the SAC proceedings series since 1998 in the Lecture Notes in Computer Science series. Last, but not least, we are very grateful to the staff members at the Concordia Institute for Information Systems Engineering (CIISE) for their tireless work in taking care of the local arrangements.

August 2014

Antoine Joux
Amr Youssef

Organization

Program Committee

Jean-Philippe Aumasson	Kudelski Security, Switzerland
Daniel J. Bernstein	University of Illinois at Chicago, USA and Technische Universiteit Eindhoven, The Netherlands
John Black	CU Boulder, USA
Céline Blondeau	Aalto University School of Science, Finland
Christina Boura	Université de Versailles Saint-Quentin-en-Yvelines, France
Anne Canteaut	INRIA, France
Carlos Cid	Royal Holloway, University of London, UK
Joan Daemen	STMicroelectronics, Belgium
Orr Dunkelman	University of Haifa, Israel
Pierre-Alain Fouque	University of Rennes 1 and Institut Universitaire de France, France
Steven Galbraith	University of Auckland, New Zealand
Joachim von zur Gathen	University of Bonn, Germany
Guang Gong	University of Waterloo, Canada
Robert Granger	EPFL, Switzerland
Michael Jacobson	University of Calgary, Canada
Antoine Joux	Fondation Partenariale de l'UPMC, LIP6, France
Pascal Junod	HEIG-VD, Switzerland
Gregor Leander	Ruhr-Universität Bochum, Germany
Arjen Lenstra	EPFL, Switzerland
Stefan Lucks	Bauhaus-University Weimar, Germany
David M'Raihi	Perzo, USA
Alexander May	Ruhr-Universität Bochum, Germany
Shiho Moriai	NICT, Japan
María Naya-Plasencia	INRIA, France
Kaisa Nyberg	Aalto University School of Science, Finland
Christiane Peters	European Network for Cyber Security, The Netherlands
Michaël Quisquater	Université de Versailles Saint-Quentin-en-Yvelines, France
Christian Rechberger	Technical University of Denmark, Denmark
Palash Sarkar	Indian Statistical Institute, India
Yu Sasaki	NTT Corporation, Japan
Douglas Stinson	University of Waterloo, Canada
Emmanuel Thomé	INRIA, LORIA, France

Frederik Vercauteren   KU Leuven – ESAT/COSIC, Belgium
Marion Videau   Université de Lorraine, France
Vanessa Vitse   Institut Fourier, University of Grenoble, France
Amr Youssef   Concordia University, Canada

Additional Reviewers

Abdel Khalek, Ahmed	Jetchev, Dimitar	Rodríguez-Henríquez,
Abed, Farzaneh	Jovanovic, Philipp	Francisco
Afshar, Arash	Järvinen, Kimmo	Roy, Arnab
Albrecht, Martin	Karpman, Pierre	Scheidler, Renate
Altawy, Riham	Kleinjung, Thorsten	Seurin, Yannick
Aoki, Kazumaro	Knellwolf, Simon	Shallue, Andrew
Aranha, Diego de Freitas	Koelbl, Stefan	Shibutani, Kyoji
Ashur, Tomer	Krasnova, Anna	Shimoyama, Takeshi
Babbage, Steve	Labrande, Hugo	Sibborn, Dale
Bai, Shi	Lallemand, Virginie	Singh, Shashank
Bauer, Aurélie	Lange, Tanja	Sinha Roy, Sujoy
Becker, Anja	Leurent, Gaëtan	Soleimany, Hadi
Bertoni, Guido	List, Eik	Stehle, Damien
Blazy, Olivier	Loebenberger, Daniel	Steinberger, John
Blömer, Johannes	Longa, Patrick	Suder, Valentin
Bogdanov, Andrey	Meier, Willi	Szepieniec, Alan
Bos, Joppe	Mendel, Florian	Takahashi, Junko
Castryck, Wouter	Mennink, Bart	Tan, Yin
Chakraborty, Debrup	Miele, Andrea	Thomae, Enrico
Chen, Yao	Murphy, Sean	Tibouchi, Mehdi
Chuengsatiansup,	Naehrig, Michael	Tiessen, Tyge
Chitchanok	Neves, Samuel	Tischhauser, Elmar
Derbez, Patrick	Niederhagen, Ruben	Van Assche, Gilles
Detrey, Jérémie	Nüsken, Michael	Wang, Lei
Dimitrov, Vassil	Onete, Cristina	Wenzel, Jakob
Dudeanu, Alina	Oswald, Elisabeth	Whitnall, Carolyn
Fan, Xinxin	Peeters, Michael	Wu, Teng
Feix, Benoit	Preneel, Bart	Yang, Bo-Yin
Forler, Christian	Procter, Gordon	Yang, Bohan
Gama, Nicolas	Quedenfeld, Frank	Yasuda, Kan
Granger, Rob	Raddum, Håvard	Yu, Mandel
Handschuh, Helena	Ramanna, Somindu	Zapalowicz,
Hermans, Jens	Reparaz, Oscar	Jean-Christophe
Isobe, Takanori	Roche, Thomas	Ziegler, Konstantin
Janson, Christian		

Contents

Malicious Hashing: Eve's Variant of SHA-1

Ange Albertini[1], Jean-Philippe Aumasson[2], Maria Eichlseder[3(✉)],
Florian Mendel[3], and Martin Schläffer[3]

[1] Corkami, Ravensburg, Germany
ange.albertini@gmail.com
[2] Kudelski Security, Cheseaux-sur-Lausanne, Switzerland
jeanphilippe.aumasson@gmail.com
[3] Graz University of Technology, Graz, Austria
{maria.eichlseder,florian.mendel,martin.schlaeffer}@iaik.tugraz.at

Abstract. We present collisions for a version of SHA-1 with modified
constants, where the colliding payloads are valid binary files. Examples are given of colliding executables, archives, and images. Our malicious SHA-1 instances have round constants that differ from the original ones in only 40 bits (on average). Modified versions of cryptographic standards are typically used on closed systems (e.g., in pay-TV, media and gaming platforms) and aim to differentiate cryptographic components across customers or services. Our proof-of-concept thus demonstrates the exploitability of custom SHA-1 versions for malicious purposes, such as the injection of user surveillance features. To encourage further research on such malicious hash functions, we propose definitions of malicious hash functions and of associated security notions.

1 Introduction

In 2013, cryptography made the headlines following the revelation that NSA may not only have compromised cryptographic software and hardware, but also cryptographic algorithms. The most concrete example is the "key escrow" [11] or "master key" [5] property of the NSA-designed Dual_EC_DRBG [27]. The alleged backdoor is the number e such that $eQ = P$, where P and Q are two points on the elliptic curve specified as constants in Dual_EC_DRBG. Knowing e allows one to determine the internal state and thus to predict all future outputs. Despite other issues [7,35] (see also [16,17]), Dual_EC_DBRG was used as default pseudorandom number generator in EMC/RSA's BSAFE library, allegedly following a $10M deal with NSA [24].

It is also speculated that NSA may have "manipulated constants" [34] of other algorithms, although no hard evidence has been published. This series of revelations prompted suspicions that NIST-standardized cryptography may be compromised by NSA. It also raised serious doubts on the security of commercial cryptography software, and even of open-source software. Several projects have been started to address those concerns, like #youbroketheinternet [43] and the Open Crypto Audit Project [29].

© Springer International Publishing Switzerland 2014
A. Joux and A. Youssef (Eds.): SAC 2014, LNCS 8781, pp. 1–19, 2014.
DOI: 10.1007/978-3-319-13051-4_1

Research on cryptographic backdoors and malicious cryptography appears to have been the monopoly of intelligence agencies and of industry. Only a handful of peer-reviewed articles have been published in the open literature (see Sect. 2.1), whereas research on related topics like covert-channel communication (e.g., [26,36]) or hardware trojans (e.g., [2,20]) is published regularly.

Malicious ciphers have been investigated by Young and Yung [44] in their "cryptovirology" project, and to a lesser extent by Rijmen and Preneel [33] and Patarin and Goubin [30]. However we are unaware of any research about *malicious hash functions*—that is, hash functions designed such that the designer knows a property that allows her to compromise one or more security notions. Note that we distinguish backdoors (covert) from trapdoors (overt); for example VSH [9] is a trapdoor hash function, such that collisions can be found efficiently if the factorization of the RSA modulus is known.

This paper thus investigates malicious hash functions: first their definition and potential applications, then a proof-of-concept by constructing a malicious version of SHA-1 with modified constants, for which two binary files (e.g., executables) collide. We have chosen SHA-1 as a target because it is (allegedly) the most deployed hash function and because of its background as an NSA/NIST design. We exploit the freedom of the four 32-bit round constants of SHA-1 to efficiently construct 1-block collisions such that two valid executables collide for this malicious SHA-1. Such a backdoor could be trivially added if a new constant is used in every step of the hash function. However, in SHA-1 only four different 32-bit round constants are used within its 80 steps, which significantly reduces the freedom of adding a backdoor. Actually our attack only modifies at most 80 (or, on average, 40) of the 128 bits of the constants.

Our malicious SHA-1 can readily be exploited in applications that use custom hash functions (e.g., for customers' segmentation) to ensure that a legitimate application can surreptitiously be replaced with a malicious one, while still passing the integrity checks such as secure boot or application code signing.

Outline. Section 2 attempts to formalize intuitive notions of malicious hash functions. We first define a malicious hash function as a pair of algorithms: a *malicious generator* (creating the function and its backdoor) and an *exploit algorithm*. Section 3 then presents a novel type of collision attack, exploiting the freedom degrees of the SHA-1 constants to efficiently construct collisions. We describe the selection of a dedicated disturbance vector that minimizes the complexity and show examples of collisions. Section 4 discusses the application to structured file formats, and whether the constraints imposed by the attack can be satisfied with common file formats. We present examples of valid binary files that collide for our malicious SHA-1: executables (e.g., master boot records or shell scripts), archives (e.g., rar), and images (jpg).

2   Malicious Hashing

We start with an overview of previous work related to malicious cryptography. Then we formalize intuitive notions of malicious hashing, first with a general definition of a malicious hash function, and then with specific security notions.

2.1   Malicious Cryptography and Backdoors

The open cryptographic literature includes only a handful of works related to malicious applications of cryptography:

- In 1997, Rijmen and Preneel [33] proposed to hide linear relations in S-boxes and presented "backdoor versions" of CAST and LOKI. These were broken in [41] as well as the general strategy proposed. Rijmen and Preneel noted that "[besides] the obvious use by government agencies to catch dangerous terrorists and drug dealers, trapdoor block ciphers can also be used for public key cryptography." [33]. Indeed, [5] previously argued that a backdoor block cipher is equivalent to a public key cryptosystem.
- In 1997, Patarin and Goubin [30] proposed an S-box-based asymmetric scheme constructed as a 2-round SPN but publicly represented as the corresponding equations – keeping the S-boxes and linear transforms secret. This was broken independently by Ye et al. and Biham [3,42]. This can be seen as an ancestor of white-box encryption schemes.
- In 1998 and later, Young and Yung designed backdoor *blackbox* malicious ciphers, which assume that the algorithm is not known to an adversary. Such ciphers exploit low-entropy plaintexts to embed information about the key in ciphertexts through a covert channel [45,46]. Young and Yung coined the term *cryptovirology* [44] and cited various malicious applications of cryptography: ransomware, deniable data stealing, etc.
- In 2010, Filiol [15] proposed to use malicious pseudorandom generators to assist in the creation of executable code difficult to reverse-engineer. Typical applications are the design of malware that resist detection methods that search for what looks like obfuscated code (suggesting the hiding of malicious instructions).

Note that we are concerned with backdoors in *algorithms*, regardless of its representation (pseudocode, assembly, circuit, etc.), as opposed to backdoors in software implementations (like Wagner and Biondi's sabotaged RC4 [38]) or in hardware implementations (like bug attacks [4] and other hardware trojans).

2.2   Definitions

We propose definitions of malicious hash functions as *adversaries* composed of a pair of algorithms: a (probabilistic) *malicious generator* and an *exploit algorithm*. Based on this formalism, we define intuitive notions of undetectability and undiscoverability.

Malicious Hash Function. Contrary to typical security definitions, our adversary is not the attacker, so to speak: instead, the adversary Eve *creates* the primitive and knows the secret (i.e. the backdoor and how to exploit it), whereas honest parties (victims) attempt to cryptanalyze Eve's design. We thus define a malicious hash function (or adversary) as a pair of efficient algorithms, modeling the ability to create malicious primitives and to exploit them:

- A *malicious generator*, i.e. a probabilistic algorithm returning a hash function and a backdoor;
- An *exploit algorithm*, i.e. a deterministic or probabilistic algorithm that uses the knowledge of the backdoor to bypass some security property of the hash function.

We distinguish two types of backdoors: *static*, which have a deterministic exploit algorithm, and *dynamic*, which have a probabilistic one.

Below, the hash algorithms and backdoors returned as outputs of a malicious generator are assumed to be encoded as bitstring in some normal form (algorithm program, etc.), and to be of reasonable length. The generator and exploit algorithms, as well as the backdoor string, are kept secret by the malicious designer.

Static Backdoors Adversaries. Eve is a *static collision adversary* (SCA) if she designs a hash function for which she knows one pair of colliding messages.

Definition 1 (SCA). *A static collision adversary is a pair* (GenSC, ExpSC) *such that*

- *The* malicious generator GenSC *is a probabilistic algorithm that returns a pair* (H, b), *where H is a hash function and b is a backdoor.*
- *The* exploit algorithm ExpSC *is a deterministic algorithm that takes a hash function H and a backdoor b and that returns distinct m and m' such that* $H(m) = H(m')$.

This definition can be generalized to an adversary producing a small number of collisions, through the definition of several ExpSC algorithms $\mathsf{ExpSC}_1, \ldots, \mathsf{ExpSC}_n$.

As a static second-preimage adversary would not differ from that of static collision, our next definition relates to (first) preimages:

Definition 2 (SPA). *A static preimage adversary is a pair* (GenSP, ExpSP) *such that*

- *The* malicious generator GenSP *is a probabilistic algorithm that returns a pair* (H, b), *where H is a hash function and b is a backdoor.*
- *The* exploit algorithm ExpSP *is a deterministic algorithm that takes a hash function H and a backdoor b and that returns m such that $H(m)$ has low entropy.*

In the above definition "low entropy" is informally defined as digest having a pattern that will convince a third party that "something is wrong" with the hash function; for example, the all-zero digest, a digest with all bytes identical, etc.

Dynamic Backdoors. Dynamic backdoors extend static backdoors from one or a few successful attacks to an arbitrary number. In some sense, dynamic backdoors are to static backdoors what universal forgery is to existential and selective forgery for MACs.

Definition 3 (DCA). *A dynamic collision adversary is a pair* (GenDC, ExpDC) *such that*

- *The* malicious generator GenDC *is a probabilistic algorithm that returns a pair* (H, b), *where* H *is a hash function and* b *is a backdoor.*
- *The* exploit algorithm ExpDC *is a probabilistic algorithm that takes a hash function* H *and a backdoor* b *and that returns distinct* m *and* m' *such that* $H(m) = H(m')$.

In this definition, ExpDC should be seen as an efficient sampling algorithm choosing the pair (m, m') within a large set of colliding pairs, as implicitly defined by GenDC. The latter may be created in such a way that sampled messages satisfy a particular property, e.g. have a common prefix.

    The definitions of dynamic second-preimage and preimage adversaries follow naturally:

Definition 4 (DSPA). *A dynamic second-preimage adversary is a pair* (GenDSP, ExpDSP) *such that*

- *The* malicious generator GenDSP *is a probabilistic algorithm that returns a pair* (H, b), *where* H *is a hash function and* b *is a backdoor.*
- *The* exploit algorithm ExpDSP *is a probabilistic algorithm that takes a hash function* H, *a backdoor* b, *and a message* m *and that returns an* m' *distinct from* m *such that* $H(m) = H(m')$.

Definition 5 (DPA). *A dynamic preimage adversary is a pair* (GenDP, ExpDP) *such that*

- *The* malicious generator GenDP *is a probabilistic algorithm that returns a pair* (H, b), *where* H *is a hash function and* b *is a backdoor.*
- *The* exploit algorithm ExpDP *is a probabilistic algorithm that takes a hash function* H, *a backdoor* b, *and a digest* d *and that returns* m *such that* $H(m) = d$.

In the definitions of DSPA and DPA, the challenge values m and d are assumed sampled at random (unrestricted to uniform distributions).

    One may consider "subset" versions of (second) preimage backdoors, i.e. where the backdoor only helps if the challenge value belongs to a specific subset. For example, one may design a hash for which only preimages of short strings—as passwords—can be found by the exploit algorithm.

    Our last definition is that of a key-recovery backdoor, for some keyed hash function (e.g. HMAC):

Definition 6 (KRA). *A dynamic key-recovery adversary is a pair* (GenKR, ExpKR) *such that*

- *The* malicious generator GenKR *is a probabilistic algorithm that returns a pair* (H, b), *where* H *is a hash function and* b *is a backdoor.*

– *The* exploit algorithm ExpKR *is a probabilistic algorithm that takes a hash function H and a backdoor b and that has oracle-access to* $H_K(\cdot)$ *for some key K and that returns K.*

The definition of KRA assumes K to be secret, and may be relaxed to subsets of "weak keys". This definition may also be relaxed to model forgery backdoors, i.e. adversaries that can forge MAC's (existentially, selectively, or universally) without recovering K.

Stealth Definitions. We attempt to formalize the intuitive notions of undetectability ("Is there a backdoor?") and of undiscoverability ("What is the backdoor?"). It is tempting to define undetectability in terms of indistinguishability between a malicious algorithm and a legit one. However, such a definition does not lend itself to a practical evaluation of hash algorithms.

We thus relax the notion to define undetectablity as the inability to determine the exploit algorithm (that is, how the backdoor works, regardless of whether one knows the necessary information, b). In other words, it should be difficult to reverse-engineer the backdoor. We thus have the following definition, applying to both collision and preimage backdoors:

Definition 7. *The backdoor in a malicious hash* (Gen, Exp) *is* undetectable *if given a H returned by* Gen *it is difficult to find* Exp.

Subtleties may lie in the specification of H: one can imagine a canonical-form description that directly reveals the presence of the backdoor, while another description or implementation would make detection much more difficult. This issue is directly related to the notion of obfuscation (be it at the level of the algorithm, source code, intermediate representation, etc.). For example, malware (such as ransomware, or jailbreak kits) may use obfuscation to dissimulate malicious features, such as cryptographic components of 0-day exploits.

Furthermore, backdoors may be introduced as sabotaged versions of legitimate designs. In that case, undetectability can take another sense, namely distinguishability from the original design. For example, in our malicious SHA-1, it is obvious that the function differs from the original SHA-1, and one may naturally suspect the existence of "poisonous" inputs, although those should be hard to determine.

Undiscoverability is more easily defined than undetectability: it is the inability to find the backdoor b given the exploit algorithm. A general definition is as follows:

Definition 8. *The backdoor in a malicious hash* (Gen, Exp) *is* undiscoverable *if given* Exp *and H returned by* Gen *it is difficult to find* b.

In our proof-of-concept of a malicious SHA-1, undiscoverability is the hardness to recover the colliding pair, given the knowledge that a pair collides (and even the differential used).

3   Eve's Variant of SHA-1

As a demonstration of the above concepts, we present an example of a *static collision backdoor*: Eve constructs a custom variant of SHA-1 that differs from the standardized specification only in the values of some round constants (up to 80 bits). Eve can use the additional freedom gained from choosing only four 32-bit constants to find a practical collision for the full modified SHA-1 function during its design. We show that Eve even has enough freedom to construct a meaningful collision block pair which she can, at a later point, use to build multiple colliding file pairs of a particular format (e.g., executable or archive format) with almost arbitrary content.

The backdoor does not exploit any particular "weaknesses" of specific round constants, nor does it weaken the logical structure of the hash function. Instead, it only relies on the designer's freedom to choose the constants during the attack. This freedom can be used to improve the complexity of previous attacks [37,40] and thus makes it feasible to find collisions for the full hash function.

For an attacker who only knows the modified constants but cannot choose them, collisions are as hard to find as for the original SHA-1. Thus, in terms of the definitions of the previous section, this backdoor is *undiscoverable*. It is, however, *detectable* since constants in hash functions are normally expected to be identifiable as nothing-up-your-sleeve numbers. This is hardly achievable in our attack.

Below, we first give a short description of SHA-1 in Sect. 3.1 and briefly review previous differential collision attacks on SHA-1 in Sect. 3.2. Then, we build upon these previous differential attacks and describe how the freedom of choosing constants can be used to improve the attack complexity in Sect. 3.3.

3.1   Short Description of SHA-1

SHA-1 is a hash function designed by the NSA and standardized by NIST in 1995. It is an iterative hash function based on the Merkle-Damgård design principle [10,25], processes 512-bit message blocks and produces a 160-bit hash value by iterating a compression function f. For a detailed description of SHA-1 we refer to [28].

The compression function f uses the Davies-Meyer construction which consists of two main parts: the message expansion and the state update transformation. The message expansion of SHA-1 is a linear expansion of the 16 message words (denoted by M_i) to 80 expanded message words W_i,

$$W_i = \begin{cases} M_i & \text{for } 0 \leq i \leq 15, \\ (W_{i-3} \oplus W_{i-8} \oplus W_{i-14} \oplus W_{i-16}) \lll 1 & \text{for } 16 \leq i \leq 79 . \end{cases}$$

The state update transformation of SHA-1 consists of 4 rounds of 20 steps each. In each step, the expanded message word W_i is used to update the 5 chaining variables as depicted in Fig. 1. In each round, the step update uses different Boolean functions f_r and additive constants K_r, which are shown in Table 1.

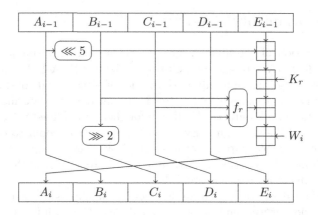

Fig. 1. The step function of SHA-1.

Table 1. The round constants K_r and Boolean functions f_r used in each step of SHA-1.

Round r	Step i	K_r	f_r
1	$0 \le i \le 19$	5a827999	$f_{\text{IF}}(B,C,D) = B \wedge C \oplus \neg B \wedge D$
2	$20 \le i \le 39$	6ed9eba1	$f_{\text{XOR}}(B,C,D) = B \oplus C \oplus D$
3	$40 \le i \le 59$	8f1bbcdc	$f_{\text{MAJ}}(B,C,D) = B \wedge C \oplus B \wedge D \oplus C \wedge D$
4	$60 \le i \le 79$	ca62c1d6	$f_{\text{XOR}}(B,C,D) = B \oplus C \oplus D$

For Eve's modified, malicious hash function, we only change the values of K_2, K_3 and K_4. The remaining definition is left unchanged. Note that the original SHA-1 constants are chosen as the square roots of 2, 3, 5 and 10:

$$K_1 = \lfloor \sqrt{2} \cdot 2^{30} \rfloor, \quad K_2 = \lfloor \sqrt{3} \cdot 2^{30} \rfloor, \quad K_3 = \lfloor \sqrt{5} \cdot 2^{30} \rfloor, \quad K_4 = \lfloor \sqrt{10} \cdot 2^{30} \rfloor.$$

3.2 Differential Attack Strategy for SHA-1

At CRYPTO 2005, Wang et al. presented the first collision attack on full SHA-1 with a theoretical complexity of about 2^{69} [40]. This was later improved to 2^{63} by the same authors [39]. Since then, several papers on the cryptanalysis of SHA-1 have been published [1,13,14,19,37]. Nevertheless, no practical collision has been shown for full SHA-1 to date.

Our practical and meaningful collision attacks on malicious SHA-1 are based on the differential attacks by Wang et al. in [40] and its improvements. In a differential collision attack, we first construct a high-probability differential characteristic that yields a zero output difference, i.e., a collision. In the second stage, we probabilistically try to find a confirming message pair for this differential characteristic.

By using a differential characteristic with a lower probability at the beginning of the hash function (first round of SHA-1), the probability of the remaining characteristic can be further improved. Since the message can be chosen freely in a hash function attack, we can significantly improve the complexity of finding confirming message pairs at the beginning of the hash function using message modification techniques [40]. A high-level overview of such a differential attack on SHA-1 is given as follows:

1. **Find a differential characteristic**
   (a) Construct the high-probability part
   (b) Determine the low-probability part
2. **Find a confirming message pair**
   (a) Use message modification in low-probability part
   (b) Perform random trials in high-probability part

The high-probability part of the differential characteristic for SHA-1 covers round 2 to round 4. It has been shown in [8,21,40] that for SHA-1, the best way to construct these high-probability characteristics is to interleave so-called local collisions (one disturbing and a set of correcting differences). These characteristics can be easily constructed by using a linearized variant of the hash function and tools from coding theory [31,32]. The probability of this characteristic determines the complexity of the attack on SHA-1.

The low-probability part and message modification take place in round 1 and are typically performed using automated non-linear equation solving tools [14,22,23]. We stress that the total complexity is still above 2^{60} for all published collision attacks so far, which is only practical for attackers with large computing power (NSA, Google, etc.).

3.3 Malicious Collision Attack

In SHA-1, a new 32-bit constant $K_1, \ldots, K_4$ is used in each of the four rounds. In our malicious collision attack, we use the freedom of these four constants to reduce the complexity of the attack. Similar to message modification, we choose the constants during the search for a confirming message pair. We modify the constants in a round-by-round strategy, always selecting a round constant such that the differential characteristic for the steps of the current round can be satisfied. Since we have to choose the constants when processing the first block, we can only improve the complexity of this block. Hence, we need to use a differential characteristic that results in a single-block collision. Note that all the collisions attacks on SHA-1 so far use a 2-block characteristic.

To find the high-probability part of a differential characteristic for round 2–4 resulting in a 1-block collision, a linearized variant of the hash function can be used. However, using algorithms from coding theory, we only find differential characteristics that maximize the overall probability and do not take the additional freedom we have in the choice of the constants in SHA-1 into account. Therefore, to minimize the overall attack complexity, we did not use these differential characteristics. Instead, we are interested in a differential characteristic such that the minimum of the three probabilities for round 2, 3 and 4 is maximized. To find such a characteristic, we start with the best overall characteristic and modify it to suit our needs.

In previous attacks on SHA-1, the best differential characteristics for rounds 2–4 have differences only at bit position 2 for some 16 consecutive state words A_i [21]. We assume that the best differential characteristic has the same property in our case. Hence, we only need to determine all 2^{16} possible differential characteristics with differences only at bit position 2 in 16 consecutive state words A_i and linearly expand them backward and forward. A similar approach has also been used to attack SHA-0 [8] and SHA-1 [21,40].

For each of these 2^{16} differential characteristics, we estimate the cost of finding a malicious single-block collision. These costs are roughly determined by the number of differences (disturbances) in A_i in each round. For details on the cost computations, we refer to [31]. The estimated costs for the best differential characteristics suited for our attack are given in Table 2, and the correspond message differences are given in Table 3.

Table 2. Probabilities for rounds 2–4 of the differential characteristics suitable for our attack.

Candidate	$r = 2$	$r = 3$	$r = 4$	Total
MD_1	2^{-40}	2^{-40}	2^{-15}	2^{-95}
MD_2	2^{-39}	2^{-42}	2^{-13}	2^{-94}
MD_3	2^{-39}	2^{-42}	2^{-11}	2^{-92}

Table 3. List of message differences suitable for our attack

MD_1	00000003	20000074	88000000	e8000062	c8000043	28000004	40000042	48000046
	88000002	00000014	08000002	a0000054	88000002	80000000	a8000003	a8000060
MD_2	20000074	88000000	e8000062	c8000043	28000004	40000042	48000046	88000002
	00000014	08000002	a0000054	88000002	80000000	a8000003	a8000060	00000003
MD_3	88000000	e8000062	c8000043	28000004	40000042	48000046	88000002	00000014
	08000002	a0000054	88000002	80000000	a8000003	a8000060	00000003	c0000002

The high-probability differential characteristic with message difference MD_1 is best suitable for our intended file formats (see Sect. 4) and used as the starting point to search for a low-probability differential characteristic for the first round of SHA-1. We use an automatic search tool [22,23] to find the low-probability part of the characteristic. The result is shown in Table 5 in the appendix. Overall, the complexity of finding a colliding message pair using malicious constants for this differential characteristic in our attack is approximately 2^{48}, which is feasible in practice as demonstrated below and in Sect. 4.

After the differential characteristic is fixed, we probabilistically search for a confirming message pair. We start with only the first constant K_1 fixed (e.g., to the standard value) and search for a message pair that confirms at least the first round (20 steps) of the characteristic, and is also suitable for our file format. This is easier than finding a message pair that works for all four rounds (with fixed constants), since fewer constraints need to be satisfied. The complexity of this step is negligible.

Now, we can exhaustively search through all 2^{32} options for K_2 until we find one that confirms round 2. Only if no such constant is found, we backtrack and modify the message words. Since the differential characteristic for message difference MD_1 holds with probability 2^{-40} in round 2 and we can test 2^{32} options for K_2, this step of the attack will only succeed with a probability of 2^{-8}. Hence, completing this step alone has a complexity of approximately 2^{40}.

Once we have found a candidate for K_2 such that the differential characteristic holds in round 2, we proceed in the same way with K_3. Again, the differential characteristic will hold with only a probability of 2^{-40} in round 3 and we can test only 2^{32} options for K_3. Therefore, we need to repeat the previous steps of the attack 2^8 times to find a solution. Including the expected 2^8 tries for the previous step to reach the current one, completing this step has an expected complexity of roughly 2^{48}.

Finally, we need to find K_4. Since the last round of the characteristic has a high probability, such a constant is very likely to exist and this step of the attack only adds negligible cost to the final attack complexity of about 2^{48}.

Normally, with fixed constants, an attacker would have to backtrack in the case of a contradiction in the later steps. Eve as the designer, on the other hand, has a chance that choosing a different constant might repair the contradictions for another round. This significantly improves the complexity of the differential attack. For predefined constants, the complexity of the attack for this particular disturbance vector would be roughly 2^{95}.

Note that we do not need the whole freedom of all 4 constants. The first constant in round 1 can be chosen arbitrarily (e.g., we keep it as in the original SHA-1 specification). For the last constant in round 4, we can fix approximately 16 bits of the constant. That is, 80 bits of the constants need to be changed compared to the original values. More freedom in choosing the constants is possible if we increase the attack complexity. An example of a colliding message pair for our malicious SHA-1 variant with modified constants is given in Table 4. The constants differ from the original values by 45 (of 128) bits. In the following section, we will show how this pair can be used to construct meaningful collisions.

4   Building Meaningful Collisions

To exploit the malicious SHA-1 described in Sect. 3, we propose several types of executable, archive and image file formats for which two colliding files can be created, and such that the behavior of the two files can be fully controlled by the attacker.

Table 4. Example of a collision for SHA-1 with modified constants $K_{1...4}$.

$K_{1...4}$	5a827999 4eb9d7f7 bad18e2f d79e5877						
IV	67452301 efcdab89 98badcfe 10325476 c3d2e1f0						
m	ffd8ffe1	e2001250	b6cef608	34f4fe83	ffae884f	afe56e6f	fc50fae6 28c40f81
	1b1d3283	b48c11bc	b1d4b511	a976cb20	a7a929f0	2327f9bb	ecde01c0 7dc00852
m^*	ffd8ffe2	c2001224	3ecef608	dcf4fee1	37ae880c	87e56e6b	bc50faa4 60c40fc7
	931d3281	b48c11a8	b9d4b513	0976cb74	2fa929f2	a327f9bb	44de01c3 d5c00832
Δm	00000003	20000074	88000000	e8000062	c8000043	28000004	40000042 48000046
	88000002	00000014	08000002	a0000054	88000002	80000000	a8000003 a8000060
$h(m)$	1896b202 394b0aae 54526cfa e72ec5f2 42b1837e						

Below, we first discuss the constraints that the files have to satisfy, in order to collide with our malicious SHA-1. We then investigate common binary file formats to determine whether they could allow us to construct a malicious SHA-1 for which two *valid* files collide. Finally, we present actual collisions, and characterize the associated instances of malicious SHA-1.

4.1 Constraints

The attack strategy and the available message differences impose several constraints for possible applications. Most importantly, the exact hash function definition with the final constants is only fixed during the attack. This implies that the differences between the two final files will be limited to a single block. In addition, this block must correspond to the first 512 bits of the final files. After this initial collision block, the file pair can be extended arbitrarily with a common suffix. Due to this limitation, for example, the method that was used to find colliding PostScript files for MD5 [12] cannot be applied here.

For the exact values of the first block, the attack allows a certain freedom. The attacker can fix the values of a few bits in advance. However, fixing too many bits will increase the attack complexity. Additionally, choosing the bits is constrained by the fixed message difference. In all our example files, we use message difference MD_1 from Table 3, which offers a slightly better expected attack complexity than MD_2 and MD_3. All of the available message differences have a difference in the first word, as well as the last byte.

4.2 Binary File Format Overview

Binary file formats typically have a predefined structure and in particular a "magic signature" in their first bytes, which is used to identify the type of binary and to define diverse metadata. As a preliminary to the construction of colliding binaries, we provide basic information on binary files so as to understand the obstacles posed for the construction of colliding files.

We also discuss both our failed and successful attempts to build colliding binary executables. Note that once a collision can be created—that is, if the block difference can be introduced without fatally altering the file structure—the programs executed in each of the two colliding files can be fully controlled. In practice, both programs may execute a legitimate application, but one of the two colliding files prepends the execution of a trojan that will persistently compromise the machine.

Magic Signatures. Most binary file formats enforce a "magic signature" at offset 0, to enable programs to recognize the type of file, its version, etc., in order to process it according to its specific format. For example, the utilities file and binwalk rely mostly on magic signatures to identify files and their type. Some formats, notably most archive formats, also allow the signature to start later in the file, at a higher offset.

Signatures are typically 4 bytes long. Some are longer, such as that of the PNG format (89504e470d0a1a0a), or the RAR archive format (526172211a0700), and some are smaller (PE's 2-byte "MZ", TIFF's 2-byte "MM" and "II"). Note that none of our colliding blocks offer four unmodified consecutive bytes. This implies that collisions for our malicious SHA-1 cannot be files with a fixed 4-byte signature at offset 0.

Executables: PE. The PE (Portable Executable) format is the standard format for Windows executables (.exe files). The PE format, as defined in 1993, is based on the older DOS EXE format (from 1981). PE thus retains the MZ signature (4d5a) from DOS, however in PE it is mostly useless: the only components of the header used are the MZ signature and the last component, which is a pointer to the more modern PE header. This leaves an entirely controllable buffer of 58 bytes near the top of the file, which is tempting to use to build colliding PEs.

PE thus seems an interesting candidate for malicious collisions: it is very commonly used, and its header provides freedom degrees to introduce differences. The only restrictions in the header are in the first two bytes (which must be set to the MZ string) and in the four bytes at offset 60, where the 4-byte pointer to the PE header is encoded.

Unfortunately, the structure of the differential attack forces the most significant byte of the PE header to be (at least) 40. This gives a minimal pointer of 40000000, that is, 1 GiB. Such a file, even if syntaxically correct, is not supported by Windows: it is correctly parsed, but then the OS fails to load it (In practice, the biggest accepted value for this pointer in a working PE is around 9000000).

Due to this limitation, we could not construct valid compact PE executables that collide for a malicious SHA-1. Note that the Unix and OS X counterpart of PEs (ELF and Mach-O files, respectively) fix at least the first 4 bytes, and thus cannot be exploited for malicious collisions either.

Headerless Executables: MBR and COM. Some older formats like master boot records (MBR) and DOS executables (COM) do not include a magic signature or any header. Instead, code execution starts directly at offset 0. By introducing a jump instruction to the subsequent block, we can have total control of the first block and thus create collisions (as long as the difference allows for the jump instruction with distinct reasonable addresses). Running in 16-bit x86 code, the block can start with a jump, encoded as `eb XX`, where `XX` is a signed char that should be positive. Both blocks will immediately jump to different pieces of code of colliding MBR or colliding COM. To demonstrate the feasibility of this approach, example files are given in the appendix.

Compressed Archives: RAR and 7z. Like any other archive file format, the RAR archive allows to start at any offset. However, unlike the ZIP, it is parsed top-down. So if a block creates a valid Rar signature that is broken by its twin, then both files can be valid Rars yet different. We could thus create two colliding archives, which can each contain arbitrary content. A very similar method can be used to build colliding 7z archives (and probably other types of compressed archives).

Images: JPEG. The JPEG file format is organized in a chunk-based manner: chunks are called segments, and each segment starts with a 2 bytes marker. The first byte of the marker is always `ff`, the second is anything but `00` or `ff`. A JPEG file must start with a "Start Of Image" (SOI) marker, `ffd8`. Segments have a variable size, encoded directly after the marker, on 2 bytes, in little endian format. Typically, right after the SOI segment starts the APP0 segment, with marker `ffe0`. This segment contains the familiar "JFIF" string.

However, most JPEG viewers do not require the second segment to start right after SOI. Adding megabytes of garbage data between the SOI marker and the APP0 segment of a JPEG file will still make it valid for most tools – as long as this data does not contain any valid marker, `ff(01-fe)`.

Not only we can insert almost-random data before the first segment, but we can insert any dummy segment that has an encoded length – this will enable us to control the parser, to give it the data we want. If each of our colliding files contains a valid segment marker with a different size and offset, each of them can have a valid APP0 segment at a different offset (provided that the sum of segment size and segment offset differs). To make the bruteforcing phase easier, we can use any of the following segments:

1. the APPx segments, with marker `ffe(0-f)`
2. the COM segment, with marker `fffe`

So, after getting 2 colliding blocks, creating 2 JPEG headers with a suitable dummy segment, we can start the actual data of file J_1 after the second dummy segment (with larger sum of segment size and offset). Right after the first dummy segment, we start another dummy segment to cover the actual data of file J_1. After this second dummy segment, the data of the file J_2 can start. If the length of any of the JPEG file cannot fit on 2 bytes, then several dummy segments need to be written consecutively. Thus, we are able to get a colliding pair of valid files, on a modern format, still used daily by most computers.

Combining Formats: Polyglots. Since the formats discussed above require their magical signatures at different positions in the file, it is possible to construct a first block (of 64 bytes) that suits multiple file formats. For instance, JPEG requires fixed values in the first few words, while archives like RAR can start their signature at a higher offset. Thus, we can construct colliding block pairs for a fixed selection of constants that can later be used to construct colliding files of multiple types with almost arbitrary content. Examples are given in Sect. 4.3 (JPEG-RAR) and in the appendix (MBR-RAR-Script).

4.3   Example Files

We use the attack strategy from Sect. 3 to build a colliding pair of JPEG images and one of RAR archives, both for the same set of malicious SHA-1 constants. Since JPEG requires the file to start with ffd8, MD_1 is the only one of the message differences given in Table 3 that is suitable for a collision between two JPEG files. The following bytes are set to ffe?, where ? differs between the two files and can be any value. Additionally, the last byte of this first 64-byte-block is fixed to also allow the block to be used for RAR collisions: It is set to the first byte of the RAR signature, 52, in one of the blocks, and to a different value as determined by MD_1 in the other block. Using these constraints as a starting point, we search for a differential characteristic. The result is given in Table 5 in the appendix. Note that at this point, the first round constant K_1 is fixed to an arbitrary value (we use the original constant), while K_2, K_3, K_4 are still free. They are determined together with the full first 64-byte block in the next phase. The result is the message pair already given in Table 4. The malicious SHA-1

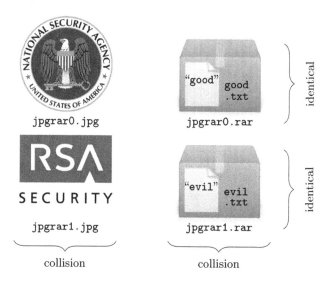

Fig. 2. Colliding JPEG/RAR polyglot file pair for malicious SHA-1 (cf. Table 4).

Table 5. Characteristic corresponding to message difference MD_1, with additional constraints for colliding JPEG/RAR polyglot files (cf. Sect. 4.3).

i	A_i	W_i
0	1001111110001101100110001001010m	1111111111011000111111111000nu
1	100n101001100----	-0011110n00 11u00010000000000---- 0100uuu0n00
2	n111n00----1----	-u000nu00 u011n1---- 000001000
3	0uuuu111--0----0--uu-	-0-0un11nn nun-n-1-- nu000n1
4	1n01u1110---u-n---	-001001n0 uu1---- 00-0011uu
5	50010011u1n00--0-uu0101-10n1u0u0 10n0-0-	-1101u11
6	n1n1n1n01000--1-100101-00n000011 1u111-	--001u0
7	nu1mmmmmmmmmmmmmmmmmmmmm000n1 0n10u00-	-000nu1
8	101111-100110000000100001111nu0n1 n001n--	--000n1 01u1u00
9	0-10101010000000000000001uu001 10110100----	01n1n00
10	n1s00----	-01n 1011n--
11	1-00-0----	-100000 1u0u-- 01n0u00
12	-0010----	-00-1 101-n-- 0-00n1
13	-1----	-1100 u0100-- n1s000n0
14	----	-0000nm -11011
15	-1-1----	-0-n1u--u-- 0un10010
16	-0----	-1110 n1u-u-- -1000uu
17	----	nn001-- 10111u1
18	----	uu001-- n-010nu
19	----	uu1- un111n1
20	----	n11 -0011nu
21	n-	0u01-- 01110n1
22	n-	0u0-- 01100nu
23	----	nn1-- uu001n1
24	----	-n10-- --100nu
25	n-	1n1-- -01001n0
26	----	0n1-- -u-110nu
27	----	nn00-- nu0n0n0
28	n-	nn101-- -11001u
29	n-	uu1n-- -0101n0
30	----	1n1n-- u1n01u0
31	----	un0-- 11101u0
32	----	01n-- -10110m
33	----	01n-- 10mm00001
34	u-	10n-- -10100nu
35	u-	1nn0-- 1001n1n1
36	n-	1n0-- -0111n0
37	----	nnn-- -0n1100u0
38	1-	n0u-- --110001
39	----	10n0-0-- 10m010111

i	A_i	W_i
40	n-	01-- -10000u0
41	----	101-- -01u1001
42	u-	n10-0-- -011um
43	----	n0u-- -01mm0010
44	----	n1u-- -10000nu
45	----	nn0-- --0111n1
46	----	uuuu-- -1111u1
47	n-	n00-- 10-0000n0
48	----	n00-- -10101u00
49	----	100-- -11-010100
50	----	010-- -1-0100010
51	----	010-- -01u100010
52	----	u01-- -11100u0
53	----	101-- -10110101
54	----	n11-- -101100m
55	----	u00-- -110u11000
56	----	100-- -00011uu
57	u-	u-0-- -11u101u0
58	----	1u1-- -001001n
59	----	nu0-- -10un001n
60	----	101-- -110000nu
61	----	un1-- -0010000n1
62	n-	1n-1-- -10u011n1
63	n-	uu1-- -11u011u0
64	----	u10-1-- -0100000u0
65	----	0-- -11110001011
66	----	111-- --0-000011n1
67	----	u1-- -0110-100100
68	----	nu-0-- --011000100
69	----	0-- -01-000010
70	----	n-1-- -111011101
71	u-	-0-- -1100-01100-
72	----	1-- -000001110
73	----	-1-- -0011-111011
74	u-	-- -101111--
75	u-	-- -0000-011101
76	u-	-- -1110001u--
77	----	-- -101000-1
78	n-	-- -000011101--
79	u-	n-- -111000101--

variant differs from the standardized SHA-1 in the values of 45 bits of the round constants.

Now, we can append suitable followup blocks to create valid JPEG or RAR file pairs, both with arbitrary content. As an example, both images in Fig. 2 hash to $h(m) = $ 1896b202 394b0aae 54526cfa e72ec5f2 42b1837e using the malicious round constants $K_1 = $ 5a827999, $K_2 = $ 4eb9d7f7, $K_3 = $ bad18e2f, $K_4 = $ d79e5877.

In a similar fashion, we were able to construct another example block pair for a different set of SHA-1 constants that is suitable for master boot records, shell scripts and RAR archives. All example file pairs and code for verification can be found online at http://maliciousshal.github.io/.

Acknowledgments. The work has been supported by the Austrian Government through the research program FIT-IT Trust in IT Systems (Project SePAG, Project Number 835919).

A   Full Characteristic for Malicious SHA-1

Table 5 shows a full differential characteristic corresponding to message difference MD_1 using the notation of generalized conditions of [14]. The message block pair given in Table 4 is based on this differential characteristic, as is the example file pair in Sect. 4.3 that uses this block pair as a first block.

The table shows the message expansion words W_i on the right-hand side and the state words A_i on the left-hand side. Note that the state words $B_i, \ldots, E_i$ can be easily derived from this representation.

The characteristic already specifies the necessary format fragments required in Sect. 4.3: The first 28 bits of word W_0 are set to ffd8ffe to accommodate the JPEG format, and the last 8 bits of word W_{15} are fixed as 52 (in one message m, for the RAR header) or 32 (in the other message m^*). Additionally, the first round constant is already fixed to the original value $K_1 = $ 5a827999, while K_2, K_3, K_4 are still free to be chosen during message modification.

References

1. Adinetz, A.V., Grechnikov, E.A.: Building a collision for 75-round reduced SHA-1 using GPU clusters. In: Kaklamanis, C., Papatheodorou, T., Spirakis, P.G. (eds.) Euro-Par 2012. LNCS, vol. 7484, pp. 933–944. Springer, Heidelberg (2012)
2. Becker, G.T., Regazzoni, F., Paar, C., Burleson, W.P.: Stealthy dopant-level hardware trojans. In: Bertoni, G., Coron, J.-S. (eds.) CHES 2013. LNCS, vol. 8086, pp. 197–214. Springer, Heidelberg (2013)
3. Biham, E.: Cryptanalysis of Patarin's 2-round public key system with S boxes (2R). In: Preneel, B. (ed.) EUROCRYPT 2000. LNCS, vol. 1807, pp. 408–416. Springer, Heidelberg (2000)
4. Biham, E., Carmeli, Y., Shamir, A.: Bug attacks. In: Wagner, D. (ed.) CRYPTO 2008. LNCS, vol. 5157, pp. 221–240. Springer, Heidelberg (2008)

5. Blaze, M., Feigenbaum, J., Leighton, T.: Master key cryptosystems. CRYPTO 1995 rump session (1995). http://www.crypto.com/papers/mkcs.pdf
6. Brassard, G. (ed.): CRYPTO 1989. LNCS, vol. 435. Springer, Heidelberg (1990)
7. Brown, D.R.L., Gjøsteen, K.: A security analysis of the NIST SP 800-90 elliptic curve random number generator. Cryptology ePrint Archive, Report 2007/048 (2007)
8. Chabaud, F., Joux, A.: Differential collisions in SHA-0. In: Krawczyk, H. (ed.) CRYPTO 1998. LNCS, vol. 1462, pp. 56–71. Springer, Heidelberg (1998)
9. Contini, S., Lenstra, A.K., Steinfeld, R.: VSH, an efficient and provable collision-resistant hash function. In: Vaudenay, S. (ed.) EUROCRYPT 2006. LNCS, vol. 4004, pp. 165–182. Springer, Heidelberg (2006)
10. Damgård, I.: A design principle for hash functions. In: Brassard, G., [6], pp. 416–427
11. Daniel R. L. Brown, S.A.V.: Elliptic curve random number generation. Patent. US 8396213 B2 (2006). http://www.google.com/patents/US8396213
12. Daum, M., Lucks, S.: Hash collisions (the poisoned message attack). CRYPTO 2005 rump session (2005). http://th.informatik.uni-mannheim.de/people/lucks/HashCollisions/
13. De Cannière, C., Mendel, F., Rechberger, C.: Collisions for 70-Step SHA-1: on the full cost of collision search. In: Adams, C., Miri, A., Wiener, M. (eds.) SAC 2007. LNCS, vol. 4876, pp. 56–73. Springer, Heidelberg (2007)
14. De Cannière, C., Rechberger, C.: Finding SHA-1 characteristics: general results and applications. In: Lai, X., Chen, K. (eds.) ASIACRYPT 2006. LNCS, vol. 4284, pp. 1–20. Springer, Heidelberg (2006)
15. Filiol, E.: Malicious cryptography techniques for unreversable (malicious or not) binaries. CoRR abs/1009.4000 (2010)
16. Green, M.: A few more notes on NSA random number generators. Blog post, December 2013. http://blog.cryptographyengineering.com/2013/12/a-few-more-notes-on-nsa-random-number.html
17. Green, M.: The many flaws of Dual_EC_DRBG. Blog post, September 2013. http://blog.cryptographyengineering.com/2013/09/the-many-flaws-of-dualecdrbg.html
18. Johansson, T., Nguyen, P.Q. (eds.): EUROCRYPT 2013. LNCS, vol. 7881. Springer, Heidelberg (2013)
19. Joux, A., Peyrin, T.: Hash functions and the (amplified) boomerang attack. In: Menezes, A. (ed.) CRYPTO 2007. LNCS, vol. 4622, pp. 244–263. Springer, Heidelberg (2007)
20. Lin, L., Kasper, M., Güneysu, T., Paar, C., Burleson, W.: Trojan side-channels: lightweight hardware trojans through side-channel engineering. In: Clavier, C., Gaj, K. (eds.) CHES 2009. LNCS, vol. 5747, pp. 382–395. Springer, Heidelberg (2009)
21. Manuel, S.: Classification and generation of disturbance vectors for collision attacks against SHA-1. Des. Codes Crypt. 59(1–3), 247–263 (2011)
22. Mendel, F., Nad, T., Schläffer, M.: Finding SHA-2 characteristics: searching through a minefield of contradictions. In: Lee, D.H., Wang, X. (eds.) ASIACRYPT 2011. LNCS, vol. 7073, pp. 288–307. Springer, Heidelberg (2011)
23. Mendel, F., Nad, T., Schläffer, M.: Improving local collisions: new attacks on reduced SHA-256. In: Johansson, T., Nguyen, P.Q., [18], pp. 262–278
24. Menn, J.: Exclusive: secret contract tied NSA and security industry pioneer. Reuters, December 2013. http://www.reuters.com/article/2013/12/20/us-usa-security-rsa-idUSBRE9BJ1C220131220
25. Merkle, R.C.: One way hash functions and DES. In: Brassard, G., [6], pp. 428–446

26. Murdoch, S.J., Lewis, S.: Embedding covert channels into TCP/IP. In: Barni, M., Herrera-Joancomartí, J., Katzenbeisser, S., Pérez-González, F. (eds.) IH 2005. LNCS, vol. 3727, pp. 247–261. Springer, Heidelberg (2005)

27. NIST: Recommendation for random number generation using deterministic random bit generators (revised). NIST Special Publication 800–90 (2007)

28. NIST: Secure hash standard (SHS). FIPS PUB 180–4 (2012)

29. Open crypto audit. http://opencryptoaudit.org. Accessed 28 May 2014

30. Patarin, J., Goubin, L.: Trapdoor one-way permutations and multivariate polynomials. In: Han, Y., Quing, S. (eds.) ICICS 1997. LNCS, vol. 1334, pp. 356–368. Springer, Heidelberg (1997)

31. Pramstaller, N., Rechberger, C., Rijmen, V.: Exploiting coding theory for collision attacks on SHA-1. In: Smart, N.P. (ed.) Cryptography and Coding 2005. LNCS, vol. 3796, pp. 78–95. Springer, Heidelberg (2005)

32. Rijmen, V., Oswald, E.: Update on SHA-1. In: Menezes, A. (ed.) CT-RSA 2005. LNCS, vol. 3376, pp. 58–71. Springer, Heidelberg (2005)

33. Rijmen, V., Preneel, B.: A family of trapdoor ciphers. In: Biham, E. (ed.) FSE 1997. LNCS, vol. 1267, pp. 139–148. Springer, Heidelberg (1997)

34. Schneier, B.: The NSA is breaking most encryption on the internet. Blog post, September 2013. https://www.schneier.com/blog/archives/2013/09/the_nsa_is_brea.html

35. Schoenmakers, B., Sidorenko, A.: Cryptanalysis of the dual elliptic curve pseudorandom generator. Cryptology ePrint Archive, Report 2006/190 (2006)

36. Shah, G., Molina, A., Blaze, M.: Keyboards and covert channels. In: USENIX Security Symposium, pp. 59–75 (2006)

37. Stevens, M.: New collision attacks on SHA-1 based on optimal joint local-collision analysis. In: Johansson, T., Nguyen, P.Q., [18], pp. 245–261

38. Wagner, D., Bionbi, P.: Misimplementation of RC4. Submission for the Third Underhanded C Contest (2007). http://underhanded.xcott.com/?page_id=16

39. Wang, X., Yao, A.C., Yao, F.: Cryptanalysis on SHA-1. NIST - First Cryptographic Hash Workshop, October 31–November 1 (2005). http://csrc.nist.gov/groups/ST/hash/documents/Wang_SHA1-New-Result.pdf

40. Wang, X., Yin, Y.L., Yu, H.: Finding collisions in the full SHA-1. In: Shoup, V. (ed.) CRYPTO 2005. LNCS, vol. 3621, pp. 17–36. Springer, Heidelberg (2005)

41. Wu, H., Bao, F., Deng, R.H., Ye, Q.-Z.: Cryptanalysis of Rijmen-Preneel trapdoor ciphers. In: Ohta, K., Pei, D. (eds.) ASIACRYPT 1998. LNCS, vol. 1514, pp. 126–132. Springer, Heidelberg (1998)

42. Ding-Feng, Y., Kwok-Yan, L., Zong-Duo, D.: Cryptanalysis of 2R schemes. In: Wiener, M. (ed.) CRYPTO 1999. LNCS, vol. 1666, pp. 315–325. Springer, Heidelberg (1999)

43. You broke the internet. http://youbroketheinternet.org. Accessed 28 May 2014

44. Young, A., Yung, M.: Malicious Cryptography: Exposing Cryptovirology. Wiley, Chichester (2004)

45. Young, A., Yung, M.: Monkey: black-box symmetric ciphers designed for MONopolizing KEYs. In: Vaudenay, S. (ed.) FSE 1998. LNCS, vol. 1372, pp. 122–133. Springer, Heidelberg (1998)

46. Young, A.L., Yung, M.: Backdoor attacks on black-box ciphers exploiting low-entropy plaintexts. In: Safavi-Naini, R., Seberry, J. (eds.) ACISP 2003. LNCS, vol. 2727, pp. 297–311. Springer, Heidelberg (2003)

Binary Elligator Squared

Diego F. Aranha[1], Pierre-Alain Fouque[2,3], Chen Qian[4], Mehdi Tibouchi[5], and Jean-Christophe Zapalowicz[6]([✉])

[1] Institute of Computing, University of Campinas, Campinas, Brazil
dfaranha@ic.unicamp.br
[2] Université de Rennes 1, Rennes, France
[3] Institut Universitaire de France, Paris, France
fouque@irisa.fr
[4] ENS Rennes, Bruz, France
chen.qian@ens-rennes.fr
[5] NTT Secure Platform Laboratories, Tokyo, Japan
tibouchi.mehdi@lab.ntt.co.jp
[6] Inria, Paris, France
jean-christophe.zapalowicz@inria.fr

Abstract. Two efficient approaches have been recently proposed to make random points on elliptic curves representable as uniform random strings (a useful property for anonymity and censorship circumvention applications): the "Elligator" technique due to Bernstein et al. (ACM CCS 2013), which is simple but supports a somewhat limited set of elliptic curves, and its variant "Elligator Squared" suggested by Tibouchi (FC 2014), which is slightly more complex but supports arbitrary curves. Despite that complexity, it was speculated that Elligator Squared could have an efficiency edge in some contexts, as it avoids a rejection sampling step necessary for Elligator, and can be used with a larger class of point encoding functions, some of them very efficient.

In this paper, we show that Elligator Squared can indeed be implemented very efficiently with a suitable choice of point encoding function. More precisely, we consider the binary curve setting, and implement the Elligator Squared bit string representation algorithm based on a suitably optimized version of the Shallue–van de Woestijne characteristic 2 encoding. On the fast binary curve of Oliveira et al. (CHES 2013), our implementation runs in an average of only 22850 Haswell cycles.

We also compare implementations of Elligator and Elligator Squared on a curve supported by Elligator, namely Curve25519, and find that generating a random point and its uniform bitstring representation is around 35–40 % faster with Elligator for protocols using a fixed base point, but 30–35 % faster with Elligator Squared in the case of a variable base point. Both are significantly slower than our binary curve implementation.

Keywords: Elligator · Binary elliptic curves · Efficient implementation · PCLMULQDQ · Anonymity and privacy

© Springer International Publishing Switzerland 2014
A. Joux and A. Youssef (Eds.): SAC 2014, LNCS 8781, pp. 20–37, 2014.
DOI: 10.1007/978-3-319-13051-4_2

1   Introduction

Elliptic curves offer many advantages for public-key cryptography compared to more traditional settings like RSA and finite field discrete logarithms, including higher efficiency, a much smaller key size that scales gracefully with security requirements, and a rich geometric structure that enables the construction of additional primitives like bilinear pairings. On the Internet, adoption of elliptic curve cryptography is growing in general-purpose protocols like TLS, SSH and S/MIME, as well as anonymity and privacy-enhancing tools like Tor (which favors ECDH key exchange in recent versions) and Bitcoin (which is based on ECDSA).

For censorship circumvention applications, however, ECC presents a weakness: points on a given elliptic curve, when represented in a usual way (even in compressed form) are easy to distinguish from random bit strings. For example, the usual compressed bit string representation of an elliptic curve point is essentially the x-coordinate of the point, and only about half of all possible x-coordinates correspond to valid points (the other half being x-coordinates of points of the quadratic twist). This makes it relatively easy for an attacker to distinguish ECC traffic (the transcripts of multiple ECDH key exchanges, say) from random traffic, and then proceed to intercept, block or otherwise tamper with such traffic.

To alleviate that problem, one possible approach is to modify protocols so that transmitted points randomly lie either on the given elliptic curve or on its quadratic twist (and the curve parameters must therefore be chosen to be twist-secure). This is the approach taken by Möller [24], who constructed a CCA-secure KEM with uniformly random ciphertexts using an elliptic curve and its twist. This approach has also been used in the context of kleptography, as considered by Young and Yung [32,33], and has already been deployed in circumvention tools, including StegoTorus [30], a camouflage proxy for Tor, and Telex [31], an anticensorship technology that uses a covert channel in TLS handshakes to securely communicate with friendly proxy servers. However, since protocols and security proofs have to be adapted to work on both a curve and its twist, this approach is not particularly versatile, and it imposes additional security requirements (twist-security) on the choice of curve parameters.

A different approach, called "Elligator", was presented at ACM CCS 2013 by Bernstein, Hamburg, Krasnova and Lange [6]. Their idea is to leverage an efficiently computable, efficiently invertible algebraic function that maps the integer interval $S = \{0, \dots, (p-1)/2\}$, p prime, *injectively* to the group $E(\mathbb{F}_p)$ where E is an elliptic curve over $\mathbb{F}_p$. Bernstein et al. observe that, since ι is injective, a uniformly random point P in $\iota(S) \subset E(\mathbb{F}_p)$ has a uniformly random preimage $\iota^{-1}(P)$ in S, and use that observation to represent an elliptic curve point P as the bit string representation of the unique integer $\iota^{-1}(P)$ if it exists. If the prime p is close to a power of 2, a uniform point in $\iota(S)$ will have a close to uniform bit string representation.

This method has numerous advantages over Möller's twisted curve method: it is easier to adapt to existing protocols using elliptic curves, since there is no need to modify them to also deal with the quadratic twist; it avoids the need to publish a twisted curve counterpart of each public key element, hence allowing a more compact public key; and it doesn't impose additional security requirements like twist-security. But it crucially relies on the existence of an injective encoding ι, only a few examples of which are known [6,13,17], all of them for elliptic curves of non-prime order over large characteristic fields. This makes the method inapplicable to implementations based on curves of prime order or on binary fields, which rules out most standardized ECC parameters [1, 11,15,23], in particular. Moreover, the rejection sampling involved (when a point P is picked outside $\iota(S)$, the protocol has to start over) can impose a significant performance penalty.

To overcome these limitations, Tibouchi [29] recently proposed a variant of Elligator, called "Elligator Squared", in which a point $P \in E(\mathbb{F}_q)$ is represented not by a preimage under an injective encoding ι, but by a randomly sampled preimage under an essentially surjective map $\mathbb{F}_q^2 \to E(\mathbb{F}_q)$ with good statistical properties, known as an *admissible encoding* following a terminology introduced by Brier et al. [10]. By results due to Farashahi et al. [14], such admissible encodings are known to exist for all isomorphism classes of elliptic curves, including curves of prime order and binary curves. Since admissible encodings are essentially surjective, the approach also eliminates the need for rejection sampling at the protocol level.

Our Contributions. While the Elligator Squared approach is quite versatile, its efficiency is highly dependent on how fast the underlying admissible encoding can be computed and sampled, and the same can be said of Elligator in the settings where it can be used. Since, to the best of our knowledge, no detailed implementation results or concrete performance numbers have been published so far for the underlying encodings, one only has some rough estimates to go by. For Elligator, Bernstein et al. give ballpark Westmere cycle count figures based on earlier implementation results [7], and for Elligator Squared, Tibouchi provides some average operation counts in [29] for a few selected encoding functions. No performance-oriented implementation is available for either approach.

In this paper, we provide the first such implementation for Elligator Squared, and do so in the binary curve setting, which had not been considered by Tibouchi. Binary curves provide a major advantage for algorithms like Elligator Squared due to the existence of a point encoding function, the binary Shallue–van de Woestijne encoding [27], that can be computed without base field exponentiations. Using the framework of Farashahi et al. [14], one can obtain an admissible encoding from that function, and hence use it to implement Elligator Squared.

We propose various algorithmic improvements and computation tricks to obtain a fast evaluation of the binary Shallue–van de Woestijne encoding and of the associated Elligator Squared sampling algorithm. In particular, our description is much more efficient than the one given in [9, Appendix E].

Based on these algorithmic improvements, we performed software implementations of Elligator Squared on the record-setting binary GLS curve of

Oliveira et al., defined over $\mathbb{F}_{2^{254}}$ [25]. We dedicate special attention to optimizing the performance-critical operations and introduce corresponding novel techniques, namely a new point addition formula in λ-affine coordinates and a faster approach for constant-time half-trace computation over quadratic extensions of $\mathbb{F}_{2^m}$. Moreover, timings are presented for both variable-time and constant-time field arithmetic.[1] The resulting timings compare very favorably to previously suggested estimates.

Finally, as a side contribution, we also propose concrete cycle counts on Ivy Bridge and Haswell for both Elligator and Elligator Squared on the Edwards curve Curve25519 [4] based on the publicly available implementation of Ed25519 [5]. We find that, on this curve, the Elligator approach is roughly 35–40 % faster than Elligator Squared for protocols that rely on fixed-base scalar multiplication, but conversely, for protocols that rely on variable-base scalar multiplication, Elligator Squared is 30–35 % faster. Both approaches are significantly slower than what we achieve on the same CPU with our binary curve implementation.

2   Preliminaries

Let E be an elliptic curve over a finite field $\mathbb{F}_q$.

2.1   Well-Bounded Encodings

Some technical definitions are required to describe the conditions under which an "encoding function" $f \colon \mathbb{F}_q \to E(\mathbb{F}_q)$ can be used in the Elligator Squared constructions. See [14,29] for details.

Definition 1. *A function $f : \mathbb{F}_q \to E(\mathbb{F}_q)$ is said to be a B-well-distributed encoding for a certain constant $B > 0$ if for any nontrivial character χ of $E(\mathbb{F}_q)$, the following holds:*

$$\left| \sum_{u \in \mathbb{F}_q} \chi(f(u)) \right| \le B\sqrt{q}.$$

Definition 2. *We call a function $f : \mathbb{F}_q \to E(\mathbb{F}_q)$ a (d, B)-well-bounded encoding, for positive constants d, B, when f is B-well-distributed and all points in $E(\mathbb{F}_q)$ have at most d preimages under f.*

2.2   Elligator Squared

Let $f : \mathbb{F}_q \to E(\mathbb{F}_q)$ be a (d, B)-well-bounded encoding and let $f^{\otimes 2}$ the tensor square defined by:

---

[1] We point out that using constant-time arithmetic for Elligator Squared is not required in most realistic adversarial models, but it does offer protection against very powerful distinguishing attackers, so the paranoid may prefer that option nonetheless.

$$f^{\otimes 2} : \mathbb{F}_q^2 \to E(\mathbb{F}_q)$$
$$(u, v) \mapsto f(u) + f(v).$$

Tibouchi shows in [29] that if we sample a uniformly random preimage under $f^{\otimes 2}$ of a uniformly random point P on the curve, we get a pair $(u, v) \in \mathbb{F}_q^2$ which is statistically close to uniform. Moreover he proves that sampling uniformly random preimages under $f^{\otimes 2}$ can be done efficiently for all points $P \in E(\mathbb{F}_q)$ except possibly a negligible fraction of them [29, Theorem 1]. The sampling algorithm Tibouchi proposed is described as Algorithm 1. The idea is to randomly pick a random u and then to compute a correct candidate v such that $P = f(u) + f(v)$. The last steps of the algorithm (step 5 to 7) are also needed in order to ensure the uniform distribution of the output (u, v).

Algorithm 1. Preimage sampling algorithm for $f^{\otimes 2}$.

1: **function** SAMPLEPREIMAGE(P)
2:    **repeat**
3:        $u \xleftarrow{\$} \mathbb{F}_q$
4:        $Q \leftarrow P - f(u)$
5:        $i \leftarrow \#f^{-1}(Q)$
6:        $j \xleftarrow{\$} \{1, \cdots, d\}$
7:    **until** $j \leq i$
8:    $\{v_1, \cdots, v_t\} \leftarrow f^{-1}(Q)$
9:    **return** (u, v_j)
10: **end function**

2.3 Shallue–van de Woestijne in Characteristic 2

In this section, we recall the Shallue–van de Woestijne algorithm in characteristic 2 [27], following the more explicit presentation given in [9, Appendix E]. An elliptic curve over a field $\mathbb{F}_{2^n}$ is a set of points $(x, y) \in (\mathbb{F}_{2^n})^2$ verifying the equation:

$$E_{a,b} : Y^2 + X \cdot Y = X^3 + a \cdot X^2 + b$$

where $a, b \in (\mathbb{F}_{2^n})^2$. Let g be the rational function $x \mapsto x^{-2} \cdot (x^3 + a \cdot x^2 + b)$. Letting $Z = Y/X$, the equation for $E_{a,b}$ can be rewritten as $Z^2 + Z = g(X)$.

Theorem 1. *Let $g(x) = x^{-2} \cdot (x^3 + a \cdot x^2 + b)$ where $a, b \in (\mathbb{F}_{2^n})^2$. Let*

$$X_1(t, u) = \frac{t \cdot c}{1 + t + t^2} \qquad X_2(t, u) = t \cdot X_1(t, u) + c \qquad X_3(t, u) = \frac{X_1(t, u) \cdot X_2(t, u)}{X_1(t, u) + X_2(t, u)}$$

where $c = a + u + u^2$. Then $g(X_1(t, u)) + g(X_2(t, u)) + g(X_3(t, u)) \in h(\mathbb{F}_{2^n})$ where h is the map $h : z \mapsto z^2 + z$.

From Theorem 1, we have that at least one of the $g(X_i(t, u))$ must be in $h(\mathbb{F}_{2^n})$, which leads to a point in $E_{a,b}(\mathbb{F}_{2^n})$. Indeed, we have that $h(\mathbb{F}_{2^n}) = \{z \in \mathbb{F}_{2^n} \mid \mathrm{Tr}(z) = 0\}$, where Tr is the trace operator $\mathrm{Tr} : \mathbb{F}_{2^n} \to \mathbb{F}_2$ with:

$$\mathrm{Tr}\, z = \sum_{i=0}^{n-1} z^{2^i}$$

(one inclusion is obvious and the other one follows from the fact that the kernel of the $\mathbb{F}_2$-linear map h is $\{0,1\}$, hence its image is a hyperplane). As a result, $\sum_{i=1}^{3} \mathrm{Tr}(g(X_i)) = 0$ and therefore at least one of the X_i must satisfy $\mathrm{Tr}(g(X_i)) = 0$ since Tr is $\mathbb{F}_2$-valued. Such an X_i is indeed the abscissa of a point in $E_{a,b}(\mathbb{F}_{2^n})$, and we can find its y-coordinate by solving the quadratic equation $Z^2 + Z = g(X_i)$. That equation is $\mathbb{F}_2$-linear, so finding Z amounts to solve a linear system over $\mathbb{F}_2$. This yields the point-encoding function described in Algorithm 2.

In the description of that algorithm, the solution of the quadratic equation is expressed in terms of a linear map QS : KerTr $\to \mathbb{F}_{2^n}$ ("quadratic solver"), which is a right inverse of $z \mapsto z^2 + z$. It is chosen among such right inverses in such a way that membership in its image is computed efficiently using a single trace computation. For example, when n is odd, it is customary to choose $\mathrm{QS}(x)$ as the trace zero solution of $z^2 + z = x$, in which case QS is simply the half-trace map HTr defined as:

$$\mathrm{HTr} : z \mapsto \sum_{i=0}^{(n-1)/2} z^{2^{2i}}.$$

When $n = 2m$ with m odd, we have $\mathbb{F}_{2^n} = \mathbb{F}_{2^m}[w]/(w^2 + w + 1)$ and we can define $\mathrm{QS}(x)$ as the solution $z = z_0 + z_1 w$ of $z^2 + z = x$ such that $\mathrm{Tr}\, z_0 = 0$ (and this clearly generalizes to extension degrees with higher 2-adic valuation). The efficient computation of QS in that case is discussed in Sect. 4.

Algorithm 2. Shallue–van de Woestijne algorithm in characteristic 2.

Require: $a, b \in \mathbb{F}_{2^n}$ and $t, u \in \mathbb{F}_{2^n}$
Ensure: $(x, y) \in E_{a,b}$
1: $c \leftarrow a + u + u^2$
2: $X_1 \leftarrow t \cdot c/(1 + t + t^2)$
3: $X_2 \leftarrow t \cdot X_1 + c$
4: $X_3 \leftarrow X_1 \cdot X_2/(X_1 + X_2)$
5: **for** $j = 1$ to 3 **do**
6:      $h_j \leftarrow (X_j^3 + a \cdot X_j^2 + b)/X_j^2$
7:      **if** $\mathrm{Tr}(h_j) = 0$ **then return** $(X_j, \mathrm{QS}(h_j) \cdot X_j)$
8:      **end if**
9: **end for**

Algorithm 2 actually maps two parameters t, u to a rational point on the curve $E_{a,b}$. One can obtain a map $f : \mathbb{F}_q \to E_{a,b}(\mathbb{F}_q)$ by picking one of the two parameters as a suitable constant and letting the other one vary. In what follows, for efficiency reasons, we fix t and use u as the variable parameter.

One can check that the resulting function is well-bounded in the sense of Sect. 2.1. Indeed, the framework of Farashahi et al. [14] can be used to establish that it is a well-distributed encoding: the proof is easily adapted from the one given in [18] for the odd characteristic version of the Shallue–van de Woestijne algorithm. Moreover, each curve point has at most 6 preimages under the corresponding function: there are at most two values of u that yield a given value of X_1, and similarly for X_2, X_3. Thus, we obtain a (d, B)-well-bounded encoding for an explicitly computable constant B and $d = 6$.

2.4   Lambda Affine Coordinates

In order to have more efficient binary elliptic curve arithmetic, we will use lambda coordinates [22,25,26]. Given a point $P = (x, y) \in E_{a,b}(\mathbb{F}_{2^n})$, with $x \neq 0$, its λ-affine representation of P is defined as (x, λ) where $\lambda = x + y/x$. The λ-affine equation of the Weierstrass Equation of the curve $y^2 + xy = x^3 + ax^2 + b$ is $(\lambda^2 + \lambda + a)x^2 = x^4 + b$. Note that the condition $x \neq 0$ is not restrictive in practice since the only point $x = 0$ satisfying Weierstrass equation is $(0, \sqrt{b})$.

3   Algorithmic Aspects

We focus on Algorithm 1 proposed by Tibouchi in [29], which we adapt for the specific characteristic 2 finite field. More precisely, we consider an elliptic curve over a field $\mathbb{F}_{2^n}$ that satisfies the equation in λ-coordinates:

$$E_{a,b} : (\lambda^2 + \lambda + a)x^2 = x^4 + b$$

where $a, b \in (\mathbb{F}_{2^n})^2$. The $(6, B)$-well-bounded encoding we consider for our efficient Elligator Squared implementation is the binary Shallue–van de Woestijne algorithm recalled in Sect. 2.3.

One of its properties is that among three candidates denoted X_1, X_2, X_3, either exactly one of them or all three are x-coordinate of a rational point over the binary elliptic curve $E_{a,b}$, and the algorithm outputs the first correct one. Owing to this property, some additional verifications are needed during preimage computation, since it is not always true that $\mathrm{SWCHAR2}_X(\mathrm{SWCHAR2}_X^{-1}(X_i)) = X_i$ for $i = 2, 3$ when it is true for $i = 1$, where we denote by $\mathrm{SWCHAR2}_X$ the x-coordinate of the binary Shallue–van de Woestijne algorithm, and by $\mathrm{SWCHAR2}_X^{-1}$ an arbitrary preimage thereof (see the discussion on the subroutine PREIMAGESSW in Sect. 3.2 for more details). We also have to consider another property of this algorithm, concerning the output. Indeed the y-coordinate has a specific form and thus, before searching for some preimages of the point Q, one has to test whether this property is verified (see the discussion on the overall complexity in Sect. 3.3 for more details).

The details of our preimage sampling algorithm in characteristic 2 are described in Algorithm 3 with t fixed to a constant such that $t(t+1)(t^2+t+1) \neq 0$, i.e. $t \notin \mathbb{F}_4$. Note that we make the choice to use the λ-coordinates for efficiency reasons justified in Sect. 3.2. The rest of the section consists in describing the two subroutines SWCHAR2 and PREIMAGESSW, as well as in evaluating the overall complexity of Algorithm 3.

Algorithm 3. Preimage Sampling Algorithm in Characteristic 2

1: Precomputed: $t_1 = \frac{t}{1+t+t^2}, t_2 = \frac{1+t}{1+t+t^2}, t_3 = \frac{t(1+t)}{1+t+t^2}$
2: **function** SAMPLEPREIMAGE$(E_{a,b}, P)$
3:     **repeat**
4:         **repeat**
5:             $u \xleftarrow{\$} \mathbb{F}_{2^n}$
6:             $R \leftarrow$ SWCHAR2$(E_{a,b}, u, t_1, t_2, t_3)$
7:             $Q \leftarrow P - R$
8:         **until** $\lambda_Q + x_Q \in \mathrm{ImQS}$ ▷ Test fails by convention for Q at infinity
9:         $k, S = \{v_1, \cdots, v_k\} \leftarrow$ PREIMAGESSW$(E_{a,b}, Q, t_1, t_2, t_3)$
10:         $j \xleftarrow{\$} \{1, \cdots, 6\}$
11:     **until** $j \leq k$
12:     **return** (u, v_j)
13: **end function**

3.1 The Subroutine SWCHAR2

The first subroutine represents the binary Shallue–van de Woestijne algorithm and its pseudocode for our case is given as Algorithm 4. Given a value $u \in \mathbb{F}_{2^n}$, it outputs the lambda coordinates of a point over the binary elliptic curve $E_{a,b}$.

Algorithm 4. Efficient Binary Shallue–van de Woestijne Algorithm

---

1: **function** SWCHAR2$(E_{a,b}, u, t_1, t_2, t_3)$
2:     $c \leftarrow u^2 + u + a$
3:     $c_{-1} \leftarrow 1/c$
4:     **for** $j = 1$ to 3 **do**   ▷ Compute h_j and perform a trace test
5:         $X_j \leftarrow t_j \cdot c$ ▷ or $X_3 \leftarrow X_1 + X_2 + c$
6:         $X_{-j} \leftarrow 1/t_j \cdot c_{-1}$ ▷ $1/t_j$ can also be precomputed
7:         $h_j \leftarrow (X_{-j})^2 \cdot b + X_j + a$
8:         **if** $\mathrm{Tr}(h_j) = 0$ **then** ▷ At least one of the three potential tests will succeed
9:             $x \leftarrow X_j$
10:             $\lambda \leftarrow \mathrm{QS}(h_j) + x$
11:             **break**   ▷ Only take into account the first correct solution
12:         **end if**
13:     **end for**
14:     **return** (x, λ) ▷ Lambda coordinates of a point over $E_{a,b}$
15: **end function**

Since the field inversion is by far the most expensive field operation (see [25] for experimental timings and Table 2 below), we have modified Algorithm 2 so that we have a single inversion of c to perform. Indeed Algorithm 2 requires at most 4 field inversions: the first one at step 4 and the three others at step 6. However the parameters X_i and $1/X_i$ for $j = 1, 2, 3$ can be expressed using c, $1/c$ and some constants depending on t which can be precomputed (see Table 1). Note that X_3 can be computed as $c \cdot t_3$, or more efficiently as $X_1 + X_2 + c$

Table 1. Efficient computation of values X_i and $1/X_i$ for $i = 1, \cdots 3$. The values $t_1 = \frac{t}{1+t+t^2}, 1/t_1, t_2 = \frac{1+t}{1+t+t^2}, 1/t_2$ and $1/t_3 = \frac{1+t+t^2}{t(1+t)}$ can be precomputed, with t a constant such that $t \notin \mathbb{F}_4$.

$X_1 \leftarrow t_1 \cdot c$	$1/X_1 \leftarrow 1/t_1 \cdot 1/c$
$X_2 \leftarrow t_2 \cdot c$	$1/X_2 \leftarrow 1/t_2 \cdot 1/c$
$X_3 \leftarrow X_1 + X_2 + c$	$1/X_3 \leftarrow 1/t_3 \cdot 1/c$

Table 2. Timings for Elligator Squared and underlying field arithmetic in two Intel platforms. Results are in clock cycles and were taken as the average of 10^4 executions with random inputs. FB/VB results refer to generating a random point with fixed-base and variable-base scalar multiplication respectively, using the constant-time, timing-attack protected scalar multiplication from [25], and computing its Elligator Squared representation with variable-time arithmetic.

Operation	Ivy Bridge	Haswell
Field squaring	13	15
Sparse multiplication	80	44
Multiplication	94	48
Inversion	959	734
Constant-time inversion	1,783	1,610
Quadratic solver	55	50
Constant-time quadratic solver	1,213	1,245
Point addition	1,500	1,026
Constant-time point addition	2,367	2,137
Elligator Squared	23,680	22,850
Constant-time Elligator Squared	52,850	51,750
FB with Elligator Squared	127,430	80,180
VB with Elligator Squared	138,480	83,680

but this requires to keep in memory X_1 and X_2. Finally this algorithm requires a single field inversion, a QS computation and some negligible field operations (multiplications, squarings and trace computations).

3.2   The Subroutine PREIMAGESSW

The second subroutine is useful to compute the number of preimages of the point $Q = (x_Q, \lambda_Q)$ by Algorithm 4. Its pseudocode is detailed as Algorithm 5 and refers to the steps 5 and 8 of Algorithm 1.

   This subroutine is more complex due to the properties of the Shallue–van de Woestijne algorithm. More precisely, there is an order relation in Algorithm 4: if X_1 corresponds to a x-coordinate of a point over the elliptic curve, then it will output this point, even if X_2 and X_3 also correspond to a possible x-coordinate. Thus, the equality $\text{SWCHAR2}(\text{SWCHAR2}^{-1}(X_j)) = X_j$ is true for $j = 1$ but not necessarily for $j = 2, 3$. In others words, for $j = 2, 3$ a solution of $\text{SWCHAR2}^{-1}(X_j)$ is not necessarily a preimage of X_j by SWCHAR2.

Starting from the equations $x_Q = X_j(t, u) = c(u) \cdot t_j$ for $j = 1, 2, 3$, with $c(u) = u^2 + u + a$, the main idea of Algorithm 5 consists in testing if there exists some values of u which satisfy these equations. If one finds some candidates for u, one also has to verify if they really correspond to preimages by Algorithm 4. From an equation $x_Q = X_j(t, u)$ we can obtain an equation $u + u^2 = x_Q/t_j + a = \alpha_j(a, t)$ which has two solutions if $\text{Tr}(\alpha_j(a, t)) = 0$ and no solution otherwise. As an example $\alpha_1(a, t)$ is equal to $x_Q \cdot (1 + t + t^2)/t + a$. The solutions are then $u_0^1 = \text{QS}(\alpha_j(a, t))$ and $u_1^1 = u_0^1 + 1$. There are thus at most 6 possible solutions for all values of j. Now for the cases $x_Q = X_2(t, u)$ and $x_Q = X_3(t, u)$, it remains to perform a verification. Actually, denoting u_0^2 one of both solutions of the equation $x_Q = X_2(t, u)$ if it exists, the computation of $\text{SWCHAR2}(u_0^2)$ can result in $X_1(t, u_0^2)$ instead of $X_2(t, u_0^2)$, and this happens with probability $1/2$ which is the probability that $\text{Tr}(h_1) = 0$. The same result holds for $x_Q = X_3(t, u)$,

---

Algorithm 5. Preimages Computation by Algorithm 4

1: **function** PREIMAGESSW$(E_{a,b}, Q = (x_Q, \lambda_Q), t_1, t_2, t_3)$
2:     $k \leftarrow 0$
3:     $S \leftarrow \{\}$
4:     **for** $j = 1$ to 3 **do**   ▷ From $x_Q = X_j(t, u)$...
5:         $\alpha_j \leftarrow x_Q \cdot 1/t_j + a$
6:         **if** $\text{Tr}(\alpha_j) = 0$ **then**   ▷ ...Test if there are some solutions
7:             **if** $j = 1$ **then**   ▷ For X_1, a solution is a preimage
8:                 $u_0 \leftarrow \text{QS}(\alpha_j)$
9:                 $u_1 \leftarrow w_0 + 1$
10:                $k \leftarrow 2$
11:                $S \leftarrow \{u_0, u_1\}$
12:            **else**   ▷ For X_2, X_3, a solution is not necessarily a preimage
13:                $X_1 \leftarrow t_1/t_j \cdot x_Q$
14:                $tmp \leftarrow [(\lambda_Q + x_Q)^2 + (\lambda_Q + x_Q) + x_Q + a] \cdot (t_j/t_1)^2$ ▷ $tmp = b/X_1^2$
15:                $h_1 \leftarrow tmp + X_1 + a$
16:                **if** $\text{Tr}(h_1) \neq 0$ **then**   ▷ Test if X_1 would also be a correct
                                                                  x-coordinate
17:                    $u_0 \leftarrow \text{QS}(\alpha_j)$
18:                    $u_1 \leftarrow u_0 + 1$
19:                    $k \leftarrow k + 2$
20:                    $S \leftarrow S \cup \{u_0, u_1\}$
21:                **end if**
22:            **end if**
23:        **end if**
24:    **end for**
25:    **return** k, S ▷ k: number of preimages, S: set of preimages
26: **end function**

however note that if X_3 is solution but not X_1 then X_2 cannot be a solution since $\sum_{i=1}^{3} \mathrm{Tr}(g(X_i)) = 0$ according to Theorem 1. Thus the verification can focus only on X_1.

Naive implementation of the verification. A simple way for implementing the verification would consist in computing $\mathrm{QS}(\alpha_j(a,t))$ for $j = 2,3$ and then calling twice the subroutine SWCHAR2 (without the steps referring to X_2 and X_3) for testing if the test on the trace is true or not. However this would require an additional inversion per call to compute SWCHAR2. Moreover, with this naive implementation we have to compute the half trace before testing if the result will be a preimage.

Efficient implementation of the verification. Since the verification focuses only on X_1 as explained above, we propose an efficient way to compute b/X_1^2, which is required in order to perform the test $\mathrm{Tr}(h_1) = \mathrm{Tr}(X_1 + a + b/X_1^2)$, without any field inversion. This trick is valuable when we are working in lambda coordinates. Our proposal has another advantage: we do not need to compute the solutions, i.e. $u_0 = \mathrm{QS}(\alpha_j(a,t))$ and $u_1 = u_0 + 1$, before to be sure that we will get two preimages. We thus save some quite expensive half trace computations.

Consider the equation:

$$x_Q = X_2 = t_2 \cdot c = t_2 \cdot X_1/t_1 \quad \text{with} \quad c = \mathrm{QS}(\alpha_2(a,t))^2 + \mathrm{QS}(\alpha_2(a,t)) + a.$$

X_1 can be expressed as $t_1/t_2 \cdot x_Q$, whose computation is negligible for t_1/t_2 a precomputed value. Now starting from the equation of the elliptic curve in affine coordinates, i.e. $E_{a,b} : Y^2 + X \cdot Y = X^3 + a \cdot X^2 + b$, we divide each term by X^2 and we evaluate the equation in the point Q. We then obtain:

$$\left(\frac{y_Q}{x_Q}\right)^2 + \frac{y_Q}{x_Q} = x_Q + a + \frac{b}{x_Q^2},$$

and finally:

$$\frac{b}{X_1^2} = \left(\frac{t_2}{t_1}\right)^2 \cdot \left[\left(\frac{y_Q}{x_Q}\right)^2 + \frac{y_Q}{x_Q} + x_Q + a\right].$$

Assuming that $(t_2/t_1)^2$ is a precomputed constant, the computation of b/X_1^2 is not costly if y_Q/x_Q does not require an expensive operation. That is the case when we are working in λ-coordinates since $\lambda_Q = y_Q/x_Q + x_Q$. The same result obviously holds for the equation $x_Q = X_3$ by replacing t_2 with t_3.

To conclude, Algorithm 5 requires at most 3 QS computations and some negligible field operations (multiplications, squarings and trace computations).

3.3   Operation Counts

We conclude this section by evaluating the average number of operations needed to evaluate Algorithm 3.

Proposition 1. *An evaluation of Algorithm 3 on uniformly random curve points requires, on average and with an error term of up to $O(2^{-n/2})$, 6 field inversions, 6 point additions, 9 quadratic solver computations and some negligible operations such as field multiplications, field squares and trace computations.*

Proof. The proof consists in evaluating the probability for exiting the two loops. First note that the output (x, λ) of Algorithm 4 has a specific property, namely $\lambda + x$ is in the image of QS. Since we want to retrieve the preimages of a point Q, we have to be sure that $\lambda_Q + x_Q$ is indeed in that image, which we test for by verifying whether $\mathrm{Tr}(\lambda_Q + x_Q) = 0$. Indeed, all elements of the form $QS(z)$ have zero trace by definition, and the converse is true for reasons of dimensions. The success probability of this test is exactly $1/2$ since Q is a uniformly random curve point. We thus have on average 2 field inversions, 2 point additions and 2 quadratic solver computations for the internal loop (steps 4 to 8).

The complexity of the external loop demands to evaluate the probabilities for having 0, 2, 4 or 6 preimages of Q. Since all tests on the trace in Algorithm 5 succeed, independently, with probability $1/2 + O(2^{-n/2})$,[2] these probabilities are then, again with an error term of $O(2^{-n/2})$, $9/32$ for 0 preimage, $15/32$ for 2 preimages, $7/32$ for 4 preimages, and $1/32$ for 6 preimages. Thus, the probability for exiting the external loop is equal to $0 \cdot 9/32 + 1/3 \cdot 15/32 + 2/3 \cdot 7/32 + 1 \cdot 1/32 = 1/3$. These probabilities also hold for evaluating the average cost of an iteration of PREIMAGESSW in term of quadratic computations. With probability $15/32$ one such computation will be performed and so on. As a consequence, one iteration of PREIMAGESSW cost on average $\frac{15 \cdot 1 + 7 \cdot 2 + 1 \cdot 3}{32} = 1$ quadratic solver computation.

To sum up, Algorithm 3 requires on average $3 \cdot 2$ field inversions, $3 \cdot 2$ additions of points and $3 \cdot (2 + 1)$ quadratic solver computations, up to a $O(2^{-n/2})$ error term. □

Note that the efficiency of this algorithm can be improved further by choosing a sparse value of b and a value of t that yields sparse precomputed constants. Many of the field multiplications will then be computed faster.

4   Implementation Aspects

Our software implementation targets modern Intel Desktop-based processors, making extensive use of the recently introduced AVX instruction set [16] accessible through compiler intrinsics. The curve choice is the GLS binary curve $(\lambda^2 + \lambda + a)x^2 = x^4 + b$ represented in λ-coordinates and defined over the quadratic extension $\mathbb{F}_{2^{254}}$. The extension is built by choosing the irreducible trinomial $g(w) = w^2 + w + 1$ over the base field $\mathbb{F}_{2^{127}}$ defined with the irreducible trinomial $f(z) = z^{127} + z^{63} + 1$. In this set of parameters, a field element a is

---

[2] This can be justified rigorously using the fact that the corresponding function field extensions are pairwise linearly disjoint, exactly as in the image size computations of [18, Sect. 4]. For simplicity, we do not include the tedious Galois extension computations involved.

represented as $a = a_0 + a_1 w$, with $a_0, a_1 \in \mathbb{F}_{2^{127}}$. For simplicity, the parameter t is chosen to be a random subfield element, allowing the computational savings by sparse multiplications described in the previous section.

Squaring and multiplication. Field squaring closely mirrors the vector formulation proposed in [3], with coefficient expansion implemented by table lookups performed through byte-shuffling instructions. The table lookups operate on registers only, allowing a very efficient constant-time implementation. Field multiplication is natively supported by the carry-less multiplier (PCLMULQDQ instruction), with the number of word multiplications reduced through application of Karatsuba formulae, as described in [28]. Modular reduction is implemented with a shift-and-add approach, with careful choice of aligning vector word shifts on multiples of 8, to explore the faster memory alignment instructions available in the target platform.

Quadratic solver. For an odd extension degree m, the half-trace function HTr : $\mathbb{F}_{2^m} \to \mathbb{F}_{2^m}$ is defined by $\mathrm{HTr}(c) = \sum_{i=0}^{(m-1)/2} c^{2^{2i}}$ and computes a solution $c \in \mathbb{F}_{2^m}$ to the quadratic equation $\lambda^2 + \lambda = c + \mathrm{Tr}(c)$. Let $\mathrm{Tr}' : \mathbb{F}_{2^{2m}} \to \mathbb{F}_2$ denote the trace function in a quadratic extension. The equation $\lambda^2 + \lambda = c$ can be solved for a trace zero element $c = c_0 + c_1 w \in \mathbb{F}_{2^{2m}}$ by computing two half-traces in $\mathbb{F}_{2^m}$, as described in [20]. First, solve $\lambda_1^2 + \lambda_1 = c_1$ to obtain λ_1, and then solve $\lambda_0^2 + \lambda_0 = c_0 + c_1 + \lambda_1 + \mathrm{Tr}(c_0 + c_1 + \lambda_1)$ to obtain the solution $\lambda = \lambda_0 + (\lambda_1 + \mathrm{Tr}(c_0 + c_1 + \lambda_1))w$. This approach is very efficient for variable-time implementations and only requires two half-trace computations in the base field, where each half-trace computation employs a large precomputed table of $2^8 \cdot \lceil \frac{m}{8} \rceil$ field elements [25].

A more naive approach evaluates the function by alternating $m - 1$ consecutive squarings and $(m - 1)/2$ additions, with the advantage of taking constant-time (if squaring and addition are also constant-time, as in the case here). We derive a faster way to compute the half-trace function in constant-time over quadratic extension fields. Applying the naive approach to a quadratic extension allows a significant speedup due to the linear property of half-trace, by reducing the cost to essentially one constant-time half-trace computation over the base field. Since $\mathrm{Tr}'(c) = 0$, we have $\mathrm{Tr}(c_1) = 0$ and $\mathrm{Tr}(\lambda_1) = 0$ for the choice of λ_1 as the half-trace of c_1 as solution of $\lambda_1^2 + \lambda_1 = c_1$. This simplifies the expression above to $\lambda_0^2 + \lambda_0 = c_0 + c_1 + \lambda_1 + \mathrm{Tr}(c_0)$. Substituting $d = c_0 + \mathrm{Tr}(c_0)$, the expression for λ_0 becomes:

$$\lambda_0 = \sum_{i=0}^{(m-1)/2} (d + c_1 + \lambda_1)^{2^{2i}} = \sum_{i=0}^{(m-1)/2} \left(d + c_1 + \sum_{j=0}^{(m-1)/2} c_1^{2^{2j}} \right)^{2^{2i}}.$$

The expansion of the inner sum allows the interleaving of the consecutive squarings. The analysis can be split in two cases, depending on the format of the extension degree m:

$$\lambda_0 = \begin{cases} c_0 + \displaystyle\sum_{i=0}^{\lfloor m/4 \rfloor - 1} (c_0^{16} + d^4 + c_1^4 + c_1^8)^{2^{4i}} & \text{if } m \equiv 1 \pmod 4 \\[2em] \displaystyle\sum_{i=0}^{\lfloor m/4 \rfloor} (c_0 + d^4 + c_1^2 + c_1^4)^{2^{4i}} & \text{if } m \equiv 3 \pmod 4. \end{cases}$$

The value λ_1 can then be computed as $\lambda_1 = \lambda_0^2 + \lambda_0 + d + c_1$, for a total of approximately m squarings and $m/4$ additions, a cost comparable to a single constant-time half-trace in the base field.

Inversion. Field inversion is implemented by two different approaches based on the Itoh-Tsuji algorithm [21]. This algorithm computes $a^{-1} = a^{(2^{m-1}-1)2}$, as proposed in [19], with the cost of $m-1$ squarings and a number of multiplications determined by the length of an addition chain for $m - 1$. For a variable-time implementation, the squarings for each 2^i-power involved can be converted into multi-squarings [8], implemented as a trade-off between space consumption and execution time. Each multi-squaring table requires the storage of $2^4 \cdot \lceil \frac{m}{4} \rceil$ field elements. A constant-time implementation must perform consecutive squarings and cannot benefit considerably from a precomputed table of field elements without introducing variance in memory latency, potentially exploitable by an intrusive attacker.

Point addition. The last performance-critical operation to be described is the point addition in λ-affine coordinates. A formula for adding points $P = (x_P, y_P)$ and $Q = (x_Q, y_Q)$ on the curve is proposed in [25], with associated cost of 2 inversions, 4 multiplications and 2 squarings:

$$x_{P+Q} = \frac{x_P \cdot x_Q (\lambda_P + \lambda_Q)}{(x_P + x_Q)^2}, \qquad \lambda_{P+Q} = \frac{x_Q \cdot (x_{P+Q} + x_P)^2}{x_{P+Q} \cdot x_P} + \lambda_P + 1.$$

Simple substitution of x_{P+Q} in the computation of λ_{P+Q} gives faster new formulas. By unifying the denominators, one field inversion can be traded for 2 multiplications in the formulas below, with associated cost of 1 inversion, 6 multiplications and 2 squarings:

$$x_{P+Q} = \frac{x_P \cdot x_Q (\lambda_P + \lambda_Q)^2}{(x_P + x_Q)^2 (\lambda_P + \lambda_Q)}$$

$$\lambda_{P+Q} = \frac{\left[(x_P + x_Q)^2 + x_Q \cdot (\lambda_P + \lambda_Q) \right]^2}{(x_P + x_Q)^2 (\lambda_P + \lambda_Q)} + \lambda_P + 1.$$

5   Experimental Results

The implementation was completed with help of the latest version of the RELIC toolkit [2]. Random number generation was implemented with the recently introduced RDRAND instruction [12]. Software was compiled with a prerelease version of GCC 4.9 available in the Arch Linux distribution with flags for loop unrolling, aggressive optimization (-O3 level) and specific tuning for the Sandy/Ivy Bridge microarchitectures. Table 2 presents timings in clock cycles for field arithmetic and Elligator Squared in two different platforms – an Intel Ivy Bridge Core i5 3317U 1.7 GHz and a Haswell Core i7 4770 K 3.5 GHz. The timings were taken as the average of 10^4 executions, with TurboBoost and HyperThreading disabled to reduce randomness in the results.

The constant-time implementation results are mostly for reference: indeed, since the Elligator Squared operation is efficiently invertible, there is no strong reason to compute it in constant time: timing information does not leak secret key data like in the case of a scalar multiplication. However, timing information could conceivably help an active distinguishing attacker; the corresponding attack scenarios are far-fetched, but the paranoid may prefer to choose constant-time arithmetic as a matter of principle.

6   Comparison of Elligator 2 and Elligator Squared on Prime Finite Fields

We have implemented Elligator 2 [6] and the corresponding Elligator Squared construction on Curve25519 [4] using the fast arithmetic provided by Bernstein et al. as part of the publicly available implementation of Curve25519 and Ed25519 [5] in SUPERCOP, in order to compare the two proposed methods on Edwards curves in large characteristic (and to see how they both perform compared to our binary implementation).

To generate a random point and compute the corresponding bitstring representation, the Elligator method requires, on average, 2 scalar multiplications, 2 tests for the existence of preimages and 1 preimage computation. On the other hand, for the same computation, Elligator Squared requires, on average, 1 scalar multiplication, 2 tests for the existence of preimages, 1 preimage computation and 2 computations of the Elligator 2 map function. As a result, compared to the Elligator approach, the Elligator Squared approach requires one scalar multiplication less, but two map function computations more. Therefore, Elligator will be faster than Elligator Squared in contexts where a scalar multiplication is cheaper than two map function evaluations and conversely. Elligator will thus tend to have an edge for protocols using fixed base point scalar multiplication, whereas Elligator Squared will perform better for protocols using variable base point scalar multiplication.

This is confirmed by our implementation results, as reported in Table 3, which are 35–40 % in favor of Elligator in the fixed-base case (FB) but 30–35 % in favor of Elligator Squared in the variable-base case (VB). Note that the variable-base scalar multiplication results are estimates based on the SUPERCOP performance numbers on haswell and hydra2. A comparison with Table 2 shows that the binary curve approach is 25 % to 200 % times faster than the fastest Curve25519 implementation. Observe that our results were obtained using a binary GLS curve with efficient arithmetic implemented in processors with native support

Table 3. Timings for Elligator Squared and Elligator 2 on Curve25519. Results are in clock cycles and were taken as the average of 10^4 executions with random inputs. FB/VB are as in Table 2.

Operation	Ivy Bridge	Haswell
Scalar multiplication (fixed-base)	42,570	42,180
Scalar multiplication (variable-base, est.)	182,490	162,460
Map function	38,420	36,590
FB with Elligator Squared	157,500	141,200
FB with Elligator 2	114,800	100,200
VB with Elligator Squared (est.)	297,420	261,480
VB with Elligator 2 (est.)	394,640	340,760

to binary field arithmetic and may not translate directly to different parameter choices or computing platforms.

References

1. ANSSI: Publication d'un paramétrage de courbe elliptique visant des applications de passeport électronique et de l'administration électronique française, November 2011. http://www.ssi.gouv.fr/fr/anssi/publications/publications-scientifiques/autres-publications/publication-d-un-parametrage-de-courbe-elliptique-visant-des-applications-de.html
2. Aranha, D.F., Gouvêa, C.P.L.: RELIC is an efficient library for cryptography. http://code.google.com/p/relic-toolkit/
3. Aranha, D.F., López, J., Hankerson, D.: Efficient software implementation of binary field arithmetic using vector instruction sets. In: Abdalla, M., Barreto, P.S.L.M. (eds.) LATINCRYPT 2010. LNCS, vol. 6212, pp. 144–161. Springer, Heidelberg (2010)
4. Bernstein, D.J.: Curve25519: new Diffie-Hellman speed records. In: Yung, M., Dodis, Y., Kiayias, A., Malkin, T. (eds.) PKC 2006. LNCS, vol. 3958, pp. 207–228. Springer, Heidelberg (2006)
5. Bernstein, D.J., Duif, N., Lange, T., Schwabe, P., Yang, B.-Y.: High-speed high-security signatures. J. Crypt. Eng. **2**(2), 77–89 (2012)
6. Bernstein, D.J., Hamburg, M., Krasnova, A., Lange, T.: Elligator: elliptic-curve points indistinguishable from uniform random strings. In: Gligor, V., Yung, Y. (eds.) ACM CCS (2013)
7. Bernstein, D.J., Hamburg, M., Krasnova, A., Lange, T.: Elligator: software, August 2013. http://elligator.cr.yp.to/software.html
8. Bos, J.W., Kleinjung, T., Niederhagen, R., Schwabe, P.: ECC2K-130 on cell CPUs. In: Bernstein, D.J., Lange, T. (eds.) AFRICACRYPT 2010. LNCS, vol. 6055, pp. 225–242. Springer, Heidelberg (2010)
9. Brier, E., Coron, J.-S., Icart, T., Madore, D., Randriam, H., Tibouchi, M.: Efficient indifferentiable hashing into ordinary elliptic curves. Cryptology ePrint Archive, Report 2009/340 (2009). http://eprint.iacr.org/. (Full version of [10])

10. Brier, E., Coron, J.-S., Icart, T., Madore, D., Randriam, H., Tibouchi, M.: Efficient indifferentiable hashing into ordinary elliptic curves. In: Rabin, T. (ed.) CRYPTO 2010. LNCS, vol. 6223, pp. 237–254. Springer, Heidelberg (2010)

11. Certicom Research. SEC 2: recommended elliptic curve domain parameters, version 2.0, January 2010

12. Intel Corporation: Intel Digital Random Number Generator (DRNG). https://software.intel.com/sites/default/files/managed/4d/91/DRNG_Software_Implementation_Guide_2.0.pdf

13. Farashahi, R.R.: Hashing into Hessian curves. In: Nitaj, A., Pointcheval, D. (eds.) AFRICACRYPT 2011. LNCS, vol. 6737, pp. 278–289. Springer, Heidelberg (2011)

14. Farashahi, R., Fouque, P.-A., Shparlinski, I., Tibouchi, M., Voloch, J.F.: Indifferentiable deterministic hashing to elliptic and hyperelliptic curves. Math. Comput. 82(281), 491–512 (2013)

15. FIPS PUB 186–3. Digital Signature Standard (DSS). NIST, USA (2009)

16. Firasta, N., Buxton, M., Jinbo, P., Nasri, K., Kuo, S.: Intel AVX: new frontiers in performance improvement and energy efficiency. White paper. http://software.intel.com/

17. Fouque, P.-A., Joux, A., Tibouchi, M.: Injective encodings to elliptic curves. In: Boyd, C., Simpson, L. (eds.) ACISP. LNCS, vol. 7959, pp. 203–218. Springer, Heidelberg (2013)

18. Fouque, P.-A., Tibouchi, M.: Indifferentiable hashing to Barreto–Naehrig curves. In: Hevia, A., Neven, G. (eds.) LatinCrypt 2012. LNCS, vol. 7533, pp. 1–17. Springer, Heidelberg (2012)

19. Guajardo, J., Paar, C.: Itoh-Tsujii inversion in standard basis and its application in cryptography and codes. Des. Codes Crypt. 25(2), 207–216 (2002)

20. Hankerson, D., Karabina, K., Menezes, A.: Analyzing the Galbraith-Lin-Scott point multiplication method for elliptic curves over binary fields. IEEE Trans. Comput. 58(10), 1411–1420 (2009)

21. Itoh, T., Tsujii, S.: A fast algorithm for computing multiplicative inverses in $GF(2^m)$ using normal bases. Inf. Comput. 78(3), 171–177 (1988)

22. Knudsen, E.W.: Elliptic scalar multiplication using point halving. In: Lam, K.-Y., Okamoto, E., Xing, C. (eds.) ASIACRYPT 1999. LNCS, vol. 1716, pp. 135–149. Springer, Heidelberg (1999)

23. Lochter, M., Merkle, J.: Elliptic curve cryptography (ECC) Brainpool standard curves and curve generation. RFC 5639 (Informational), March 2010

24. Möller, B.: A public-key encryption scheme with pseudo-random ciphertexts. In: Samarati, P., Ryan, P.Y.A., Gollmann, D., Molva, R. (eds.) ESORICS 2004. LNCS, vol. 3193, pp. 335–351. Springer, Heidelberg (2004)

25. Oliveira, T., López, J., Aranha, D.F., Rodríguez-Henríquez, F.: Two is the fastest prime: lambda coordinates for binary elliptic curves. J. Crypt. Eng. 4(1), 3–17 (2014)

26. Schroeppel, R.: Elliptic curves: twice as fast! Presentation at the CRYPTO 2000 Rump Session (2000)

27. Shallue, A., van de Woestijne, C.E.: Construction of rational points on elliptic curves over finite fields. In: Hess, F., Pauli, S., Pohst, M. (eds.) ANTS 2006. LNCS, vol. 4076, pp. 510–524. Springer, Heidelberg (2006)

28. Taverne, J., Faz-Hernández, A., Aranha, D.F., Rodríguez-Henríquez, F., Hankerson, D., López, J.: Speeding scalar multiplication over binary elliptic curves using the new carry-less multiplication instruction. J. Crypt. Eng. 1(3), 187–199 (2011)

29. Tibouchi, M.: Elligator Squared: uniform points on elliptic curves of prime order as uniform random strings. In: Christin, N., Safavi-Naini, R. (eds.) Financial Cryptography. LNCS. Springer, Heidelberg (2014). (To appear)

30. Weinberg, Z., Wang, J., Yegneswaran, V., Briesemeister, L., Cheung, S., Wang, F., Boneh, D.: StegoTorus: a camouflage proxy for the Tor anonymity system. In: Yu, T., Danezis, G., Gligor, V.D. (eds.) ACM CCS, pp. 109–120. ACM (2012)

31. Wustrow, E., Wolchok, S., Goldberg, I., Halderman, J.A.: Telex: anticensorship in the network infrastructure. In: USENIX Security Symposium. USENIX Association (2011)

32. Young, A.L., Yung, M.: Space-efficient kleptography without random oracles. In: Furon, T., Cayre, F., Doërr, G., Bas, P. (eds.) IH 2007. LNCS, vol. 4567, pp. 112–129. Springer, Heidelberg (2008)

33. Young, A., Yung, M.: Kleptography from standard assumptions and applications. In: Garay, J.A., De Prisco, R. (eds.) SCN 2010. LNCS, vol. 6280, pp. 271–290. Springer, Heidelberg (2010)

Batch NFS

Daniel J. Bernstein[1,2]($\boxtimes$) and Tanja Lange[2]($\boxtimes$)

[1] Department of Computer Science, University of Illinois at Chicago,
Chicago, IL 60607-7045, USA
djb@cr.yp.to
[2] Department of Mathematics and Computer Science,
Technische Universiteit Eindhoven, P.O. Box 513,
5600 MB Eindhoven, The Netherlands
tanja@hyperelliptic.org

Abstract. This paper shows, assuming standard heuristics regarding the number-field sieve, that a "batch NFS" circuit of area $L^{1.181...+o(1)}$ factors $L^{0.5+o(1)}$ separate B-bit RSA keys in time $L^{1.022...+o(1)}$. Here $L = \exp((\log 2^B)^{1/3}(\log \log 2^B)^{2/3})$. The circuit's area-time product (price-performance ratio) is just $L^{1.704...+o(1)}$ per key. For comparison, the best area-time product known for a single key is $L^{1.976...+o(1)}$.

This paper also introduces new "early-abort" heuristics implying that "early-abort ECM" improves the performance of batch NFS by a super-polynomial factor, specifically $\exp((c + o(1))(\log 2^B)^{1/6}(\log \log 2^B)^{5/6})$ where c is a positive constant.

Keywords: Integer factorization · Number-field sieve · Price-performance ratio · Batching · Smooth numbers · Elliptic curves · Early aborts

1 Introduction

The cryptographic community reached consensus a decade ago that a 1024-bit RSA key can be broken in a year by an attack machine costing significantly less than 10^9 dollars. See [51], [38], [24], and [23]. The attack machine is an optimized version of the number-field sieve (NFS), a factorization algorithm that has been intensively studied for twenty years, starting in [36]. The run-time analysis of NFS relies on various heuristics, but these heuristics have been confirmed in a broad range of factorization experiments using several independent NFS software implementations: see, e.g., [29], [30], [31], and [4].

Despite this threat, 1024-bit RSA remains the workhorse of the Internet's "DNS Security Extensions" (DNSSEC). For example, at the time of this writing (September 2014), the IP address of the domain dnssec-deployment.org is

This work was supported by the National Science Foundation under grant 1018836 and by the Netherlands Organisation for Scientific Research (NWO) under grant 639.073.005. Permanent ID of this document: 4f99b1b911984e501c099f514d8fd2ce. Date: 2014.09.17.

© Springer International Publishing Switzerland 2014
A. Joux and A. Youssef (Eds.): SAC 2014, LNCS 8781, pp. 38–58, 2014.
DOI: 10.1007/978-3-319-13051-4_3

- signed by that domain's 1024-bit "zone-signing key", which in turn is
- signed by that domain's 2048-bit "key-signing key", which in turn is
- signed by .org's 1024-bit zone-signing key, which in turn is
- signed by .org's 2048-bit key-signing key, which in turn is
- signed by the DNS root's 1024-bit zone-signing key, which in turn is
- signed by the DNS root's 2048-bit key-signing key.

An attacker can forge this IP address by factoring any of the three 1024-bit RSA keys in this chain.

A report [41] last year indicated that, out of the 112 top-level domains using DNSSEC, 106 used the same key sizes as .org. We performed our own survey of zone-signing keys in September 2014, after many new top-level domains were added. We found 286 domains using 1024-bit keys; 4 domains using 1152-bit keys; 192 domains using 1280-bit keys; and just 22 domains using larger keys. Almost all of the 1280-bit keys are for obscure domains such as .boutique and .rocks; high-volume domains practically always use 1024-bit keys.

Evidently DNSSEC users find the attacks against 1024-bit RSA less worrisome than the obvious costs of moving to larger keys. There are, according to our informal surveys of these users, three widespread beliefs supporting the use of 1024-bit RSA:

- A typical RSA key is believed to be worth less than the cost of the attack machine.
- Building the attack machine means building a huge farm of application-specific integrated circuits (ASICs). Standard computer clusters costing the same amount of money are believed to take much longer to perform the same calculations.
- It is believed that switching RSA signature keys after (e.g.) a month will render the attack machine useless, since the attack machine requires a full year to run.

Consider, for example, the following quote from the latest "DNSSEC operational practices" recommendations [32, Sect. 3.4.2], published December 2012:

> DNSSEC signing keys should be large enough to avoid all known cryptographic attacks during the effectivity period of the key. To date, despite huge efforts, no one has broken a regular 1024-bit key; in fact, the best completed attack is estimated to be the equivalent of a 700-bit key. An attacker breaking a 1024-bit signing key would need to expend phenomenal amounts of networked computing power in a way that would not be detected in order to break a single key. Because of this, it is estimated that most zones can safely use 1024-bit keys for at least the next ten years.

This quote illustrates the first and third beliefs reported above: the attack cost would be "phenomenal" and would break only "a single key"; furthermore, the attack would have to be completed "during the effectivity period of the key". A typical DNSSEC key is valid for just one month and is then replaced by a new key.

1.1 Contents of this paper. This paper analyzes the *asymptotic* cost, specifically the *price-performance ratio*, of breaking *many* RSA keys. We emphasize several words here:

- "Many": The attacker is faced not with a single target, but with many targets. The algorithmic task here is not merely to break, e.g., a single 1024-bit RSA key; it is to break more than one hundred 1024-bit RSA keys for DNSSEC top-level domains, many more 1024-bit RSA keys at lower levels of DNSSEC, millions of 1024-bit RSA keys in SSL (as in [25], and [35]; note that upgrading SSL to 2048-bit RSA does nothing to protect the confidentiality of previously recorded SSL traffic), etc. This is important if there are ways to share attack work across the keys.
- "Price-performance ratio": As in [53], [18], [50], [15], [54], [7], [51], [56], [23], [24], etc., our main interest is not in the number of "operations" carried out by an algorithm, but in the actual price and performance of a machine carrying out those operations. Parallelism increases price but often improves performance; large storage arrays are a problem for both price and performance. We use price-performance ratio as our primary cost metric, but we also report time separately since signature-key rotation puts a limit upon time.
- "Asymptotic": The cost improvements that we present are superpolynomial in the size of the numbers being factored. We thus systematically suppress all polynomial factors in our cost analyses, simplifying the analyses.

This paper presents a new "batch NFS" circuit of area $L^{1.181...+o(1)}$ that, assuming standard NFS heuristics, factors $L^{0.5+o(1)}$ separate B-bit RSA keys in total time just $L^{1.022...+o(1)}$. The area-time product is $L^{1.704...+o(1)}$ for each key; i.e., the price-performance ratio is $L^{1.704...+o(1)}$. Here (as usual for NFS) L means $\exp((\log N)^{1/3}(\log \log N)^{2/3})$ where $N = 2^B$.

For comparison (see Table 1.4), the best area-time product known for factoring a single key (without quantum computers) is $L^{1.976...+o(1)}$, even if nonuniform precomputations such as Coppersmith's "factorization factory" are allowed. The literature is reviewed below.

This paper also looks more closely at the $L^{o(1)}$. The main bottleneck in batch NFS is not traditional sieving, but rather low-memory factorization, motivating new attention to the complexity of low-memory factorization. Traditional ECM, the elliptic-curve method of recognizing y-smooth integers, works in low memory and takes time $\exp(\sqrt{(2+o(1))}\log y \log \log y)$. One can reasonably guess that, compared to traditional ECM, "early-abort ECM" saves a subexponential factor here, but the complexity of early-abort ECM has never been analyzed. Section 3 of this paper introduces new early-abort heuristics implying that the complexity of early-abort ECM is $\exp\left(\sqrt{\left(\frac{8}{9}+o(1)\right)}\log y \log \log y\right)$. Using early aborts increases somewhat the number of auxiliary integers that need to be factored, producing a further increase in cost, but the cost is outweighed by the faster factorization.

The ECM cost is obviously bounded by $L^{o(1)}$: more precisely, the cost is $\exp(\Theta((\log N)^{1/6}(\log \log N)^{5/6}))$ in the context of batch NFS, since $y \in L^{\Theta(1)}$.

This cost is invisible at the level of detail of $L^{1.704\ldots+o(1)}$. The speedup from ECM to early-abort ECM is nevertheless superpolynomial and directly translates into the same speedup in batch NFS.

1.2 Security consequences. We again emphasize that our results are asymptotic. This prevents us from directly drawing any conclusions about 1024-bit RSA, or 2048-bit RSA, or any other specific RSA key size. Our results are nevertheless sufficient to undermine all three of the beliefs described above:

- Users comparing the value of an RSA key to the cost of an attack machine need to know the *per-key* cost of batch NFS. This has not been seriously studied. What the literature has actually studied in detail is the cost of NFS attacking *one key at a time*; this is not the same question. Our asymptotic results do not rule out the possibility that these costs are the same for 1024-bit RSA, but there is also no reason to be confident about any such possibility.
- Most of the literature on single-key NFS relies heavily on operations that—for large key sizes—are not handled efficiently by current CPUs and that become much more efficient on ASICs: consider, for example, the routing circuit in [51]. Batch NFS relies much more heavily on massively parallel elliptic-curve scalar multiplication, exactly the operation that is shown in [12], [11], and [17] to fit very well into off-the-shelf graphics cards. The literature supports the view that off-the-shelf hardware is much less cost-effective than ASICs for single-key NFS, but there is no reason to think that the same is true for batch NFS.
- The natural machine size for batch NFS (i.e., the circuit area if price-performance ratio is optimized) is larger than the natural machine size for single-key NFS, but the natural *time* is considerably smaller. As above, these asymptotic results undermine any confidence that one can obtain from comparing the natural time for single-key NFS to the rotation interval for signature keys: there is no reason to think that the latency of batch NFS will be as large as the latency of single-key NFS. Note that, even though this paper emphasizes optimal price-performance ratio for simplicity, there are also techniques to further reduce the time below the natural time, hitting much lower latency targets without severely compromising price-performance ratio: in particular, for the core sorting subroutines inside linear algebra, one can replace time T with T/f at the expense of replacing area A with Af^2.

The standard measure of security is the total cost of attacking *one* key. For example, this is what NIST is measuring in [6] when it reports "80-bit security" for 1024-bit RSA, "112-bit security" for 2048-bit RSA, "128-bit security" for 3072-bit RSA, etc. What batch NFS illustrates is that, when there are many user keys, the attacker's cost *per key* can be smaller than the attacker's total cost for one key. It is much more informative to measure the attacker's total cost of attacking U user keys, as a function of U. It is even more informative to measure the attacker's chance of breaking exactly K out of U simultaneously attacked keys in time T using a machine of cost A, as a function of (K, U, T, A).

There are many other examples of cryptosystems where the attack cost does not grow linearly with the number of targets. For example, it is well known that exhaustive search finds preimages for U hash outputs in about the same time as a preimage for a single hash output; furthermore, the first preimage that it finds appears after only $1/U$ of the total time, reducing actual security by $\lg U$ bits. However, most cryptosystems have moved up to at least a "128-bit" security level, giving them a buffer against losing some bits of security. RSA is an exception: its poor performance at high security levels has kept it at a bleeding-edge "80-bit security" level. Even when users can be convinced to move away from 1024-bit keys, they normally move to $\leq$2048-bit keys. We question whether it is appropriate to view 1024-bit keys as "80-bit" security and 2048-bit keys as "112-bit" security if the attacker's costs *per key* are not so high.

1.3   Previous work. In the NFS literature, as in the algorithm literature in general, there is a split between traditional analyses of "operations" (adding two 64-bit integers is one "operation"; looking up an element of a 2^{64}-byte array is one "operation") and modern analyses of more realistic models of computation. We follow the terminology of our paper [14]: the "RAM metric" counts traditional operations, while the "AT metric" multiplies the area of a circuit by the time taken by the same circuit.

Buhler, H. Lenstra, and Pomerance showed in [19] (assuming standard NFS heuristics, which we now stop mentioning) that NFS factors a single key N with RAM cost $L^{1.922...+o(1)}$. As above, L means $\exp((\log N)^{1/3}(\log\log N)^{2/3})$. This exponent $1.922...$ is the most frequently quoted cost exponent for NFS.

Coppersmith in [20] introduced two improvements to NFS. The first, "multiple number fields", reduces the exponent $1.922... + o(1)$ to $1.901... + o(1)$. The second, the "factorization factory", is a *non-uniform* algorithm that reduces $1.901...+o(1)$ to just $1.638...+o(1)$. Recall that (size-)non-uniform algorithms are free to perform arbitrary amounts of precomputation as functions of the *size* of the input, i.e., the number of bits of N. A closer look shows that Coppersmith's precomputation costs $L^{2.006...+o(1)}$, so if it is applied to more than $L^{0.368...+o(1)}$ inputs then the precomputation cost can quite reasonably be ignored.

Essentially all of the subsequent NFS literature has consisted of analysis and optimization of algorithms that cost $L^{1.922...+o(1)}$, such as the algorithm of [19]. The ideas of [20] have been dismissed for three important reasons:

- The bottleneck in [19] is sieving, while the bottleneck in [20] is ECM. Both of these algorithms use $L^{o(1)}$ operations in the RAM metric, but the $o(1)$ is considerably smaller for sieving than for ECM.
- Even if the $o(1)$ in [20] were as small as the $o(1)$ in [19], there would not be much benefit in $1.901... + o(1)$ compared to $1.922... + o(1)$. For example, $(2^{50})^{1.922} \approx 2^{96}$ while $(2^{50})^{1.901} \approx 2^{95}$.
- The change from $1.901... + o(1)$ to $1.638... + o(1)$ is much larger, but it comes at the cost of massive memory consumption. Specifically, [20] requires space $L^{1.638...+o(1)}$, while [19] uses space just $L^{0.961...+o(1)}$. This is not visible in the RAM metric but is obviously a huge problem in reality, and it becomes

Table 1.4. Asymptotic exponents for several variants of NFS, assuming standard heuristics. "Exponent" e means asymptotic cost $L^{e+o(1)}$ per key factored. "Precomp" 2θ means that there is a precomputation involving integer pairs (a, b) up to $L^{\theta+o(1)}$, for total precomputation cost $L^{2\theta+o(1)}$; algorithms without precomputation have $2\theta = 0$. "Batch" β means batch size $L^{\beta+o(1)}$; algorithms handling each key separately have $\beta = 0$. See Sect. 2 for further details.

metric	exponent	precomp	batch	source
AT	1.976...	0	0	2001 Bernstein [7]
RAM (unrealistic)	1.922...	0	0	1993 Buhler–H. Lenstra–Pomerance [19]
RAM (unrealistic)	1.901...	0	0	1993 Coppersmith [20]
AT	1.900...	0	0.1	**batch NFS**; this paper
AT	1.829...	0	0.2	**batch NFS**; this paper
AT	1.763...	0	0.3	**batch NFS**; this paper
AT	1.710...	0	0.4	**batch NFS**; this paper
AT	1.704...	0	0.5	**batch NFS**; this paper
RAM (unrealistic)	1.638...	2.006...	0	1993 Coppersmith [20]

increasingly severe as computations grow larger. As a concrete illustration of the real-world costs of storage and computation, paying for 2^{70} bytes of slow storage (about $30 \cdot 10^9$ USD in hard drives) is much more troublesome than paying for 2^{80} floating-point multiplications (about $0.02 \cdot 10^9$ USD in GPUs plus $0.005 \cdot 10^9$ USD for a year of electricity).

We quote A. Lenstra, H. Lenstra, Manasse, and Pollard [37]: "There is no indication that the modification proposed by Coppersmith has any practical value."

At the time there was already more than a decade of literature showing how to analyze algorithm asymptotics in more realistic models of computation that account for memory consumption, communication, etc.; see, e.g., [18]. Bernstein in [7] analyzed the circuit performance of NFS, concluding that an optimized circuit of area $L^{0.790...+o(1)}$ would factor N in time $L^{1.18...+o(1)}$, for price-performance ratio $L^{1.976...+o(1)}$. [7] did not analyze the factorization factory but did analyze multiple number fields, concluding that they did not reduce AT cost. The gap between the RAM exponent $1.901... + o(1)$ from [20] and the AT exponent $1.976... + o(1)$ from [7] is explained primarily by communication overhead inside linear algebra, somewhat moderated by parameter choices that reduce the cost of linear algebra at the expense of relation collection.

We pointed out in [14] that the factorization factory does not reduce AT cost. In Sect. 2 we review the reason for this and explain how batch NFS works around it. We also presented in [14] a superpolynomial improvement to the factorization factory in the RAM metric, by eliminating ECM in favor of batch trial division, but this is not useful in the AT metric.

2 Exponents

This section reviews NFS and then explains how to drastically reduce the AT cost of NFS through batching. The resulting cost exponent, $1.704\ldots$ in Table 1.4, is new. All costs in this section are expressed as $L^{e+o(1)}$ for various exponents e. Section 3 looks more closely at the $L^{o(1)}$ factor.

2.1 QS: the Quadratic sieve (1982). As a warmup for NFS we briefly review the general idea of combining congruences, using QS as an example.

QS writes down a large collection of congruences modulo the target integer N and tries to find a nontrivial subcollection whose product is a congruence of squares. One can then reasonably hope that the difference of square roots has a nontrivial factor in common with N.

Specifically, QS computes $s \approx \sqrt{N}$ and writes down the congruences $s^2 \equiv s^2 - N$, $(s+1)^2 \equiv (s+1)^2 - N$, etc. The left side of each congruence is already a square. The main problem is to find a nontrivial set of integers a such that the product of $(s+a)^2 - N$ is a square.

If $(s+a)^2 - N$ is divisible by a very large prime then it is highly unlikely to participate in a square: the prime would have to appear a second time. QS therefore focuses on **smooth** congruences: congruences where $(s+a)^2 - N$ factors completely into small primes. Applying linear algebra modulo 2 to the matrix of exponents in these factorizations is guaranteed to find nonempty subsets of the congruences with square product once the number of smooth congruences exceeds the number of small primes.

The integers a such that $(s+a)^2 - N$ is divisible by a prime p form a small number of arithmetic progressions modulo p. "Sieving" means jumping through these arithmetic progressions to mark divisibility, the same way that the sieve of Eratosthenes jumps through arithmetic progressions to mark non-primality.

2.2 NFS: the number-field sieve (1993). NFS applies the same idea, but instead of congruences modulo N it uses congruences modulo a related algebraic number $m - \alpha$. This algebraic number is chosen to have norm N (divided by a certain denominator shown below), and one can reasonably hope to obtain a factorization of N by obtaining a random factorization of this algebraic number.

Specifically, NFS chooses a positive integer m, and writes N as a polynomial in radix m: specifically, $N = f(m)$ where f is a degree-d polynomial with coefficients $f_d, f_{d-1}, \ldots, f_0 \in \{0, 1, \ldots, m-1\}$. NFS then takes α as a root of f. The norm of $a - b\alpha$ is then $f_d a^d + f_{d-1} a^{d-1} b + \cdots + f_0 b^d$ (divided by f_d), and in particular the norm of $m - \alpha$ is N (again divided by f_d).

It is not difficult to see that optimizing NFS requires d to grow slowly with N, so m is asymptotically on a much smaller scale than N, although not as small as L. More precisely, NFS takes

$$m \in \exp((\mu + o(1))(\log N)^{2/3}(\log \log N)^{1/3})$$

where μ is a positive real constant, optimized below. Note that the inequalities $m^d \le N < m^{d+1}$ imply

$$d \in (1/\mu + o(1))(\log N)^{1/3}(\log \log N)^{-1/3}.$$

NFS uses the congruences $a - bm \equiv a - b\alpha$ modulo $m - \alpha$. There are now two numbers, $a - bm$ and $a - b\alpha$, that both need to be smooth. Smoothness of the algebraic number $a - b\alpha$ is defined as smoothness of the (scaled) norm $f_d a^d + f_{d-1} a^{d-1} b + \cdots + f_0 b^d$, and smoothness of an integer is defined as having no prime divisors larger than y. Here $y \in L^{\gamma + o(1)}$ is another parameter chosen by NFS; $\gamma > 1/(6\mu)$ is another real constant, optimized below.

The range of pairs (a, b) searched for smooth congruences is the set of integer pairs in the rectangle $[-H, H] \times [1, H]$. Here H is chosen so that there will be enough smooth congruences to produce squares at the end of the algorithm. Standard heuristics state that $a - bm$ has smoothness probability $L^{-\mu/(3\gamma) + o(1)}$ if a and b are on much smaller scales than m; in particular, if $H \in L^{\theta + o(1)}$ for some positive real number θ then the number of congruences with $a - bm$ smooth is $L^{\phi + o(1)}$ with $\phi = 2\theta - \mu/(3\gamma)$. Standard heuristics also state the simultaneous smoothness probability of $a - bm$ and $a - b\alpha$, implying that to obtain enough smooth congruences one can take $H \in L^{\theta + o(1)}$ with $\theta = (3\mu\gamma^2 + 2\mu^2)/(6\mu\gamma - 1)$ and $\phi = (18\mu\gamma^3 + 6\mu^2\gamma + \mu)/(18\mu\gamma^2 - 3\gamma)$. See, e.g., [19]. We henceforth assume these formulas for θ and ϕ in terms of μ and γ.

2.3 RAM cost analysis (1993).

Sieving for y-smoothness of $H^{2+o(1)}$ polynomial values uses $H^{2+o(1)}$ operations, provided that y is bounded by $H^{2+o(1)}$. The point here is that the pairs (a, b) with congruences divisible by p form a small number of shifted lattices of determinant p, usually with basis vectors of length $O(\sqrt{p})$, making it easy to find all the lattice points inside the rectangle $[-H, H] \times [1, H]$. The number of operations is thus essentially the number of points marked, and each point is marked just $\sum_{p \le y} 1/p \approx \log \log y$ times.

Sparse techniques for linear algebra involve $y^{1+o(1)}$ matrix-vector multiplications, each involving $y^{1+o(1)}$ operations, for a total of $y^{2+o(1)}$ operations. Other subroutines in NFS take negligible time, so the overall RAM cost of NFS is $L^{\max\{2\theta, 2\gamma\} + o(1)}$.

It is not difficult to see that the exponent $\max\{2\theta, 2\gamma\}$ achieves its minimum value $(64/9)^{1/3} = 1.922\ldots$ with $\mu = (1/3)^{1/3} = 0.693\ldots$ and $\theta = \gamma = (8/9)^{1/3} = 0.961\ldots$. This exponent $1.922\ldots$ is the NFS exponent from [19], and as mentioned earlier is the most frequently quoted NFS exponent. We do not review the multiple-number-fields improvement to $1.901\ldots$ from [20]; as far as we know, multiple number fields do not improve any of the exponents analyzed below.

2.4 AT cost analysis (2001).

In the AT metric there is an important obstacle to cost $H^{2+o(1)}$ for sieving: namely, communicating across area $H^{2+o(1)}$

takes time at least $H^{1+o(1)}$. One can efficiently split the sieving problem into $H^{2+o(1)}/y^{1+o(1)}$ tasks, running one task after another on a smaller array of size $y^{1+o(1)}$, but communicating across this array still takes time at least $y^{0.5+o(1)}$, so AT is at least $H^{2+o(1)}y^{0.5+o(1)}$.

Fortunately, there is a much more efficient alternative to sieving: ECM, explained in Appendix A of the full version of this paper online. What matters in this section is that ECM tests y-smoothness in time $y^{o(1)}$ on a circuit of area $y^{o(1)}$. A parallel array of ECM units, each handling a separate number, tests y-smoothness of $H^{2+o(1)}$ polynomial values in time $H^{2+o(1)}/y^{1+o(1)}$ on a circuit of area $y^{1+o(1)}$, achieving $AT = H^{2+o(1)}$.

Unfortunately, the same obstacle shows up again for linear algebra, and this time there is no efficient alternative. Multiplying a sparse matrix by a vector requires time $y^{0.5+o(1)}$ on a circuit of area $y^{1+o(1)}$, and must be repeated $y^{1+o(1)}$ times. The overall AT cost of NFS is $L^{\max\{2\theta,2.5\gamma\}+o(1)}$.

The exponent $\max\{2\theta, 2.5\gamma\}$ achieves its minimum value $1.976\ldots$ with $\mu = 0.702\ldots$, $\gamma = 0.790\ldots$, and $\theta = 0.988\ldots$. This exponent $1.976\ldots$ is the NFS exponent from [7]. Notice that γ is much smaller here than it was in the RAM optimization: y has been reduced to keep the cost of linear algebra under control, but this also forced θ to increase.

2.5 The factorization factory (1993).

Coppersmith in [20] precomputes "tables which will be useful for factoring any integers in a large range … after the precomputation, an individual integer can be factored in time $L[1/3, 1.639]$", i.e., $L^{\approx 1.639+o(1)}$.

Coppersmith's table is simply the set of (a, b) such that $a - bm$ is smooth. One reuses m, and thus this table, for any integer N between (e.g.) m^d and m^{d+1}.

Coppersmith's method to factor "an individual integer" is to test smoothness of $a - b\alpha$ for each (a, b) in the table. At this point Coppersmith has found the same smooth congruences as conventional NFS, and continues with linear algebra in the usual way.

Coppersmith uses ECM to test smoothness. The problem with sieving here is not efficiency, as in the (subsequent) paper [7], but functionality: sieving can handle polynomial values only at regularly spaced inputs, and the pairs (a, b) in this table are not regularly spaced.

Recall that the size of this table is $L^{\phi+o(1)}$ with $\phi = 2\theta - \mu/(3\gamma)$. ECM uses $L^{o(1)}$ operations per number, for a total smoothness cost of $L^{\phi+o(1)}$, asymptotically a clear improvement over the $L^{2\theta+o(1)}$ for conventional NFS.

The overall RAM cost of the factorization factory is $L^{\max\{\phi,2\gamma\}+o(1)}$. The exponent achieves its minimum value $1.638\ldots$ with $\mu = 0.905\ldots$, $\gamma = 0.819\ldots$, $\theta = 1.003\ldots$, and $\phi = 1.638\ldots$. This is the exponent from [20].

The AT metric tells a completely different story, as we pointed out in [14]. The area required for the table is $L^{\phi+o(1)}$. This area is easy to reuse for very fast parallel smoothness detection, finishing in time $L^{o(1)}$. Unfortunately, collecting the smooth results then takes time $L^{0.5\phi+o(1)}$, for an AT cost of at least

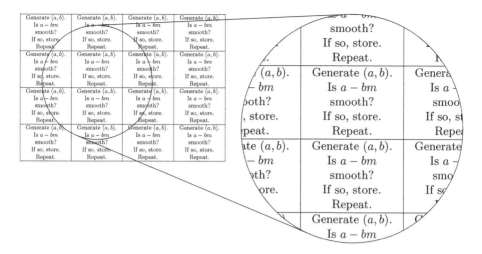

Fig. 2.7. Relation-search mesh finding pairs (a, b) where $a - bm$ is smooth. The following exponents are optimized for factoring a batch of $L^{0.5+o(1)}$ B-bit integers: The mesh has height $L^{0.25+o(1)}$, width $L^{0.25+o(1)}$, and area $L^{0.5+o(1)}$. The mesh consists of $L^{0.5+o(1)}$ small parallel processors (illustration contains 16). Each processor has area $L^{o(1)}$. Each processor knows the same $m \in \exp((0.92115 + o(1))(\log 2^B)^{2/3}(\log\log 2^B)^{1/3})$. Each processor generates its own $L^{0.200484+o(1)}$ pairs (a, b), where a and b are bounded by $L^{1.077242+o(1)}$. Each processor tests each of its own $a - bm$ for smoothness using ECM, using smoothness bound $L^{0.681600+o(1)}$. Together the processors generate $L^{0.700484+o(1)}$ separate pairs (a, b), of which $L^{0.25+o(1)}$ have $a - bm$ smooth.

$L^{\max\{1.5\phi, 2.5\gamma\}+o(1)}$, never mind the problem of matching the table area with the linear-algebra area. The minimum exponent here is above 2.4.

2.6   Batch NFS (new). We drastically reduce AT cost by sharing work across many N's in a different way: we process a *batch* of N's in parallel, rather than performing precomputation to be used for one N at a time. We dynamically enumerate the pairs (a, b) with $a - bm$ smooth, distribute each pair across all the N's in the batch, and remove each pair as soon as possible, rather than storing a complete table of the pairs. To avoid excessive communication costs we completely reorganize data in the middle of the computation: at the beginning each N is repeated many times to bring N close to the pairs (a, b), while at the end the pairs (a, b) relevant to each N are moved much closer together. The rest of this subsection presents the details of the algorithm.

Consider as input a batch of $L^{\beta+o(1)}$ simultaneous targets N within the large range described above. We require $\beta \le \min\{2\phi - 2\gamma, 4\theta - 2\phi\}$; if there are more targets available at once then we actually process those targets in batches of size $L^{\min\{2\phi-2\gamma, 4\theta-2\phi\}+o(1)}$, storing no data between runs.

Consider a square mesh of $L^{\beta+o(1)}$ small parallel processors. This mesh is large enough to store all of the targets N. Use each processor in parallel to test

Is $a - b\alpha_1$ smooth? If so, store. Send (a,b) right. Repeat.	Is $a - b\alpha_2$ smooth? If so, store. Send (a,b) right. Repeat.	Is $a - b\alpha_3$ smooth? If so, store. Send (a,b) right. Repeat.	Is $a - b\alpha_4$ smooth? If so, store. Send (a,b) down. Repeat.
Is $a - b\alpha_5$ smooth? If so, store. Send (a,b) up. Repeat.	Is $a - b\alpha_6$ smooth? If so, store. Send (a,b) left. Repeat.	Is $a - b\alpha_7$ smooth? If so, store. Send (a,b) left. Repeat.	Is $a - b\alpha_8$ smooth? If so, store. Send (a,b) left. Repeat.
Is $a - b\alpha_9$ smooth? If so, store. Send (a,b) right. Repeat.	Is $a - b\alpha_{10}$ smooth? If so, store. Send (a,b) right. Repeat.	Is $a - b\alpha_{11}$ smooth? If so, store. Send (a,b) right. Repeat.	Is $a - b\alpha_{12}$ smooth? If so, store. Send (a,b) down. Repeat.
Is $a - b\alpha_{13}$ smooth? If so, store. Send (a,b) up. Repeat.	Is $a - b\alpha_{14}$ smooth? If so, store. Send (a,b) left. Repeat.	Is $a - b\alpha_{15}$ smooth? If so, store. Send (a,b) left. Repeat.	Is $a - b\alpha_{16}$ smooth? If so, store. Send (a,b) left. Repeat.

Fig. 2.8. Relation-search mesh from Fig. 2.7, now finding pairs (a,b) where both $a - bm$ and $a - b\alpha_i$ are smooth. The mesh knows $L^{0.25+o(1)}$ pairs (a,b) with $a - bm$ smooth from Fig. 2.7. Each (a,b) is copied $L^{0.25+o(1)}$ times (2 times in the illustration) so that it appears in the first two rows, the next two rows, etc. Each (a,b) visits each mesh position within $L^{0.25+o(1)}$ steps (8 steps in the illustration). Each processor knows its own target N_i and the corresponding α_i, and in each step tests each $a - b\alpha_i$ for smoothness using ECM. Together Figs. 2.7 and 2.8 take time $L^{0.25+o(1)}$ to search $L^{0.700484+o(1)}$ pairs (a,b).

smoothness of $a - bm$ for $L^{2\theta-\phi-0.5\beta+o(1)}$ pairs (a,b) using ECM; by hypothesis $2\theta - \phi - 0.5\beta \geq 0$. The total number of pairs here is $L^{2\theta-\phi+0.5\beta+o(1)}$. Each smoothness test takes time $L^{o(1)}$. Overall the mesh takes time $L^{2\theta-\phi-0.5\beta+o(1)}$ and produces a total of $L^{0.5\beta+o(1)}$ pairs (a,b) with $a - bm$ smooth, i.e., only $L^{o(1)}$ pairs for each column of the mesh. See Fig. 2.7.

Move these pairs to the top row of the mesh (spreading them evenly across that row) by a standard sorting algorithm, say the Schnorr–Shamir algorithm from [50], taking time $L^{0.5\beta+o(1)}$. Then broadcast each pair to its entire column, taking time $L^{0.5\beta+o(1)}$. Actually, it will suffice for each pair to appear once somewhere in the first two rows, once somewhere in the next two rows, etc.

Now consider a pair at the top-left corner. Send this pair to its right until it reaches the rightmost column, then down one row, then repeatedly to its left, then back up. In parallel move all the other elements in the first two rows on the same path. In parallel do the same for the third and fourth rows, the fifth and sixth rows, etc. Overall this takes time $L^{0.5\beta+o(1)}$.

Observe that each pair has now visited each position in the mesh. When a pair (a,b) visits a mesh position holding a target N, use ECM to check whether $a - b\alpha$ is smooth, taking time $L^{o(1)}$. The total time to check all $L^{0.5\beta+o(1)}$ pairs against all $L^{\beta+o(1)}$ targets is just $L^{0.5\beta+o(1)}$, plus the time $L^{2\theta-\phi-0.5\beta+o(1)}$ to generate the pairs in the first place. See Fig. 2.8.

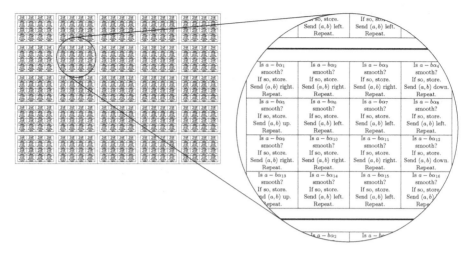

Fig. 2.9. $L^{0.681600+o(1)}$ copies (25 copies in the illustration) of the mesh from Figs. 2.7 and 2.8. Each copy has the same $L^{0.5+o(1)}$ target integers to factor. The total area of this circuit is $L^{1.181600+o(1)}$. In time $L^{0.25+o(1)}$ this circuit searches $L^{1.382084+o(1)}$ pairs (a, b). In time $L^{1.022400+o(1)}$ this circuit searches all $L^{2.154484+o(1)}$ pairs (a, b) and finds, for each target N_i and the corresponding α_i, all $L^{0.681600+o(1)}$ pairs (a, b) for which $a - bm$ and $a - b\alpha_i$ are both smooth.

Repeat this entire procedure $L^{\phi-\gamma-0.5\beta+o(1)}$ times; by hypothesis $\phi - \gamma - 0.5\beta \geq 0$. This covers a total of $L^{2\theta-\gamma+o(1)}$ pairs (a, b), of which $L^{\phi-\gamma+o(1)}$ have $a - bm$ smooth, so for each N there are $L^{o(1)}$ pairs (a, b) for which $a - bm$ and $a - b\alpha$ are both smooth. The total number of smooth congruences found this way across all N is $L^{\beta+o(1)}$. Store each smooth congruence as (N, a, b); all of these together fit into a mesh of area $L^{\beta+o(1)}$. The time spent is $L^{\max\{\phi-\gamma, 2\theta-\gamma-\beta\}+o(1)}$.

Build $L^{\gamma+o(1)}$ copies of the same mesh, all operating in parallel, for a total circuit area of $L^{\beta+\gamma+o(1)}$. Each copy of the mesh has its own copy of the entire list of N's; distributing the N's from an input port through the total circuit area takes time $L^{0.5\beta+0.5\gamma+o(1)}$. The total circuit covers all $L^{2\theta+o(1)}$ pairs (a, b) and obtains, for each N, all of the $L^{\gamma+o(1)}$ smooth congruences required to factor that N. See Fig. 2.9.

We are not done yet: we still need to perform linear algebra for each N. To keep the communication costs of linear algebra under control we pack the linear algebra for each N into the smallest possible area. Allocate a separate square of area $L^{\gamma+o(1)}$ to each N, and route each smooth congruence (N, a, b) in parallel to the corresponding square; this is another standard sorting step, taking total time $L^{0.5\beta+0.5\gamma+o(1)}$ for all $L^{\beta+\gamma+o(1)}$ smooth congruences. Finally, perform linear algebra separately in each square, and complete the factorization of each N as usual. This takes time $L^{1.5\gamma+o(1)}$. See Fig. 2.10.

The overall time exponent is $\max\{\phi - \gamma, 2\theta - \gamma - \beta, 0.5\beta + 0.5\gamma, 1.5\gamma\}$, and the area exponent is $\beta + \gamma$. The final price-performance ratio, AT per integer factored, has exponent $\max\{\phi, 2\theta - \beta, 0.5\beta + 1.5\gamma, 2.5\gamma\}$.

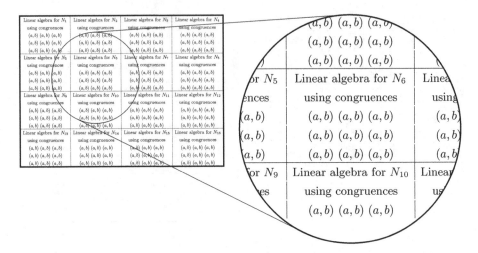

Fig. 2.10. $L^{0.5+o(1)}$ copies (16 copies in the illustration) of a linear-algebra circuit. Each circuit has area $L^{0.681600+o(1)}$. The total area is $L^{1.181600+o(1)}$. Each circuit has its own integer N_i to factor and $L^{0.681600+o(1)}$ pairs (a, b) for which $a - bm$ and $a - b\alpha_i$ are smooth. Routing all pairs (a, b) from Fig. 2.9 to an adjacent (or overlapping and reconfigured) Fig. 2.10 takes time $L^{0.590800+o(1)}$. Each circuit uses $L^{0.681600+o(1)}$ matrix-vector multiplications, and takes time $L^{0.340800+o(1)}$ for each matrix-vector multiplication. The total time is $L^{1.022400+o(1)}$.

2.11   Comparison and numerical parameter optimization.

Bernstein's AT exponent from [7] was $\max\{2\theta, 2.5\gamma\}$. Batch NFS replaces 2θ with $2\theta - \beta$, allowing γ to be correspondingly reduced, at least until β becomes large enough for 2.5γ to cross below ϕ. In principle one should also watch for 2.5γ to cross below $0.5\beta + 1.5\gamma$, but Table 2.12 shows that ϕ is more important.

Of course, even if we ignore the cost of finding the smooth $a - bm$ (the term $2\theta - \beta$), our AT exponent is not as small as Coppersmith's RAM exponent $\max\{\phi, 2\gamma\}$ from [20]. We have an extra $0.5\beta + 1.5\gamma$ term, reflecting the cost of communicating smooth congruences across a batch, and, more importantly, 2.5γ instead of 2γ, reflecting the communication cost of linear algebra.

Table 2.12 shows the smallest exponents that we obtained for various β, in each case from a brief search through 2500000000 pairs (μ, γ). The exponent of the price-performance ratio for batch NFS drops below Bernstein's $1.976\ldots$ as soon as β increases past 0, and reaches a minimum of $1.704\ldots$ as the batch size increases. (The minimum is actually very slightly below 1.704, but our table does not include enough precision to show this.) Finding all (a, b) with $a - bm$ smooth is still a slight bottleneck for $\beta = 0.4$ but disappears for $\beta = 0.5$. When there are more inputs we partition them into batches of size $L^{0.5+o(1)}$, preserving exponent $1.704\ldots$ for the price-performance ratio.

Our optimal $\gamma = 0.681\ldots$ is much smaller than Coppersmith's $\gamma = 0.819\ldots$, for the same reasons that Bernstein's $\gamma = 0.790\ldots$ is smaller than the conventional $\gamma = 0.961\ldots$. The natural time exponent for batch NFS—as above, this means

Table 2.12. Cost exponents for batch NFS in the AT metric. The batch size is $L^{\beta+o(1)}$. The AT cost is $L^{e+o(1)}$. The parameter m is chosen as $\exp((\mu + o(1))(\log N)^{2/3}(\log\log N)^{1/3})$. The prime bound y is chosen as $L^{\gamma+o(1)}$. The (a,b) bound H is chosen as $L^{\theta+o(1)}$. The number of $a - b\alpha$ smoothness tests is $L^{\phi+o(1)}$ per target. The number of $a - bm$ somothness tests is $L^{2\theta-\beta+o(1)}$ per target. The AT cost of routing is $L^{0.5\beta+1.5\gamma+o(1)}$ per target. The AT cost of linear algebra is $L^{2.5\gamma+o(1)}$ per target. All operations take place on a circuit of size $L^{\beta+\gamma+o(1)}$.

batch	AT	m size	primes	(a,b)	$a-b\alpha$	$a-bm$	route	linear
β	e	μ	γ	θ	ϕ	$2\theta-\beta$	$0.5\beta+1.5\gamma$	2.5γ
0.0	1.976052	0.702860	0.790420	0.988026	1.679645	1.976052	1.185630	1.976050
0.1	1.900575	0.705460	0.760230	1.000287	1.691256	1.900575	1.190345	1.900575
0.2	1.829615	0.712320	0.731840	1.014808	1.705173	1.829615	1.197760	1.829600
0.3	1.763034	0.718160	0.705210	1.031517	1.723580	1.763034	1.207815	1.763025
0.4	1.710375	0.820920	0.684150	1.055172	1.710374	1.710345	1.226225	1.710375
0.5	1.704000	0.921150	0.681600	1.077242	1.704000	1.654484	1.272400	1.704000

the time exponent when price-performance ratio is optimized—is just $1.022\ldots$, considerably smaller than the natural time exponent $1.185\ldots$ for single-key NFS. This means that collecting targets into batches produces not merely a drastic improvement in price-performance ratio, but also a side effect of considerably reducing latency.

3 Early-abort ECM

Section 2 used ECM as a low-area smoothness test for auxiliary integers $c = f_d a^d + \cdots + f_0 b^d$. Each curve in ECM catches a fraction of the primes $p \le y$ dividing c, and many curves in sequence catch essentially all of the primes $p \le y$.

This section analyzes a much faster smoothness-detection method, "early-abort ECM". Not all smooth numbers are detected by early-abort ECM, but new heuristics introduced in this section imply that this loss is much smaller than the speedup factor. The overall improvement grows as a superpolynomial function of $\log y$, and therefore grows as a superpolynomial function of the NFS input size.

Specifically, it is well known (see, e.g. [21, page 302]) that (assuming standard conjectures) ECM uses $\exp(\sqrt{(2 + o(1))\log y \log\log y})$ multiplications modulo c to find essentially all primes $p \le y$ dividing c. Here $o(1)$ is some function of y that converges to 0 as $y \to \infty$. Consequently, if a fraction $1/S$ of the ECM inputs are smooth, then ECM uses

$$S \cdot \exp(\sqrt{(2 + o(1))\log y \log\log y})$$

modular multiplications for each smooth integer that it finds. This section's heuristics imply that early-abort ECM uses only

$$S \cdot \exp\left(\sqrt{\left(\frac{8}{9} + o(1)\right) \log y \log \log y}\right)$$

modular multiplications for each smooth integer that it finds. Notice the change from $2 + o(1)$ to $8/9 + o(1)$ in the exponent.

We emphasize again that this paper's analyses are asymptotic. We do not claim that early-abort ECM is better than ECM for any particular value of y.

The rest of this section uses the word "time" to count simple arithmetic operations, such as multiplication and division, on integers with $O(\lg c)$ bits. Each of these operations actually takes time $(\lg c)^{1+o(1)}$, but this extra factor is absorbed into other $o(1)$ terms when c is bounded by the usual functions of y.

3.1   Early-abort trial division. Early aborts predate ECM. They became popular in the 1970s as a component of CFRAC [43], a subexponential-time factorization method that, like batch NFS, generates many "random" numbers that need to be tested for smoothness.

The simplest form of early aborts is single-early-abort trial division. Trial division simply checks divisibility of c by each prime $p \leq y$, taking time $y^{1+o(1)}$. Single-early-abort trial division first checks divisibility of c by each prime $p \leq y^{1/2}$; then throws c away (this is the early abort) if the unfactored part of c is too large; and then, if c has survived the early abort, checks divisibility of c by each prime $p \leq y$.

The definition of "too large" is chosen so that $1/y^{1/2+o(1)}$ of all inputs survive the abort, balancing the cost of the stages before and after the abort. In other words, single-early-abort trial division checks divisibility of each input by each prime $p \leq \sqrt{y}$; keeps the smallest $1/y^{1/2+o(1)}$ of all inputs; and, for each of those inputs, checks divisibility by each prime $p \leq y$.

More generally, $(k-1)$-early-abort trial division removes each prime $p \leq y^{1/k}$ from each input (by dividing by factors found); reduces the number of inputs by a factor of $y^{1/k}$, keeping the smallest inputs; removes each prime $p \leq y^{2/k}$ from each remaining input; reduces the number of inputs by another factor of $y^{1/k}$, keeping the smallest inputs; and so on through $y^{k/k} = y$.

The time per input for $(k-1)$-early-abort trial division is only $y^{1/k+o(1)}$, saving a factor $y^{1-1/k+o(1)}$, if k is limited to a slowly growing function of y. The method does not detect all smooth numbers, but Pomerance's analysis in [45, Sect. 4] shows that the loss factor is only $y^{(1-1/k)/2+o(1)}$, i.e., that the method detects 1 out of every $y^{(1-1/k)/2+o(1)}$ smooth numbers. The overall improvement factor in price-performance ratio is $y^{(1-1/k)/2+o(1)}$; if k is chosen so that $k \to \infty$ as $y \to \infty$ then the improvement factor is $y^{1/2+o(1)}$.

3.2   Early aborts in more generality. One can replace trial division with any method, or combination of methods, of checking for primes $\leq y^{1/k}$, primes $\leq y^{2/k}$, etc.

In particular, Pomerance considered an early-abort version of Pollard's rho method. The original method takes time $y^{1/2+o(1)}$ to find all primes $p \leq y$. Early-abort rho takes time only $y^{1/(2k)+o(1)}$, and Pomerance's analysis shows that it has a loss factor of only $y^{(1-1/k)/4+o(1)}$.

Pomerance actually considered a different method by Pollard and Strassen. The Pollard–Strassen method takes essentially the same amount of time as Pollard's rho method, and has the advantage of a proof of speed without any conjectures, but has the disadvantage of using much more memory.

Pomerance's paper was published in 1982, so of course it did not analyze the elliptic-curve method. After seeing early aborts improve trial division from y to $y^{1/2}$, and improve Pollard's rho method from $y^{1/2}$ to $y^{1/4}$, one might guess that early aborts improve ECM from $\exp(\sqrt{(2+o(1))}\log y \log \log y)$ to $\exp((1/2)\sqrt{(2+o(1))}\log y \log \log y)$, but our heuristics do not agree with this guess.

3.3 Performance of early aborts.

Recall that ECM takes time $T(y)^{1+o(1)}$ to find primes $p \leq y$, where $T(y) = \exp(\sqrt{2 \log y \log \log y})$. We actually consider, in much more generality, any factorization method M taking time $T(y)^{1+o(1)}$ to find primes $p \leq y$, where T is any sufficiently smooth function.

Our early-abort heuristics state that the price-performance ratio of $(k-1)$-early-abort M is the geometric average

$$T(y^{1/k})^{1/k} T(y^{2/k})^{1/k} T(y^{3/k})^{1/k} \cdots T(y)^{1/k}$$

to the power $1 + o(1)$. More generally, cutoffs $y_1, y_2, y_3, \ldots$ produce a geometric average of $T(y_1), T(y_2), T(y_3), \ldots$ with weights $\log y_1, \log y_2 - \log y_1, \log y_3 - \log y_2, \ldots$.

In particular, for any purely exponential $T(y) = y^C$, the price-performance ratio is

$$\left(T(y^{1/k})T(y^{2/k}) \cdots T(y^{(k-1)/k})T(y)\right)^{1/k} = \left(y^{C/k} y^{2C/k} \cdots y^{(k-1)C/k} y^C\right)^{1/k}$$

$$= \left(y^{C \sum_{i=1}^{k} i/k}\right)^{1/k} = y^{C(k+1)/(2k)}$$

which converges to $y^{C/2} = T(y)^{1/2}$ as k increases, matching Pomerance's analyses of early-abort trial division and early-abort rho. More generally, if $T(y) = \exp(C(\log y)^{1/f})$ then $T(y^{i/k}) = T(y)^{(i/k)^{1/f}}$ so

$$\left(T(y^{1/k})T(y^{2/k}) \cdots T(y^{(k-1)/k})T(y)\right)^{1/k} = T(y)^{(\sum_{i=1}^{k}(i/k)^{1/f})/k} \to T(y)^{f/(f+1)}.$$

To prove that $(\sum_{i=1}^{k}(i/k)^{1/f})/k \to f/(f+1)$ as $k \to \infty$, observe that $\sum_{i=1}^{k} i^{1/f}$ is within $k^{1/f}$ of $\int_0^k z^{1/f} dz = (f/(f+1))k^{(f+1)/f}$. ECM is essentially the case $f = 2$: the geometric average is $T(y)^{2/3+o(1)}$.

3.4 Understanding the heuristics. Let y and u be real numbers larger than 1, define $x = y^u$, and define $S_0 = \{1, 2, \ldots, \lfloor x \rfloor\}$. Define $\Psi(x, y)$ as the number of y-smooth integers in S_0. Then $\Psi(x, y)$ is approximately x/u^u. See [45, Theorem 2.1] for a precise statement. The same approximation is still valid for $\Psi(x, y, z)$, the number of y-smooth integers in S_0 having no prime factor $\leq z$, assuming that $z < y^{1-1/\log u}$; see [45, Theorem 2.2].

Let k be a positive integer. Let $y_0, y_1, y_2, \ldots, y_k$ be real numbers with $1 = y_0 < y_1 < y_2 < \cdots < y_k = y$. Let $x_1, x_2, \ldots, x_k$ be positive real numbers with $x = x_1 x_2 \cdots x_k$. Define

$$S_1 = \{c \in S_0 : c/(y_1\text{-smooth part of } c) \leq x/x_1\};$$
$$S_2 = \{c \in S_1 : c/(y_2\text{-smooth part of } c) \leq x/(x_1 x_2)\};$$
$$\vdots$$
$$S_k = \{c \in S_{k-1} : c/(y_k\text{-smooth part of } c) \leq x/(x_1 x_2 \cdots x_k)\}.$$

Note that each element $c \in S_k$ is y-smooth, since c divided by its y-smooth part is bounded by $x/(x_1 x_2 \cdots x_k) = 1$.

Consider any vector $(s_1, s_2, \ldots, s_k)$ such that each s_i is a y_i-smooth positive integer $\leq x_i$ having no prime factors $\leq y_{i-1}$. For any such $(s_1, s_2, \ldots, s_k)$, the product $c = s_1 s_2 \cdots s_k$ is a positive integer bounded by $x_1 x_2 \cdots x_k = x$, so $c \in S_0$. Dividing c by its y_1-smooth part produces $s_2 \cdots s_k \leq x/x_1$, so $c \in S_1$. Similarly $c \in S_2$ and so on through $c \in S_k$.

The map from $(s_1, s_2, \ldots, s_k)$ to $s_1 s_2 \cdots s_k \in S_k$ is injective: the y_1-smooth part of $s_1 s_2 \cdots s_k$ is exactly s_1, the y_2-smooth part is exactly $s_1 s_2$, etc. Hence $\#S_k$ is at least the number of such vectors $(s_1, s_2, \ldots, s_k)$, which is exactly $\Psi(x_1, y_1, y_0)\Psi(x_2, y_2, y_1)\Psi(x_3, y_3, y_2) \cdots \Psi(x_k, y_k, y_{k-1})$. Pomerance's early-abort analysis in [45] says, in some cases, that $\#S_k$ is not much larger than this. We heuristically assume that this is true in more generality.

The approximation $\Psi(x_i, y_i, y_{i-1}) \approx x_i/u_i^{u_i}$, where $u_i = (\log x_i)/\log y_i$, now implies that $\#S_k$ is approximately $x/(u_1^{u_1} \cdots u_k^{u_k})$. More generally, $\#S_i$ is approximately $x/(u_1^{u_1} \cdots u_i^{u_i})$.

Write T_i for the cost of finding the y_i-smooth part of an integer. The early-abort factorization method, applied to a uniform random element of S_0, always takes time T_1 to find primes $\leq y_1$; with probability $\#S_1/\#S_0 \approx 1/u_1^{u_1}$ takes additional time T_2 to find primes $\leq y_2$; with probability $\#S_2/\#S_0 \approx 1/(u_1^{u_1} u_2^{u_2})$ takes additional time T_3 to find primes $\leq y_3$; and so on. With probability $\#S_k/\#S_0 \approx 1/(u_1^{u_1} \cdots u_k^{u_k})$ an integer survives all aborts and is y-smooth.

Balancing the time for the early-abort stages, i.e., ensuring that each stage takes time approximately T_1, requires choosing x_1 (depending on y_1) so that $u_1^{u_1} \approx T_2/T_1$, choosing x_2 (depending on y_2) so that $u_2^{u_2} \approx T_3/T_2$, and so on through choosing x_{k-1} (depending on y_{k-1}) so that $u_{k-1}^{u_{k-1}} \approx T_k/T_{k-1}$. Then x_k is determined as $x/(x_1 \cdots x_{k-1})$, and u_k is determined as $(\log x_k)/\log y_k = u - (\log x_1 \cdots x_{k-1})/\log y = u - (\theta_1 u_1 + \theta_2 u_2 + \cdots + \theta_{k-1} u_{k-1})$ where $\theta_i = (\log y_i)/\log y$.

As a special case (including the cases considered by Pomerance), if all u_i are in $u^{1+o(1)}$, then $T_{i+1}/T_i \approx u^{u_i}$ is a $1 + o(1)$ power of u^{u_i}, so $u_1^{u_1} \cdots u_k^{u_k}$ is a $1 + o(1)$ power of $u^{u_1 + \cdots + u_k} = u^{u + u_1(1-\theta_1) + \cdots + u_{k-1}(1-\theta_{k-1})}$, which is a $1 + o(1)$ power of

$$u^u (T_2/T_1)^{1-\theta_1} \cdots (T_k/T_{k-1})^{1-\theta_{k-1}}$$
$$= u^u T_1^{\theta_1} T_2^{\theta_2-\theta_1} T_3^{\theta_3-\theta_2} \cdots T_{k-1}^{\theta_{k-1}-\theta_{k-2}} T_k^{1-\theta_{k-1}}/T_1.$$

In other words, compared to the original smoothness probability $1/u^u$ of integers in S_0, the found-by-early-abort-factorization probability is smaller by a factor $T_1^{\theta_1} T_2^{\theta_2-\theta_1} \cdots T_{k-1}^{\theta_{k-1}-\theta_{k-2}} T_k^{1-\theta_{k-1}}/T_1$. The time for all stages of early-abort factorization is essentially T_1. For example, for $\theta_i = i/k$, the product of the time and the loss factor is $(T_1 T_2 \cdots T_k)^{1/k}$.

We see two obstacles to proving the formula $(T_1 T_2 \cdots T_k)^{1/k}$ for early-abort ECM. First, the assumption $u_i \in u^{1+o(1)}$ is correct for exponential-time smoothness tests for standard ranges of x and y; but $u_i \in u^{0.5+o(1)}$ for ECM, except for $i = k$. Second, the error factor $u^{o(u)}$ in the standard u^u approximation is larger than the entire ECM running time. Despite these caveats we conjecture that the heuristics apply beyond the case of exponential-time smoothness tests, and in particular apply to early-abort ECM.

Even when smoothness theorems are available, one should not overstate the extent to which they constitute rigorous analyses of NFS. There is no proof that NFS congruences have similar smoothness probability to uniform random integers; this is one of the NFS heuristics. There is no proof that ECM finds all small primes at similar speed; this is another heuristic. As mentioned earlier, Pomerance's analysis in [45] actually uses the provable Pollard–Strassen smoothness-detection method, and Bernstein's batch trial-division method [8] is proven to run in polynomial time per input; but both of these methods perform poorly in the AT metric. Similarly, Pomerance proved in [45] that Dixon's random squares have similar smoothness probability to uniform random integers; but Dixon's method is much slower than NFS, and proving something similar about NFS is an open problem.

3.5   Impact of early aborts on smoothness probabilities. Because early-abort ECM does not find *all* smooth values, it forces batch NFS to consider more pairs (a, b), and therefore slightly larger pairs (a, b). This increase means that the auxiliary integers c are larger and less likely to be smooth. We conclude by showing that this effect does not eliminate the (heuristic) asymptotic gain produced by early aborts.

Recall that the smoothness probability of c is heuristically $1/v^v$, where v is the ratio of the number of bits in $(|f_d| + \cdots + |f_0|)H^d$ and the number of bits in y. The derivative of v with respect to $\log H$ is $d/\log y$, so the derivative of $\log(v^v)$ with respect to $\log H$ is $d(1 + \log v)/\log y \in 1/(3\gamma\mu) + o(1)$; here we have used the asymptotics $d \in (1/\mu + o(1))(\log N)^{1/3}(\log \log N)^{-1/3}$, $\log y \in (\gamma + o(1))(\log N)^{1/3}(\log \log N)^{2/3}$, and $\log v \in (1/3 + o(1)) \log \log N$.

Write $\delta = (2/3)/(2 - 1/(3\gamma\mu))$. Multiplying H by a factor $T^{\delta+o(1)}$ means multiplying the number of pairs (a, b) by a factor $T^{2\delta+o(1)}$ and thus multiplying the number of smoothness tests by a factor $T^{2\delta+o(1)}$. Meanwhile it multiplies v^v by a factor $T^{\delta/(3\gamma\mu)+o(1)}$, and thus multiplies the final number of smooth congruences by a factor $T^{(2-1/(3\gamma\mu))\delta+o(1)} = T^{2/3+o(1)}$. Our heuristics state that switching from ECM to early-abort ECM reduces the number of smooth congruences found by a factor $T^{2/3+o(1)}$, producing just enough smooth congruences for a successful factorization, while decreasing the cost of each smoothness test by a factor $T^{1+o(1)}$. The overall speedup factor is $T^{1-2\delta+o(1)}$.

For example, [7] took $\gamma \approx 0.790420$ and $\mu \approx 0.702860$, so the speedup factor is $T^{0.047\ldots+o(1)}$. As another example, batch NFS with $\beta = 0.5$ takes $\gamma \approx 0.681600$ and $\mu \approx 0.921150$, so the speedup factor is $T^{0.092\ldots+o(1)}$.

A   ECM

See [10] and the full version of this paper.

References

[4] Bai, S., Bouvier, C., Filbois, A., Gaudry, P., Imbert, L., Kruppa, A., Morain, F., Thomé, E., Zimmermann, P.: CADO-NFS—Crible Algébrique: Distribution, Optimisation—Number Field Sieve (2013). http://cado-nfs.gforge.inria.fr/. Citations in this document: §1

[6] Barker, E., Barker, W., Burr, W., Polk, W., Smid, M.: Recommendation for key management—part 1: general (revision 3) (2012). http://csrc.nist.gov/publications/nistpubs/800-57/sp800-57_part1_rev3_general.pdf. Citations in this document: §1.2

[7] Bernstein, D.J.: Circuits for integer factorization: a proposal (2001). http://cr.yp.to/papers.html#nfscircuit. Citations in this document: §1.1, §1.3, §1.3, §1.3, §1.3, §2.4, §2.5, §2.11, §3.5

[8] Bernstein, D.J.: How to find small factors of integers (2002). http://cr.yp.to/papers.html#sf. Citations in this document: §3.4

[10] Bernstein, D.J., Birkner, P., Lange, T., Peters, C.: ECM using Edwards curves. Math. Comput. **82**, 1139–1179 (2013). Citations in this document: §A

[11] Bernstein, D.J., Chen, H.-C., Chen, M.-S., Cheng, C.-M., Hsiao, C.-H., Lange, T., Lin, Z.-C. , Yang, B.-Y.: The billion-mulmod-per-second PC. In: Workshop Record of SHARCS'09: Special-Purpose Hardware for Attacking Cryptographic Systems, pp. 131–144 (2009). http://www.hyperelliptic.org/tanja/SHARCS/record2.pdf. Citations in this document: §1.2

[12] Bernstein, D.J., Chen, T.-R., Cheng, C.-M., Lange, T., Yang, B.-Y.: ECM on graphics cards. In: Joux, A. (ed.) EUROCRYPT 2009. LNCS, vol. 5479, pp. 483–501. Springer, Heidelberg (2009). Citations in this document: §1.2

[14] Bernstein, D.J., Lange, T.: Non-uniform cracks in the concrete: the power of free precomputation. In: Sako, K., Sarkar, P. (eds.) ASIACRYPT 2013, Part II. LNCS, vol. 8270, pp. 321–340. Springer, Heidelberg (2013). Citations in this document: §1.2, §1.3, §1.3, §2.5

[15] Bilardi, G., Preparata, F.P.: Horizons of parallel computation. J. Parallel Distrib. Comput. **27**, 172–182 (1995). Citations in this document: §1.1

[17] Bos, J.W., Kleinjung, T.: ECM at work. In: Wang, X., Sako, K. (eds.) ASIACRYPT 2012. LNCS, vol. 7658, pp. 467–484. Springer, Heidelberg (2012). Citations in this document: §1.2

[18] Brent, R.P., Kung, H.T.: The area-time complexity of binary multiplication. J. ACM **28**, 521–534 (1981). Citations in this document: §1.1, §1.3

[19] Buhler, J.P., Lenstra Jr., H.W., Pomerance, C.: Factoring integers with the number field sieve. See [36], pp. 50–94 (1993). Citations in this document: §1.3, §1.3, §1.3, §1.3, §1.3, §1.3, §2.2, §2.3

[20] Coppersmith, D.: Modifications to the number field sieve. J. Cryptol. **6**, 169–180 (1993). Citations in this document: §1.3, §1.3, §1.3, §1.3, §1.3, §1.3, §1.3, §1.3, §2.3, §2.5, §2.5, §2.11

[21] Crandall, R., Pomerance, C.: Prime numbers: A Computational Perspective. Springer, New York (2001). Citations in this document: §3

[23] Franke, J., Kleinjung, T., Paar, C., Pelzl, J., Priplata, C., Stahlke, C.: SHARK: a realizable special hardware sieving device for factoring 1024-bit integers. In: Rao, J.R., Sunar, B. (eds.) CHES 2005. LNCS, vol. 3659, pp. 119–130. Springer, Heidelberg (2005). Citations in this document: §1, §1.1

[24] Geiselmann, W., Shamir, A., Steinwandt, R., Tromer, E.: Scalable hardware for sparse systems of linear equations, with applications to integer factorization. In: Rao, J.R., Sunar, B. (eds.) CHES 2005. LNCS, vol. 3659, pp. 131–146. Springer, Heidelberg (2005). Citations in this document: §1, §1.1

[25] Heninger, N., Durumeric, Z., Wustrow, E., Halderman, J.A.: Mining your Ps and Qs: detection of widespread weak keys in network devices. In: USENIX Security Symposium (2012). Citations in this document: §1.1

[29] Kleinjung, T.: On polynomial selection for the general number field sieve. Math. Comput. **75**, 2037–2047 (2006). Citations in this document: §1

[30] Kleinjung, T.: Polynomial selection. Slides presented at the CADO workshop, Nancy, France (2008). http://cado.gforge.inria.fr/workshop/slides/kleinjung.pdf. Citations in this document: §1

[31] Kleinjung, T., Aoki, K., Franke, J., Lenstra, A.K., Thomé, E., Bos, J.W., Gaudry, P., Kruppa, A., Montgomery, P.L., Osvik, D.A., te Riele, H.J.J., Timofeev, A., Zimmermann, P.: Factorization of a 768-bit RSA modulus. In: Rabin, T. (ed.) CRYPTO 2010. LNCS, vol. 6223, pp. 333–350. Springer, Heidelberg (2010). Citations in this document: §1

[32] Kolkman, O.M., Mekking, M., Gieben, M.: RFC 6781: DNSSEC operational practices, version 2 (2012). http://tools.ietf.org/html/rfc6781. Citations in this document: §1

[35] Lenstra, A.K., Hughes, J.P., Augier, M., Bos, J.W., Kleinjung, T., Wachter, C.: Public keys. In: Safavi-Naini, R., Canetti, R. (eds.) CRYPTO 2012. LNCS, vol. 7417, pp. 626–642. Springer, Heidelberg (2012). Citations in this document: §1.1

[36] Lenstra, A.K., Lenstra Jr., H.W. (eds.): The Development of the Number Field Sieve. Lecture Notes in Mathematics, vol. 1554. Springer, Berlin (1993). Citations in this document: §1. See [19]

[37] Lenstra, A.K., Lenstra Jr., H.W., Manasse, M.S., Pollard, J.M.: The factorization of the ninth Fermat number. Math. Comput. **61**, 319–349 (1993). Citations in this document: §1.3

[38] Lenstra, A.K., Tromer, E., Shamir, A., Kortsmit, W., Dodson, B., Hughes, J., Leyland, P.: Factoring estimates for a 1024-bit RSA modulus. In: Laih, C.-S.

(ed.) ASIACRYPT 2003. LNCS, vol. 2894, pp. 55–74. Springer, Heidelberg (2003). Citations in this document: §1

[40] Lenstra Jr., H.W., Tijdeman, R. (eds.): Computational Methods in Number Theory I. Mathematical Centre Tracts, vol. 154. Mathematisch Centrum, Amsterdam (1982). See [45]

[41] Lewis, E.: DNSSEC at TLDs, start of 4Q 2013 (2013). https://elists.isoc.org/pipermail/dnssec-coord/2013-October/000172.html. Citations in this document: §1

[43] Morrison, M.A., Brillhart, J.: A method of factoring and the factorization of F_7. Math. Comput. **29**, 183–205 (1975). Citations in this document: §3.1

[45] Pomerance, C.: Analysis and comparison of some integer factoring algorithms. In: [40], pp. 89–139 (1982). http://cr.yp.to/bib/1982/pomerance.html. Citations in this document: §3.1, §3.4, §3.4, §3.4, §3.4, §3.4

[50] Schnorr, C.P., Shamir, A.: An optimal sorting algorithm for mesh-connected computers. In: STOC 1986, pp. 255–261 (1986). Citations in this document: §1.1, §2.6

[51] Shamir, A., Tromer, E.: Factoring large numbers with the TWIRL device. In: Boneh, D. (ed.) CRYPTO 2003. LNCS, vol. 2729, pp. 1–26. Springer, Heidelberg (2003). Citations in this document: §1, §1.1, §1.2

[53] Thompson, C.D., Kung, H.T.: Sorting on a mesh-connected parallel computer. Commun. ACM **20**, 263–271 (1977). Citations in this document: §1.1

[54] van Oorschot, P.C., Wiener, M.: Parallel collision search with cryptanalytic applications. J. Cryptol. **12**, 1–28 (1999). Citations in this document: §1.1

[56] Wiener, M.J.: The full cost of cryptanalytic attacks. J. Cryptol. **17**, 105–124 (2004). Citations in this document: §1.1

An Improvement of Linear Cryptanalysis with Addition Operations with Applications to FEAL-8X

Eli Biham and Yaniv Carmeli[✉]

Computer Science Department, Technion - Israel Institute of Technology,
3200003 Haifa, Israel
{biham,yanivca}@cs.technion.ac.il
http://www.cs.technion.ac.il/~{biham,yanivca}/

Abstract. FEAL is a Feistel cipher that uses addition operations. Since its introduction 26 years ago it played a key role in the development of many cryptanalytic techniques, including differential and linear cryptanalysis. For its 25th anniversary Mitsuru Matsui announced a challenge for an improved known plaintext attack on FEAL-8X. In this paper we describe our attack and introduce several improvements to linear cryptanalysis that allowed us to recover the key given 2^{14} known plaintexts in about 14 h of computation, and led us to win the challenge. An especially interesting improvement considers the approximation of addition-based S-boxes by partitioning into several sets in a way that amplifies the bias, and therefore allows for a reduction in the number of required known plaintexts as well as saving computation time. We also describe attacks that require only a few (even 2 or 3) known plaintexts that recover the key much faster than exhaustive search.

Keywords: FEAL · Linear cryptanalysis · Partitioning · Meet in the middle

1 Introduction

FEAL [13] was introduced in 1987 as a fast encryption algorithm which combines the simplicity of software-based operations with an improved security over prior designs. Over the years FEAL inspired the development of many cryptanalytic techniques, including differential and linear cryptanalysis [3,7,8]. The best known attacks on FEAL required (until recently) a few hundreds of chosen plaintexts [4] or 16 million known plaintexts [2,6].

In CRYPTO 2012 Mitsuru Matsui announced a year-long challenge [6] for developing improved attacks on FEAL-8X [9], and an award which will be given to the best attack capable of recovering the key of given sets of known plaintexts with various amounts of data. The attack recovering the key using the smallest number of known plaintexts would be declared the winner. In the course of this year we developed an improved attack capable of recovering the key of

© Springer International Publishing Switzerland 2014
A. Joux and A. Youssef (Eds.): SAC 2014, LNCS 8781, pp. 59–76, 2014.
DOI: 10.1007/978-3-319-13051-4_4

FEAL-8X, and three weeks before the deadline we submitted our solution for the challenge set with a million known plaintexts, and were the first to submit a correct solution. A few days later another group submitted a solution for a smaller set of 2^{15} known plaintexts. It took us another two weeks to finalize our program with all the additional tricks and to submit the solution for the set of 2^{14} known plaintexts, which became the winning solution. The secret key is 5681891EEC34CE1241ED0F52C9C23F65.

In this paper we present the cryptanalytic attacks that we developed for this challenge, and the techniques that we used to improve linear cryptanalysis. We first describe a linear attack which uses a 6-round approximation and analyzes both the first and last rounds simultaneously, recovering 37 subkey bits in total. We then describe how running it a second time with a different approximation can reduce the number of required plaintexts and find 44 bit of the subkeys. We describe the rest of the steps needed in order to recover the remaining subkeys and show how the FEAL-8X key can be reconstructed from those subkeys. The above mentioned techniques can find the FEAL-8X key given 2^{15} known plaintexts in about 26 h on our computer.

We then present our main contribution – a new partitioning method that can amplify the bias of a linear approximation of addition. The data is partitioned into two sets such that in one of the sets the bias of the linear approximation is stronger than it is when all the messages are considered. Interestingly, we cannot tell in advance which of the two sets is the one with the increased bias, and therefore we try both of them. The amplified bias allows us to reduce the number of plaintexts needed for the attack while keeping the analysis time per plaintext the same. Due to the smaller number of required plaintexts the attack time when using this method even decreases. Incorporating this technique with our previous methods allowed us to find the key given 2^{14} known plaintexts in about 14 h. In the summary of this paper (Sect. 7) we discuss the differences between our technique and *Partitioning Cryptanalysis* [5].

In addition to the practical attacks we also discuss attacks that can find the key with fewer plaintexts faster than exhaustive search. We describe an attack that can recover the key given 2^{10} known plaintexts in time of 2^{62} FEAL-8X encryptions. In addition, we describe attacks in which given only 11–21 known or chosen plaintexts the FEAL-8X key can be recovered with complexity about 2^{80} and given 2 or 3 known plaintexts the FEAL-8X key can be recovered with complexity about 2^{96}. These attacks combine linear cryptanalysis and differential cryptanalysis with exhaustive search of many subkeys, as well as meet in the middle attacks. These attacks exploit the fact that the total size of the subkeys is not sufficiently larger than the size of the key.

The structure of the paper is as follows: In Sect. 2 we describe FEAL-8X, give two equivalent descriptions of the cipher, and define notations. In Sect. 3 we describe the linear attack that recovers the key given 2^{15} known plaintexts. In Sect. 4 we present the new partitioning method and how to recover the key given 2^{14} known plaintexts. In Sect. 5 we extend the methods from the previous sections, and describe an attack on 2^{10} known plaintexts faster than exhaustive search. Finally, in Sect. 6 we describe the attacks that require only a few known ciphertexts or a few chosen plaintexts. In Appendix A we describe an efficient implementation of our attacks which is able to save a factor of about 2^6 in the attack time. Appendix B illustrates the linear approximations used in our attacks. In the full version of the paper we also show how to find the key of FEAL-8X given the subkeys that are found by our attacks.

2   The Cipher FEAL-8X

The block size of FEAL-8X is 64 bits and the key size is 128 bits. The key processing algorithm of FEAL-8X takes the 128-bit key and generates 16 subkeys, denoted by $K0$–Kf, each of length 16 bits.

FEAL-8X is an 8-round Feistel cipher. Before the first round the plaintext is mixed with a 64-bit whitening subkey ($K89ab$) which is followed by XORing the left half of the data into the right half. The inverse of this operation is performed after the last round, i.e., the left half of the data is XORed into the right half and the result is mixed with a 64-bit whitening key ($Kcdef$). In each round a function F is computed on the right half of the data and a 16-bit subkey (one of $K0$–$K7$), and the output is XORed into the left half. The two halves are then swapped.

The function F takes four bytes as input, and starts by XORing the first and last bytes into the two middle bytes, and then XORs the subkey into the same bytes. It then applies four S-boxes in the order described in Fig. 1. Each S-box adds two bytes and an index (0 or 1) and rotates the output by two bits to the left. FEAL-8X and the F-function are outlined in Fig. 1.

2.1   An Equivalent Description of FEAL-8X

In order to simplify the analysis we prefer to eliminate the whitening keys. This is possible on one end of the cipher by extending the size of the subkeys to 32 bits in each round and by XORing the eliminated whitening key information into all the subkeys. We consider two equivalent descriptions of the cipher. In the first we eliminate the whitening at the beginning of the cipher, and in the second we eliminate the whitening at the end (this latter version is outlined in Fig. 2). The 32-bit subkeys of the equivalent description are called *actual subkeys*. We call the actual subkeys of the version with eliminated whitening key at beginning *encryption actual subkeys* and denote them by $EK0$–$EK7$, while we call the actual subkeys of the version with eliminated whitening key at the end *decryption actual subkeys*, and denote them by $DK0$–$DK7$. To simplify the description we define the function

$$mw(X, Y) = (Y_0, Y_0 \oplus Y_1 \oplus X_0, Y_2 \oplus Y_3 \oplus X_1, Y_3)$$

where X is a 16-bit value, Y is a 32-bit value, and $X_0, X_1, Y_0, Y_1, Y_2, Y_3$ are their individual bytes. Note that $mw(X, Y)$ is just the first part of the F-function before the S-boxes (see Fig. 1). The mapping between the subkeys of all three

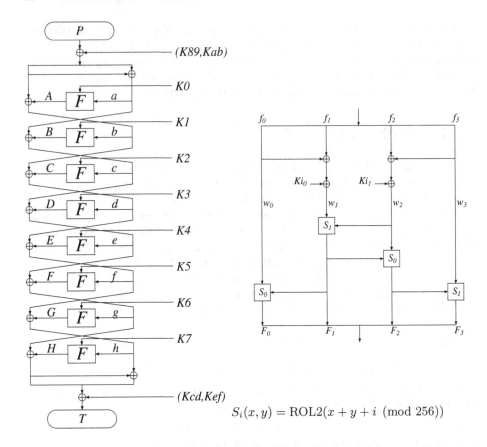

Fig. 1. The outline of FEAL-8 and of the F-function

descriptions of the cipher (the subkeys of FEAL and the two equivalent descriptions) is summarized in Table 1.

In our attacks when we analyze the last rounds of the cipher we assume the whitening at the end is zero, and therefore retrieve the bits of the decryption actual subkeys DK. Similarly, when we analyze the first rounds of the cipher we retrieve the bits of the encryption actual subkeys EK.

Note that since there is a linear relation between the subkeys of all three descriptions of FEAL it is possible to target actual subkeys of different descriptions in the same linear attack. For example, the attack presented in Sect. 3.2 targets both $EK0$ and $DK7$.

3   First Attack – Finding the Key Using 2^{15} Known Plaintexts

In this section we describe a linear attack that requires 2^{15} known plaintexts and finds the key in about 26 h on a server with an Intel(R) Xeon(R) X5650

Table 1. The Subkeys of FEAL-8X and the actual subkeys of the equivalent descriptions

Subkeys of FEAL-8X	Equivalent description without whitening at the beginning	Equivalent description without whitening at the end
$K89ab$	0	$(K89 \oplus Kcd \oplus Kef, Kab \oplus Kef)$
$K0$	$EK0 = mw(K0, K89 \oplus Kab)$	$DK0 = mw(K0, Kcd)$
$K1$	$EK1 = mw(K1, K89)$	$DK1 = mw(K1, Kcd \oplus Kef)$
$K2$	$EK2 = mw(K2, K89 \oplus Kab)$	$DK2 = mw(K2, Kcd)$
$K3$	$EK3 = mw(K3, K89)$	$DK3 = mw(K3, Kcd \oplus Kef)$
$K4$	$EK4 = mw(K4, K89 \oplus Kab)$	$DK4 = mw(K4, Kcd)$
$K5$	$EK5 = mw(K5, K89)$	$DK5 = mw(K5, Kcd \oplus Kef)$
$K6$	$EK6 = mw(K6, K89 \oplus Kab)$	$DK6 = mw(K6, Kcd)$
$K7$	$EK7 = mw(K7, K89)$	$DK7 = mw(K7, Kcd \oplus Kef)$
$Kcdef$	$(K89 \oplus Kab \oplus Kcd, Kab \oplus Kef)$	0

2.67 GHz processor with 12 cores. We first describe a 6-round linear approximation and then the basic attack which performs the analysis on both ends of the cipher simultaneously. We then describe how to use it with reduced number of plaintexts, and how to recover the rest of the actual subkeys and the full key.

3.1 The Linear Approximations

In [1,11] eight 7-round linear approximations with a bias of about 2^{-9} were presented. The attack we present in this paper uses two 6-round approximations with bias of about 2^{-6}, which we got by truncating two of the 7-round approximations of [1,11] by one round. These approximations are outlined in Figs. 4 and 5 in Appendix B.

3.2 The Basic Attack

The attack we present targets both the encryption actual subkey of the first round ($EK0$), and the decryption actual subkey of the last round ($DK7$). The six-round linear approximation covers the six middle rounds of the cipher (rounds 1–6), while the first and last rounds are used for analysis. We found that when using Approximation 1 there are only 37 bits of $EK0$ and $DK7$ that affect the parity of the bits in the approximation: 22 bits in the last actual subkey ($DK7$, given by the mask 03 FF FF 0F), and 15 bits of the first actual subkey ($EK0$, given by the mask 00 7F 7F 00 and the parity of the two bits 00 80 80 00). The remaining 27 bits of $EK0$ and $DK7$ have no impact on the parity of the bits in the linear approximation of the six middle rounds. It is therefore that this basic attack finds the 37 bits of the two actual subkeys.

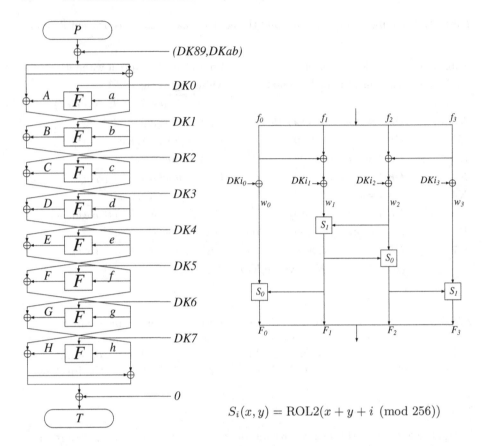

Fig. 2. Equivalent description of FEAL-8X without whitening at the end

$$S_i(x, y) = \text{ROL2}(x + y + i \pmod{256})$$

The attack is as follows:

1. For each of the 2^{15} candidates for the 16 bits of $EK0$:
   (a) For each of the 2^{22} candidates for the 22 bits of $DK7$:
      i. For each known plaintext P, C:
         A. Decrypt C by one round using $DK7$.
         B. Encrypt one round of P using $EK0$.
         C. Compute the parity of the approximated bits.
      ii. Count the number of messages for which the linear approximation holds and compute the bias.
2. The correct key is expected to be the one with the highest bias.

We also observed that not all 37 bits of the subkeys have the same impact on the bias. While some bits completely throw off the observed bias if guessed incorrectly, others have only a minor impact. We can take advantage of this observation to reduce the running time of the attack by excluding a few such bits with a minor impact on the bias, and to search for them only when the rest of the bits are already known. For example, instead of guessing 15 bits of $EK0$ with a bias of about 2^{-6}, we may guess only 13 bits (the 12 bits whose mask is 00 6E 7F 00, and the parity of the two bits 00 80 80 00) with a slightly lower expected bias of $2^{-6.5}$, and save a factor of 4 in computation time.

Clearly, the more data we have at our disposal the more accurate the results are (since it is easier to detect the linear bias). If the available data is a lot larger than required in order to detect the bias then we have more freedom to exclude such minor-impact bits (as the measurement of the bias is only slightly inaccurate). As the number of known plaintexts decreases, the identification of the correct key becomes harder (as the bias is harder to detect), and in this case we usually cannot afford to reduce the bias in return for speeding up the attack.

3.3   Matching Subkeys from the Backward and Forward Directions

As noted above, the basic attack does not suffice to find the correct key using 2^{15} known plaintexts. In this section we apply the basic attack twice: once in the forward direction, and once in the backward direction.

We first generate a list $L1$ of the N (for some parameter N) keys which exhibit the highest bias according to Approximation 1 in the forward direction, as described in Sect. 3.2. Recall that for each such key we get 15 bits of the first encryption actual subkey $EK0$, and 22 bits of the last decryption actual subkey $DK7$.

We now run the attack again in the backward direction, i.e., we use the reverse of Approximation 1. In this run we guess 22 bits of $EK0$ and 15 bits of $DK7$. We generate a second list $L2$ of the N keys that exhibit the highest biases.

There is an overlap of 15 bits between the bits we guess in $EK0$ in both runs, and similarly, an overlap of 15 bits in $DK7$. Seven bits of $EK0$ are available only in $L2$, and seven bits of $DK7$ are available only in $L1$. The correct value of these 30 overlapping bits is expected to be in both lists. In such a case, we can easily find the correct value of $30 + 7 + 7 = 44$ bits of the actual subkeys as the (usually single) value that has a match in those 30 bits in both lists.

As we noted earlier, some of the bits of the key only have a minor impact on the measured bias if they are guessed incorrectly. If we cannot find a match between an entry in $L1$ and an entry in $L2$, we can try looking for entries that have a low Hamming distance in the overlapping bits, and between these prefer entries that differ in bits that are known to have a minor impact on the bias.

This is the most time-consuming part of our attack. When we ran it[1] on the server mentioned above it found the 44 bits of the actual subkeys within 24 h using 2^{15} known plaintexts (12 h for each call to the basic attack). The correct key bits were among the top $N = 3200$ keys in each list.

3.4   Retrieving the Rest of the Subkeys

In the previous section we found 44 bits of the actual subkeys. In this section we briefly describe additional steps for finding the rest of the bits of $EK0$ and $DK7$,

---

[1] With the implementation improvement described in Appendix A.

as well as the rest of the actual subkeys. The steps are described in the order in which they are performed, as each step assumes knowledge of the subkey bits that are retrieved in the preceding steps.

Finding 8 additional bits of $EK0$ and $DK7$. This step is similar to the attack presented in Sect. 3.2, but uses Approximation 2 instead of Approximation 1. Since the linear approximation is different, there are also different bits of the subkeys of Rounds 0 and 7 that affect the parity. The bits of $EK0$ that affect the parity are given by the mask 3F FF FF 00 and the bits of $DK7$ are given by the mask 0F FF FF 03. Most of those bits are already known, except for eight bits. The correct values of these remaining eight bits can be identified by standard linear cryptanalysis techniques, similarly to the attack of Sect. 3.2. After this step is performed we know 26 bits in each of $EK0$, $DK7$, a total of 52 actual subkey bits.

Finding 4 additional bits of $DK7$ and 15 bits of $DK6$. At this point there are still 6   bits missing in the subkey $DK7$, which are difficult to retrieve by analyzing Round 7. We therefore move on to analyze Round 6 by using a shorter linear Approximation. We use the first five rounds of Approximation 1 with a bias of 2^{-3}, and use it to cover rounds 1–5. In order to compute the parity of the approximated bits in Round 5 we need to guess the values of four more bits of $DK7^2$, and 15 bits of $DK6$. After this step is performed we know a total of 30 bits of $DK7$ (given by the mask 7F FF FF 7F) and 15 bits of $DK6$ (given by the mask 00 7F 7F 00 and in addition the parity of bits 00 80 80 00).

Finding 7 additional bits of $DK6$. This step is similar to the previous step, but this time we use a 5-round approximation obtained from the last five rounds of Approximation 1, which covers Rounds 1–5. There are 22 bits in $DK6$ that affect the parity of this linear approximation. We already found 15 of them in the previous step, and we should now search for the remaining seven.

Finding 4 additional bits of $DK6$. We use a 5-round approximation comprised of the last five rounds of Approximation 2, which covers Rounds 1–5 We can obtain four more bits of $DK6$, and get a total of 26 bits of $DK6$.

Finding the rest of the subkeys $DK1$–$DK7$. In a similar way, we can attack the rest of the rounds until we have all the actual subkeys $DK1$–$DK7$. Note that as we progress in the attack, analyzing each additional round becomes easier for two main reasons: First, we use shorter approximations with higher biases, which significantly decrease the chances of errors. Second, since the actual subkeys $DK0$, $DK2$, $DK4$ and $DK6$ have 16 bits in common (and similarly for $DK1$, $DK3$, $DK5$ and $DK7$) there are only 16 bits to retrieve in each of those actual subkeys once $DK6$ and $DK7$ are fully known.

Finding $EK0$–$EK6$. Once we finish recovering the decryption actual subkeys, we can repeat the entire process in the reverse direction in order to find the

---

[2] The value of the two remaining bits of $DK7$ can only be determined when we analyze round 3. Until then those bits have only a linear effect on the parity of the approximation, and therefore cannot be discovered by methods of linear cryptanalysis.

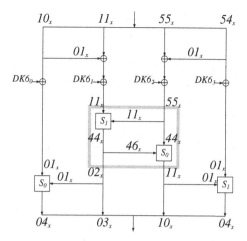

Fig. 3. The approximation of the seventh round

encryption actual subkeys $EK0$–$EK6$.[3] These actual subkeys depend on the whitening key of the plaintext, and are needed in order to retrieve the FEAL-8X key.

Finding The Key Itself. Given $DK1$–$DK7$ and $EK0$–$EK6$ we find the FEAL-8X key within a fraction of a second. The details are omitted here due to space constraint, but the algorithm is described in the full version of the paper.

4 The Partitioning Technique – Finding the Key Using 2^{14} Known Plaintexts

In this section we describe a technique that can reduce the number of known plaintexts by a factor of 3.1 compared to the algorithm of Sect. 3.2. In this technique we partition the data into several sets, such that the bias of the approximation in some of them is higher than when measured across all the data, with a ratio that overcomes the smaller number of messages in those sets. Therefore, fewer messages are required in order to detect the amplified bias.

4.1 A Simplified Example

We apply this technique to Round 6 of the cipher in the inner loop of the algorithm, after the output of the last F-function is already (partially) computed. It is therefore that most bits of the inputs to the S-boxes of Round 6 are known up to an XOR with $DK6$.

---

[3] We note that instead of searching for $EK0$–$EK6$, we can continue the analysis in the decryption direction and retrieve the actual subkey $DK0$ and the whitening key. Once all the decryption actual subkeys $DK0$–$DK7$ and the whitening key are known, the encryption actual subkeys $EK0$–$EK7$ can easily be computed (see Table 1).

At Round 6 we approximate the first S-box by 11 11 → 44 (see Fig. 3). The input mask 11 11 is approximated to the output mask 44 through the addition operations in the S-boxes (and the rotation), and therefore the quality of the approximation is determined by the carry bits from lower bits into the approximated bits.

We are interested in improving our control on the carry bits, which in turn will improve our approximations.

For that we identified that some of the bits in the inputs to this S-box (denoted by w_1 and w_2 in Fig. 2) in this round are known to us up to an XOR with the actual subkey $DK6$ (as mentioned above).

The approximation 11 11 → 44 approximates two bits through the addition operations. One of them involves the addition of the least significant bits of the inputs (mask 01 01 or $w_{1,0} + w_{2,0} = F_{1,2}$, where w_i are the input bytes to the S-boxes and F_i is the output, as denoted in Fig. 2, and $w_{i,j}$ is bit j of w_i). The approximation of this bit has probability 1, as there cannot be a carry into the LSB. The other approximates Bit 4 of both inputs (mask 10 10), the carry to which involves Bit 3 of both inputs ($w_{1,3}$ and $w_{2,3}$, identified by the mask 08 08). If we would know in advance that the unknown values of these two bits are both 0 then it is certain that there cannot be any carry into Bit 4, which would ensure that this approximation will also have probability 1 (bias +0.5). Similarly, when both bits are 1, a carry from this bit to the next one is guaranteed, and therefore we would also be able to make the approximation with probability 1 (knowing that the carry always flips the approximated output, thus the bias is −0.5). In the other cases (where the bits $w_{1,3}$ and $w_{2,3}$ are either 0,1 or 1,0), we have no idea what the carry is, but we expect that it would occur in about half of the inputs, which would cause the bias to be much closer to zero. We refer to the four possible cases by the values of $w_{1,3}$, $w_{2,3}$ as cases 00, 11, 01, and 10, respectively. The bias of the S-box (on all inputs) is close to 0.25, and therefore the bias of the entire Approximation 1 is 0.25α, for some α that depends on the other parts of the approximation.

If we could choose only plaintexts of cases 00 and 11 and run the attack only on these plaintexts, we would need fewer messages due to the larger bias. Unfortunately, the values of $w_{1,3}$ and $w_{2,3}$ are only known up to a XOR with two missing bits of $DK6$ (see Fig. 2):

$$w_{1,3} = f_{0,3} \oplus f_{1,3} \oplus DK6_{1,3}, \qquad w_{2,3} = f_{2,3} \oplus f_{3,3} \oplus DK6_{2,3},$$

and therefore they clearly cannot be chosen or known directly. Nevertheless, the corresponding bits $f_{0,3}$, $f_{1,3}$, $f_{2,3}$ and $f_{3,3}$ in the input of the F-function are all known as a result of the partial guess of the actual subkey $DK7$. We observe that we can still partition all the data into the same four sets according to $f_{0,3} \oplus f_{1,3}$ and $f_{2,3} \oplus f_{3,3}$, instead of $w_{1,3}$ and $w_{2,3}$, but we do not know which of the four sets have the amplified biases.

Though we cannot identify the two sets with an amplified bias, we can run this inner part of the attack four times, once on each of the sets. We expect the following results: In each set we would have about a quarter of the known plaintexts but in two of them we would have a bias twice as large as we had originally (meaning $\pm 0.5\alpha$).[4] Therefore the number of required plaintexts in these sets is about 4 times $(0.5^2/0.25^2)$ smaller than would have been needed without applying this technique.

A more careful analysis shows that we can merge the two sets with bias ± 0.5 (with an appropriate sign coefficient) and partition the plaintexts only to two sets. This merges the sets of cases 00 and 11 into one set, and the sets of cases 01 and 10 into another set according to the parity of the two bits $f_{0,3} \oplus f_{1,3}$ and $f_{2,3} \oplus f_{3,3}$. Denote the number of known plaintexts required for the original attack by m. As discussed above, the amplified bias can be detected with $m/4$ plaintexts. Since each of the two unified sets has about half of the plaintexts, we deduce that $m/2$ known plaintexts suffice for the partitioning technique.

4.2   The Attack

The attack follows the lines of the above example, but considers that the details of the approximation of the S-boxes are more complicated than described so far. While for a single S-box and appropriate independence assumptions the technique would work as described, in practice there is a correlation between the approximation of the two middle S-boxes of F. We give the combination of both middle S-boxes the name *T-box* (marked by a rectangle in Fig. 3). The joint approximation of the two S-boxes in the T-box cannot be described as a combination of two independent approximations since the input bits to the second S-box are all either inputs of the first S-box or its output. Therefore, a closer examination of the joint distribution is in order.

We computed the joint approximation of the S-boxes (the T-box) with the approximation 11 55 → 02 11 and observed that the partition to two sets (by the value of $f_{0,3} \oplus f_{1,3} \oplus f_{2,3} \oplus f_{3,3}$) has the following effect: In the cases 01 and 10 the bias is increased by a factor of about 2.49 compared to the original bias, while the absolute value of the bias in the other cases (00 and 11) is halved. It is therefore that the number of known plaintexts needed by the attack is reduced by a factor of about $2.49^2/2 \approx 3.1$.

We also note that there are other possible partitions (by other control bits) that yield an increased bias in one or more of the sets, that can be used for alternative implementations of this technique.

We applied this improvement to the attack of Sect. 3 and successfully reduced the number of required known plaintexts from 2^{15} to 2^{14}. Applying this technique did not add a noticeable overhead to the running time of the attack. In fact, the time it took to recover the 44 bits of the actual subkeys using 2^{14} plaintexts was 12 h – which is about half the time that was required using 2^{15} plaintexts (without using this technique). The rest of the attack took about two more hours, and the key was found after 14 h of computation. The key that was found for the challenge with 2^{14} plaintexts is 5681891EEC34CE1241ED0F52C9C23F65.

---

[4] For the purpose of this simplified example we assume that the linear approximation of this S-box is independent of the rest of Approximation 1. We will see later that this is not the case.

5   Attacking FEAL-8X Using 2^{10} Known Plaintexts with Complexity 2^{62}

The methods we described in the previous sections can be used to break FEAL-8X with even fewer known plaintexts in time which is still faster than exhaustive search. In particular, the key can be found given 2^{10} known plaintexts in time of about 2^{62} FEAL encryptions.

To justify the above claim, we describe an attack on seven rounds of FEAL, which is based on the attack of Sect. 3.2, and then extend it to 8 rounds by exhaustively searching for the subkey of the last round.

The attack on seven rounds of FEAL uses the first five rounds of Approximation 1, with a bias of 2^{-3}. Similarly to the attack of Sect. 3.2, the approximation covers the five middle rounds and the analysis is performed on the first and last rounds. In each of the first and last rounds there are 15 bits that we need to guess in order to compute the parity of the linear approximation, and therefore the attack requires encrypting/decrypting an equivalent of $2^{15} \cdot 2^{15} \cdot 2^{10} \cdot 2 = 2^{41}$ rounds of FEAL.

In order to extend the attack to eight rounds, we also guess 30 bits of the actual subkey $DK7$ of the last round (recall that two of the 32 bits have no effect on the parity of the linear approximation). For each candidate for these 30 bits of $DK7$ we decrypt the last round of all the inputs, and then apply the above attack to the remaining seven rounds. The attack on seven rounds is performed 2^{30} times, and therefore the total time complexity is equivalent to computing $2^{30} \cdot 2^{41} = 2^{71}$ rounds of FEAL (or 2^{68} encryptions of the full cipher), which is much faster than exhaustively searching for the 128-bit key.

When applying the optimization improvements described in Appendix A we get an even lower complexity of about 2^{62} FEAL encryptions.[5]

6   Attacks with a Few Known or Chosen Plaintexts

In this section we describe several attacks that require only a few (even 2 or 3) known or chosen plaintexts, which are based on linear cryptanalysis or differential cryptanalysis combined with exhaustive search of most subkeys, as well as meet in the middle attacks.

6.1   Differential and Linear Exhaustive Search Attacks

During the work on this paper we noticed that the actual subkeys of FEAL-8X are mixed very slowly through the encryption function. In particular, we observed that only 112 bits of the actual subkeys are needed in order to decrypt a ciphertext by 5 rounds and compute the data after the third round of the cipher from the ciphertext. In addition, we recalled that there are four independent

---

[5] Recall that the key size of FEAL-8X is 128 bit.

3-round linear approximations with probability 1 (creating a total of 15 non-trivial approximations) and two independent 3-round differential characteristics with probability 1 (creating a total of 3 characteristics). These approximations and characteristics can be found in [2,4].

In the case of the linear approximations with probability 1, each allows us to test one parity bit of the data after the third round and to compare to a parity bit of the plaintext. Therefore, a total of 4 bits can be tested on each plaintext (except for the first known plaintext to whose parities we compare). Given 5 known plaintexts the attack would be:

1. For each value of the set of subkeys $DK3$, $DK4$, $DK5$, $DK6$, $DK7$ (in total these 160 bits only contain 112 independent bits).

   (a) For each plaintext-ciphertext pair (P, C) decrypt the ciphertext by 5 rounds to D_3 and compute the parity of each approximation $P\lambda_P^3 \oplus D_3\lambda_T^3$, where $\lambda_P^3 \rightarrow \lambda_T^3$ is the mask of the linear approximation in use.

   (b) Discard any guess for which the five results (each of 4 bits, one for each approximation) are not the same.

   (c) Note that at this point only about 2^{96} of the guesses of the subkeys remain.

   (d) For each value of the subkey $DK2$ (16 more bits)

      i. Note that at this point we have about 2^{112} guesses of the subkeys.

      ii. We will now use four 2-round approximations $\lambda_P^2 \rightarrow \lambda_T^2$ which are based on the last two rounds of the prior ones.

      iii. For each plaintext-ciphertext pair (P, C) decrypt the ciphertext by 6 rounds to D_2 and compute the parity of each approximation $P\lambda_P^2 \oplus D_2\lambda_T^2$.

      iv. Discard any guess for which the five results are not the same.

      v. Note that at this point we are left again with only about 2^{96} guesses of the subkeys.

      vi. For each value of the subkey $DK1$ (16 more bits)

         A. Note that at this point we have about 2^{112} guesses of the subkeys.

         B. For each plaintext-ciphertext pair (P, C) decrypt the ciphertext by 7 rounds to D_1 and compute the XOR of both halves of the whitening key $DK89 \oplus DKab$ (32 bits in total).

         C. Discard any guess for which the five results are not the same.

         D. Note that at this point we expect that only the correct values of all the above guesses remain.

         E. Complete the rest of the subkeys by guessing $DK0$ and comparing the resulting $DK89$ in 2^{16} time.

         F. Recover the original key. The algorithm is described in the full version of the paper (note that given all the decryption actual subkeys and the whitening key it is easy to compute the encryption actual subkeys needed by that algorithm).

The complexity of this attack is 2^{112}, taking into consideration that the various decryptions need not be computed several times (once by 5 rounds, then by 6, then by 7), but that the intermediate values can be cached to save computation time. A careful implementation would require an average computation of only two rounds in each guess for each of the three guessing loops. Thus the total complexity is about $3 \cdot 2 \cdot 2^{112}$ round computations $= 0.75 \cdot 2^{112}$ encryption of FEAL-8X.

A similar attack that uses the 3-round differential characteristics with probability 1 requires only three chosen plaintexts (whose plaintexts differ by the two plaintext differences of the two characteristics). Since each differential characteristic predicts 64 bits of the intermediate difference, we have a much better elimination of wrong guesses, and thus we need only three chosen plaintexts. The complexity of the attack is also 2^{112}.

6.2   Meet in the Middle Attacks

The attack that requires the least number of known plaintexts is a meet in the middle attack. We observe that the number of (independent) bits of the actual subkeys that are required to partially encrypt (or decrypt) four rounds of the cipher is 96. Therefore, a meet in the middle attack using two (or three) known plaintexts computes 2^{96} 4-round partial encryptions of two blocks plus 2^{96} 4-round partial decryptions of two blocks. This attack also requires 2^{96} memory words of size 128 bits (or even 96 bits). The list of about 2^{64} (or 2^{96}) colliding values should then be checked by auxiliary techniques, and be completed to a full key with the same known plaintexts.

An improvement of this attack may reduce the complexity to 2^{80}, by encrypting or decrypting only three rounds from each end, using 11 known plaintexts. This improvement considers that the F-function in the fourth round can be approximated by the four independent linear approximations with probability 1 (each one is represented by a single parity bit in the output of encryption and a single parity bit in the output of decryption). The fifth round can be approximated similarly. This way, each known plaintext contributes 8 bits to the colliding values (except for the first, whose 8 parity bits are XORed into the parity bits of all the other ones), and thus in order to collide on 80 bits, we need 11 known plaintexts. Each of the 2^{80} colliding values can then be checked by auxiliary techniques, and be completed to the full key.

We also note that these meet in the middles attacks can be transformed to memoryless meet in the middle attacks by standard techniques [10,12]. The simplest implementation of the former encrypts/decrypts three blocks at a time, each encrypted or decrypted by four rounds, resulting in a collision on 192 intermediate data bits, which ensures that the real value of the subkeys are easily identified in time 2^{96}. The simplest implementation of the latter encrypts/decrypts 21 blocks at a time, each encrypted or decrypted by three rounds, resulting in a collision on 160 intermediate data bits, which ensures that the real value of the subkeys be easily identified in time $21 \cdot (3/8 + 3/8) \cdot 2^{80} \approx 2^{84}$.

7 Summary

We presented the techniques which allowed us to break FEAL-8X with only 2^{14} known plaintexts and recover the secret key. This is an improvement of the best known-plaintext attacks prior to this paper. Our attack is based on a few improvements and optimizations to linear cryptanalysis, the most important of which is the new partitioning technique which allowed us to reduce the amount of known plaintexts needed for the attack.

In addition to the practical attacks on FEAL-8X we also presented a few attacks which are based on linear and differential cryptanalysis in combination with meet-in-the-middle techniques. Those attacks can find the secret key given only a few messages in time which is faster than exhaustive search.

We wish to discuss the similarities and differences between our partitioning technique and partitioning cryptanalysis [5]. They both partition the data into several sets based on functions that take the plaintexts or ciphertexts and guessed key bits, where each set of the input-partition is related to some linear approximation and expected biases. In that sense, our technique is a variant of partitioning cryptanalysis. However, in partitioning cryptanalysis the expected biases are known in advance for each input block of the partition, and thus the attacker can select the best block and choose all the *chosen plaintexts* to be in that block. In our case we succeed (in the particular case of the addition operation) to take one step further and divide to partitions such that we do not know which set should have which bias. The identification of the sets is part of the attack, and it is therefore that our technique is a *known plaintext attack*. But perhaps the most significant improvement of our technique stems directly from the motivation that is the basis of our partition – we use the partition in order to discard (or rather ignore) messages that do not contribute to the linear bias. By doing so the bias in the remaining set is higher, which allows us to reduce the number of messages needed for the attack. We also note that our technique may in some cases be applied both on the plaintext side and on the ciphertext side simultaneously, and gain the extra factor in cases that partitioning cryptanalysis may not.

Acknowledgements. The authors would like to thank Mitsuru Matsui for initiating the FEAL 25 Years challenge. We would also like to thank Orr Dunkelman for his insightful comments and helpful suggestions.

A Efficient Implementation

We describe an optimization to the implementation of the attack of Sect. 3.2 which saves a factor of about 2^6 in the computation time of the attack. This optimization can also be applied to other attacks presented in this paper that are based on the attack of Sect. 3.2.

Recall that in the attack of Sect. 3.2 we iterate over 2^{15} possible values for (16 bits of) the encryption actual subkey of the first round ($EK0$), and 2^{22} possible values for (22 bits of) the decryption actual subkey of the last round ($DK7$). For each of the 2^{37} combinations, two rounds of FEAL are encrypted/decrypted for each known plaintext. We denote the number of known plaintexts by m.

We observe that given a known plaintext-ciphertext pair P, C, the parity of the approximated bits can be written as $b_P \oplus b_C$, where b_P is a bit that depends only on the plaintext and the actual subkey of the first round, and b_C is a bit that depends only on the ciphertext and the actual subkey of the last round. Therefore, we can change the attack as follows:

1. For each of the 2^{15} candidates for the 16 bits of $EK0$:
   (a) Compute a vector B_P of length D bits, where $(B_P)_i = b_{P_i}$. Save all the vectors in a table.
2. For each of the 2^{22} candidates for the 22 bits of $DK7$:
   (a) Compute a vector B_C of length m bits, where $(B_C)_i = b_{C_i}$.
   (b) For each of the 2^{15} candidates for the 16 bits of $EK0$, get the vector B_P from the table, and compute the number of plaintexts for which the parity of the Approximations is 1 by $H(B_P \oplus B_C)$, where H is the Hamming weight function.
   (c) Compute the bias for approximation.
3. The correct key is expected to be the one with the highest bias.

Assuming a processor with a word size of 64 bits, this optimization lets us compute the parity of 64 plaintexts at the same time, and therefore saves a factor of about 2^6 in the attack.

We note that this optimized implementation also works with the partitioning technique described in Sect. 4. In Step 2a of the algorithm above, in addition to generating the vector B_C we generate a third vector W. The i-th bit of W determines to which set of the partition the i-th plaintext belongs. We can compute the number of plaintexts with a parity of 1 in the bits of the approximation in each of the sets as $H((B_P \oplus B_C)\&W)$ and $H((B_P \oplus B_C)\&\overline{W})$, where $\&$ is the bitwise-and operator, and $\overline{W}$ denotes the binary complement of W.

B   The Linear Approximations Used in Our Attacks

The appendix lists the two linear approximations from [1,11] which we use in our attacks. Approximation 1 is presented in Fig. 4 and Approximation 2 is in Fig. 5.

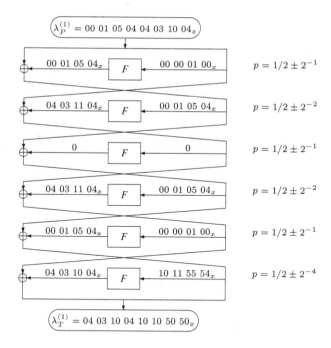

Fig. 4. Approximation 1 – A six round approximation with bias 2^{-6}

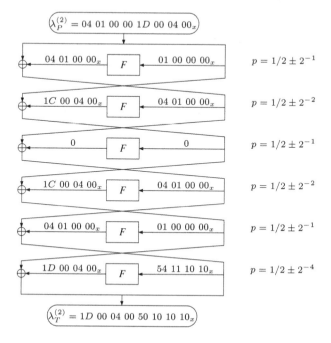

Fig. 5. Approximation 2 – A six round approximation with a bias 2^{-6}

References

1. Aoki, K., Ohta, K., Moriai, S., Matsui, M.: Linear cryptanalysis of FEAL. IEICE Trans. Fundam. Electron. Commun. Comput. Sci. **E81–A**(1), 88–97 (1998)
2. Biham, E.: On matsui's linear cryptanalysis. In: De Santis, A. (ed.) EUROCRYPT 1994. LNCS, vol. 950, pp. 341–355. Springer, Heidelberg (1995)
3. Biham, E., Shamir, A.: Differential cryptanalysis of DES-like cyptosystems. J. Cryptol. **4**(1), 3–72 (1991)
4. Biham, E., Shamir, A.: Differential cryptanalysis of feal and N-Hash. In: Davies, D.W. (ed.) EUROCRYPT 1991. LNCS, vol. 547, pp. 1–16. Springer, Heidelberg (1991)
5. Harpes, C., Massey, J.L.: Partitioning cryptanalysis. In: Biham, E. (ed.) FSE 1997. LNCS, vol. 1267, pp. 13–27. Springer, Heidelberg (1997)
6. Matsui, M.: Celebrating the 25th year of FEAL - A new prize problem, rump session of CRYPTO'12. http://crypto.2012.rump.cr.yp.to/19997d5a295baee62 c05ba73534745ef.pdf
7. Matsui, M.: Linear cryptanalysis method for DES cipher. In: Helleseth, T. (ed.) EUROCRYPT 1993. LNCS, vol. 765, pp. 386–397. Springer, Heidelberg (1994)
8. Matsui, M., Yamagishi, A.: A new method for known plaintext attack of FEAL cipher. In: Rueppel, R.A. (ed.) EUROCRYPT 1992. LNCS, vol. 658, pp. 81–91. Springer, Heidelberg (1993)
9. Miyaguchi, S.: News on FEAL Cipher, talk at the rump session at CRYPTO'90 (1990)
10. Morita, H., Ohta, K., Miyaguchi, S.: A switching closure test to analyze cryptosystems. In: Feigenbaum, J. (ed.) CRYPTO 1991. LNCS, vol. 576, pp. 183–193. Springer, Heidelberg (1992)
11. Ohta, K., Aoki, K.: Linear cryptanalysis of fast data encipherment algorithm. Technical Report of IEICE (1994)
12. Quisquater, J.-J., Delescaille, J.-P.: How easy is collision search. New results and applications to DES. In: Brassard, G. (ed.) CRYPTO 1989. LNCS, vol. 435, pp. 408–413. Springer, Heidelberg (1990)
13. Shimizu, A., Miyaguchi, S.: Fast data encipherment algorithm FEAL. In: Price, W.L., Chaum, D. (eds.) EUROCRYPT 1987. LNCS, vol. 304, pp. 267–278. Springer, Heidelberg (1988)

Colliding Keys for *SC2000-256*

Alex Biryukov[1] and Ivica Nikolić[2]([✉])

[1] University of Luxembourg, Walferdange, Luxembourg
alex.biryukov@uni.lu
[2] Nanyang Technological University, Singapore, Singapore
inikolic@ntu.edu.sg

Abstract. In this work we present analysis for the block cipher *SC2000*, which is in the Japanese CRYPTREC portfolio for standardization. In spite of its very complex and non-linear key-schedule we have found a property of the full *SC2000-256* (with 256-bit keys) which allows the attacker to find many pairs of keys which generate identical sets of subkeys. Such colliding keys result in identical encryptions. We designed an algorithm that efficiently produces colliding key pairs in 2^{39} time, which takes a few hours on a PC. We show that there are around 2^{68} colliding pairs, and the whole set can be enumerated in 2^{58} time. This result shows that *SC2000-256* cannot model an ideal cipher. Furthermore we explain how practical collisions can be produced for both Davies-Meyer and Hiroses hash function constructions instantiated with *SC2000-256*.

Keywords: SC2000 · Block cipher · Key collisions · Equivalent keys · CRYPTREC · Hash function

1 Introduction

The block cipher *SC2000* [15] was designed by researchers from Fujitsu and the Science University of Tokyo, and submitted to the open call for 128-bit encryption standards organized by Cryptography Research and Evaluation Committees (CRYPTREC). Started in 2000, CRYPTREC is a program of the Japanese government set up to evaluate and recommend cryptographic algorithms for use in industry and institutions across Japan. An algorithm becomes a CRYPTREC recommended standard after two stages of evaluations. Unlike AES, eSTREAM and SHA-3 competitions, the evaluation stages of CRYPTREC do not have strictly defined time limits, but an algorithm progresses to the next stage (or becomes a standard), when its security level has been confirmed by a substantial amount of cryptanalysis. CRYPTREC takes into account all published cryptanalysis in academia and, as well, hires experts to evaluate the security of the algorithm. *SC2000* has passed the first two stages, and for a decade it was among the recommended standards.

Cryptanalysis of the full 6.5–7.5 round (depending on the key size) *SC2000* is still unknown, however, single-key attacks on round-reduced *SC2000* were presented in several papers: boomerang and rectangle attacks on 3.5 rounds

© Springer International Publishing Switzerland 2014
A. Joux and A. Youssef (Eds.): SAC 2014, LNCS 8781, pp. 77–91, 2014.
DOI: 10.1007/978-3-319-13051-4_5

by Dunkelman and Keller [7] and Biham et al. [2], high probability 3.5-round differential characteristics were used in 4.5-round attack by Raddum and Knudsen [13], iterative differential and linear characteristics resulting in attacks on 4.5 rounds by Yanami et al. [17], and a differential attack on 5 rounds by Lu [10].

In spite of considerable evaluation effort by world leading analysts, the cryptanalytic progress on the cipher was slow. A possible reason is given in one of the evaluation reports [16] – the authors state that "... the design is complicated and uses components which do not facilitate for easy analysis". Indeed, *SC2000* uses surprisingly large number of different operations: modular additions, subtractions and multiplications, bitwise additions (XOR), two S-boxes of different size (5 bits and 6 bits), diffusion layers based on multiplications by binary matrices, and rotations. Compared to the widely used design methods such as substitution-permutations (SP) networks (only S-boxes and diffusion layers), or ARX (additions, rotations and XOR), *SC2000* seems too complex, which in turn makes the analysis hard to perform. Moreover, in *SC2000* there are more operations in the key schedule than in the state – this may explain the absence of the key schedule attacks. This paper is the first analysis on the key schedule – we find a weakness in the complex key schedule that we exploit to find colliding keys, i.e. two different master keys that result in the same subkeys. Our result works on the full cipher and independently of the number of its rounds.

In [11] Matsui investigates the behavior of colliding key pairs for the stream cipher RC4. He shows that even in the case of a key size as small as 24 bytes, there are related keys that create the same initial state, hence they generate the same pseudo-random byte stream. In other words, the streams collide. Matsui's discovery is rather interesting and unexpected as the number of possible distinct initial states in RC4 is $256! \approx 2^{1684}$ while the number of states generated from 24-byte key is only 2^{192}. No key collisions should occur in any cipher (the key schedule should be injective), in particular in ciphers that have strictly expandable key schedule, where the accumulative size of the subkeys is larger than the size of the master key. The ratio of the expanded key size/master key size usually depends on the number of rounds and on the length of subkey input in each round. For example, in AES-256 this ratio is 7.5 as there are 15 128-bit subkeys produced from the 256-bit master key. Colliding keys are often called equivalent keys and the existence of such keys is known for a few ciphers. For instance, Robshaw [14] has shown that another CRYPTREC candidate, the block cipher CIPHERUNICORN-A, has equivalent keys. Kelsey et al. [9] found trivial equivalent keys for the Tiny Encryption Algorithm (TEA) block cipher. Furthermore, Aumasson et al. [1] have discovered that the ISDB Scrambling Algorithm, the cipher MULTI2, allows such keys as well.

For *SC2000-256*, despite the fact that the total size of the subkeys is 8 times larger than the size of the master key, we show that this cipher does not have an injective key schedule. There exists a set of 2^{68} pairs of colliding master keys – each pair is composed of two different master keys that after the key schedule lead to the same set of subkeys. Therefore encryptions of *any* plaintext under the first and under the second key produce the same ciphertext, hence the two master keys are equivalent. We achieve the collisions in the subkeys by exploiting weaknesses in the two-stage key schedule: in the first stage we efficiently find a key pair that results in two intermediate keys with a special relation, which in turn is a sufficient condition for the second stage to produce the same subkeys.

Our algorithm for finding a colliding key pair requires only 2^{39} operations, and we have tested our analysis in practice by implementing a search on a regular PC. The produced collisions (see Table 1) confirm the correctness of the analysis and the complexity of the algorithm. We show how an attacker can use the colliding key pairs in order to construct practical collisions in hash functions instantiated with *SC2000-256*. In both single-block-length Davies-Meyer hash, and in the double-block-length Hirose's hash [8] the level of collisions resistance drops from 64,128 to only 39 bits, if instantiated with *SC2000-256*. This suggests that *SC2000-256*, although possibly secure for encryption, has a serious key-schedule weakness and cannot model an ideal cipher.

2   Description of *SC2000-256*

SC2000 is 128-bit block cipher that supports 128, 192, and 256-bit keys. In this work we focus on 256-bit key cipher, further denoted as *SC2000-256*. This cipher has 7.5 rounds, but our analysis is independent of the number of rounds and the round function, as it is valid for any number and for any function. Therefore, in the sequel we describe only the key schedule.

Most of the operations in the key schedule are word-oriented. The only exception is S_{func}, which is a bijective non-linear operation that applies in parallel 5-bit and 6-bit S-boxes (see Fig. 1). The 32-bit input word is split into six chunks of sizes 6,5,5,5,5, and 6 bits, respectively, then 6-bit or 5-bit S-boxes (depending on the size of the chunk) are applied to the chunks, and finally the outputs of the S-boxes are concatenated to produce the final output of S_{func}. The remaining operations in the key schedule are all word-oriented, and include:

1. M_{func} : bijective linear transformation which is a multiplication by a 32×32 matrix. The input is seen as a vector of 32 elements, and it is multiplied by a binary matrix.
2. $+, \boxplus$: addition mod 2^{32}.
3. $-, \boxminus$: subtraction mod 2^{32}.
4. $\times, \boxtimes$: multiplication mod 2^{32}.

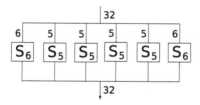

Fig. 1. The operation S_{func} used in the key schedule of *SC2000-256*.

5. $\oplus$: XOR (bitwise addition).
6. $\lll 1$: rotation by 1 bit to the left of 32-bit words.

The key schedule needs two steps (or phases) to produce the subkey words (used in the round functions) from the master key. At the beginning, it starts by dividing the 256-bit master key into eight 32-bit words $uk_j, j = 0, 1, \ldots, 7$, called master key words.

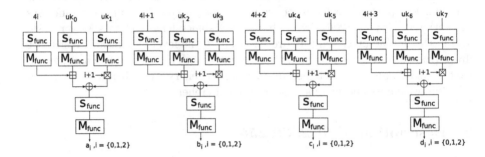

Fig. 2. The intermediate key generation used in *SC2000-256*.

The first phase, called intermediate key generation, takes the 8 words uk_j and outputs 12 intermediate key words $a_i, b_i, c_i, d_i, i = 0, 1, 2$ (see Fig. 2). It applies four similar transformations, called branches, to the four pairs of master key words: the first branch operates on uk_0, uk_1 and produces a_0, a_1, a_2, the second branch on uk_2, uk_3 and outputs b_0, b_1, b_2, etc. Hence each triplet of intermediate key words depends only on two master key words. In a pseudo code, this phase can be described as:

for $i = 0 \rightarrow 2$ **do**
$\quad a_i \leftarrow M_f(S_f((M_f(S_f(uk_0)) + M_f(S_f(4i))) \oplus M_f(S_f(uk_1))))$
$\quad b_i \leftarrow M_f(S_f((M_f(S_f(uk_2)) + M_f(S_f(4i + 1))) \oplus M_f(S_f(uk_3))))$
$\quad c_i \leftarrow M_f(S_f((M_f(S_f(uk_4)) + M_f(S_f(4i + 2))) \oplus M_f(S_f(uk_5))))$
$\quad d_i \leftarrow M_f(S_f((M_f(S_f(uk_6)) + M_f(S_f(4i + 3))) \oplus M_f(S_f(uk_7))))$
end for

The second phase, called extended key generation, takes the 12 intermediate key words and produces 64 subkey words $ek_i, i = 0, 1, \ldots, 63$, called extended key words. Each subkey word is obtained with a non-symmetric transformation (see Fig. 3) of four intermediate key words that come from different branches, hence every subkey word depends on all master key words. For each subkey word, to determine which four intermediate key words should be taken, and in what order, this phase requires two lookup tables. The first table *Order* specifies the order (recall that the transformation is non-symmetric, so the order matters) in which the words are put into the transformation. The second table *Index* determines which word within a branch should be taken. As a result, no two subkey words depend on the same intermediate key words put in the same order. Refer to Fig. 3

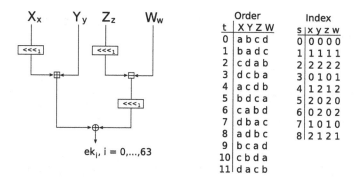

Fig. 3. The non-symmetric transformation used in the extended key generation (left), and the values of the lookup tables (right).

for a pictorial view of the non-symmetric transformation and for the values of the lookup tables. In a pseudo code, the second phase can be described as:

for $i = 0 \rightarrow 63$ **do**
  $s \leftarrow i \ (mod \ 9)$
  $t \leftarrow (i + \lfloor \frac{i}{36} \rfloor) \ (mod \ 12)$
  $X \leftarrow Order[t][0]$
  $x \leftarrow Index[s][0]$
  $Y \leftarrow Order[t][1]$
  $y \leftarrow Index[s][1]$
  $Z \leftarrow Order[t][2]$
  $z \leftarrow Index[s][2]$
  $W \leftarrow Order[t][3]$
  $w \leftarrow Index[s][3]$
  $ek_i \leftarrow (X_h \lll 1 + Y_y) \oplus ((Z_z \lll 1 - W_w) \lll 1)$
end for

As the description of the key schedule suffices to understand our attack, for a full specification of the cipher we refer the interested reader to [15].

3 Key Collisions for *SC2000-256*

For *SC2000-256*, we show how to find two distinct master keys that produce the same subkey words and hence we obtain key collisions. The core idea of our analysis is a weakness in the second phase (extended key generation) – it can cancel a particular input difference (i.e. a particular difference in each pair of intermediate key words), resulting in subkey collisions. If we are able to deterministically find two master keys that after the first phase produce the particular difference, then the second phase will cancel the difference, and we will end up with collisions. Therefore, to present the analysis we focus on:

1. **(Second phase)** Specify the difference in intermediate key words, and prove that it leads to collisions after the second phase.
2. **(First phase)** Give an algorithm that finds two master keys that lead to the difference in the intermediate key words after the first phase.

This seemingly upside-down approach (at the beginning we analyze the second phase, and then the first), is taken to understand why the algorithm at step 2 has to target the specific difference (and not some other).

Notations. With superscripts 1 and 2 we denote various master, intermediate and extended keys for the first and respectively the second master key, e.g. a_0^2 is the first intermediate key produced from the second master key, ek_{10}^1 is the eleventh subkey word produced from the first master key. The subscript h stands for hexadecimal number, for instance 80000000_h is 2^{31}. With $\overline{X}$ we denote the bitwise negation of the word X, i.e. $\overline{X} = X \oplus (-1)$, while $\wedge$ stands for bitwise AND.

3.1 Specifying the Difference for the Second Phase

Let us start our analysis by focusing on the second phase. The following Lemma defines the required difference (and the additional conditions) in the intermediate key words, that plays the main role in the analysis:

Lemma 1. *Let each pair (X_1, X_2) in the set of pairs of intermediate key words (a_i^1, a_i^2), (b_i^1, b_i^2), (c_i^1, c_i^2), (d_i^1, d_i^2), $i = 0, 1, 2$ satisfy the following two conditions:*

$$\textbf{Condition 1} \quad X_2 = \overline{X_1} + 3,$$
$$\textbf{Condition 2} \quad X_1 \wedge 8000000f_h = 80000003_h.$$

Then the extended key generation will produce the same extended keys (subkey words), i.e. $ek_i^1 = ek_i^2, i = 0, \ldots, 63$.

The Lemma claims that if: (Condition 1) the pairs of intermediate words produced from the first and the second master key have the special relation, and (Condition 2) the intermediate key words produced from the first master key have particular values in five bits (the most significant, and the four least significant), then after the second phase they will lead to the same subkeys. The Condition 2 becomes clear in the proof, and when the five bits have this specific value, the probability that the subkey words collide is 1, otherwise it is less than 1.

Proof. To prove the Lemma we focus on the extended key generation function

$$f(X, Y, Z, W) = (X \lll 1 + Y) \oplus [(Z \lll 1 - W) \lll 1].$$

We claim that if X, Y, Z, W are randomly chosen words with the MSB fixed to 1 and the four LSBs fixed to 3, then

$$f(X, Y, Z, W) = f(\overline{X} + 3, \overline{Y} + 3, \overline{Z} + 3, \overline{W} + 3).$$

Let us rewrite f as an XOR of two functions g, h, i.e.

$$f(X, Y, Z, W) = g(X, Y) \oplus h(Z, W),$$
$$g(X, Y) = X \lll 1 + Y,$$
$$h(Z, W) = (Z \lll 1 - W) \lll 1.$$

We will prove that

$$g(X, Y) \oplus g(\overline{X} + 3, \overline{Y} + 3) = \texttt{ffffffff7}_h, \qquad (1)$$
$$h(Z, Y) \oplus h(\overline{Z} + 3, \overline{W} + 3) = \texttt{ffffffff7}_h, \qquad (2)$$

and thus

$$f(X, Y, Z, W) \oplus f(\overline{X} + 3, \overline{Y} + 3, \overline{Z} + 3, \overline{W} + 3) =$$
$$g(X, Y) \oplus h(Z, W) \oplus g(\overline{X} + 3, \overline{Y} + 3) \oplus h(\overline{Z} + 3, \overline{W} + 3) = 0.$$

We need to following supplementary facts:

Fact 1 *Let X, Y be 32-bit words. If $X \wedge \texttt{7fffffff}_h + Y \wedge \texttt{7fffffff}_h < 2^{31}$ then*

$$(X + Y) \lll 1 = X \lll 1 + Y \lll 1.$$

Proof. The fact can be seen as corollary of Theorem 4.11 from [5]. □

Fact 2 *For any values X, Y*

$$\overline{X + Y} = \overline{X} + \overline{Y} + 1.$$

Proof. Note that for any value V, $V \oplus \overline{V} = V + \overline{V} = -1$, and thus $\overline{V} = -1 - V$. Therefore:

$$\overline{X + Y} = -1 - (X + Y) = (-1 - X) + (-1 - Y) + 1 = \overline{X} + \overline{Y} + 1$$

□

Fact 3 *If $U \wedge m = 0$ then $\overline{U} \oplus (U + m) = \texttt{ffffffff}_h \oplus m$.*

Proof. When $U \wedge m = 0$, then $U + m = U \oplus m$. Therefore

$$\overline{U} \oplus (U + m) = \overline{U} \oplus U \oplus m = \texttt{ffffffff}_h \oplus m.$$

□

Now we are ready to present the proof of the Lemma. We will prove only the part for g – the part for h is similar and instead of modular addition we have to work with modular subtraction. Let us focus on (1). We get:

$$g(X, Y) \oplus g(\overline{X} + 3, \overline{Y} + 3) = \qquad (3)$$
$$= (X \lll 1 + Y) \oplus ((\overline{X} + 3) \lll 1 + \overline{Y} + 3) = \qquad (4)$$
$$= (X \lll 1 + Y) \oplus ((\overline{X}) \lll 1 + 3 \lll 1 + \overline{Y} + 3) = \qquad (5)$$
$$= (X \lll 1 + Y) \oplus ((\overline{X \lll 1} + \overline{Y}) + 9) = \qquad (6)$$
$$= (X \lll 1 + Y) \oplus ((\overline{X \lll 1 + Y}) + 8) = \qquad (7)$$
$$= \overline{U} \oplus (U + 8), \qquad (8)$$

where $U = \overline{X} \lll 1 + Y$. The transition (4) to (5) is due to Fact 1 – the two least significant bits of $\overline{X}$ are 00 thus $\overline{X} \wedge \texttt{7fffffff}_h + 3 < 2^{31}$. Note, this is where we actually use the requirement of Condition 2: the two least significant bits of X must be '11'. The transition (6) to (7) is due to Fact 2. Finally, the four least significant bits of U are 0101 (again use of Condition 2!) and thus by Fact 3, $g(X,Y) \oplus g(\overline{X} + 3, \overline{Y} + 3) = \texttt{ffffff7}_h$. This concludes the proof. $\square$

We have discovered the strange conditions of the Lemma (and then provided a formal proof), when we analyzed the behavior of the non-symmetric function f – it became clear that f cancels some modular differences. The similarity of the left and the right side (the function g and the function h without the final rotation) of f, and the fact that the rotations are only on 1 bit, suggested that there may exist a universal difference for the intermediate words, such that cancellation in f would occur when all four words have this difference. We started experimenting with various differences between X and $\overline{X}$, and various values for the two most significant bit (as in h we have twice rotation on 1 to the left, we took 2 bits), and several least significant bits. The experiments were implemented as an exhaustive computer search that tries all such differences and bit values, and for each combination checks the probability that f cancels the difference. The results of our experiment led to the actual Conditions 1,2.

3.2   Finding Pairs in the First Phase

Let us focus on the first phase and produce a pair of master key words that after this phase result in pairs of intermediate key words that comply with Conditions 1 and 2 of the Lemma. For the sake of simplicity, first we take into account only Condition 1, and later we consider Condition 2.

Let $u_i^1, i = 0, \ldots, 7$ be the words of the first master key K_1, and $u_i^2, i = 0, \ldots, 7$ be the words of the second master key K_2. Let U_i be the corresponding words of the master keys after the application of S_{func} and M_{func}, i.e. $U_i^j = M_{func}(S_{func}(u_i^j))), i = 0, \ldots, 7, j = 1, 2$. Also, let I_i be the constants $I_{i+1} = M_{func}(S_{func}(4 \cdot i)), i = 0, 1, 2$. Then, taking into account the intermediate key generation procedure, Condition 1 for the pairs $(a_i^1, a_i^2), i = 0, 1, 2$ is equivalent to solving the following system of equations (refer to Fig. 4):

$$(U_0^1 + I_1) \oplus U_1^1 = A^1 \tag{9}$$

$$(U_0^1 + I_2) \oplus 2 \cdot U_1^1 = B^1 \tag{10}$$

$$(U_0^1 + I_3) \oplus 3 \cdot U_1^1 = C^1 \tag{11}$$

$$(U_0^2 + I_1) \oplus U_1^2 = A^2 \tag{12}$$

$$(U_0^2 + I_2) \oplus 2 \cdot U_1^2 = B^2 \tag{13}$$

$$(U_0^2 + I_3) \oplus 3 \cdot U_1^2 = C^2 \tag{14}$$

$$A^2 = S_{func}^{-1}(M_{func}^{-1}(\overline{M_{func}(S_{func}(A^1))} + 3)) \tag{15}$$

$$B^2 = S_{func}^{-1}(M_{func}^{-1}(\overline{M_{func}(S_{func}(B^1))} + 3)) \tag{16}$$

$$C^2 = S_{func}^{-1}(M_{func}^{-1}(\overline{M_{func}(S_{func}(C^1))} + 3)) \tag{17}$$

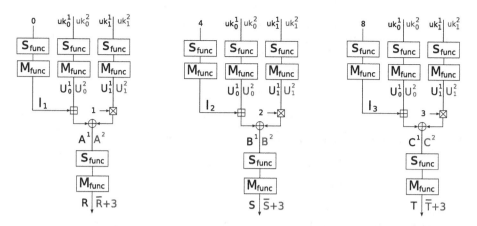

Fig. 4. The intermediate values used to describe the algorithm in the first branch (with outputs a_0, a_1, a_2). The values produced from the first master key are on the left and have a dark red color, while from the second are on the right and have blue color. R, S, T are 32-bit words that do not have specified values (Color figure online).

Let $G(x) = S_{func}^{-1}(M_{func}^{-1}(\overline{M_{func}(S_{func}(x))} + 3))$. Then the system can be rewritten as:

$$(U_0^1 + I_1) \oplus U_1^1 = G((U_0^2 + I_1) \oplus U_1^2)$$
$$(U_0^1 + I_2) \oplus 2 \cdot U_1^1 = G((U_0^2 + I_2) \oplus 2 \cdot U_1^2)$$
$$(U_0^1 + I_3) \oplus 3 \cdot U_1^1 = G((U_0^2 + I_3) \oplus 3 \cdot U_1^2)$$

The system has three equations and four unknowns $(U_0^1, U_1^1, U_0^2, U_1^2)$ – theoretically, for any values of I_1, I_2, I_3 and a bijective function G^1, it has 2^{32} solutions. To produce one solution, we find a good pair of master keys for the first two intermediate words (the first two equations of the system), and then we check if the pair is good as well for the third. The algorithm is as follows:

1. Fix random A_1, B_1.
2. Find U_0^1, U_1^1 that satisfy Eqs. (9) and (10).
3. Compute C^1 from U_0^1, U_1^1 with (11).
4. Produce A_2, B_2 from A_1, B_1 with the function G by (15),(16).
5. Find U_0^2, U_1^2 that satisfy Eqs. (12) and (13).
6. Compute C^2 from U_0^2, U_1^2 with (14).
7. Compute $\tilde{C}^2$ from C^1 with the function G by (17).
8. If C^2 is not equal to $\tilde{C}^2$ go to Step 1.
9. The quartet $(U_0^1, U_1^1, U_0^2, U_1^2)$ is the solution for the system.

---

[1] This is not always the case as the authors have tried to launch a much simpler attack with $A^1 = A^2, B^1 = B^2, C^1 = C^2$ and failed due to the fact that no solutions exist for such system.

The values of C^2 and $\tilde{C}^2$ coincide with probability 2^{-32}, hence to find a solution, we need to repeat around 2^{32} times the Steps 1–7. The complexity of each step is constant (just an application of a formula), with the exception of Steps 2 and 5 – here, we need to find the unknown (U_0, U_1) given the two equations:

$$(U_0 + I_1) \oplus U_1 = A$$
$$(U_0 + I_2) \oplus 2 \cdot U_1 = B$$

After basic algebraic transformations, they are reduced to the form:

$$U_1 = ((2 \cdot U_1 \oplus B) + (I_1 - I_2)) \oplus A \tag{18}$$
$$U_0 = (A \oplus U_1) - I_1 \tag{19}$$

Thus we want to efficiently solve Eq. $(18)^2$. The complexity of finding the value of U_1 is given by the following Lemma. We note that the algorithm relies on solving a word equation and to a certain extend is similar to the algorithms from [3].

Lemma 2. *There is an algorithm that, with complexity linear in the size of the words, finds the unique solution of the equation:*

$$X = ((2 \cdot X \oplus B) + C) \oplus A, \tag{20}$$

where A, B, C are some word constants.

Proof. Let us use subscripts to denote the bits of a word, e.g. X_5 is the sixth least significant bit of the word X. The multiplication $2 \cdot X$ is a shift to the left of X by one position, and therefore the $(s + 1)$-th bit of $2 \cdot X$ is indeed the s-th bit of X, or in our notation $(2 \cdot X)_s = X_{s-1}$. We solve the above Eq. (20) bit by bit, starting from the least significant bit, and moving towards the most significant bit. In other words, we use a recursive algorithm: first show how to find the least significant (i.e. 0-th) bit, and assuming that we have found the t-th bit, describe how we can find the $(t + 1)$-th bit. The equation involves modular addition, therefore for each bit we should keep track of the carry – with cr_i we denote the carry bit of $(2 \cdot X \oplus B) + C$ at i-th bit position.

– *Bit 0.* For the least significant bit, Eq. (20) takes the form:

$$X_0 = B_0 \oplus C_0 \oplus A_0,$$

hence the least significant bit of X_0 can be uniquely determined with a simple XOR of three bits, while $cr_0 = B_0 \cdot C_0$.

– *Bit $t + 1$.* We assume we have the previous carry cr_t, and we have found the value for X_t. Then for the bit $t + 1$, we have:

$$X_{t+1} = X_t \oplus B_{t+1} \oplus C_{t+1} \oplus cr_t \oplus A_{t+1}$$

and for the carry we get $cr_{t+1} = m(X_t \oplus B_{t+1}, C_{t+1}, cr_t)$, where $m(x, y, z) = xy \oplus xz \oplus yz$. Again, X_{t+1}, cr_{t+1} are determined uniquely with a constant number of operations.

---

^{2} Once we have the value of U_1, we can easily find U_0 by (19).

As each step of the algorithm requires constant number of operations, and there are n steps in total (n is the word size), we can claim that the complexity of finding the unique solution is linear in the size of the words.   □

The Lemma gives us the complexity for the Steps 2 and 5 of the algorithm, i.e. we can solve the system for any A, B with a constant complexity (since $n = 32$). As a result, the total complexity of the algorithm is 2^{32}.

Now we are ready to consider Condition 2. To satisfy this condition (as well as Condition 1), we have to slightly tweak our algorithm and make sure that we have the precise value in the 5 bits of each intermediate key word produced from the first master key. Further we present the full algorithm for computing the pair of master key words that produces the required intermediate key words in the first branch:

1. Fix random R, S such that the most significant bits of R, S are 1, and the values of the four least significant bits equal 3. Compute $A_1 = S_{func}^{-1}(M_{func}^{-1}(R))$, $B_1 = S_{func}^{-1}(M_{func}^{-1}(S))$.
2. Find U_0^1, U_1^1 that satisfy Eqs. (9) and (10).
3. Compute C^1 from U_0^1, U_1^1 with (11).
4. Produce A_2, B_2 from A_1, B_1 with the function G by (15),(16).
5. Find U_0^2, U_1^2 that satisfy Eqs. (12) and (13).
6. Compute C^2 from U_0^2, U_1^2 with (14).
7. Compute $\tilde{C}^2$ from C^1 with the function G by (17).
8. If C^2 is not equal to $\tilde{C}^2$, or in the word T such that $T = M_{func}(S_{func}(C^2))$, the most significant bit of T is not 1, or the value of the four least significant bits is not 3, go to Step 1.
9. The quartet $(U_0^1, U_1^1, U_0^2, U_1^2)$ is the solution for the system.

The new method of defining the values of A_1, B_1 introduced at Step 1, does not change the complexity of the algorithm (compared to the previous). On the other hand, the additional filter at Step 8 (condition on 5 bits) increases the frequency of repeating Steps 1–7 by a factor of 2^5. Hence, the total complexity of the algorithm is $2^{32} \cdot 2^5 = 2^{37}$.

The above complexity is required to find the values of $U_0^1, U_1^1, U_0^2, U_1^2$, which are the 4 words produced from a pair of two master key words in the first branch only. To find the precise value of the pair of two master key words, i.e. to find $(u_0^1, u_1^1), (u_1^1, u_1^2)$, we just need to invert S_{func}, M_{func} (see Fig. 4), hence $u_i^j = S_{func}^{-1}(M_{func}^{-1}(U_i^j)), i = 0, 1, j = 1, 2$. Computing the values of the master key words for all four branches can be done similarly, and with complexity $4 \cdot 2^{37} = 2^{39}$. Therefore, in 2^{39} we can find a pair of master key words, that after the first phase result in a pair of intermediate key words that comply with the conditions of Lemma 1, and thus after the second phase lead to colliding subkey words. Therefore, encryption of any plaintext under the first, and under the second key, produces the same ciphertext.

4   Results and Applications

We have implemented the above search for colliding keys on a PC and, in a matter of hours, we were able to find a pair of master keys (K_1, K_2) that produces the same subkeys. The words of the master keys are given in Table 1. These practical results confirm our analysis and the complexity of the algorithm.

Table 1. Example of colliding pair of master keys for *SC2000-256*

K_1	0x59d0d459	0x4473d8dd	0xcc7d3064	0xd3bbda93
	0x8ff60b58	0xe9dc073d	0x8776c115	0x743c9cfe
K_2	0x10672240	0xb94214ff	0x2bc72c50	0x539cdd3e
	0xf9e9f251	0x921811fa	0x35bf5b7f	0x82ab8bdd
ek^1, ek^2	0xff582ab3	0x4d261f23	0xcb9f9ad3	0x7c81f9c2
	0x0997d523	0xc42fc563	0x2172df72	0x95d8dcb3
	0x18121223	0x9d034e02	0x1baa1423	0xe9190113
	0x4d148522	0xd9247b13	0xb49e6723	0xa393b3e3
	0x3953dbc3	0xb2f85ee2	0x0c17c0a2	0x29d7a162
	0x45ba8593	0x14eb6423	0xe4780213	0xdf8f8b23
	0xd7b48013	0xb5a368a3	0xc47fffc3	0xdee3ff23
	0x4f279343	0xb4a34873	0xe2881a63	0x0c1b8372
	0xae1a47e3	0x3285cd02	0x96418533	0x8a904d03
	0xf1633b43	0x0664d382	0x35fb0a83	0xe246b6c2
	0x8fc44d93	0x2fe1e763	0xd2823073	0x530dffc2
	0xe7dd8fe3	0xe4503972	0xad5f9022	0xdebed232
	0x10a9a642	0x9db60612	0x3ea3de03	0x5ed728a2
	0x3941d142	0xd961e823	0x43df53b2	0x7d7f7a82
	0x766512c3	0x6d9e3863	0xaaccccc73	0xf74a2b92
	0x9ca25a32	0xd6a613e2	0x94819ca3	0xc98a4542

Our next task is to find the number of colliding key pairs. A careful look at the proposed algorithm reveals the number. At Step 1, we can choose $2^{32-5} = 2^{27}$ possible values for A_1, and the same number for B_1. The equations at Steps 2 and 5 can be solved always, thus there are $2^{2 \cdot 27} = 2^{54}$ different values for the tuple $(U_0^1, U_1^1, C^1, A^2, B^2, U_0^2, U_1^2, C^2, \tilde{C}^2)$ obtained at Steps 2–7. The condition at Step 8 filters $2^{32+5} = 2^{37}$ tuples, therefore we end up with $2^{54-37} = 2^{17}$ possible different values for $(U_0^1, U_1^1, U_0^2, U_1^2)$, and thus there are 2^{17} values for $(uk_0^1, uk_1^1, uk_0^2, uk_1^2)$. This is for the first branch only – if we take into account the four branches, in total there are $2^{4 \cdot 17} = 2^{68}$ colliding key pairs in *SC2000-256*. It is interesting to note that the collisions are found independently for each branch. Thus, to find all 2^{68} colliding key pairs we need only around $4 \cdot 2^{54} = 2^{56}$ operations.

Application to hash functions. The key collision attack on *SC2000-256* leads to practical collisions for single-block-length hash function instantiated with *SC2000-256*. Assume that the compression function $C(M, H)$ is based on the Davies-Meyer construction[3], i.e. $C(M, H) = E_M(H) \oplus H$. For the cipher $E_M(H)$, we find two colliding keys $M_1, M_2, M_1 \neq M_2$, such that for any plaintext H, it holds $E_{M_1}(H) = E_{M_2}(H)$. Therefore, in 2^{39} time we can find collisions for the compression function as

$$C(M_1, H) \oplus C(M_2, H) = E_{M_1}(H) \oplus H \oplus E_{M_2}(H) \oplus H = 0.$$

The collisions do not depend on the chaining value H, hence the above result holds for the hash function as well.

Double-block-length hash constructions based on *SC2000-256*, are not collision resistant as well. Let us take Hirose's construction [8], where the 256-bit compression function $C(g, h, M)$ (here g, h are two 128-bit chaining values, M is 128-bit message block) is based upon a cipher $E_K(P)$ with 256-bit key K and 128-bit state, and it is defined as

$$C(g, h, M) = E_{h||M}(g) \oplus g||E_{h||M}(g \oplus c) \oplus g \oplus c,$$

where $||$ is concatenation of two 128-bit words, and c is some non-zero constant. Hirose proved the collision resistance level of this construction to be around 2^{128}, when the underlying cipher is ideal. However, if we use *SC2000-256*, and two colliding keys $h_1||M_1, h_2||M_2$, then for any 128-bit chaining value g we obtain

$C(g, h_1, M_1) \oplus C(g, h_2, M_2) =$
$= [E_{h_1||M_1}(g) \oplus g||E_{h_1||M_1}(g \oplus c) \oplus g \oplus c] \oplus [E_{h_2||M_2}(g) \oplus g||E_{h_2||M_2}(g \oplus c) \oplus g \oplus c] =$
$= E_{h_1||M_1}(g) \oplus g \oplus E_{h_2||M_2}(g) \oplus g||E_{h_1||M_1}(g \oplus c) \oplus g \oplus c \oplus E_{h_2||M_2}(g \oplus c) \oplus g \oplus c =$
$= 0^{128}||0^{128}.$

Therefore, instead of 128-bit collision level achieved when using an ideal cipher, we obtain only 39 bits in the case of *SC2000-256*.

Application to SC2000-128 and SC2000-192. For the cases of 128-bit and 192-bit key *SC2000*, the last four, respectively two, words entering the intermediate key generation are copies of the original master key words. Hence in these cases two, respectively one, branches of the intermediate key generation has to be satisfied probabilistically. As there are 96 conditions per branch, and the remaining freedom per branch is 2^{32}, and the branches are cross dependent, i.e. for 128-bit key, the third branch depends on the keys of the first branch, and for 128-bit and 192-bit keys, the fourth branch depends on the second branch, the analysis for *SC2000-256* cannot be extended to *SC2000* with 128-bit and 192-bit keys.

---

[3] Among the analyzed by Preneel-Govaerts-Vandewalle [12] secure single- block-length modes, only Davies-Meyer mode, i.e. $C(M, H) = E_M(H) \oplus H$ allows 256-bit messages inputs and 128-bit chaining values.

5   Conclusion

We have shown that the key schedule of *SC2000-256* is not injective, and the cipher has 2^{68} pairs of colliding keys. These pairs are due to the two weaknesses in the key schedule: the non-symmetric function used in the second phase easily cancels a particular difference, and the four branches in the first phase are independent and can be attacked separately. Based on the combination of the two weaknesses, we have derived an algorithm that in 2^{39} operations finds one pair of colliding keys, and in 2^{56} finds all of them. As a proof of concept we have produced colliding keys in a matter of a few hours on a PC. Thus the collision resistance level of hash functions based on *SC2000-256* is very low.

SC2000-256 suffers from a practically exploitable security weakness and cannot model an ideal cipher. In spite of the *SC2000-256* cipher being in the CRYPTREC portfolio for more than 10 years and in spite of considerable previous evaluation, this paper is the first to discover this design flaw. This is probably due to complexity of the design and is an example in favor of clean and easy to analyze design strategies.

Acknowledgement. The authors would like to thank Jérémy Jean for proposing improvements to the results and the anonymous reviewers of SAC 2014 for their helpful comments. Ivica Nikolić is supported by the Singapore National Research Foundation Fellowship 2012 NRF-NRFF2012-06.

References

1. Aumasson, J.-P., Nakahara Jr., J., Sepehrdad, P.: Cryptanalysis of the ISDB scrambling algorithm (MULTI2). In: Dunkelman [6], pp. 296–307
2. Biham, E., Dunkelman, O., Keller, N.: New results on boomerang and rectangle attacks. In: Daemen and Rijmen [4], pp. 1–16
3. Biryukov, A., Gauravaram, P., Guo, J., Khovratovich, D., Ling, S., Matusiewicz, K., Nikolić, I., Pieprzyk, J., Wang, H.: Cryptanalysis of the LAKE hash family. In: Dunkelman [6], pp. 156–179
4. Daemen, J., Rijmen, V. (eds.): FSE 2002. LNCS, vol. 2365. Springer, Heidelberg (2002)
5. Daum, M.: Cryptanalysis of hash functions of the MD4-Family. Ph.D. Thesis, Ruhr-Universität Bochum, May 2005
6. Dunkelman, O. (ed.): FSE 2009. LNCS, vol. 5665. Springer, Heidelberg (2009)
7. Dunkelman, O., Keller, N.: Boomerang and rectangle attacks on SC2000. In: NESSIE 2nd Workshop (London) (2001)
8. Hirose, S.: Some plausible constructions of double-block-length hash functions. In: Robshaw, M. (ed.) FSE 2006. LNCS, vol. 4047, pp. 210–225. Springer, Heidelberg (2006)
9. Kelsey, J., Schneier, B., Wagner, D.: Key-schedule cryptanalysis of IDEA, G-DES, GOST, SAFER, and Triple-DES. In: Koblitz, N. (ed.) CRYPTO 1996. LNCS, vol. 1109, pp. 237–251. Springer, Heidelberg (1996)
10. Lu, J.: Differential attack on five rounds of the SC2000 block cipher. In: Bao, F., Yung, M., Lin, D., Jing, J. (eds.) Inscrypt 2009. LNCS, vol. 6151, pp. 50–59. Springer, Heidelberg (2010)

11. Matsui, M.: Key collisions of the RC4 stream cipher. In: Dunkelman [6], pp. 38–50
12. Preneel, B., Govaerts, R., Vandewalle, J.: Hash functions based on block ciphers: A synthetic approach. In: Stinson, D.R. (ed.) CRYPTO 1993. LNCS, vol. 773, pp. 368–378. Springer, Heidelberg (1994)
13. Raddum, H., Knudsen, L.R.: A differential attack on reduced-round SC2000. In: Vaudenay, S., Youssef, A.M. (eds.) SAC 2001. LNCS, vol. 2259, pp. 190–198. Springer, Heidelberg (2001)
14. Robshaw, M.: A cryptographic review of Cipherunicorn-A. Technical Report, CRYPTRECT Technical report (2001)
15. Shimoyama, T., Yanami, H., Yokoyama, K., Takenaka, M., Itoh, K., Yajima, J., Torii, N., Tanaka, H.: The block cipher SC2000. In: Matsui, M. (ed.) FSE 2001. LNCS, vol. 2355, pp. 312–327. Springer, Heidelberg (2002)
16. Technical Report 1087. Analysis of SC2000 (2000). http://www.ipa.go.jp/security/enc/CRYPTREC/fy15/doc/1087_sc2000.pdf
17. Yanami, H., Shimoyama, T., Dunkelman, O.: Differential and linear cryptanalysis of a reduced-round SC2000. In: Daemen and Rijmen [4], pp. 34–48

Faster Binary-Field Multiplication
and Faster Binary-Field MACs

Daniel J. Bernstein[1,2] and Tung Chou[2]

[1] Department of Computer Science, University of Illinois at Chicago,
Chicago, IL 60607-7053, USA
djb@cr.yp.to
[2] Department of Mathematics and Computer Science,
Technische Universiteit Eindhoven, P.O. Box 513,
5600 MB Eindhoven, The Netherlands
blueprint@crypto.tw

Abstract. This paper shows how to securely authenticate messages using just 29 bit operations per authenticated bit, plus a constant overhead per message. The authenticator is a standard type of "universal" hash function providing information-theoretic security; what is new is computing this type of hash function at very high speed.

At a lower level, this paper shows how to multiply two elements of a field of size 2^{128} using just $9062 \approx 71 \cdot 128$ bit operations, and how to multiply two elements of a field of size 2^{256} using just $22164 \approx 87 \cdot 256$ bit operations. This performance relies on a new representation of field elements and new FFT-based multiplication techniques.

This paper's constant-time software uses just 1.89 Core 2 cycles per byte to authenticate very long messages. On a Sandy Bridge it takes 1.43 cycles per byte, without using Intel's PCLMULQDQ polynomial-multiplication hardware. This is much faster than the speed records for constant-time implementations of GHASH without PCLMULQDQ (over 10 cycles/byte), even faster than Intel's best Sandy Bridge implementation of GHASH with PCLMULQDQ (1.79 cycles/byte), and almost as fast as state-of-the-art 128-bit prime-field MACs using Intel's integer-multiplication hardware (around 1 cycle/byte).

Keywords: Performance · FFTs · Polynomial multiplication · Universal hashing · Message authentication

1 Introduction

NIST's standard AES-GCM authenticated-encryption scheme uses GHASH to authenticate ciphertext (produced by AES in counter mode) and to authenticate

This work was supported by the National Science Foundation under grant 1018836 and by the Netherlands Organisation for Scientific Research (NWO) under grant 639.073.005. Permanent ID of this document: 393655eb413348f4e17c7ec451b9e159. Date: 2014.09.14.

A. Joux and A. Youssef (Eds.): SAC 2014, LNCS 8781, pp. 92–111, 2014.
DOI: 10.1007/978-3-319-13051-4_6

additional data. GHASH converts its inputs into a polynomial and evaluates that polynomial at a secret element of $\mathbb{F}_{2^{128}} = \mathbb{F}_2[x]/(x^{128}+x^7+x^2+x+1)$, using one multiplication in $\mathbb{F}_{2^{128}}$ for each 128-bit input block. The cost of GHASH is an important part of the cost of GCM, and it becomes almost the entire cost when large amounts of non-confidential data are being authenticated without being encrypted, or when a denial-of-service attack is sending a flood of forgeries to consume all available processing time.

Most AES-GCM software implementations rely heavily on table lookups and presumably leak their keys to cache-timing attacks. Käsper and Schwabe [35] (CHES 2009) addressed this problem by introducing a constant-time implementation of AES-GCM using 128-bit vector instructions. Their GHASH implementation takes 14.4 cycles/byte on one core of an Intel Core 2 processor. On a newer Intel Sandy Bridge processor the same software takes 13.1 cycles/byte. For comparison, HMAC-SHA1, which is widely used in Internet applications, takes 6.74 Core 2 cycles/byte and 5.18 Sandy Bridge cycles/byte.

1.1 Integer-Multiplication Hardware

Much better speeds than GHASH were already provided by constant-time MACs that used integer multiplication rather than multiplication of polynomials mod 2. Examples include UMAC [15], Poly1305 [5], and VMAC [36]. Current Poly1305 software from [22] runs at 1.89 Core 2 cycles/byte and 1.22 Sandy Bridge cycles/byte. VMAC, which uses "pseudo dot products" (see Sect. 4), is even faster than Poly1305.

CPUs include large integer-multiplication units to support many different applications, so it is not a surprise that these MACs are much faster in software than GHASH (including non-constant-time GHASH software; see [35]). However, integer multiplication uses many more bit operations than multiplication of polynomials mod 2, so for hardware designers these MACs are much less attractive. MAC choice is a continuing source of tension between software designers and hardware designers.

1.2 New Speeds for Binary-Field MACs

This paper introduces Auth256, an $\mathbb{F}_{2^{256}}$-based MAC at a 2^{255} security level; and a constant-time software implementation of Auth256 running at just 1.89 cycles/byte on a Core 2. We also tried our software on a Sandy Bridge; it runs at just 1.43 cycles/byte. We also have a preliminary Cortex-A8 implementation below 14 cycles/byte.

This new binary-field MAC is not quite as fast as integer-multiplication MACs. However, the gap is quite small, while the hardware advantages of binary fields are quite important. We plan to put our software into the public domain.

Caveat: All of the above performance figures ignore short-message overhead, and in particular our software has very large overhead, tens of thousands of cycles. For 32-, 64-, 128-kilobyte messages, our software takes 3.07, 2.44, 2.14 Core 2 cycles per byte, and 2.85, 2.09, 1.74 Sandy Bridge cycles per byte. This software is designed primarily for authenticating large files, not for authenticating network packets. However, a variant of Auth256 ($b = 1$ in Sect. 6) takes only 0.81 additional cycles/byte and has much smaller overhead. We also expect that, compared to previous MAC designs, this variant will allow significantly lower area for high-throughput hardware, as explained below.

1.3   New Bit-Operation Records for Binary-Field Multiplication

The software speed advantage of Auth256 over GHASH, despite the much higher security level of Auth256, is easily explained by the following comparison. Schoolbook multiplication would take 128^2 ANDs and approximately 128^2 XORs for each 128 bits of GHASH input, i.e., approximately 256 bit operations per authenticated bit. Computing a 256-bit authenticator in the same way would use approximately 512 bit operations per authenticated bit. Auth256 uses just 29 bit operations per authenticated bit.

Of course, Karatsuba's method saves many bit operations at this size. See, e.g., [4,8,9,18,21,41,42,45,48]. Bernstein's Karatsuba/Toom combination in [9] multiplies 256-bit polynomials using only about $133 \cdot 256$ bit operations. Multiplying 256-bit field elements has only a small overhead. However, 133 bit operations is still much larger than 29 bit operations.

Our improved number of bit operations is a combination of four factors. The first factor is faster multiplication: we reduce the cost of multiplication in $\mathbb{F}_{2^{256}}$ from $133 \cdot 256$ bit operations to just $22292 \approx 87 \cdot 256$ bit operations. The second factor, which we do not take credit for, is the use of pseudo dot products to reduce the number of multiplications by a factor of 2, reducing 87 below 44. The third factor, which reduces 44 to 32, is an extra speedup from an interaction between the structure of pseudo dot products and the structure of the multiplication algorithms that we use. The fourth factor, which reduces 32 to just 29, is to use a different field representation for the input to favor the fast multiplication algorithm we use.

Specifically, we use a fast Fourier transform (FFT) to multiply polynomials in $\mathbb{F}_{2^8}[x]$. The FFT is advertised in algorithm courses as using an essentially linear number of field additions and multiplications but is generally believed to be much slower than other multiplication methods for cryptographic sizes. Bernstein, Chou, and Schwabe showed at CHES 2013 [11] that a new "additive FFT" saves time for decryption in McEliece's code-based public-key cryptosystem, but the smallest FFT sizes in [11] were above 10000 bits (evaluation at every element in $\mathbb{F}_{2^m}$, where $m \geq 11$). We introduce an improved additive FFT that uses fewer bit operations than any previously known multiplier for fields as small as $\mathbb{F}_{2^{64}}$, provided that the fields contain $\mathbb{F}_{2^8}$. Our additive FFT, like many AES hardware implementations, relies heavily on a tower-field representation of $\mathbb{F}_{2^8}$, but benefits from this representation in different ways from AES. The extra speedup inside pseudo dot products comes from merging inverse FFTs, which requires breaking the standard encapsulation of polynomial multiplication; see Sect. 4.

The fact that we are optimizing bit operations is also the reason that we expect our techniques to produce low area for high-throughput hardware. Optimizing the area of a fully unrolled hardware multiplier is highly correlated with optimizing the number of bit operations. We do not claim relevance to very small serial multipliers.

1.4 Polynomial-Multiplication Hardware: PCLMULQDQ

Soon after [35], in response to the performance and security problems of AES-GCM software, Intel added "AES New Instructions" to some of its CPUs. These instructions include PCLMULQDQ, which computes a 64-bit polynomial multiplication in $\mathbb{F}_2[x]$.

Krovetz and Rogaway reported in [37] that GHASH takes 2 Westmere cycles/byte using PCLMULQDQ. Intel's Shay Gueron reported in [26] that heavily optimized GHASH implementations using PCLMULQDQ take 1.79 Sandy Bridge cycles/byte. Our results are faster at a higher security level, although they do require switching to a different authenticator.

Of course, putting sufficient resources into a hardware implementation will beat any software implementation. To quantify this, consider what is required for GHASH to run faster than 1.43 cycles/byte using PCLMULQDQ. GHASH performs one multiplication for every 16 bytes of input, so it cannot afford more than 22.88 cycles for each multiplication. If PCLMULQDQ takes t cycles and t is not very small then presumably Karatsuba is the best approach to multiplication in $\mathbb{F}_{2^{128}}$, taking $3t$ cycles plus some cycles for latency, additions, and reductions. The latest version of Fog's well-known performance survey [23] indicates that $t = 7$ for AMD Bulldozer, Piledriver, and Steamroller and that $t = 8$ for Intel Sandy Bridge and Ivy Bridge; on the other hand, $t = 2$ for Intel Haswell and $t = 1$ for AMD Jaguar. Gueron, in line with this analysis, reported 0.40 Haswell cycles/byte for GHASH.

It is quite unclear what to expect from future CPUs. Intel did not put PCLMULQDQ hardware into its low-cost "Core i3" lines of Sandy Bridge, Ivy Bridge, and Haswell CPUs; and obviously Intel is under pressure from other manufacturers of small, low-cost CPUs. To emphasize the applicability of our techniques to a broad range of CPUs, we have avoided PCLMULQDQ in our software.

2 Field Arithmetic in $\mathbb{F}_{2^8}$

This section reports optimized circuits for field arithmetic in $\mathbb{F}_{2^8}$. We write "circuit" here to mean a fully unrolled combinatorial circuit consisting of AND gates and XOR gates. Our main cost metric is the total number of bit operations, i.e., the total number of AND gates and XOR gates, although as a secondary metric we try to reduce the number of registers required in our software.

Subsequent sections use these circuits as building blocks. The techniques also apply to larger $\mathbb{F}_{2^s}$, but $\mathbb{F}_{2^8}$ is large enough to support the FFTs that we use in this paper.

2.1 Review of Tower Fields

We first construct $\mathbb{F}_{2^2}$ in the usual way as $\mathbb{F}_2[x_2]/(x_2^2 + x_2 + 1)$. We write α_2 for the image of x_2 in $\mathbb{F}_{2^2}$, so $\alpha_2^2 + \alpha_2 + 1 = 0$. We represent elements of $\mathbb{F}_{2^2}$ as linear combinations of 1 and α_2, where the coefficients are in $\mathbb{F}_2$. Additions in $\mathbb{F}_{2^2}$ use 2 bit operations, namely 2 XORs.

We construct $\mathbb{F}_{2^4}$ as $\mathbb{F}_{2^2}[x_4]/(x_4^2 + x_4 + \alpha_2)$, rather than using a polynomial basis for $\mathbb{F}_{2^4}$ over $\mathbb{F}_2$. We write α_4 for the image of x_4 in $\mathbb{F}_{2^4}$. We represent elements of $\mathbb{F}_{2^4}$ as linear combinations of 1 and α_4, where the coefficients are in $\mathbb{F}_{2^2}$. Additions in $\mathbb{F}_{2^4}$ use 4 bit operations.

Finally, we construct $\mathbb{F}_{2^8}$ as $\mathbb{F}_{2^4}[x_8]/(x_8^2 + x_8 + \alpha_2\alpha_4)$; write α_8 for the image of x_8 in $\mathbb{F}_{2^8}$; and represent elements of $\mathbb{F}_{2^8}$ as $\mathbb{F}_{2^4}$-linear combinations of 1 and α_8. Additions in $\mathbb{F}_{2^8}$ use 8 bit operations.

2.2 Variable Multiplications

A variable multiplication is the computation of ab given $a, b \in \mathbb{F}_{2^8}$ as input. We say "variable multiplication" to distinguish this operation from multiplication by a constant; we will optimize constant multiplication later.

For variable multiplication in $\mathbb{F}_{2^2}$, we perform a multiplication of $a_0 + a_1 x, b_0 + b_1 x \in \mathbb{F}_2[x]$ and reduction modulo $x^2 + x + 1$. Here is a straightforward sequence of 7 operations using schoolbook polynomial multiplication: $c_0 \leftarrow a_0 \otimes b_0; c_1 \leftarrow a_0 \otimes b_1; c_2 \leftarrow a_1 \otimes b_0; c_3 \leftarrow a_1 \otimes b_1; c_4 \leftarrow c_1 \oplus c_2; c_5 \leftarrow c_0 \oplus c_3; c_6 \leftarrow c_4 \oplus c_3$. The result is c_5, c_6.

For $\mathbb{F}_{2^4}$ and $\mathbb{F}_{2^8}$ we use 2-way Karatsuba. Note that since the irreducible polynomials are of the form $x^2 + x + \alpha$ the reductions involve a different type of multiplication described below: multiplication of a field element with a constant.

We end up with just 110 bit operations for variable multiplication in $\mathbb{F}_{2^8}$. For comparison, Bernstein [9] reported 100 bit operations to multiply 8-bit polynomials in $\mathbb{F}_2[x]$, but reducing modulo an irreducible polynomial costs many extra operations. A team led by NIST [19], improving upon various previous results such as [31], reported 117 bit operations to multiply in $\mathbb{F}_2[x]$ modulo the AES polynomial $x^8 + x^4 + x^3 + x + 1$.

2.3 Constant Multiplications

A constant multiplication in $\mathbb{F}_{2^s}$ is the computation of αb given $b \in \mathbb{F}_{2^s}$ as input for some constant $\alpha \in \mathbb{F}_{2^s}$. This is trivial for $\alpha \in \mathbb{F}_2$ so we focus on $\alpha \in \mathbb{F}_{2^s} \setminus \mathbb{F}_2$. One can substitute a specific α into our 110-gate circuit for variable multiplication to obtain a circuit for constant multiplication, and then shorten the circuit by eliminating multiplications by 0, multiplications by 1, additions of 0, etc.; but for small fields it is much better to use generic techniques to optimize the cost of multiplying by a constant matrix.

Our linear-map circuit generator combines various features of Paar's greedy additive common-subexpression elimination algorithm [40] and Bernstein's two-operand "xor-largest" algorithm [10]. For $\alpha \in \mathbb{F}_{2^8} \setminus \mathbb{F}_2$ our constant-multiplication circuits use 14.83 gates on average. Compared to Paar's results, this is slightly more gates but is much better in register use; compared to Bernstein's results, it is considerably fewer gates.

The real importance of the tower-field construction for us is that constant multiplications become much faster when the constants are in subfields. Multiplying an element of $\mathbb{F}_{2^8}$ by a constant $\alpha \in \mathbb{F}_{2^4} \backslash \mathbb{F}_2$ takes only 7.43 gates on average, and multiplying an element of $\mathbb{F}_{2^8}$ by a constant $\alpha \in \mathbb{F}_{2^2} \backslash \mathbb{F}_2$ takes only 4 gates on average. The constant multiplications in our FFT-based multiplication algorithms for $\mathbb{F}_{2^{256}}$ (see Sect. 3) are often in subfields of $\mathbb{F}_{2^8}$, and end up using only 9.02 gates on average.

2.4 Subfields and Decomposability

A further advantage of the tower-field construction, beyond the number of bit operations, is that it allows constant multiplications by subfield elements to be decomposed into independent subcomputations. For example, when an $\mathbb{F}_{2^8}$ element in this representation is multiplied by a constant in $\mathbb{F}_{2^2}$, the computation decomposes naturally into 4 independent subcomputations, each of which takes 2 input bits to 2 output bits.

Decomposability is a nice feature for software designers; it guarantees a smaller working set, which in general implies easier optimization, fewer memory operations and cache misses, etc. The ideal case is when the working set can fit into registers; in this case the computation can be done using the minimum number of memory accesses. Section 5 gives an example of how decomposability can be exploited to help optimization of a software implementation.

The decomposition of multiplication by a constant in a subfield has the extra feature that the subcomputations are identical. This allows extra possibilities for efficient vectorization in software, and can also be useful in hardware implementations that reuse the same circuit several times. Even when subcomputations are not identical, decomposability increases flexibility of design and is desirable in general.

3   Faster Additive FFTs

Given a 2^{m-1}-coefficient polynomial f with coefficients in $\mathbb{F}_{2^8}$, a size-2^m additive FFT computes $f(0), f(\beta_m), f(\beta_{m-1}), f(\beta_m + \beta_{m-1}), f(\beta_{m-2})$, etc., where $\beta_m, \ldots, \beta_2, \beta_1$ are $\mathbb{F}_2$-linearly independent elements of $\mathbb{F}_{2^8}$ specified by the algorithm. We always choose a "Cantor basis", i.e., elements $\beta_m, \ldots, \beta_2, \beta_1$ satisfying $\beta_{i+1}^2 + \beta_{i+1} = \beta_i$ and $\beta_1 = 1$; specifically, we take $\beta_1 = 1$, $\beta_2 = \alpha_2$, $\beta_3 = \alpha_4 + 1$, $\beta_4 = \alpha_2\alpha_4$, $\beta_5 = \alpha_8$, and $\beta_6 = \alpha_2\alpha_8 + \alpha_2\alpha_4 + \alpha_2 + 1$. We do not need larger FFT sizes in this paper.

Our additive FFT is an improvement of the Bernstein–Chou–Schwabe [11] additive FFT, which in turn is an improvement of the Gao–Mateer [24] additive FFT. This section presents details of our size-4, size-8, and size-16 additive FFTs over $\mathbb{F}_{2^8}$. All of our improvements are already visible for size 16. At the end of the section gate counts for all sizes are collected and compared with state-of-the-art Karatsuba/Toom-based methods.

3.1   Size-4 FFTs: The Lowest Level of Recursion

Given a polynomial $f = a + bx \in \mathbb{F}_{2^8}[x]$, the size-4 FFT computes $f(0) = a, f(\beta_2) = a + \beta_2 b, f(1) = a + b, f(\beta_2 + 1) = a + (\beta_2 + 1)b$. Recall that $\beta_2 = \alpha_2$ so $\beta_2^2 + \beta_2 + 1 = 0$. The size-4 FFT is of interest because it serves as the lowest level of recursion for larger-size FFTs.

As mentioned in Sect. 2, since $\beta_2 \in \mathbb{F}_{2^2}$, the size-4 FFT can be viewed as a collection of 4 independent pieces, each dealing with only 2 out of the 8 bits.

Let a_0, a_1 be the first 2 bits of a; similarly for b. Then a_0, a_1 and b_0, b_1 represent $a_0 + a_1\beta_2, b_0 + b_1\beta_2 \in \mathbb{F}_{2^2}$. Since $\beta_2(a_0 + a_1\beta_2) = a_1 + (a_0 + a_1)\beta_2$, a 6-gate circuit that carries out the size-4 FFTs operations on the first 2 bits is $c_{00} \leftarrow a_0; c_{01} \leftarrow a_1; c_{20} \leftarrow a_0 \oplus b_0; c_{21} \leftarrow a_1 \oplus b_1; c_{10} \leftarrow a_0 \oplus b_1; c_{31} \leftarrow a_1 \oplus b_0; c_{11} \leftarrow c_{31} \oplus b_1; c_{30} \leftarrow c_{10} \oplus b_0$. Then c_{00}, c_{01} is the 2-bit result of a; c_{10}, c_{11} is the 2-bit result of $a + \beta_2 b$; similarly for c_{20}, c_{21} and c_{30}, c_{31}. In conclusion, a size-4 FFT can be carried out using a $6 \cdot 4 = 24$-gate circuit.

The whole computation costs the same as merely 3 additions in $\mathbb{F}_{2^8}$. This is the result of having evaluation points to be in the smallest possible subfield, namely $\mathbb{F}_{2^2}$, and the use of tower field construction for $\mathbb{F}_{2^8}$.

3.2   The Size-8 FFTs: The First Recursive Case

Given a polynomial $f = f_0 + f_1 x + f_2 x^2 + f_3 x^3 \in \mathbb{F}_{2^8}[x]$, the size-8 FFT computes $f(0), f(\beta_3), f(\beta_2), f(\beta_2 + \beta_3), f(1), f(\beta_3 + 1), f(\beta_2 + 1), f(\beta_2 + \beta_3 + 1)$. Recall that $\beta_3 = \alpha_4 + 1$ so $\beta_3^2 + \beta_3 + \beta_2 = 0$. The size-8 FFT is of interest because it is the smallest FFT that involves recursion.

In general, a recursive size-2^m FFT starts with a radix conversion that computes $f^{(0)}$ and $f^{(1)}$ such that $f = f^{(0)}(x^2 + x) + xf^{(1)}(x^2 + x)$. When f is a 2^{m-1}-coefficient polynomial we call this a size-2^{m-1} radix conversion. Since the size-4 radix conversion can be viewed as a change of basis in $\mathbb{F}_2^4$, each coefficient in $f^{(0)}$ and $f^{(1)}$ is a subset sum of f_0, f_1, f_2, and f_3. In fact, $f^{(0)} = f_0 + (f_2 + f_3)x$ and $f^{(1)} = (f_1 + f_2 + f_3) + f_3 x$ can be computed using exactly 2 additions.

After the radix conversion, 2 size-4 FFTs are invoked to evaluate $f^{(0)}, f^{(1)}$ at $0^2 + 0 = 0$, $\beta_3^2 + \beta_3 = \beta_2$, $\beta_2^2 + \beta_2 = 1$, and $(\beta_2 + \beta_3)^2 + (\beta_2 + \beta_3) = \beta_2 + 1$. Each of these size-4 FFTs takes 24 bit operations.

Note that we have

$$f(\alpha) = f^{(0)}(\alpha^2 + \alpha) + \alpha f^{(1)}(\alpha^2 + \alpha),$$
$$f(\alpha + 1) = f^{(0)}(\alpha^2 + \alpha) + (\alpha + 1)f^{(1)}(\alpha^2 + \alpha).$$

Starting from $f^{(0)}(\alpha^2 + \alpha)$ and $f^{(1)}(\alpha^2 + \alpha)$, Gao and Mateer multiply $f^{(1)}(\alpha^2 + \alpha)$ by α and add $f^{(0)}(\alpha^2 + \alpha)$ to obtain $f(\alpha)$, and then add $f^{(1)}(\alpha^2 + \alpha)$ with $f(\alpha)$ to obtain $f(\alpha + 1)$. We call this a **muladddadd** operation.

The additive FFT thus computes all the pairs $f(\alpha), f(\alpha + 1)$ at once: given $f^{(0)}(0)$ and $f^{(1)}(0)$ apply muladdadd to obtain $f(0)$ and $f(1)$, given $f^{(0)}(\beta_2)$ and $f^{(1)}(\beta_2)$ apply muladdadd operation to obtain $f(\beta_3)$ and $f(\beta_3 + 1)$, and so on.

The way that the output elements form pairs is a result of having 1 as the last basis element. In general the Gao–Mateer FFT is able to handle the case where 1 is not in the basis with some added cost, but here we avoid the cost by making 1 the last basis element.

Generalizing this to the case of size-2^m FFTs implies that the i-th output element of $f^{(0)}$ and $f^{(1)}$ work together to form the ith and $(i + 2^{m-1})$th output element for f. We call the collection of muladdadds that are used to combine 2 size-2^{m-1} FFT outputs to form a size-2^m FFT output a **size-2^m combine routine**.

We use our circuit generator introduced in Sect. 2 to generate the circuits for all the constant multiplications. The muladdadds take a total of 76 gates. Therefore, a size-8 FFT can be carried out using $2 \cdot 8 + 2 \cdot 24 + 76 = 140$ gates.

Note that for a size-8 FFT we again benefit from the special basis and the $\mathbb{F}_{2^8}$ construction. The recursive calls still use the good basis $\beta_2, 1$ so that there are only constant multiplications by $\mathbb{F}_{2^2}$ elements. The combine routine, although not having only constant multiplications by $\mathbb{F}_{2^2}$ elements, at least has only constant multiplications by $\mathbb{F}_{2^4}$ elements.

3.3   The Size-16 FFTs: Saving Additions for Radix Conversions

The size-16 FFT is the smallest FFT in which non-trivial radix conversions happen in recursive calls. Gao and Mateer presented an algorithm performing a size-2^n radix conversion using $(n - 1)2^{n-1}$ additions. We do better by combining additions across levels of recursion.

The size-8 radix conversion finds $f^{(0)}$, $f^{(1)}$ such that $f = f^{(0)}(x^2 + x) + xf^{(1)}(x^2 + x)$. The two size-4 radix conversion in size-8 FFT subroutines find $f^{(i0)}$, $f^{(i1)}$ such that $f^{(i)} = f^{(i0)}(x^2+x)+xf^{(i1)}(x^2+x)$ for $i \in \{0,1\}$. Combining all these leads to $f = f^{(00)}(x^4 + x) + (x^2 + x)f^{(01)}(x^4 + x) + xf^{(10)}(x^4 + x) + x(x^2 + x)f^{(11)}(x^4 + x)$.

In the end the size-8 and the two size-4 radix conversions together compute from f the following: $f^{(00)} = f_0+(f_4+f_7)x$, $f^{(01)} = (f_2+f_3+f_5+f_6)+(f_6+f_7)x$, $f^{(10)} = (f_1+f_2+f_3+f_4+f_5+f_6+f_7)+(f_5+f_6+f_7)x$, and $f^{(11)} = (f_3+f_6)+f_7x$. The Gao–Mateer algorithm takes 12 additions for this computation, but one sees by hand that 8 additions suffice. One can also obtain this result by applying the circuit generator introduced in Sect. 2. Here is an 8-addition sequence generated by the circuit generator: $f_0^{(00)} \leftarrow f_0; f_1^{(11)} \leftarrow f_7; f_1^{(00)} \leftarrow f_4 + f_7; f_0^{(01)} \leftarrow f_2 + f_5; f_0^{(11)} \leftarrow f_3 + f_6; f_1^{(01)} \leftarrow f_6 + f_7; f_0^{(10)} \leftarrow f_1 + f_1^{(00)}; f_1^{(10)} \leftarrow f_5 + f_1^{(01)}; f_0^{(01)} \leftarrow f_0^{(01)} + f_0^{(11)}; f_0^{(10)} \leftarrow f_0^{(10)} + f_0^{(01)}$.

We applied the circuit generator for larger FFTs and found larger gains. A size-32 FFT, in which the input is a size-16 polynomial, requires 31 rather than 48 additions for radix conversions. A size-64 FFT, in which the input is a size-32 polynomial, requires 82 rather than 160 additions for radix conversions.

We also applied our circuit generator to the muladdadds, obtaining a 170-gate circuit for the size-16 combined routine and thus a size-16 FFT circuit using $8 \cdot 8 + 4 \cdot 24 + 2 \cdot 76 + 170 = 482$ gates.

3.4 Size-16 FFTs Continued: Decomposition at Field-Element Level

The size-16 FFT also illustrates the decomposability of the combine routines of a FFT. Consider the size-16 and size-8 combine routines; the computation takes as input the FFT outputs for the $f^{(ij)}$'s to compute the FFT output for f.

Let the output for f be $a_0, a_1, \ldots, a_{15}$, the output for $f^{(i)}$ be $a_0^{(i)}, a_1^{(i)}, \ldots, a_7^{(i)}$, and similarly for $f^{(ij)}$. For $k \in \{0, 1, 2, 3\}$, a_k, a_{k+8} are functions of $a_k^{(0)}$ and $a_k^{(1)}$, which in turn are functions of $a_k^{(00)}$, $a_k^{(01)}$, $a_k^{(10)}$, and $a_k^{(11)}$; a_{k+4}, a_{k+12} are functions of $a_{k+4}^{(0)}$ and $a_{k+4}^{(1)}$, which in turn are functions of the same 4 elements. We conclude that $a_k, a_{k+4}, a_{k+8}, a_{k+12}$ depend only on $a_k^{(00)}$, $a_k^{(01)}$, $a_k^{(10)}$, and $a_k^{(11)}$. In this way, the computation is decomposed into 4 independent parts; each takes as input 4 field elements and outputs 4 field elements. Note that here the decomposition is at the field-element level, while Sect. 2 considered decomposability at the bit level.

More generally, for size-2^m FFTs we suggest decomposing k levels of combine routines into 2^{m-k} independent pieces, each taking 2^k $\mathbb{F}_{2^8}$ elements as input and producing 2^k $\mathbb{F}_{2^8}$ elements as output.

3.5 Improvements: A Summary

We have two main improvements to the additive FFT: reducing the cost of multiplications and reducing the number of additions in radix conversion. We also use these ideas to accelerate size-32 and size-64 FFTs, and obviously they would also save time for larger FFTs.

The reduction in the cost of multiplications is a result of (1) choosing a "good" basis for which constant multiplications use constants in the smallest possible subfield; (2) using a tower-field representation to accelerate those constant multiplications; and (3) searching for short sequences of additions. The reduction of additions for radix conversion is a result of (1) merging radix conversion at different levels of recursion and again (2) searching for short sequences of additions.

3.6 Polynomial Multiplications: A Comparison with Karatsuba and Toom

Just like other FFT algorithms, any additive FFT can be used to multiply polynomials. Given two 2^{m-1}-coefficient polynomials in $\mathbb{F}_{2^s}$, we apply a size-2^m additive FFT to each polynomial, a pointwise multiplication consisting of 2^m variable multiplications in $\mathbb{F}_{2^s}$, and a size-2^m inverse additive FFT, i.e., the inverse of an FFT with both input and output size 2^m. An FFT (or inverse FFT) with input

Table 1. Cost of multiplying $b/8$-coefficient polynomials over $\mathbb{F}_{2^8}$. "Forward" is the cost of two size-$b/4$ FFTs with size-$b/8$ inputs. "Pointwise" is the cost of pointwise multiplication. "Inverse" is the cost of an inverse size-$b/4$ FFT. "Total" is the sum of forward, pointwise, and inverse. "Competition" is the cost from [9] of an optimized Karatsuba/Toom multiplication of b-coefficient polynomials over $\mathbb{F}_2$; note that slight improvements appear in [21].

b	forward	pointwise	inverse	total	competition
16	$2 \cdot 24$	$4 \cdot 110$	60	$448 \approx 14 \cdot 2 \cdot 16$	$350 \approx 10.9 \cdot 2 \cdot 16$
32	$2 \cdot 140$	$8 \cdot 110$	228	$1388 \approx 21.7 \cdot 2 \cdot 32$	$1158 \approx 18.1 \cdot 2 \cdot 32$
64	$2 \cdot 482$	$16 \cdot 110$	746	$3470 \approx 27.1 \cdot 2 \cdot 64$	$3682 \approx 28.8 \cdot 2 \cdot 64$
128	$2 \cdot 1498$	$32 \cdot 110$	2066	$8582 \approx 33.5 \cdot 2 \cdot 128$	$11486 \approx 44.9 \cdot 2 \cdot 128$
256	$2 \cdot 4068$	$64 \cdot 110$	5996	$21172 \approx 41.4 \cdot 2 \cdot 256$	$34079 \approx 66.6 \cdot 2 \cdot 256$

and output size 2^m is slightly more expensive than an FFT with input size 2^{m-1} and output size 2^m: input size 2^{m-1} is essentially input size 2^m with various 0 computations suppressed.

Table 1 summarizes the number of bit operations required for multiplying b-bit (i.e., $b/8$-coefficient) polynomials in $\mathbb{F}_{2^8}[x]$. Field multiplication is slightly more expensive than polynomial multiplication. For $\mathbb{F}_{2^{256}}$ we use the polynomial $x^{32} + x^{17} + x^2 + \alpha_8$; reduction costs 992 bit operations. However, as explained in Sect. 4, in the context of Auth256 we can almost eliminate the inverse FFT and the reduction, and eliminate many operations in the forward FFTs, making the additive FFT even more favorable than Karatsuba.

4 The Auth256 Message-Authentication Code: Major Features

Auth256, like GCM's GHASH, follows the well-known Wegman–Carter [47] recipe for building an MAC with (provably) information-theoretic security. The recipe is to apply a (provably) "δ-xor-universal hash" to the message and encrypt the result with a one-time pad. Every forgery attempt then (provably) has success probability at most δ, no matter how much computer power the attacker used.

Of course, real attackers do not have unlimited computer power, so GCM actually replaces the one-time pad with counter-mode AES output to reduce key size. This is safe against any attacker who cannot distinguish AES output from uniform random; see, e.g., [33, comments after Corollary 3]. Similarly, it is safe to replace the one-time pad in Auth256 with cipher output.

This section presents two important design decisions for Hash256, the hash function inside Auth256. Section 4.1 describes the advantages of the Hash256 output size. Section 4.2 describes the choice of pseudo dot products inside Hash256, and the important interaction between FFTs and pseudo dot products. Section 4.3 describes the use of a special field representation for inputs to reduce the cost of FFTs.

Section 6 presents, for completeness, various details of Hash256 and Auth256 that are not relevant to this paper's performance evaluation.

4.1   Output Size: Bigger-Birthday-Bound Security

Hash256 produces 256-bit outputs, as its name suggests, and Auth256 produces 256-bit authenticators. Our multiplication techniques are only slightly slower per bit for $\mathbb{F}_{2^{256}}$ than for $\mathbb{F}_{2^{128}}$, so Auth256 is only slightly slower than an analogous Auth128 would be. An important advantage of an increased output size is that one can safely eliminate nonces.

Encrypting a hash with a one-time pad, or with a stream cipher such as AES in counter mode, requires a nonce, and becomes insecure if the user accidentally repeats a nonce; see, e.g., [30]. Directly applying a PRF (as in HMAC) or PRP (as in WMAC) to the hash, without using a nonce, is much more resilient against misuse but becomes insecure if hashes collide, so b-bit hashes are expected to be broken within $2^{b/2}$ messages (even with an optimal $\delta = 2^{-b}$) and already provide a noticeable attack probability within somewhat fewer messages.

This problem has motivated some research into "beyond-birthday-bound" mechanisms for authentication and encryption that can safely be used for more than $2^{b/2}$ messages. See, e.g., [38]. Hash256 takes a different approach, which we call "bigger-birthday-bound" security: simply increasing b to 256 (and correspondingly reducing δ) eliminates all risk of collisions. For the same reason, Hash256 provides extra strength inside other universal-hash applications, such as wide-block disk encryption; see, e.g., [28].

In applications with space for only 128-bit authenticators, it is safe to simply truncate the Hash256 and Auth256 output from 256 bits to 128 bits. This increases δ from 2^{-255} to 2^{-127}.

4.2   Pseudo Dot Products and FFT Addition

Hash256 uses the same basic construction as NMH [29, Sect. 5], UMAC [15], Badger [16], and VMAC [36]: the hash of a message with blocks $m_1, m_2, m_3, m_4, \ldots$ is $(m_1 + r_1)(m_2 + r_2) + (m_3 + r_3)(m_4 + r_4) + \cdots$. Halevi and Krawczyk [29] credit this hash to Carter and Wegman; Bernstein [6] credits it to Winograd and calls it the "pseudo dot product". The pseudo-dot-product construction of Hash256 gives $\delta < 2^{-255}$; see Appendix A for the proof.

A simple dot product $m_1 r_1 + m_2 r_2 + m_3 r_3 + m_4 r_4 + \cdots$ uses one multiplication per block. The same is true for GHASH and many other polynomial-evaluation hashes. The basic advantage of $(m_1 + r_1)(m_2 + r_2) + (m_3 + r_3)(m_4 + r_4) + \cdots$ is that there are only 0.5 multiplications per block.

For Auth256 each block contains 256 bits, viewed as an element of the finite field $\mathbb{F}_{2^{256}}$. Our cost of Auth256 per 512 authenticated bits is $29 \cdot 512 = 58 \cdot 256$ bit operations, while the our cost for a multiplication in $\mathbb{F}_{2^{256}}$ is $87 \cdot 256$ bit operations. We now explain one of the two sources of this gap.

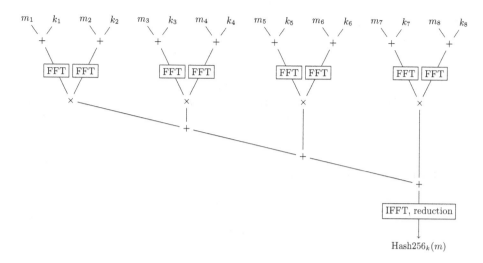

Fig. 1. Hash256 flowchart

FFT-based multiplication of two polynomials $f_1 f_2$ has several steps: apply an FFT to evaluate f_1 at many points; apply an FFT to evaluate f_2 at many points; compute the corresponding values of the product $f_1 f_2$ by pointwise multiplication; and apply an inverse FFT to reconstruct the coefficients of $f_1 f_2$. FFT-based multiplication of field elements has the same steps plus a final reduction step.

These steps for $\mathbb{F}_{2^{256}}$, with our optimizations from Sect. 3, cost 4068 bit operations for each forward FFT, $64 \cdot 110$ bit operations for pointwise multiplication, 5996 bit operations for the inverse FFT (the forward FFT is less expensive since more polynomial coefficients are known to be 0), and 992 bit operations for the final reduction. Applying these steps to each 512 bits of input would cost approximately 15.89 bit operations per bit for the two forward FFTs, 13.75 bit operations per bit for pointwise multiplication, 11.71 bit operations per bit for the inverse FFT, and 1.94 bit operations per bit for the final reduction, plus 1.5 bit operations per bit for the three additions in the pseudo dot product.

We do better by exploiting the structure of the pseudo dot product as a sum of the form $f_1 f_2 + f_3 f_4 + f_5 f_6 + \cdots$. Optimizing this computation is not the same problem as optimizing the computation of $f_1 f_2$. Specifically, we apply an FFT to each f_i and compute the corresponding values of $f_1 f_2$, $f_3 f_4$, etc., but we then add these values *before* applying an inverse FFT. See Fig. 1. There is now only one inverse FFT (and one final reduction) per message, rather than one inverse FFT for every two blocks. Our costs are now 15.89 bit operations per bit for the two forward FFTs, 13.75 bit operations per bit for pointwise multiplication, 1 bit operation per bit for the input additions in the pseudo dot product, and 1 bit operation per bit for the pointwise additions, for a total of 31.64 bit operations per bit, plus a constant (not very large) overhead per message.

This idea is well known in the FFT literature (see, e.g., [7, Sect. 2]) but we have never seen it applied to message authentication. It reduces the cost of

FFT-based message authentication by a factor of nearly 1.5. Note that this also reduces the cutoff between FFT and Karatsuba.

UMAC and VMAC actually limit the lengths of their pseudo dot products, to limit key size. This means that longer messages produce two or more hashes; these hashes are then further hashed in a different way (which takes more time per byte but is applied to far fewer bytes). For simplicity we instead use a key as long as the maximum message length. We have also considered the small-key hashes from [6] but those hashes obtain less benefit from merging inverse FFTs.

4.3   Embedding Invertible Linear Operations into FFT Inputs

Section 4.2 shows how to achieve 31.64 bit operations per message bit by skipping the inverse FFTs for almost all multiplications in the pseudo dot product. Now we show how Auth256 achieves 29 bit operations per message bit by skipping operations in the forward FFTs.

Section 3.3 shows that the radix conversions can be merged into one invertible $\mathbb{F}_{2^8}$-linear (actually $\mathbb{F}_2$-linear) map, which takes place before all other operations in the FFT. The input is a $\mathbb{F}_{2^{256}}$ element which is represented as coefficients in $\mathbb{F}_{2^8}$ with respect to a polynomial basis. Applying an invertible linear map on the coefficients implies a change of basis. In other words, the radix conversions convert the input into another 256-bit representation. If we define the input to be elements in this new representation, all the radix conversions can simply be skipped. Note that the authenticator still uses the original representation. See Sect. 6.2 for a definition of the new representation.

This technique saves a significant fraction of the operations in the forward FFT. As shown in Sect. 3, one forward FFT takes 4068 bit operations, where $82 \cdot 8 = 656$ of them are spent on radix conversions. Eliminating all radix conversions then gives the 29 bit operations per message bit.

The additive FFTs described so far are "2-way split" FFTs since they require writing the input polynomial $f(x)$ in the form $f^{(0)}(x^2 + x) + x f^{(1)}(x^2 + x)$. It is easy to generalize this to a "2^k-way split" in which $f(x)$ is written as $\sum_{i=0}^{2^k-1} x^i f^{(i)}(\psi^k(x))$, where $\psi(x) = x^2 + x$. In particular, Gao and Mateer showed how to perform $2^{2^{k-1}}$-way-split FFTs for polynomials in $\mathbb{F}_{2^{2^k}}[x]$. The technique of changing input representation works for any 2^k-way split. In fact we found that with 8-way-split FFTs, the number of bit operations per message bit can be slightly better than 29. However, for simplicity, Auth256 is defined in a way that favors 2-way-split FFTs.

5   Software Implementation

Our software implementation uses bitslicing. This means that we convert each bit in previous sections into w bits, where w is the register width on the machine; we convert each AND into a bitwise w-bit AND instruction; and we convert each XOR into a bitwise w-bit XOR instruction.

Bitslicing is efficient only if there is adequate parallelism in the algorithm. Fortunately, the pseudo-dot-product computation is naturally parallelizable: we let the jth bit position compute the sum of all products $(m_{2i+1}+r_{2i+1})(m_{2i+2}+r_{2i+2})$ where $i \equiv j \pmod{w}$. After all the products are processed, the results in all bit positions are summed up to get the final value.

The detailed definition of Auth256 (see Sect. 6) has a parameter b. Our software takes $b = w$, allowing it to simply pick up message blocks as vectors. If b is instead chosen as 1 then converting to bitsliced form requires a transposition of message blocks; in our software this transposition costs an extra 0.81 cycles/byte.

5.1 Minimizing Memory Operations in Radix Conversions

We exploit the decomposability of additions to minimize memory operations for a radix conversion. When dealing with size-2^k radix conversions with $k \leq 4$, we decompose at bit level the computation into 2^k parts, each of which deals with $16/2^k$ bit positions. This minimizes the number of loads and stores. The same technique applies for a radix conversion combined with smaller-size radix conversions in the FFT subroutines.

Our implementation uses the size-16 FFT as a subroutine. Inside a size-16 FFT the size-8 radix conversion is combined with the 2 size-4 radix conversions in FFT subroutines. Our bit-operation counts handle larger radix conversions in the same way, but in our software we sacrifice some of the bit operations saved here to improve instruction-level parallelism and register utilization. For size-16 radix conversion the decomposition method is adopted. For size-32 radix conversion the decomposition method is used only for the size-16 recursive calls.

5.2 Minimizing Memory Operations in Muladdadd Operations

For a single muladdadd operation $a \leftarrow a + \alpha b; b \leftarrow b + a$, each of a and b consumes 8 vectors; evidently at least 16 loads and 16 stores are required. While we showed how the sequence of bit operations can be generated, it does not necessarily mean that there are enough registers to carry out the bit operations using the minimum number of loads and stores.

Here is one strategy to maintain both the number of bit operations and the lower bound on number of loads and stores. First load the 8 bits of b into 8 registers $t_0, t_1, \ldots, t_7$. Use the sequence of XORs generated by the code generator, starting from the t_i's, to compute the 8 bits of αb, placing them in the other 8 registers $s_0, s_1, \ldots, s_7$. Then perform $s_i \leftarrow s_i \oplus a[i]$, where $a[i]$ is the corresponding bit of a in memory, to obtain $a + \alpha b$. After that overwrite a with the s_i's. Finally, perform $t_i \leftarrow t_i \oplus s_i$ to obtain $a + (\alpha + 1)b$, and overwrite b with the t_i's.

In our software muladdadd operations are handled one by one in size-64 and size-32 combine routines. See below for details about how muladdadds in smaller size combine routines are handled.

5.3   Implementing the Size-16 Additive FFT

In our size-16 FFT implementation the size-8 radix conversion is combined with the two size-4 ones in the FFT subroutines using the decomposition method described earlier in this section. Since the size-4 FFTs deal with constants in $\mathbb{F}_{2^2}$, we further combine the radix conversions with size-4 FFTs.

At the beginning of one of the 4 rounds of the whole computation, the $2 \cdot 8 = 16$ bits of the input for size-8 radix conversion are loaded. Then the logic operations are carried out in registers, and eventually the result is stored in $2 \cdot 16 = 32$ bits of the output elements. The same routine is repeated 4 times to cover all the bit positions.

The size-16 and size-32 combine routines are also merged as shown in Sect. 3. The field-level decomposition is used together with a bit-level decomposition: in size-16 FFT all the constants are in $\mathbb{F}_{2^4}$, so it is possible to decompose any computation that works on field elements into a 2-round procedure and handle 4 bit positions in each round. In conclusion, the field-level decomposition turns the computation into 4 pieces, and the bit-level decomposition further decomposes each of these into 2 smaller pieces. In the end, we have an 8-round procedure.

At the beginning of one of the 8 rounds of the whole computation, the $4 \cdot 4 = 16$ bits of the outputs of the size-4 FFTs are loaded. Then the logic operations are carried out in registers, and eventually the result is stored in $4 \cdot 4 = 16$ bits of the output elements. The same routine is repeated 8 times to cover all the bit positions.

6   Auth256: Minor Details

To close we fill in, for completeness, the remaining details of Hash256 and Auth256.

6.1   Review of Wegman–Carter MACs

Wegman–Carter MACs work as follows. The authenticator of the nth message m is $H(r, m) \oplus s_n$. The key consists of independent uniform random $r, s_1, s_2, s_3, \ldots$; the pad is $s_1, s_2, s_3, \ldots$.

The hash function H is designed to be "δ-xor-universal", i.e., to have "differential probability at most δ". This means that, for every message m, every message $m' \neq m$, and every difference Δ, a uniform random r has $H(r, m) \oplus H(r, m') = \Delta$ with probability at most δ.

6.2   Field Representation

We represent an element of $\mathbb{F}_{2^s}$ as a sequence of s bits. If we construct $\mathbb{F}_{2^s}$ as $\mathbb{F}_{2^t}[x]/\phi$ then we recursively represent the element $c_0 + c_1 x + \cdots + c_{t/s-1} x^{t/s-1}$ as the concatenation of the representations of $c_0, c_1, \ldots, c_{t/s-1}$. At the bottom of the recursion, we represent an element of $\mathbb{F}_2$ as 1 bit in the usual way. See Sects. 2 and 3.6 for the definition of ϕ for $\mathbb{F}_{2^2}$, $\mathbb{F}_{2^4}$, $\mathbb{F}_{2^8}$, and $\mathbb{F}_{2^{256}}$.

As mentioned in Sect. 4.3, we do not use the polynomial basis $1, x, \ldots, x^{31}$ for $\mathbb{F}_{2^{256}}$ inputs. Here we define the representation for them. Let $y_{(b_{k-1}b_{k-2}\cdots b_0)_2} = \prod_{i=0}^{k-1}(\psi^i(x))^{b_i}$, where $\psi(x)$ follows the definition in Sect. 4.3. Then each $\mathbb{F}_{2^{256}}$ input $c_0 y_0 + c_1 y_1 + \cdots + c_{31} y_{31}$ is defined as the concatenation of the representations of $c_0, c_1, \ldots, c_{31}$. One can verify that $y_0, y_1, \ldots, y_{31}$ is the desired basis by writing down the equation between $f(x)$ and $f^{(00000)}(x)$, $f^{(00001)}(x)$, $\ldots$, $f^{(111111)}(x)$ as in Sect. 3.3.

If $s \geq 8$ then we also represent an element of $\mathbb{F}_{2^s}$ as a sequence of $s/8$ bytes, i.e., $s/8$ elements of $\{0, 1, \ldots, 255\}$. The 8-bit sequence $b_0, b_1, \ldots, b_7$ is represented as the byte $b = \sum_i 2^i b_i$.

6.3 Hash256 Padding and Conversion

Hash256 views messages as elements of $K^0 \cup K^2 \cup K^4 \cup \cdots$, i.e., even-length strings of elements of K, where K is the finite field $\mathbb{F}_{2^{256}}$. It is safe to use a single key with messages of different lengths.

In real applications, messages are strings of bytes, so strings of bytes need to be encoded invertibly as strings of elements of K. The simplest encoding is standard "10*" padding, where a message is padded with a 1 byte and then as many 0 bytes as necessary to obtain a multiple of 64 bytes. Each 32-byte block is then viewed as an element of K.

We define a more general encoding parametrized by a positive integer b; the encoding of the previous paragraph has $b = 1$. The message is padded with a 1 byte and then as many 0 bytes as necessary to obtain a multiple of $64b$ bytes, say $64bN$ bytes. These bytes are split into $2N$ segments $M_0, M_0', M_1, M_1', \ldots, M_{N-1}$, M_{N-1}', where each segment contains $32b$ consecutive bytes. Each segment is then transposed into b elements of K: segment M_i is viewed as a column-major bit matrix with b rows and 256 columns, and row j of this matrix is labeled c_{bi+j}, while c_{bi+j}' is defined similarly using M_i'. This produces $2bN$ elements of K, namely $m_0, m_0', m_1, m_1', m_2, m_2', \ldots, m_{bN-1}, m_{bN-1}'$.

The point of this encoding is to allow a simple bitsliced vectorized implementation; see Sect. 5. Our 1.59 cycle/byte implementation uses $b = 256$. We have also implemented $b = 1$, which costs 0.81 cycles/byte extra for transposition and is compatible with efficient handling of much shorter messages. An interesting intermediate possibility is to take, e.g., $b = 8$, eliminating the most expensive (non-bytewise) transposition steps while still remaining suitable for authentication of typical network packets.

6.4 Hash256 and Auth256 Keys and Authenticators

The Hash256 key is a uniform random byte string of the same length as a maximum-length padded message, representing elements $r_0, r_0', r_1, r_1', \ldots$ of K. If the key is actually defined as, e.g., counter-mode AES output then the maximum length does not need to be specified in advance: extra key elements can be computed on demand and cached for subsequent use.

The Hash256 output is $(m_0 + r_0)(m'_0 + r'_0) + (m_1 + r_1)(m'_1 + r'_1) + \cdots$. This is an element of K.

The Auth256 key is a Hash256 key together with independent uniform random elements $s_1, s_2, \ldots$ of K. The Auth256 authenticator of the nth message m_n is $\text{Auth256}(r, m_n) \oplus s_n$.

References

1. — (no editor): Information Theory Workshop, 2006. ITW '06 Chengdu. IEEE (2006). See [45]
2. — (no editor): Design, Automation & Test in Europe Conference & Exhibition, 2007. DATE '07. IEEE (2007). See [41]
3. — (no editor): Proceedings of the 6th WSEAS World Congress: Applied Computing Conference (ACC 2013). WSEAS (2013). See [21]
4. Bernstein, D.J.: Fast multiplication (2000). http://cr.yp.to/talks.html#2000.08. 14. Citations in this document: §1.3
5. Bernstein, D.J.: The Poly1305-AES message-authentication code. In: FSE 2005 [25], pp. 32–49 (2005). http://cr.yp.to/papers.html#poly1305. Citations in this document: §1.1
6. Bernstein, D.J.: Polynomial evaluation and message authentication (2007). http://cr.yp.to/papers.html#pema. Citations in this document: §4.2, §4.2
7. Bernstein, D.J.: Fast multiplication and its applications. In: [17], pp. 325–384 (2008). http://cr.yp.to/papers.html#multapps. Citations in this document: §4.2
8. Bernstein, D.J.: Batch binary Edwards. In: Crypto 2009 [27], pp. 317–336 (2009). http://cr.yp.to/papers.html#bbe. Citations in this document: §1.3
9. Bernstein, D.J.: Minimum number of bit operations for multiplication (2009). http://binary.cr.yp.to/m.html. Citations in this document: §1.3, §1.3, §2.2, §1, §1
10. Bernstein, D.J.: Optimizing linear maps modulo 2. In: Workshop Record of SPEED-CC: Software Performance Enhancement for Encryption and Decryption and Cryptographic Compilers, pp. 3–18 (2009). http://cr.yp.to/papers.html#linearmod2. Citations in this document: §2.3
11. Bernstein, D.J., Chou, T., Schwabe, P.: McBits: fast constant-time code-based cryptography. In: CHES 2013 [12], pp. 250–272 (2013). Citations in this document: §1.3, §1.3, §3
12. Bertoni, G., Coron, J.-S. (eds.): CHES'13. LNCS, vol. 8086. Springer, Heidelberg (2013). ISBN 978-3-642-40348-4. See [11]
13. Biham, E. (ed.): FSE 1997. LNCS, vol. 1267. Springer, Heidelberg (1997). ISBN 3-540-63247-6. See [29]
14. Biham, E., Youssef, A.M. (eds.): SAC 2006. LNCS, vol. 4356. Springer, Heidelberg (2007). ISBN 978-3-540-74461-0. See [36]
15. Black, J., Halevi, S., Krawczyk, H., Krovetz, T., Rogaway, P.: UMAC: fast and secure message authentication. In: Crypto 1999 [49], pp. 216–233 (1999). http://www.cs.ucdavis.edu/~rogaway/umac/. Citations in this document: §1.1, §4.2
16. Boesgaard, M., Scavenius, O., Pedersen, T., Christensen, T., Zenner, E.: Badger—a fast and provably secure MAC. In: [32], pp. 176–191 (2005). Citations in this document: §4.2
17. Buhler, J.P., Stevenhagen, P. (eds.): Surveys in Algorithmic Number Theory. Mathematical Sciences Research Institute Publications, vol. 44. Cambridge University Press, New York (2008). See [7]

18. Chang, N.S., Kim, C.H., Park, Y.-H., Lim, J.: A non-redundant and efficient architecture for Karatsuba-Ofman algorithm. In: [50], pp. 288–299 (2005). Citations in this document: §1.3

19. Circuit Minimization Team.: Multiplication circuit for GF(256) with irreducible polynomial $X^8 + X^4 + X^3 + X + 1$ (2010). http://cs-www.cs.yale.edu/homes/peralta/CircuitStuff/slp_84310.txt. Citations in this document: §2.2

20. Clavier, C., Gaj, K. (eds.): CHES 2009. LNCS, vol. 5747. Springer, Heidelberg (2009). ISBN 978-3-642-04137-2. See [35]

21. D'Angella, D., Schiavo, C.V., Visconti, A.: Tight upper bounds for polynomial multiplication. In: [3] (2013). http://www.wseas.us/e-library/conferences/2013/Nanjing/ACCIS/ACCIS-03.pdf. Citations in this document: §1.3, §1, §1

22. Andrew, M. "Floodyberry". Optimized implementations of Poly1305, a fast message-authentication-code (2014). https://github.com/floodyberry/poly1305-opt. Citations in this document: §1.1

23. Fog, A.: Instruction tables (2014). http://www.agner.org/optimize/instruction_tables.pdf. Citations in this document: §1.4

24. Gao, S., Mateer, T.: Additive fast Fourier transforms over finite fields. IEEE Trans. Inf. Theory **56**, 6265–6272 (2010). http://www.math.clemson.edu/~sgao/pub.html. Citations in this document: §3

25. Gilbert, H., Handschuh, H. (eds.): FSE 2005. LNCS, vol. 3557. Springer, Heidelberg (2005). ISBN 3-540-26541-4. See [5]

26. Gueron, S.: AES-GCM software performance on the current high end CPUs as a performance baseline for CAESAR (2013). http://2013.diac.cr.yp.to/slides/gueron.pdf. Citations in this document: §1.4

27. Halevi, S. (ed.): CRYPTO 2009. LNCS, vol. 5677. Springer, Heidelberg (2009). See [8]

28. Halevi, S.: Invertible universal hashing and the TET encryption mode. In: Crypto 2007 [39], pp. 412–429 (2007). http://eprint.iacr.org/2007/014. Citations in this document: §4.1

29. Halevi, S., Krawczyk, H.: MMH: software message authentication in the Gbit/second rates. In: FSE 1997 [13], pp. 172–189 (1997). http://www.research.ibm.com/people/s/shaih/pubs/mmh.html. Citations in this document: §4.2, §4.2

30. Handschuh, H., Preneel, B.: Key-recovery attacks on universal hash function based MAC algorithms. In: [46], pp. 144–161 (2008). Citations in this document: §4.1

31. Imaña, J.L., Hermida, R., Tirado, F.: Low-complexity bit-parallel multipliers based on a class of irreducible pentanomials. IEEE Trans. Very Large Scale Integr. (VLSI) Syst. **14**, 1388–1393 (2006). Citations in this document: §2.2

32. Ioannidis, J., Keromytis, A.D., Yung, M. (eds.): ACNS 2005. LNCS, vol. 3531. Springer, Heidelberg (2005). ISBN 3-540-26223-7. See [16]

33. Iwata, T., Ohashi, K., Minematsu, K.: Breaking and repairing GCM security proofs. In: Crypto 2012 [44], pp. 31–49 (2012). http://eprint.iacr.org/2012/438. Citations in this document: §4

34. Joux, A. (ed.): FSE 2011. LNCS, vol. 6733. Springer, Heidelberg (2011). ISBN 978-3-642-21701-2. See [37]

35. Käsper, E., Schwabe, P.: Faster and timing-attack resistant AES-GCM. In: CHES 2009 [20], pp. 1–17 (2009). http://eprint.iacr.org/2009/129. Citations in this document: §1, §1.1, §1.4

36. Krovetz, T.: Message authentication on 64-bit architectures. In: SAC 2006 [14], pp. 327–341 (2007). Citations in this document: §1.1, §4.2

37. Krovetz, T., Rogaway, P.: The software performance of authenticated-encryption modes. In: FSE 2011 [34], pp. 306–327 (2011). http://www.cs.ucdavis.edu/~rogaway/papers/ae.pdf. Citations in this document: §1.4

38. Landecker, W., Shrimpton, T., Terashima, R.S.: Tweakable blockciphers with beyond birthday-bound security. In: Crypto 2012 [44], pp. 14–30 (2012). Citations in this document: §4.1

39. Menezes, A. (ed.): CRYPTO 2007. LNCS, vol. 4622. Springer, Heidelberg (2007). ISBN 978-3-540-74142-8. See [28]

40. Paar, C.: Optimized arithmetic for Reed-Solomon encoders (1997). http://www.emsec.rub.de/media/crypto/veroeffentlichungen/2011/01/19/cnst.ps. Citations in this document: §2.3

41. Peter, S., Langendörfer, P.: An efficient polynomial multiplier in $GF(2^m)$ and its application to ECC designs. In: DATE 2007 [2] (2007). http://ieeexplore.ieee.org/xpl/freeabs_all.jsp?isnumber=4211749&arnumber=4211979&count=305&index=229. Citations in this document: §1.3

42. Rodríguez-Henríquez, F., Koç, Ç.K.: On fully parallel Karatsuba multipliers for $GF(2^m)$. In: [43], pp. 405–410 (2003). Citations in this document: §1.3

43. Sahni, S. (ed.): Proceedings of the International Conference on Computer Science and Technology. Acta Press, Crete (2003). See [42]

44. Safavi-Naini, R., Canetti, R. (eds.): CRYPTO 2012. LNCS, vol. 7417. Springer, Heidelberg (2012). ISBN 978-3-642-32008-8. See [33,38]

45. von zur Gathen, J., Shokrollahi, J.: Fast arithmetic for polynomials over $\mathbf{F}_2$ in hardware. In: ITW 2006 [1], pp. 107–111 (2006). Citations in this document: §1.3

46. Wagner, D. (ed.): CRYPTO 2008. LNCS, vol. 5157. Springer, Heidelberg (2008). ISBN 978-3-540-85173-8. See [30]

47. Wegman, M.N., Carter, J.L.: New hash functions and their use in authentication and set equality. J. Comput. Syst. Sci. **22**, 265–279 (1981). ISSN 0022-0000, MR 82i:68017. Citations in this document: §4

48. Weimerskirch, A., Paar, C.: Generalizations of the Karatsuba algorithm for efficient implementations (2006). http://eprint.iacr.org/2006/224. Citations in this document: §1.3

49. Wiener, M. (ed.): CRYPTO 1999. LNCS, vol. 1666. Springer, Heidelberg (1999). ISBN 3-5540-66347-9, MR 2000h:94003. See [15]

50. Zhou, J., López, J., Deng, R.H., Bao, F. (eds.): ISC 2005. LNCS, vol. 3650. Springer, Heidelberg (2005). ISBN 3-540-29001-X. See [18]

A Security Proof

This appendix proves that Hash256 has differential probability smaller than 2^{-255}. This is not exactly the same as the proofs for the pseudo-dot-product portions of UMAC and VMAC: UMAC and VMAC specify fixed lengths for their pseudo dot products, whereas we allow variable lengths.

Theorem 1. *Let K be a finite field. Let ℓ, ℓ', k be nonnegative integers with $\ell \leq k$ and $\ell' \leq k$. Let $m_1, m_2, \ldots, m_{2\ell-1}, m_{2\ell}$ be elements of K. Let $m'_1, m'_2, \ldots, m'_{2\ell'-1}, m'_{2\ell'}$ be elements of K. Assume that $(m_1, m_2, \ldots, m_{2\ell}) \neq (m'_1, m'_2, \ldots,$*

$m'_{2\ell'}$). Let Δ be an element of K. Let $r_1, r_2, \ldots, r_{2k}$ be independent uniform random elements of k. Let p be the probability that $h = h' + \Delta$, where

$$h = (m_1 + r_1)(m_2 + r_2) + (m_3 + r_3)(m_4 + r_4) + \cdots$$
$$+ (m_{2\ell-1} + r_{2\ell-1})(m_{2\ell} + r_{2\ell}),$$
$$h' = (m'_1 + r_1)(m'_2 + r_2) + (m'_3 + r_3)(m'_4 + r_4) + \cdots$$
$$+ (m'_{2\ell'-1} + r_{2\ell'-1})(m'_{2\ell'} + r_{2\ell'}).$$

Then $p < 2/\#K$. If $\ell = \ell'$ then $p \leq 1/\#K$, and if $\ell \neq \ell'$ then $p < 1/\#K + 1/\#K^{|\ell-\ell'|}$.

Proof. Case 1: $\ell = \ell'$. Then $h = h' + \Delta$ if and only if

$$r_1(m_2 - m'_2) + r_2(m_1 - m'_1) + r_3(m_4 - m'_4) + r_4(m_3 - m'_3) + \cdots$$
$$= \Delta + m'_1 m'_2 - m_1 m_2 + m'_3 m'_4 - m_3 m_4 + \cdots.$$

This is a linear equation in $r_1, r_2, \ldots, r_{2k}$. This linear equation is nontrivial: by hypothesis $(m_1, m_2, \ldots, m_{2\ell}) \neq (m'_1, m'_2, \ldots, m'_{2\ell})$, so there must be some i for which $m_i - m'_i \neq 0$. Consequently there are most $\#K^{2k-1}$ solutions to the equation out of the $\#K^{2k}$ possibilities for r; i.e., $p \leq 1/\#K$ as claimed.

Case 2: $\ell < \ell'$ and $(m_1, \ldots, m_\ell) \neq (m'_1, \ldots, m'_\ell)$. Define

$$f = (m'_{2\ell+1} + r_{2\ell+1})(m'_{2\ell+2} + r_{2\ell+2}) + \cdots + (m'_{2\ell'-1} + r_{2\ell'-1})(m'_{2\ell'} + r_{2\ell'}).$$

Then $h = h' + \Delta$ if and only if

$$r_1(m_2 - m'_2) + r_2(m_1 - m'_1) + r_3(m_4 - m'_4) + r_4(m_3 - m'_3) + \cdots$$
$$+ r_{2\ell-1}(m_{2\ell} - m'_{2\ell}) + r_{2\ell}(m_{2\ell-1} - m'_{2\ell-1})$$
$$= f + \Delta + m'_1 m'_2 - m_1 m_2 + m'_3 m'_4 - m_3 m_4 + \cdots + m'_{2\ell-1} m'_{2\ell} - m_{2\ell-1} m_{2\ell}.$$

This is a linear equation in $r_1, \ldots, r_{2\ell}$, since f is independent of $r_1, \ldots, r_{2\ell}$. For each choice of $(r_{2\ell+1}, r_{2\ell+2}, \ldots, r_{2k})$, there are at most $\#K^{2\ell-1}$ choices of $(r_1, \ldots, r_{2\ell})$ satisfying this linear equation. Consequently $p \leq 1/\#K$ as above.

Case 3: $\ell < \ell'$ and $(m_1, \ldots, m_\ell) = (m'_1, \ldots, m'_\ell)$. Then $h = h' + \Delta$ if and only if $0 = f + \Delta$, where f is defined as above. This is a linear equation in $r_{2\ell+2}, r_{2\ell+4}, \ldots, r_{2\ell'}$ for each choice of $r_{2\ell+1}, r_{2\ell+3}, \ldots, r_{2\ell'-1}$. The linear equation is nontrivial except when $r_{2\ell+1} = -m'_{2\ell+1}$, $r_{2\ell+3} = -m'_{2\ell+3}$, and so on through $r_{2\ell'-1} = -m'_{2\ell'-1}$. The linear equation thus has at most $\#K^{\ell'-\ell-1}$ solutions $(r_{2\ell+2}, r_{2\ell+4}, \ldots, r_{2\ell'})$ for $\#K^{\ell'-\ell} - 1$ choices of $(r_{2\ell+1}, r_{2\ell+3}, \ldots, r_{2\ell'-1})$, plus at most $\#K^{\ell'-\ell}$ solutions $(r_{2\ell+2}, r_{2\ell+4}, \ldots, r_{2\ell'})$ for 1 exceptional choice of $(r_{2\ell+1}, r_{2\ell+3}, \ldots, r_{2\ell'-1})$, for a total of $\#K^{2\ell'-2\ell-1} - \#K^{\ell'-\ell-1} + \#K^{\ell'-\ell} < \#K^{2\ell'-2\ell}(1/\#K + 1/\#K^{\ell'-\ell})$ solutions. Consequently $p < 1/\#K + 1/\#K^{\ell'-\ell}$ as claimed.

Case 4: $\ell' < \ell$. Exchanging ℓ, m with ℓ', m' produces Case 2 or Case 3. $\square$

OMD: A Compression Function Mode of Operation for Authenticated Encryption

Simon Cogliani[1], Diana-Ştefania Maimuţ[1], David Naccache[1],
Rodrigo Portella do Canto[2], Reza Reyhanitabar[3(✉)],
Serge Vaudenay[3], and Damian Vizár[3]

[1] ENS, Paris, France
Diana.Maimut@ens.fr
[2] Université Paris II - Panthéon-Assas, Paris, France
[3] EPFL, Lausanne, Switzerland
reza.reyhanitabar@epfl.ch

Abstract. We propose the Offset Merkle-Damgård (OMD) scheme, a mode of operation to use a compression function for building a nonce-based authenticated encryption with associated data. In OMD, the parts responsible for privacy and authenticity are tightly coupled to minimize the total number of compression function calls: for processing a message of ℓ blocks and associated data of a blocks, OMD needs $\ell + a + 2$ calls to the compression function (plus a single call during the whole lifetime of the key). OMD is provably secure based on the standard pseudorandom function (PRF) property of the compression function. Instantiations of OMD using the compression functions of SHA-256 and SHA-512, called OMD-SHA256 and OMD-SHA512, respectively, provide much higher quantitative level of security compared to the AES-based schemes. OMD-SHA256 can benefit from the new Intel SHA Extensions on next-generation processors.

Keywords: Authenticated encryption · Provable security · Standard model · Intel SHA Extensions

1 Introduction

An authenticated encryption (AE) scheme delivers on two complementary data security goals: confidentiality (privacy) and integrity (authenticity). Historically, these goals were achieved by combining two separate cryptographic primitives, an encryption scheme to ensure privacy and a message authentication code (MAC) to guarantee authenticity [4,5]. This generic composition paradigm is neither most efficient (for instance, it requires processing the input stream at least twice) nor most robust to implementation errors [8,21,27]. The notion of AE, as a desirable primitive in its own right, was originally formalized in [4,6,17] (as a probabilistic algorithm) and later developed to include notions of nonce-based AE [25], nonce-based AE with associated data (AEAD) [22,24], deterministic AE

© Springer International Publishing Switzerland 2014
A. Joux and A. Youssef (Eds.): SAC 2014, LNCS 8781, pp. 112–128, 2014.
DOI: 10.1007/978-3-319-13051-4_7

(DAE) [26] (providing a solution to nonce-misuse resistance) and online DAE (online nonce-misuse resistant AE) [11].

Importance of useable AE to practice, and to some extent, difficulty of getting it right, is evident from the number of standards that were developed over the years, specifying different methods; for instance, the CCM method is specified in IEEE 802.11i, IPsec ESP and IKEv2 and NIST SP 800-38C; the GCM method is specified in NIST SP 800-38D; the EAX method is specified in ANSI C12.22; and ISO/IEC 19772:2009 defines six methods including five dedicated AE designs and one generic composition method, namely Encrypt-then-MAC. Surprisingly, the way that the latter generic method is specified in ISO/IEC 19772:2009 was very recently shown to be flawed, opening way for incorrect and insecure implementations [21].

AE schemes have been studied for over a decade, yet the topic remains a highly active and interesting area of research as evidenced by the recently initiated CAESAR competition by the cryptographic community. The competition aims to boost public discussions towards a better understanding of AE designs and to identify a portfolio of efficient and secure AE schemes by mid-December 2017 [7].

In this paper, we present a new nonce-based AEAD, called Offset Merkle-Damgård (OMD), offering several attractive features. Unlike the mainstream schemes which are either blockcipher-based or permutation-based schemes, OMD is designed as a mode of operation for a compression function. The motivation for this is manifold: (1) the cryptographic community has spent more than two decades on public research and standardization activities on hash functions resulting to development of a rich source of secure and efficient compression functions; (2) the standard SHA family of algorithms is heavily employed in many of the most common cryptographic applications and one can easily use off-the-shelf highly optimized implementations of these functions [12,13]; (3) Intel has recently introduced new instructions that support performance acceleration of SHA-1 and SHA-256 on next-generation processors [14]; (4) we believe that having a diverse set of AE schemes based on different primitives can be interesting from a practical viewpoint, providing the opportunity to choose among the AE algorithms based on what primitives have already been available and implemented and to reuse them.

Some of the interesting features of OMD, and its instantiations OMD-SHA256 and OMD-SHA512, are as follows:

Provable Security in the Standard Model. OMD achieves its security goals (privacy and authenticity) provably, based on the standard assumption that its underlying keyed compression function is PRF, an assumption which is among the most well-known and widely-used assumptions [2]. From a theoretical point of view, this is an advantage compared to permutation-based AE schemes whose security proofs rely on the *ideal permutation* assumption.

High Quantitative Security Level. When implemented with an off-the-shelf compression function such as those of the standard SHA family [1], OMD can achieve much higher security level compare to AES-based schemes. For example, the proven security of OMD-SHA256 and OMD-SHA512 falls off in about $\frac{\sigma^2}{2^{256}}$ and $\frac{\sigma^2}{2^{512}}$, respectively, where σ is the total number of calls to the compression function. In comparison, for the same key size and tag size, the proven security of all the standardized blockcipher-based AE schemes using AES (e.g. all five dedicated schemes specified in ISO/IEC 19772:2009) falls off in about $\frac{\sigma'^2}{2^{128}}$ where σ' is the total number of calls to AES. We note that it is possible to get blockcipher-based AE schemes with (high) beyond birthday-bound security, but the existing schemes with beyond birthday-bound security have a degraded efficiency [15,16,19,20].

Online. OMD encryption is online; that is, it outputs a stream of ciphertext as a stream of plaintext arrives with a constant latency and using constant memory. After receiving an indication that the plaintext is over, the final part of ciphertext together with the tag is output. OMD decryption is *internally* online: one can generate a stream of plaintext bits as the stream of ciphertext bits comes in, but no part of the plaintext stream will be returned before the whole ciphertext stream is decrypted and the tag is verified to be correct.

Flexible Parameters. OMD-SHA256 can support any key length up to 256 bits, tag length up to 256 bits, and nonce length up to 255 bits. OMD-SHA512 can support any key length up to 512 bits, tag length up to 512 bits, and nonce length up to 511 bits. These upper bounds on the parameters' length will satisfy the required security level of almost any imaginable application today and well beyond. The lower bounds on the parameters' lengths should be selected based on the specific security level sought by an application; for instance, most applications would not use keys shorter than 128 bits, tags shorter than 32 bits and nonce shorter than 64 bits.

ORGANIZATION OF THE PAPER. Notations and preliminary concepts are presented in Sect. 2. Security goals are defined in Sect. 3. Section 4 provides the specification of the OMD mode of operation. In Sect. 5, we provide the security analysis of OMD. Section 6 describes our recommended instantiations OMD-SHA256 and OMD-SHA512.

2   Preliminaries

NOTATIONS. If S is a finite set, $x \xleftarrow{\$} S$ means that x is chosen from S uniformly at random. $X \leftarrow Y$ is used for denoting the assignment statement where the value of Y is assigned to X. The set of all binary strings of length n bits (for some positive integer n) is denoted as $\{0,1\}^n$, the set of all binary strings whose lengths are variable but upper-bounded by L is denoted by $\{0,1\}^{\leq L}$ and the set of all binary strings of arbitrary but finite length is denoted by $\{0,1\}^*$. For two strings X and Y we use $X||Y$ and XY analogously to denote the string obtained by concatenating Y to X. For an m-bit binary string $X = X_{m-1} \cdots X_0$ we denote the left-most bit by $\mathsf{msb}(X) = X_{m-1}$ and the right-most bit by $\mathsf{lsb}(X) = X_0$; let $X[i \cdots j] = X_i \cdots X_j$ denote a substring of X, for $0 \leq j \leq i \leq (m-1)$. Let $1^n 0^m$ denote concatenation of n ones by m zeros. For a non-negative integer i let $\langle i \rangle_m$ denote binary representation of i by an m-bit string.

For a binary string $X = X_{m-1} \cdots X_0$, let $X \ll n$ denote the left-shift operation, where the n left-most bits are discarded and the n vacated right bits are set to 0. We let $X \gg n$ denote the (unsigned) right-shift operation where the n right-most bits are discarded and the n vacated left bits are set to 0. We let $X \gg_s n$ denote the *signed* right-shift operation where the n right-most bits are discarded and the n vacated left bits are filled with the left-most bit (which is considered as the sign bit); for example, $1001100 \gg_s 3 = 1111001$. If the left-most bit of X is 0 then we have $X \gg_s n = X \gg n$. Let $\mathtt{ntz}(i)$ denote the number of trailing zeros (i.e. the number of rightmost bits that are zero) in the binary representation of a positive integer i.

The special symbol $\perp$ means that the value of a variable is undefined; we also overload this symbol and use it to signify an error. Let $|Z|$ denote the number of elements of Z if Z is a set, and the length of Z in bits if Z is a binary string. For $X \in \{0,1\}^*$ let $X[1]\|X[2] \cdots \|X[m] \overset{b}{\leftarrow} X$ denote partitioning X into blocks $X[i]$ such that $|X[i]| = b$ for $1 \leq i \leq m-1$ and $|X[m]| \leq b$; let $m = |X|_b$ denote length of X in b-bit blocks.

For two binary strings $X = X_{m-1} \cdots X_0$ and $Y = Y_{n-1} \cdots Y_0$, the notation $X \oplus Y$ denotes bitwise xor of $X_{m-1} \cdots X_{m-1-\ell}$ and $Y_{n-1} \cdots Y_{n-1-\ell}$ where $\ell = min\{m-1, n-1\}$. Clearly, if X and Y have the same length then $X \oplus Y$ simply means their usual bitwise xor. The empty string is denoted by ε and we let $|\varepsilon| = 0$. For any string X, define $X \oplus \varepsilon = \varepsilon \oplus X = \varepsilon$.

THE FINITE FIELD WITH 2^n POINTS. Let $(GF(2^n), \oplus, \cdot)$ denote the Galois Field with 2^n points. When considering a point α in $GF(2^n)$ it can be represented in any of the following equivalent ways: (1) as an integer between 0 and 2^n, (2) as a binary string $\alpha_{n-1} \cdots \alpha_0 \in \{0,1\}^n$, or (3) as a formal polynomial $\alpha(X) = \alpha_{n-1}X^{n-1} + \cdots + \alpha_1 X + \alpha_0$ with binary coefficients. The addition "$\oplus$" and multiplication "$\cdot$" of two field elements in $GF(2^n)$ are defined as usual (e.g., see [25]). For $GF(2^{256})$ we use $P_{256}(X) = X^{256} + X^{10} + X^5 + X^2 + 1$, and for $GF(2^{512})$ we use $P_{512}(X) = X^{512} + X^8 + X^5 + X^2 + 1$ as the irreducible polynomials used in the field multiplications. It is easy to multiply an arbitrary field element α by the element 2 (i.e. X). For example, in $GF(2^{256})$ using $P_{256}(X)$ the doubling operation can be described as follows:

$$2 \cdot \alpha = \begin{cases} \alpha \ll 1 & \text{if } \mathsf{msb}\,(\alpha) = 0 \\ (\alpha \ll 1) \oplus 0^{245}10000100101 & \text{if } \mathsf{msb}\,(\alpha) = 1 \end{cases} \quad (1)$$

$$= (\alpha \ll 1) \oplus ((\alpha \gg_s 255) \wedge 0^{245}10000100101) \quad (2)$$

We note that the results computed in (1) and (2) are the same but an implementation using (2) will not be susceptible to the timing attacks unlike one which uses (1).

3 Definitions and Security Goals

As usual in the concrete-security definitions, we measure the insecurity of a scheme Π using the resource parametrized function $\mathbf{Adv}_{\Pi}^{\text{xxx}}(\mathbf{r})$, denoting the maximal value of the adversarial advantage, $\mathbf{Adv}_{\Pi}^{\text{xxx}}(\mathbf{r}) = max_{A} \{\mathbf{Adv}_{\Pi}^{\text{xxx}}(A)\}$, over all adversaries A, against the xxx property of a primitive or scheme Π, that use resources bounded by $\mathbf{r}$. Let A be an adversary that returns a binary value; by $A^{f(\cdot)}(X) \Rightarrow 1$ we refer to the event that A on input X and access to an oracle function $f(.)$ returns 1.

PSEUDORANDOM FUNCTIONS (PRFS) AND TWEAKABLE PRFS. Let $\text{Func}(m, n)$ be the set of all functions from m-bit strings to n-bit strings; i.e., $\text{Func}(m, n) = \{f : \{0,1\}^m \rightarrow \{0,1\}^n\}$. A random function (RF) R with m-bit input and n-bit output is a function selected uniformly at random from $\text{Func}(m, n)$. We denote this by $R \xleftarrow{\$} \text{Func}(m, n)$.

Let $\text{Func}^{\mathcal{T}}(m, n)$ be the set of all functions $\left\{ \widetilde{f} : \mathcal{T} \times \{0,1\}^m \rightarrow \{0,1\}^n \right\}$, where $\mathcal{T}$ is a set of tweaks. A tweakable RF with the tweak space $\mathcal{T}$, m-bit input and n-bit output is a map $\widetilde{R} : \mathcal{T} \times \{0,1\}^m \rightarrow \{0,1\}^n$ selected uniformly at random from $\text{Func}^{\mathcal{T}}(m, n)$; i.e. $\widetilde{R} \xleftarrow{\$} \text{Func}^{\mathcal{T}}(m, n)$. Clearly, if $\mathcal{T} = \{0,1\}^t$ then $|\text{Func}^{\mathcal{T}}(m, n)| = |\text{Func}(m + t, n)|$, and hence, $\widetilde{R}$ can be instantiated using a random function R with $(m + t)$-bit input and n-bit output. We use $\widetilde{R}^{\langle T \rangle}(.)$ and $\widetilde{R}(T, .)$ interchangeably, for every $T \in \mathcal{T}$. Notice that each tweak T names a random function $\widetilde{R}^{\langle T \rangle} : \{0,1\}^m \rightarrow \{0,1\}^n$ and distinct tweaks name distinct (independent) random functions.

Let $F : \mathcal{K} \times \{0,1\}^m \rightarrow \{0,1\}^n$ be a keyed function and let $\widetilde{F} : \mathcal{K} \times \mathcal{T} \times \{0,1\}^m \rightarrow \{0,1\}^n$ be a keyed and tweakable function, where the key space $\mathcal{K}$ is some nonempty set. Let $F_K(.) = F(K, .)$ and $\widetilde{F}_K^{\langle T \rangle}(.) = \widetilde{F}(K, T, .)$. Let A be an adversary. Then:

$$\mathbf{Adv}_F^{\text{prf}}(A) = \Pr \left[K \xleftarrow{\$} \mathcal{K} : A^{F_K(\cdot)} \Rightarrow 1 \right] - \Pr \left[R \xleftarrow{\$} \text{Func}(m, n) : A^{R(\cdot)} \Rightarrow 1 \right]$$

$$\mathbf{Adv}_{\widetilde{F}}^{\text{prf}}(A) = \Pr \left[K \xleftarrow{\$} \mathcal{K} : A^{\widetilde{F}_K^{\langle \cdot \rangle}(\cdot)} \Rightarrow 1 \right] - \Pr \left[\widetilde{R} \xleftarrow{\$} \text{Func}^{\mathcal{T}}(m, n) : A^{\widetilde{R}^{\langle \cdot \rangle}(\cdot)} \Rightarrow 1 \right]$$

The resource parametrized advantage functions are defined accordingly, considering that the adversarial resources of interest here are the time complexity (t) of the adversary and the total number of queries (q) asked by the adversary (note that we just consider fixed-input-length functions, so the lengths of queries are fixed and known). We say that F is $(t, q; \epsilon)$-PRF if $\mathbf{Adv}_F^{\text{prf}}(t, q) \leq \epsilon$. We say that $\widetilde{F}$ is $(t, q; \epsilon)$-tweakable PRF if $\mathbf{Adv}_{\widetilde{F}}^{\text{prf}}(t, q) \leq \epsilon$.

SYNTAX OF AN AEAD SCHEME. A nonce-based authenticated encryption with associated data, AEAD for short, is a symmetric key scheme $\Pi = (\mathcal{K}, \mathcal{E}, \mathcal{D})$. The key space $\mathcal{K}$ is some non-empty finite set. The encryption algorithm $\mathcal{E} : \mathcal{K} \times \mathcal{N} \times \mathcal{A} \times \mathcal{M} \rightarrow \mathcal{C} \cup \{\perp\}$ takes four arguments, a secret key $K \in \mathcal{K}$, a nonce $N \in \mathcal{N}$, an associated data (a.k.a. header data) $A \in \mathcal{A}$ and a message $M \in \mathcal{M}$, and returns either a ciphertext $\mathbb{C} \in \mathcal{C}$ or a special symbol $\perp$ indicating an error. The decryption algorithm $\mathcal{D} : \mathcal{K} \times \mathcal{N} \times \mathcal{A} \times \mathcal{C} \to \mathcal{M} \cup \{\perp\}$ takes four arguments $(K, N, A, \mathbb{C})$ and either outputs a message $M \in \mathcal{M}$ or an error indicator $\perp$.

For correctness of the scheme, it is required that $\mathcal{D}(K, N, A, \mathbb{C}) = M$ for any $\mathbb{C}$ such that $\mathbb{C} = \mathcal{E}(K, N, A, M)$. It is also assumed that if algorithms $\mathcal{E}$ and $\mathcal{D}$ receive parameter not belonging to their specified domain of arguments they will output $\perp$. We write $\mathcal{E}_K(N, A, M) = \mathcal{E}(K, N, A, M)$ and similarly $\mathcal{D}_K(N, A, \mathbb{C}) = \mathcal{D}(K, N, A, \mathbb{C})$.

We assume that the message and associated data can be any binary string of arbitrary but finite length; i.e. $\mathcal{M} = \{0, 1\}^*$ and $\mathcal{A} = \{0, 1\}^*$, but the key and nonce are some fixed-length binary strings, i.e. $\mathcal{N} = \{0, 1\}^{|N|}$ and $\mathcal{K} = \{0, 1\}^k$, where the positive integers $|N|$ and k are respectively the nonce length and the key length of the scheme in bits. We assume that $|\mathcal{E}_K(N, A, M)| = |M| + \tau$ for some positive fixed constant τ; that is, we will have $\mathbb{C} = C\|\mathsf{Tag}$ where $|C| = |M|$ and $|\mathsf{Tag}| = \tau$. We call C the core ciphertext and Tag the tag.

NONCE RESPECTING ADVERSARIES. Let A be an adversary. We say that A is nonce-respecting if it never repeats a nonce in its *encryption* queries. That is, if A queries the encryption oracle $\mathcal{E}_K(\cdot, \cdot, \cdot)$ on $(N^1, A^1, M^1) \cdots (N^q, A^q, M^q)$ then $N^1, \cdots, N^q$ must be distinct.

PRIVACY OF AEAD SCHEMES. Let $\Pi = (\mathcal{K}, \mathcal{E}, \mathcal{D})$ be a nonce-based AEAD scheme. Let A be a nonce-respecting adversary. A is provided with an oracle which can be either a real encryption oracle $\mathcal{E}_K(\cdot, \cdot, \cdot)$ such that on input (N, A, M) returns $\mathbb{C} = \mathcal{E}_K(N, A, M)$, or a fake encryption oracle $\$(\cdot, \cdot, \cdot)$ which on any input (N, A, M) returns $|\mathbb{C}|$ fresh random bits. The advantage of A in mounting a chosen plaintext attack (CPA) against the privacy property of Π is measured as follows:

$$\mathbf{Adv}_\Pi^{\mathrm{priv}}(A) = \Pr[K \xleftarrow{\$} \mathcal{K} : A^{\mathcal{E}_K(\cdot, \cdot, \cdot)} \Rightarrow 1] - \Pr[A^{\$(\cdot, \cdot, \cdot)} \Rightarrow 1].$$

This privacy notion, also called indistinguishability of ciphertext from random bits under CPA (IND\$-CPA), is defined originally in [25] and is a stronger variant of the classical IND-CPA notion [3,4] for conventional symmetric-key encryption schemes.

AUTHENTICITY OF AEAD SCHEMES. Let $\Pi = (\mathcal{K}, \mathcal{E}, \mathcal{D})$ be a nonce-based AEAD scheme. Let A be a nonce-respecting adversary. We stress that nonce-respecting is only regarded for the encryption queries; that is, A can repeat nonces during its decryption queries and it can also ask an encryption query with a nonce that was already used in a decryption query. Let $\mathcal{A}$ be provided with the encryption oracle $\mathcal{E}_K(\cdot, \cdot, \cdot)$ and the decryption oracle $\mathcal{D}_K(\cdot, \cdot, \cdot)$; that is, we consider adversaries that can mount chosen ciphertext attacks (CCA). We say that A forges if it makes a decryption query $(N, A, \mathbb{C})$ such that $\mathcal{D}_K(N, A, \mathbb{C}) \neq \perp$ and no previous encryption query $\mathcal{E}_K(N, A, M)$ returned $\mathbb{C}$.

$$\mathbf{Adv}_\Pi^{\mathrm{auth}}(A) = \Pr[K \xleftarrow{\$} \mathcal{K} : A^{\mathcal{E}_K(\cdot, \cdot, \cdot), \mathcal{D}_K(\cdot, \cdot, \cdot)} \text{ forges}].$$

This authenticity notion, also called integrity of ciphertext (INT-CTXT) under CCA attacks, is defined originally in [4].

RESOURCE PARAMETERS FOR THE ADVERSARY. Let an adversary A make encryption queries $(N^1, A^1, M^1) \cdots (N^{q_e}, A^{q_e}, M^{q_e})$ and decryption queries $(N'^1, A'^1, \mathbb{C}'^1) \cdots (N'^{q_v}, A'^{q_v}, \mathbb{C}'^{q_v})$. We define the resource parameters of A as $(t, q_e, q_v, \sigma_A, \sigma_M, \sigma_{A'}, \sigma_{\mathbb{C}'}, L_{max})$, where t is the time complexity, q_e and q_v are respectively the total number of encryption queries and decryption queries, L_{max} is the maximum length of each query in bits, $\sigma_A = \sum_{i=1}^{q_e} |A^i|$, $\sigma_M = \sum_{i=1}^{q_e} |M^i|$, $\sigma_{A'} = \sum_{i=1}^{q_v} |A'^i|$ and $\sigma_{\mathbb{C}'} = \sum_{i=1}^{q_v} (|\mathbb{C}'^i| - \tau)$.

We remind that absence of a resource parameter means that the parameter is irrelevant in the context and hence omitted.

The use of the aforementioned privacy (IND\$-CPA) and authenticity (INT-CTXT) goals to define security of AE schemes dates back to [4] where it was shown that if an AE scheme satisfies the combination of IND-CPA and INT-CTXT properties then it will also fulfill indistinguishability under the strongest form of chosen-ciphertext attack (IND-CCA) which, in turn, is equivalent to non-malleability under chosen-ciphertext attack (NM-CCA).

4   The OMD Mode of Operation

To use OMD one must specify a keyed compression function $F : \mathcal{K} \times (\{0,1\}^n \times \{0,1\}^m) \to \{0,1\}^n$ and fix a tag length $\tau \leq n$; where the key space $\mathcal{K} = \{0,1\}^k$ and $m \leq n$. We let $\text{OMD}[F, \tau]$ denote the OMD mode of operation using the keyed compression function F_K and the fixed tag length τ.

At first glance, imposing $m \leq n$ may look a bit odd as usually a compression function has a larger input block length than its output length, but we note that in practice, the compression function of standard hash functions (e.g. SHA-1 or the SHA-2 family) are keyless, therefore one will need to use k bits of their b-bit message block to get a keyed function. So, there will be no waste in each call to the compression function if $m = n$ and $b = n + k$; for example, when the key length is 256 bits and the compression function of SHA-256 is used.

Figure 1 depicts the encryption algorithm of $\text{OMD}[F, \tau]$. The construction of the decryption algorithm is straightforward and almost the same as the encryption algorithm except a tag comparison (verification) at the end of the decryption process. An algorithmic description of $\text{OMD}[F, \tau]$ is provided in Fig. 2.

The encryption algorithm of $\text{OMD}[F, \tau]$ inputs four arguments (secret key $K \in \{0,1\}^k$, nonce $N \in \{0,1\}^{|N|}$, associated data $A \in \{0,1\}^*$, message $M \in \{0,1\}^*$) and outputs $\mathbb{C} = C \| \text{Tag} \in \{0,1\}^{|M|+\tau}$. The decryption algorithm of $\text{OMD}[F, \tau]$ inputs four arguments (secret key $K \in \{0,1\}^k$, nonce $N \in \{0,1\}^{|N|}$, associated data $A \in \{0,1\}^*$, ciphertext $C \| \text{Tag} \in \{0,1\}^*$) and either outputs the whole $M \in \{0,1\}^{|C|-\tau}$ at once or an error message ($\bot$). Note that we have either $C = C_1 \cdots C_\ell$ or $C = C_1 \cdots C_{\ell-1} C_*$ depending on whether the message length in bits is a multiple of the block length m or not, respectively.

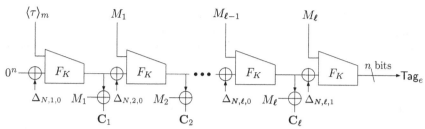

Encrypting a message whose length is a multiple of the block length. No padding is needed.

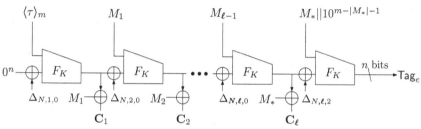

Encrypting a message whose length is not a multiple of the block length. The final message block is padded to make it a full block

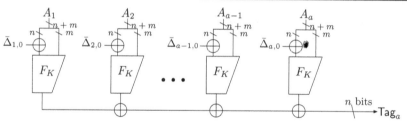

Computing Tag_a for an associate data whose length is a multiple of the input length (i.e $|A_a| = n+m$).

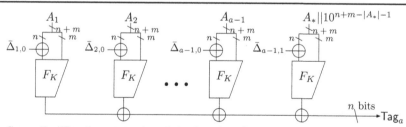

Computing Tag_a for an associate data whose length is not a multiple of the input length. The final block is padded to make it a full block .

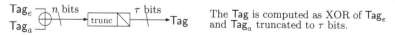

The Tag is computed as XOR of Tag_e and Tag_a truncated to τ bits.

Fig. 1. The encryption process of $\mathrm{OMD}[F, \tau]$ using a keyed compression function $F_K : (\{0,1\}^n \times \{0,1\}^m) \rightarrow \{0,1\}^n$ and a fixed tag length τ.

COMPUTING THE MASKING VALUES. As seen from the description of OMD in Fig. 1, before each call to the underlying keyed compression function we xor a masking value denoted as $\Delta_{N,i,j}$ (the top and middle parts of Fig. 1) and $\bar{\Delta}_{i,j}$ (the bottom part of Fig. 1). In the following, we describe how these masks are generated.

There are different ways to compute the masking values to satisfy both the security and efficiency criteria; for example, we refer to [9,18,23]. We use the method proposed in [18]. In the following, all multiplications (denoted by ".") are in $GF(2^n)$.

Initialization:
$\Delta_{N,0,0} = F_K(N||10^{n-1-|N|}, 0^m); \bar{\Delta}_{0,0} = 0^n; L_* = F_K(0^n, 0^m); L(0) = 4 \cdot L_*$, and $L(i) = 2 \cdot L(i-1)$ for $i \geq 1$. We note that the values $L(i)$ can be preprocessed and stored (for a fast implementation) in a table for $0 \leq i \leq \lceil \log_2(\ell_{max}) \rceil$, where ℓ_{max} is the bound on the maximum number of blocks in any input that can be encrypted or decrypted. Alternatively, (if there is a memory restriction) they can be computed on-the-fly for $i \geq 1$. It is also possible to precompute and store some values and then compute the others as needed on-the-fly.

Masking sequence for processing the message:
For $i \geq 1$: $\Delta_{N,i,0} = \Delta_{N,i-1,0} \oplus L(\text{ntz}(i)); \Delta_{N,i,1} = \Delta_{N,i,0} \oplus 2 \cdot L_*$; and $\Delta_{N,i,2} = \Delta_{N,i,0} \oplus 3 \cdot L_*$.

Masking sequence for processing the associated data:
$\bar{\Delta}_{i,0} = \bar{\Delta}_{i-1,0} \oplus L(\text{ntz}(i))$ for $i \geq 1$; and $\bar{\Delta}_{i,1} = \bar{\Delta}_{i,0} \oplus L_*$ for $i \geq 0$.

5   Security Analysis

Theorem 1 provides the security bounds of OMD.

Theorem 1. *Fix $n \geq 1$ and $\tau \in \{0, 1, \cdots, n\}$. Let $F : \mathcal{K} \times (\{0,1\}^n \times \{0,1\}^m) \rightarrow \{0,1\}^n$ be a PRF, where the key space $\mathcal{K} = \{0,1\}^k$ for $k \geq 1$ and $1 \leq m \leq n$. Then*

$$\mathbf{Adv}_{OMD[F,\tau]}^{\text{priv}}(t, q_e, \sigma_e, \ell_{max}) \leq \mathbf{Adv}_F^{\text{prf}}(t', 2\sigma_e) + \frac{3\sigma_e^2}{2^n}$$

$$\mathbf{Adv}_{OMD[F,\tau]}^{\text{auth}}(t, q_e, q_v, \sigma, \ell_{max}) \leq \mathbf{Adv}_F^{\text{prf}}(t', 2\sigma) + \frac{3\sigma^2}{2^n} + \frac{q_v \ell_{max}}{2^n} + \frac{q_v}{2^\tau}$$

where q_e and q_v are, respectively, the number of encryption and decryption queries, ℓ_{max} denotes the maximum number of m-bit blocks in an encryption or decryption query, $t' = t + cn\sigma$ for some constant c, and σ_e and σ are the total number of calls to the underlying compression function F in all queries asked by the CPA and CCA adversaries against the privacy and authenticity of the scheme, respectively.

The proof is obtained by combing Lemma 1 in Subsect. 5.1 with Lemmas 2 and 3 in Subsect. 5.2.

```
1: Algorithm INITIALIZE(K)                    12:        H ← F_K(H ⊕ Δ, M_i)
2:     L_* ← F_K(0^n, 0^m)                     13:        C_ℓ ← H ⊕ M_ℓ
3:     L(0) ← 4 · L_*                          14:        if |M_ℓ| = m then
4:     for i ← 1 to ⌈log_2(ℓ_max)⌉ do         15:            Δ ← Δ ⊕ 2 · L_*
5:         L(i) = 2 · L(i − 1)                  16:            Tag_e ← F_K(H ⊕ Δ, M_ℓ)
6:     return                                   17:        else if |M_ℓ| ≠ 0 then
                                                18:            Δ ← Δ ⊕ 3 · L_*
                                                19:            M_pad ← M_ℓ||10^{m−1−|M_ℓ|−1}
1: Algorithm HASH_K(A)                         20:            Tag_e ← F_K(H ⊕ Δ, M_pad)
2:     b ← n + m                               21:        else
3:     A_1||A_2 · · · A_{ℓ−1}||A_ℓ  ←̲ᵇ  A       22:            Tag_e ← H
4:     Tag_a ← 0^n                             23:        Tag_a ← HASH_K(A)
5:     Δ ← 0^n                                 24:        Tag ← (Tag_e ⊕ Tag_a)[n − 1 · · · n − τ]
6:     for i ← 1 to ℓ − 1 do                   25:        C ← C_1||C_2|| · · · ||C_ℓ||Tag
7:         Δ ← Δ ⊕ L(ntz(i))                   26:        return C
8:         Left ← A_i[b − 1 · · · m]
9:         Right ← A_i[m − 1 · · · 0]
10:        Φ ← F_K(Left ⊕ Δ, Right)            1: Algorithm D_K(N, A, C)
11:        Tag_a ← Tag_a ⊕ Φ                   2:     if |N| > n − 1 or |C| < τ then
12:    if |A_ℓ| = b then                       3:         return ⊥
13:        Δ ← Δ ⊕ L(ntz(ℓ))                   4:     C_1||C_2 · · · C_{ℓ−1}||C_ℓ||Tag  ←̲ᵐ  C
14:        Left ← A_ℓ[b − 1 · · · m]           5:     Δ ← F_K(N||10^{n−1−|N|}, 0^m)
15:        Right ← A_ℓ[m − 1 · · · 0]          6:     H ← 0^n
16:        Φ ← F_K(Left ⊕ Δ, Right)            7:     Δ ← Δ ⊕ L(0)
17:        Tag_a ← Tag_a ⊕ Φ                   8:     H ← F_K(H ⊕ Δ, ⟨τ⟩_m)
18:    else                                    9:     for i ← 1 to ℓ − 1 do
19:        Δ ← Δ ⊕ L_*                         10:        M_i ← H ⊕ C_i
20:        A_pad ← A_ℓ||10^{b−|A_ℓ|−1}         11:        Δ ← Δ ⊕ L(ntz(i + 1))
21:        Left ← A_pad[b − 1 · · · m]         12:        H ← F_K(H ⊕ Δ, M_i)
22:        Right ← A_pad[m − 1 · · · 0]        13:    M_ℓ ← H ⊕ C_ℓ
23:        Φ ← F_K(Left ⊕ Δ, Right)            14:    if |C_ℓ| = m then
24:        Tag_a ← Tag_a ⊕ Φ                   15:        Δ ← Δ ⊕ 2 · L_*
25:    return Tag_a                            16:        Tag_e ← F_K(H ⊕ Δ, M_ℓ)
                                                17:    else if |C_ℓ| ≠ 0 then
                                                18:        Δ ← Δ ⊕ 3 · L_*
1: Algorithm E_K(N, A, M)                      19:        M_pad ← M_ℓ||10^{m−|M_ℓ|−1}
2:     if |N| > n − 1 then                     20:        Tag_e ← F_K(H ⊕ Δ, M_pad)
3:         return ⊥                            21:    else
4:     M_1||M_2 · · · M_{ℓ−1}||M_ℓ  ←̲ᵐ  M       22:        Tag_e ← H
5:     Δ ← F_K(N||10^{n−1−|N|}, 0^m)           23:    Tag_a ← HASH_K(A)
6:     H ← 0^n                                 24:    Tag' ← (Tag_e ⊕ Tag_a)[n − 1 · · · n − τ]
7:     Δ ← Δ ⊕ L(0)                            25:    if Tag' = Tag then
8:     H ← F_K(H ⊕ Δ, ⟨τ⟩_m)                  26:        return M ← M_1||M_2|| · · · ||M_ℓ
9:     for i ← 1 to ℓ − 1 do                   27:    else
10:        C_i ← H ⊕ M_i                       28:        return ⊥
11:        Δ ← Δ ⊕ L(ntz(i + 1))
```

Fig. 2. Definition of OMD$[F, \tau]$. The function $F : \mathcal{K} \times (\{0,1\}^n \times \{0,1\}^m) \rightarrow \{0,1\}^n$ is a keyed compression function with $\mathcal{K} = \{0,1\}^k$ and $m \leq n$. The tag length is $\tau \in \{0, 1, \cdots, n\}$. Algorithms $\mathcal{E}$ and $\mathcal{D}$ can be called with arguments $K \in \mathcal{K}$, $N \in \{0,1\}^{\leq n-1}$, and $A, M, \mathbb{C} \in \{0,1\}^*$. ℓ_{max} is the bound on the maximum number of blocks in any input to the encryption or decryption algorithms.

Remark 1. Referring to Subsect. 3 for definitions of the resource parameters, it can be seen that: $\sigma_e = \lceil \sigma_M/m \rceil + \lceil \sigma_A/(n+m) \rceil + 2q_e$; $\sigma = \lceil (\sigma_M + \sigma_{\mathbb{C}'})/m \rceil + \lceil (\sigma_A + \sigma_{A'})/(n+m) \rceil + 2q$; *and* $\ell_{max} = \lceil L_{max}/m \rceil$.

5.1   Generalized OMD Using a Tweakable Random Function

Figure 3 shows the G-OMD$[\widetilde{R}, \tau]$ scheme which is a generalization of OMD$[F, \tau]$ using a tweakable random function $\widetilde{R} : \mathcal{T} \times (\{0,1\}^n \times \{0,1\}^m) \to \{0,1\}^n$. The tweak space $\mathcal{T}$ consists of five mutually exclusive sets of tweaks; namely, $\mathcal{T} = \mathcal{N} \times \mathbb{N} \times \{0\} \cup \mathcal{N} \times \mathbb{N} \times \{1\} \cup \mathcal{N} \times \mathbb{N} \times \{2\} \cup \mathbb{N} \times \{0\} \cup \mathbb{N} \times \{1\}$, where $\mathcal{N} = \{0,1\}^{|N|}$ is the set of nonces and $\mathbb{N}$ is the set of positive integers.

Lemma 1. *Let G-OMD$[\widetilde{R}, \tau]$ be the scheme shown in Fig. 3. Then*

$$\mathbf{Adv}^{priv}_{G\text{-}OMD[\widetilde{R},\tau]}(q_e, \sigma_e, \ell_{max}) = 0$$

$$\mathbf{Adv}^{auth}_{G\text{-}OMD[\widetilde{R},\tau]}(q_e, q_v, \sigma, \ell_{max}) \leq \frac{q_v \ell_{max}}{2^n} + \frac{q_v}{2^\tau}$$

where q_e and q_v are, respectively, the number of encryption and decryption queries, ℓ_{max} denotes the maximum number of m-bit blocks in an encryption or decryption query, and σ_e and σ are the total number of calls to the underlying tweakable random function $\widetilde{R}$ in all queries asked by the CPA and CCA adversaries against the privacy and authenticity of the scheme, respectively.

The proof is provided in the full version of this paper [10].

5.2   Instantiating Tweakable RFs with PRFs

Step 1. Replace the tweakable RF $\widetilde{R} : \mathcal{T} \times (\{0,1\}^n \times \{0,1\}^m) \to \{0,1\}^n$ in G-OMD with a tweakable PRF $\widetilde{F} : \mathcal{K} \times \mathcal{T} \times (\{0,1\}^n \times \{0,1\}^m) \to \{0,1\}^n$, where $\mathcal{K} = \{0,1\}^k$. The following lemma states the classical bound on the security loss induced by this replacement step. The proof is a straightforward reduction and omitted here.

Lemma 2. *Let $\widetilde{R} : \mathcal{T} \times (\{0,1\}^n \times \{0,1\}^m) \to \{0,1\}^n$ be a tweakable RF and $\widetilde{F} : \mathcal{K} \times \mathcal{T} \times (\{0,1\}^n \times \{0,1\}^m) \to \{0,1\}^n$ be a tweakable PRF. Then*

$$\mathbf{Adv}^{priv}_{G\text{-}OMD[\widetilde{F},\tau]}(t, q_e, \sigma_e, \ell_{max}) \leq \mathbf{Adv}^{priv}_{G\text{-}OMD[\widetilde{R},\tau]}(q_e, \sigma_e, \ell_{max}) + \mathbf{Adv}^{\widetilde{prf}}_{\widetilde{F}}(t', \sigma_e)$$

$$\mathbf{Adv}^{auth}_{G\text{-}OMD[\widetilde{F},\tau]}(t, q_e, q_v, \sigma, \ell_{max}) \leq \mathbf{Adv}^{auth}_{G\text{-}OMD[\widetilde{R},\tau]}(q_e, q_v, \sigma, \ell_{max}) + \mathbf{Adv}^{\widetilde{prf}}_{\widetilde{F}}(t'', \sigma)$$

where q_e and q_v are, respectively, the number of encryption and decryption queries, $q = q_e + q_v$, ℓ_{max} denotes the maximum number of m-bit blocks in an encryption or decryption query, $t' = t + cn\sigma_e$ and $t'' = t + c'n\sigma$ for some constants c, c', and σ_e and σ are the total number of calls to the underlying compression function F in all queries asked by the CPA and CCA adversaries against the privacy and authenticity of the scheme, respectively.

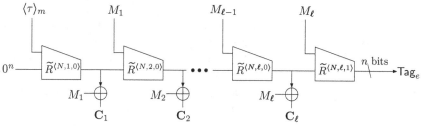

Encrypting a message whose length is a multiple of the block length. No padding is needed.

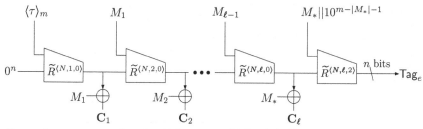

Encrypting a message whose length is not a multiple of the block length. The final message block is padded to make it a full block

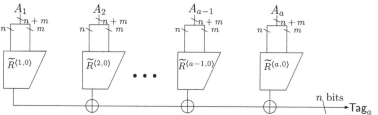

Computing Tag_a for an associate data whose length is a multiple of the input length (i.e $|A_a| = n + m$).

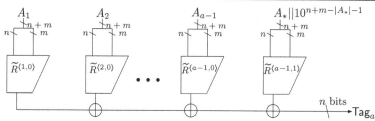

Computing Tag_a for an associate data whose length is not a multiple of the input length. The final block is padded to make it a full block .

Tag_e, Tag_a — n bits → trunc → τ bits → Tag. The Tag is computed as XOR of Tag_e and Tag_a truncated to τ bits.

Fig. 3. The G-OMD$[\widetilde{R}, \tau]$ scheme using a tweakable random function $\widetilde{R} : \mathcal{T} \times (\{0,1\}^n \times \{0,1\}^m) \to \{0,1\}^n$ (i.e. $\widetilde{R} \xleftarrow{\$} \mathrm{Func}^{\mathcal{T}}(n+m, n)$).

Step 2. We instantiate a tweakable PRF using a PRF by means of XORing (part of) the input by a mask generated as a function of the key and tweak as shown in Fig. 4. This method to tweak a PRF is (essentially) the XE method of [23]. In OMD the tweaks are of the form $T = (\alpha, i, j)$ where $\alpha \in \mathcal{N} \cup \{\varepsilon\}$, $1 \leq i \leq 2^{n-8}$ and $j \in \{0, 1, 2\}$. We note that not all combinations are used; for example, if $\alpha = \varepsilon$ (empty) which corresponds to processing of the associate data in Fig. 1 then $j \neq 2$. The masking function $\Delta_K(T) = \Delta_K(\alpha, i, j)$ outputs an n-bit mask such that the following two properties hold for any fixed string $H \in \{0, 1\}^n$:

1. $\Pr[\Delta_K(\alpha, i, j) = H] \leq 2^{-n}$ for any (α, i, j)
2. $\Pr[\Delta_K(\alpha, i, j) \oplus \Delta_K(\alpha', i', j') = H] \leq 2^{-n}$ for $(\alpha, i, j) \neq (\alpha', i', j')$

where the probabilities are taken over random selection of the key.

It is easy to verify that these two properties are satisfied by the specific masking scheme of OMD as described in Sect. 4.

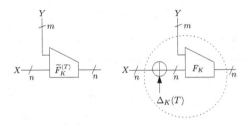

Fig. 4. Building a tweakable PRF $\widetilde{F}_K^{\langle T \rangle} : \{0, 1\}^n \times \{0, 1\}^m \rightarrow \{0, 1\}^n$ using a PRF $F_K : \{0, 1\}^n \times \{0, 1\}^m \rightarrow \{0, 1\}^n$. There are several efficient ways to define the masking function $\Delta_K(T)$ [9, 18, 23]; we use the method of [18].

Lemma 3. *Let* $F : \mathcal{K} \times (\{0, 1\}^n \times \{0, 1\}^m) \rightarrow \{0, 1\}^n$ *be a function family with key space* $\mathcal{K}$. *Let* $\widetilde{F} : \mathcal{K} \times \mathcal{T} \times (\{0, 1\}^n \times \{0, 1\}^m) \rightarrow \{0, 1\}^n$ *be defined by* $\widetilde{F}_K^{\langle T \rangle}(X, Y) = F_K((X \oplus \Delta_K(T)), Y)$ *for every* $T \in \mathcal{T}, K \in \mathcal{K}, X \in \{0, 1\}^n, Y \in \{0, 1\}^m$ *and* $\Delta_K(T)$ *is the masking function of OMD as defined in Sect. 4. If* F *is PRF then* $\widetilde{F}$ *is tweakable PRF; more precisely*

$$\mathbf{Adv}_{\widetilde{F}}^{\widetilde{\mathrm{prf}}}(t, q) \leq \mathbf{Adv}_F^{\mathrm{prf}}(t', 2q) + \frac{3q^2}{2^n}.$$

The proof is a simple adaptation of a similar result on the security of the XE construction (to tweak a blockcipher) in [18]. As we use a PRF rather than PRP, our bound has two main terms. The first term is a single birthday bound loss of $\frac{0.5q^2}{2^n}$ to take care of the case that a collision might happen when computing the initial mask $\Delta_{N,0,0} = F_K(N||10^{n-1-|N|}, 0^m)$ using a PRF (F) rather than a PRP (as in [18]). The analysis of the remaining term (i.e. $\frac{2.5q^2}{2^n}$) is essentially the same as the similar part in [18], but we note that in the context of our construction as we are directly dealing with PRFs unlike [18] in which PRPs are used, the bound obtained here does not have any loss terms caused by the switching (PRF–PRP) lemma. Therefore, instead of the $\frac{6q^2}{2^n}$ bound in [18] (from which $\frac{3.5q^2}{2^n}$ is due to using the switching lemma) our bound has only $\frac{2.5q^2}{2^n}$.

6   Instantiations

6.1   OMD-SHA256

Our primary recommendation to instantiate OMD, called OMD-SHA256, uses the underlying compression function of SHA-256 [1]. This is intended to be the appropriate choice for next-generation processors supporting Intel SHA Extensions.

The compression function of SHA-256 is a map SHA-256 $: \{0,1\}^{256} \times \{0,1\}^{512} \to \{0,1\}^{256}$. On input a 256-bit chaining block X and a 512-bit message block Y, it outputs a 256-bit digest Z, i.e. let $Z = \text{SHA-256}(X,Y)$.

To use OMD with SHA-256, we use the first 256-bit argument X for chaining values as usual. We use the 512-bit argument Y (the message block in SHA-256) to input both a 256-bit message block and the key K which can be of any length $k \leq 256$ bits. If $k < 256$ then let the key be $K||0^{256-k}$. That is, we define the keyed compression function $F_K : \{0,1\}^{256} \times \{0,1\}^{256} \to \{0,1\}^{256}$ needed in OMD as $F_K(H,M) = \text{SHA-256}(H, K||0^{256-k}||M)$.

The parameters of OMD-SHA256 are as follows:

- The message block length in bits is $m = 256$; i.e. $|M_i| = 256$. *If needed*, we pad the final block of the message with 10^* (i.e., a single 1 followed by the minimal number of 0's needed) to make its length exactly 256 bits.
- The key length in bits can be $80 \leq k \leq 256$; but $k < 128$ is not recommended. *If needed*, we pad the key K with 0^{256-k} to make its length exactly 256 bits.
- The nonce (public message number) length in bits can be $96 \leq |N| \leq 255$. We *always* pad the nonce with $10^{255-|N|}$ to make its length exactly 256 bits.
- The secret message number length in bits is 0; that is, our scheme does not support secret message numbers.
- The associated data block length in bits is $2n = 512$; i.e. $|A_i| = 512$. *If needed*, we pad the final block of the associated data with 10^* (i.e., a single 1 followed by the minimal number of 0's needed) to make its length exactly 512 bits.
- The tag length in bits can be $32 \leq \tau \leq 256$; but it must be noted that the selection of the tag length directly affects the security level.

6.2   OMD-SHA512

Our second recommendation to instantiate OMD, called OMD-SHA512, uses the underlying compression function of SHA-512 [1]. This is intended to be the appropriate choice for implementations on 64-bit machines.

The compression function of SHA-512 is a map SHA-512 : $\{0,1\}^{512} \times \{0,1\}^{1024} \rightarrow \{0,1\}^{512}$. On input a 512-bit chaining block X and a 1024-bit message block Y, it outputs a 512-bit digest Z, i.e. let $Z = \text{SHA-512}(X, Y)$.

To use OMD with SHA-512, we use the first 512-bit argument X for chaining values as usual. In our notation (see Fig. 1) this means that $n = 512$. We use the 1024-bit argument Y (the message block in SHA-512) to input both a 512-bit message block and the key K which can be of any length $k \leq 512$ bits. If $k < 512$ then let the key be $K\|0^{512-k}$. That is, we define the keyed compression function $F_K : \{0,1\}^{512} \times \{0,1\}^{512} \rightarrow \{0,1\}^{512}$ needed in OMD as $F_K(H, M) = \text{SHA-512}(H, K\|0^{512-k}\|M)$.

The parameters of OMD-SHA512 are set as follows:

- The message block length in bits is $m = 512$; i.e. $|M_i| = 512$. *If needed*, we pad the final block of the message with 10^* (i.e., a single 1 followed by the minimal number of 0's needed) to make its length exactly 512 bits.
- The key length in bits can be $80 \leq k \leq 512$; but $k < 128$ is not recommended. *If needed*, we pad the key K with 0^{512-k} to make its length exactly 512 bits.
- The nonce (public message number) length in bits can be $96 \leq |N| \leq 511$. We *always* pad the nonce with $10^{511-|N|}$ to make its length exactly 512 bits.
- The secret message number length in bits is 0; that is, our scheme does not support secret message numbers.
- The associated data block length in bits is $2n = 1024$; i.e. $|A_i| = 1024$. *If needed*, we pad the final block of the associated data with 10^* (i.e., a single 1 followed by the minimal number of 0's needed) to make its length exactly 1024 bits.
- The tag length in bits can be $32 \leq \tau \leq 512$.

6.3 Instantiating G-OMD with a Native Tweakable PRF

Considering the proof steps of Theorem 1, it can be seen that the security bound for the core of OMD, called G-OMD (Fig. 3), is free from the quadratic degradation (birthday term) in the total number of blocks in queries; namely, $\frac{3\sigma^2}{2^n}$. This term is introduced only in Lemma 3, when we instantiated a tweakable PRF $\widetilde{F}$: $\mathcal{K} \times \mathcal{T} \times (\{0,1\}^n \times \{0,1\}^m) \rightarrow \{0,1\}^n$ from a PRF $F : \mathcal{K} \times (\{0,1\}^n \times \{0,1\}^m) \rightarrow \{0,1\}^n$ using the XE masking method; i.e., $\widetilde{F}_K^{\langle T \rangle}(X, Y) = F_K(X \oplus \Delta_K(T), Y)$. This instantiation aimed to provide a large range for the allowed key length and the nonce length in instantiations of OMD using practical compression functions, regarding that most practical compression functions (e.g. those of the SHA family) do not have a dedicated key input and must be keyed by allocating some part of their input for the key.

To avoid such a degradation in the security bound, caused by the XE method, one can directly instantiate the tweakable PRF $\widetilde{F}$ from a PRF $F' : \mathcal{K} \times (\{0,1\}^n \times \{0,1\}^{m'}) \rightarrow \{0,1\}^n$, where $m' = m + |T|$, by defining $\widetilde{F}_K^{\langle T \rangle}(X, Y) = F_K'(X, T\|Y)$. In practice, using an off-the-shelf (keyless) compression function with fixed input and output sizes, this will imply supporting a more restricted range of parameters' sizes (key length and nonce length) compared to OMD. Fortunately, the allowed parameter sizes for would be still large enough for most of applications today, if one simply use the compression functions of SHA-256 or SHA-512. Let's call the G-OMD instantiated with these compression functions, G-OMD-SHA256 and G-OMD-SHA512, respectively.

For G-OMD-SHA256 and G-OMD-SHA512, we can define the tweakable function $\widetilde{F}_K^{\langle T \rangle}(X,Y)$ by $\widetilde{F}_K^{\langle T \rangle}(X,Y) = \text{SHA-256}(X, K||T||0^{256-k-|T|}||M)$ and $\widetilde{F}_K^{\langle T \rangle}(X,Y) = \text{SHA-512}(X, K||T||0^{512-k-|T|}||M)$, respectively. Considering that $T = (N, i, j)$, where N is the nonce, i is the block number, and $j \in \{0, 1, 2\}$, G-OMD-SHA256 and G-OMD-SHA512 will allow k-bit keys, $|N|$-bit nonce, ℓ-block messages and a-block associated data, subject to the constraints $k + |N| + max(\log \ell, \log a) + 2 \leq 256$ and $k + |N| + max(\log \ell, \log a) + 2 \leq 512$, respectively.

Acknowledgments. We would like to thank the anonymous reviewers of SAC 2014 for their constructive comments. The EPFL team was partially supported by Microsoft Research under MRL Contract No. 2014-006 (DP1061305).

References

1. Secure Hash Standard (SHS). NIST FIPS PUB 180–4, Mar 2012
2. Bellare, M.: New proofs for NMAC and HMAC: security without collision-resistance. IACR Cryptology ePrint Archive 2006, 43 (2006)
3. Bellare, M., Desai, A., Jokipii, E., Rogaway, P.: A concrete security treatment of symmetric encryption. In: FOCS, pp. 394–403 (1997)
4. Bellare, M., Namprempre, C.: Authenticated encryption: relations among notions and analysis of the generic composition paradigm. In: Okamoto, T. (ed.) ASI-ACRYPT 2000. LNCS, vol. 1976, pp. 531–545. Springer, Heidelberg (2000)
5. Bellare, M., Namprempre, C.: Authenticated encryption: relations among notions and analysis of the generic composition paradigm. J. Cryptol. **21**(4), 469–491 (2008)
6. Bellare, M., Rogaway, P.: Encode-then-encipher encryption: how to exploit nonces or redundancy in plaintexts for efficient cryptography. In: Okamoto, T. (ed.) ASI-ACRYPT 2000. LNCS, vol. 1976, pp. 317–330. Springer, Heidelberg (2000)
7. Bernstein, D.J.: Cryptographic competitions: CAESAR. http://competitions.cr.yp.to
8. Canvel, B., Hiltgen, A.P., Vaudenay, S., Vuagnoux, M.: Password interception in a SSL/TLS channel. In: Boneh, D. (ed.) CRYPTO 2003. LNCS, vol. 2729, pp. 583–599. Springer, Heidelberg (2003)
9. Chakraborty, D., Sarkar, P.: A general construction of tweakable block ciphers and different modes of operations. IEEE Trans. Inf. Theory **54**(5), 1991–2006 (2008)
10. Cogliani, S., Maimut, D., Naccache1, D., do Canto, R.P., Reyhanitabar, R., Vaudenay, S., Vizár, D.: Offset Merkle-Damgård (OMD) version 1.0: A CAESAR Proposal, Mar 2014. http://competitions.cr.yp.to/round1/omdv10.pdf
11. Fleischmann, E., Forler, C., Lucks, S.: McOE: a family of almost foolproof on-line authenticated encryption schemes. In: Canteaut, A. (ed.) FSE 2012. LNCS, vol. 7549, pp. 196–215. Springer, Heidelberg (2012)

12. Guilford, J., Cote, D., Gopal, V.: Fast SHA512 Implementations on Intel®
    Architecture Processors, Nov 2012. http://www.intel.com/content/www/us/en/
    intelligent-systems/intel-technology/fast-sha512-implementations-ia-processors-
    paper.html
13. Guilford, J., Yap, K., Gopal, V.: Fast SHA-256 Implementations on Intel®
    Architecture Processors, May 2012. http://www.intel.com/content/www/us/en/
    intelligent-systems/intel-technology/sha-256-implementations-paper.html
14. Gulley, S., Gopal, V., Yap, K., Feghali, W., Guilford, J., Wolrich, G.: Intel® SHA
    Extensions: New Instructions Supporting the Secure Hash Algorithm on Intel®
    Architecture Processors, Jul 2013. https://software.intel.com/sites/default/files/
    article/402097/intel-sha-extensions-white-paper.pdf
15. Iwata, T.: New blockcipher modes of operation with beyond the birthday bound
    security. In: Robshaw, M. (ed.) FSE 2006. LNCS, vol. 4047, pp. 310–327. Springer,
    Heidelberg (2006)
16. Iwata, T.: Authenticated encryption mode for beyond the birthday bound secu-
    rity. In: Vaudenay, S. (ed.) AFRICACRYPT 2008. LNCS, vol. 5023, pp. 125–142.
    Springer, Heidelberg (2008)
17. Katz, J., Yung, M.: Unforgeable encryption and chosen ciphertext secure modes of
    operation. In: Schneier, B. (ed.) FSE 2000. LNCS, vol. 1978, pp. 284–299. Springer,
    Heidelberg (2001)
18. Krovetz, T., Rogaway, P.: The software performance of authenticated-encryption
    modes. In: Joux, A. (ed.) FSE 2011. LNCS, vol. 6733, pp. 306–327. Springer,
    Heidelberg (2011)
19. Landecker, W., Shrimpton, T., Terashima, R.S.: Tweakable blockciphers with
    beyond birthday-bound security. In: Safavi-Naini, R., Canetti, R. (eds.) CRYPTO
    2012. LNCS, vol. 7417, pp. 14–30. Springer, Heidelberg (2012)
20. Lefranc, D., Painchault, P., Rouat, V., Mayer, E.: A generic method to design
    modes of operation beyond the birthday bound. In: Adams, C., Miri, A., Wiener,
    M. (eds.) SAC 2007. LNCS, vol. 4876, pp. 328–343. Springer, Heidelberg (2007)
21. Namprempre, C., Rogaway, P., Shrimpton, T.: Reconsidering generic composition.
    In: Nguyen, P.Q., Oswald, E. (eds.) EUROCRYPT 2014. LNCS, vol. 8441, pp.
    257–274. Springer, Heidelberg (2014)
22. Rogaway, P.: Authenticated-encryption with associated-data. In: ACM Conference
    on Computer and Communications Security, pp. 98–107 (2002)
23. Rogaway, P.: Efficient instantiations of tweakable blockciphers and refinements to
    modes OCB and PMAC. In: Lee, P.J. (ed.) ASIACRYPT 2004. LNCS, vol. 3329,
    pp. 16–31. Springer, Heidelberg (2004)
24. Rogaway, P.: Nonce-based symmetric encryption. In: Roy, B., Meier, W. (eds.) FSE
    2004. LNCS, vol. 3017, pp. 348–359. Springer, Heidelberg (2004)
25. Rogaway, P., Bellare, M., Black, J., Krovetz, T.: OCB: a block-cipher mode of
    operation for efficient authenticated encryption. In: ACM Conference on Computer
    and Communications Security, pp. 196–205 (2001)
26. Rogaway, P., Shrimpton, T.: A provable-security treatment of the key-wrap prob-
    lem. In: Vaudenay, S. (ed.) EUROCRYPT 2006. LNCS, vol. 4004, pp. 373–390.
    Springer, Heidelberg (2006)
27. Vaudenay, S.: Security flaws induced by CBC padding - applications to SSL,
    IPSEC, WTLS. In: Knudsen, L.R. (ed.) EUROCRYPT 2002. LNCS, vol. 2332,
    pp. 534–546. Springer, Heidelberg (2002)

Security Amplification
for the Composition of Block Ciphers:
Simpler Proofs and New Results

Benoit Cogliati[1], Jacques Patarin[1], and Yannick Seurin[2]([✉])

[1] University of Versailles, Versailles, France
benoit.cogliati@ens.uvsq.fr, jacques.patarin@uvsq.fr
[2] ANSSI, Paris, France
yannick.seurin@m4x.org

Abstract. Security amplification results for block ciphers typically state that cascading (i.e., composing with independent keys) two (or more) block ciphers yields a new block cipher that offers better security against some class of adversaries and/or that resists stronger adversaries than each of its components. One of the most important results in this respect is the so-called "two weak make one strong" theorem, first established up to logarithmic terms by Maurer and Pietrzak (TCC 2004), and later optimally tightened by Maurer, Pietrzak, and Renner (CRYPTO 2007), which states that, in the information-theoretic setting, cascading F and G^{-1}, where F and G are respectively (q, ε_F)-secure and (q, ε_G)-secure against non-adaptive chosen-plaintext (NCPA) attacks, yields a block cipher which is $(q, \varepsilon_F + \varepsilon_G)$-secure against adaptive chosen-plaintext and ciphertext (CCA) attacks. The first contribution of this work is a surprisingly simple proof of this theorem, relying on Patarin's H-coefficient method. We then extend our new proof to obtain new results (still in the information-theoretic setting). In particular, we prove a new composition theorem (which can be seen as the generalization of the "two weak make one strong" theorem to the composition of $n > 2$ block ciphers) which provides both amplification of the advantage and strengthening of the distinguisher's class in some optimal way (indeed we prove that our new composition theorem is tight up to some constant).

Keywords: Block cipher · Security amplification · Cascade · Composition · Provable security

1 Introduction

SECURITY AMPLIFICATION FOR BLOCK CIPHERS. The usual security notion for a block cipher E is *pseudorandomness*, which measures the (in-)ability of

The author 'Y. Seurin' was partially supported by the French National Agency of Research through the BLOC project (contract ANR-11-INS-011).

© Springer International Publishing Switzerland 2014
A. Joux and A. Youssef (Eds.): SAC 2014, LNCS 8781, pp. 129–146, 2014.
DOI: 10.1007/978-3-319-13051-4_8

an adversary (the *distinguisher*) which is given oracle access to a permutation (and potentially its inverse) to tell whether it is interacting with the block cipher E_K for some randomly drawn key K or with a truly random permutation. One usually classifies distinguishers according to the way they can issue their queries. A distinguisher which can only make direct (plaintext) queries to the permutation oracle is called a CPA-distinguisher, whereas it is called a CCA-distinguisher when it can make both direct and inverse (ciphertext) queries. Both types come in a non-adaptive variant (NCPA and NCCA respectively), i.e., the adversary must choose all its queries before receiving any answer from the permutation oracle. A block cipher is said to be (q, ε)-ATK secure when no distinguisher in the attack class ATK (for instance NCPA, etc.) making at most q oracle queries can distinguish E_K from a truly random permutation with advantage better than ε.

The security amplification problem is to determine whether adequately combining some mildly secure block ciphers $E_1, \ldots, E_n$ can yield a block cipher F with stronger security guarantees than each of its components. (This question naturally extends to other cryptographic primitives such as pseudorandom generators or pseudorandom functions, but in this paper we focus on pseudorandom permutations, i.e., block ciphers.) Here, "stronger" security guarantees might mean either that F has a smaller distinguishing advantage in face of some fixed class of distinguishers than each component E_i (something we will informally refer to as *ε-amplification*), or that F can withstand attacks from a stronger class of adversaries than each of its components (something we will call *class-amplification*). We clarify this distinction with a prominent example of each type of result.

The classical example of an ε-amplification result states that cascading two block ciphers F and G which are respectively (q, ε_F)- and (q, ε_G)-NCPA (resp. CPA) secure yields a block cipher which is $(q, 2\varepsilon_F \varepsilon_G)$-NCPA (resp. CPA) secure. Hence, when $\varepsilon_F, \varepsilon_G < 1/2$, the new block cipher is indeed strictly more secure than each of its components. This was proved (in the information-theoretic setting, i.e., when considering computationally unbounded adversaries) by Vaudenay (see [Vau98] for the non-adaptive case and [Vau99] for the adaptive case) using the *decorrelation theory* framework [Vau03]. (See also [KNR09, Theorem 3.8] for a different proof for self-composition in the non-adaptive case.) A computational analogue of this result was later proved by Maurer and Tessaro [MT09].

For the class-amplification type of results, one of the most notable examples is what we will refer to as the "two weak make one strong" (*2W1S* for short) theorem, which states that if F and G are resp. (q, ε_F)- and (q, ε_G)-NCPA secure, then the composition $G^{-1} \circ F$ is $(q, \varepsilon_F + \varepsilon_G)$-CCA secure (a result which is tight in general). Note that here, the resulting cipher withstands much stronger attacks than each component F and G, but its CCA advantage is strictly larger than each of the NCPA advantages of F and G. This theorem was first proved up to logarithmic terms by Maurer and Pietrzak [MP04], while the tight version was later proved by Maurer, Pietrzak, and Renner [MPR07] using the framework of random systems [Mau02]. We stress that this result only holds in the information-theoretic setting. In the computational setting, the composition of non-adaptively secure block ciphers does not, in general, yield an adaptively secure one [Mye04, Pie05a], though some partial positive results are known [LR86, Pie06].

OUR CONTRIBUTION.   The starting point of our work is a surprisingly simple proof of the 2W1S theorem. Our new technique relies on simple manipulations of transition probabilities (which are nothing else, up to some normalization factors, than the H-coefficients of Patarin [Pat08]) and eschews completely the heavy machinery of the random systems framework [Mau02] on which the only previously known proof was based [MPR07]. We think that having an elementary proof of an important result (on which a number of subsequent papers rely, notably in coupling-based security proofs [MRS09, HR10, LPS12, LS14]) is an interesting contribution in itself. To emphasize our point, we stress that a crucial lemma of the random systems framework (namely Theorem 2 of [Mau02]), to which the proof of the 2W1S theorem of [MPR07] appeals, was later found to be incorrectly stated (and also that the only known proof of this lemma in [Pie05b] was flawed) by Jetchev *et al.* [JÖS12]. Hence, the 2W1S theorem can only be considered formally proven by combining results from three different papers [Mau02, MPR07, JÖS12], a somehow unsatisfying state of affairs.

Motivated by our findings, we consider the following problem: given three (or more) block ciphers which are (q, ε)-NCPA secure, can we get both ε-amplification *and* class-amplification at the same time, i.e., a composed block cipher which is (q, ε')-CCA secure for $\varepsilon' < \varepsilon$, in some optimal manner?[1] Focusing on self-composition for simplicity, consider a block cipher E such that both E and E^{-1} are (q, ε)-NCPA secure.[2] What can we say about the CCA-security of the n-fold composition E^n? Using known results, a straightforward answer (assuming n even) can be obtained by first (recursively) applying the ε-amplification theorem for NCPA-secure block ciphers to each half of the cascade, thereby getting

$$\mathbf{Adv}^{\mathrm{ncpa}}_{E^{n/2}}(q) \leq 2^{\frac{n}{2}-1}\varepsilon^{\frac{n}{2}} \quad \text{and} \quad \mathbf{Adv}^{\mathrm{ncpa}}_{(E^{n/2})^{-1}}(q) \leq 2^{\frac{n}{2}-1}\varepsilon^{\frac{n}{2}},$$

and then the 2W1S theorem to obtain

$$\mathbf{Adv}^{\mathrm{cca}}_{E^n}(q) \leq \mathbf{Adv}^{\mathrm{ncpa}}_{E^{n/2}}(q) + \mathbf{Adv}^{\mathrm{ncpa}}_{(E^{n/2})^{-1}}(q) \leq (2\varepsilon)^{\frac{n}{2}}.$$

For n odd, a similar reasoning yields (by cutting E^n into two unbalanced halves)

$$\mathbf{Adv}^{\mathrm{cca}}_{E^n}(q) \leq \mathbf{Adv}^{\mathrm{ncpa}}_{E^{(n+1)/2}}(q) + \mathbf{Adv}^{\mathrm{ncpa}}_{(E^{(n-1)/2})^{-1}}(q) \leq 2^{\frac{n-1}{2}}\varepsilon^{\frac{n+1}{2}} + 2^{\frac{n-3}{2}}\varepsilon^{\frac{n-1}{2}}.$$

---

[1] This requires at least three block ciphers since the 2W1S theorem is tight. Hence, in general, from two (q, ε)-NCPA secure block ciphers F and G, one can at best obtain a $(q, 2\varepsilon)$-CCA secure one.

[2] A larger number of block cipher designs have similar provable security in the direct and inverse direction because of their involution-like structure, for example balanced Feistel schemes.

In particular, for $n = 3$, the best one can prove from previous results is that

$$\mathbf{Adv}_{E^3}^{\mathrm{cca}} \leq \varepsilon + 2\varepsilon^2.$$

Hence, one gets (provable) ε-amplification only for $n \geq 4$, assuming $\varepsilon < 1/4$.

In this paper, we prove that the CCA-security of E^n is actually much better, namely

$$\mathbf{Adv}_{E^n}^{\mathrm{cca}}(q) \leq (2\varepsilon)^{n-1}.$$

Hence, for $n \geq 3$, this provides both ε-amplification *and* class-amplification as soon as

$$\varepsilon < \frac{1}{2 \cdot 2^{1/(n-2)}}$$

(hence, in particular as soon as $\varepsilon < 1/4$ for any $n \geq 3$). In fact we prove a more general theorem (see Theorem 2) which also implies the following interesting corollary. Let E, F, G be three block ciphers such that E, F, F^{-1} and G^{-1} are (q, ε)-NCPA secure. Then the composition $G \circ F \circ E$ is $(q, 4\varepsilon^2)$-CCA secure.

A WORD OF INTERPRETATION.   Our new result has some interesting implications regarding the superiority of triple- versus double-encryption. This fact has already been widely analyzed in the ideal cipher model [ABCV98, BR06]. Our new theorem may be seen as yet another expression of this phenomenon in the standard, information-theoretic setting. For concreteness, assume that we have at hand a block cipher E such that E and E^{-1} are only, say, $(2^{40}, 2^{-30})$-NCPA secure, a mild security insurance by current standards. Using double-encryption, one "restores" NCPA-security (since E^2 and $(E^2)^{-1}$ are ensured to be $(2^{40}, 2^{-59})$-NCPA secure) but in general one cannot exclude that a CCA-attack will break E^2 with 2^{40} queries and advantage 2^{-30}. On the other hand, triple-encryption is good enough here, since our new result shows that E^3 is $(2^{40}, 2^{-58})$-CCA secure.

RELATED WORK.   The topic of security amplification is too broad to be entirely covered here. Restricting our attention to block cipher security amplification, we mention that a long line of work considered provable security results for cascade encryption *in the ideal cipher model* [ABCV98, BR06, GM09, Lee13], which is quite orthogonal to our setting: working in the ideal cipher model is in some sense equivalent to upper bounding the knowledge of the adversary on the underlying block cipher(s) (since it can only make a limited number of ideal cipher queries), whereas we consider computationally unbounded adversaries, in the standard, non-idealized model (in particular, the adversary has complete knowledge of the underlying block cipher(s), and may, e.g., represent them as a huge look-up table).

ORGANIZATION.   We start with useful definitions and the necessary background on transition probabilities and how these quantities are related to the advantage against different classes of distinguishers in Sect. 2. In Sect. 3, we give our new and substantially simpler proof of the 2W1S theorem. Then, in Sect. 4, we extend this result to the general case of the composition of $n \geq 2$ non-adaptively secure block ciphers (we treat the special case $n = 3$ in the full version of the paper [CPS14]). Finally, in Sect. 5, we show that our new result is tight up to some constant.

2   Preliminaries

2.1   Notation and Definitions

Given a non-empty set S, the set of all permutations of S is denoted $\mathsf{Perm}(S)$. We write $s \leftarrow_\$ S$ to mean that a value is sampled uniformly at random from S and assigned to s.

Definition 1 (Statistical Distance). *Let Ω be a finite event space and let μ and ν be two probability distributions defined on Ω. The* statistical distance *(or* total variation distance*) between μ and ν, denoted $\|\mu - \nu\|$ is defined as:*

$$\|\mu - \nu\| = \frac{1}{2} \sum_{\omega \in \Omega} |\mu(\omega) - \nu(\omega)|.$$

The following definitions can easily be seen equivalent:

$$\|\mu - \nu\| = \max_{S \subseteq \Omega} \{\mu(S) - \nu(S)\} = \max_{S \subseteq \Omega} \{\nu(S) - \mu(S)\} = \max_{S \subseteq \Omega} \{|\mu(S) - \nu(S)|\}.$$

COMPOSITION OF BLOCK CIPHERS. Let $\mathcal{M}$ and $\mathcal{K}$ be two sets. A block cipher with message space $\mathcal{M}$ and key space $\mathcal{K}$ is a mapping $E : \mathcal{K} \times \mathcal{M} \to \mathcal{M}$ such that for any $K \in \mathcal{K}$, the partial mapping $E(K, \cdot)$ is a permutation of $\mathcal{M}$. We interchangeably use the notation $E_K(x)$ for $E(K, x)$, the inverse of E_K being denoted E_K^{-1}. Given two block ciphers E and F with the same message space $\mathcal{M}$ and respective key spaces $\mathcal{K}_E$ and $\mathcal{K}_F$, we denote $F \circ E$ the block cipher with message space $\mathcal{M}$ and key space $\mathcal{K}_E \times \mathcal{K}_F$ defined as

$$F \circ E_{(K_E, K_F)}(x) = F_{K_F}(E_{K_E}(x)).$$

We call $F \circ E$ interchangeably the *composition* or the *cascade* of E and F. This definition extends straightforwardly to the composition of $n > 2$ block ciphers. We denote E^n the n-fold self-composition of E (with independent keys).

2.2   Security Definitions and Classical Lemmas

Fix some message space $\mathcal{M}$ and denote $M = |\mathcal{M}|$. We denote $(\mathcal{M})_q$ the set of all q-tuple of pairwise distinct elements of $\mathcal{M}$. Let E be a block cipher with message space $\mathcal{M}$ and key space $\mathcal{K}_E$. Given an integer $q \geq 1$ and two q-tuples $x = (x_1, \ldots, x_q) \in (\mathcal{M})_q$ and $y = (y_1, \ldots, y_q) \in (\mathcal{M})_q$ of pairwise distinct elements of $\mathcal{M}$, we denote

$$\mathsf{p}_E(x, y) = \Pr\left[K \leftarrow_\$ \mathcal{K}_E : E_K(x) = y\right] = \frac{|\{K \in \mathcal{K}_E : E_K(x) = y\}|}{|\mathcal{K}_E|},$$

where the notation $E_K(x) = y$ is a shorthand meaning that $E_K(x_i) = y_i$ for all $1 \leq i \leq q$. We also denote

$$\mathsf{p}^* = \Pr\left[P \leftarrow_\$ \mathsf{Perm}(\mathcal{M}) : P(x) = y\right] = \frac{1}{M(M-1)\cdots(M-q+1)}.$$

When x is fixed,
$$\mathsf{p}_{E,x} : y \mapsto \mathsf{p}_E(x, y)$$

is the probability distribution (over the choice of a uniformly random key $K \leftarrow_\$ \mathcal{K}_E$) of the q-tuple of ciphertexts when E receives the q-tuple of plaintexts x. Similarly, when y is fixed,

$$\mathsf{p}_{E^{-1},y} : x \mapsto \mathsf{p}_E(x, y)$$

is the probability distribution of the q-tuples of plaintexts when E^{-1} receives the q-tuple of ciphertexts y. Overloading the notation, p^* will also denote the uniform probability distribution over $(\mathcal{M})_q$. Note that for any $x = (x_1, \ldots, x_q) \in (\mathcal{M})_q$ and any $y = (y_1, \ldots, y_q) \in (\mathcal{M})_q$,

$$\sum_{z \in (\mathcal{M})_q} (\mathsf{p}_E(x, z) - \mathsf{p}^*) = \sum_{z \in (\mathcal{M})_q} (\mathsf{p}_E(z, y) - \mathsf{p}^*) = 0. \tag{1}$$

Let $\mathcal{D}$ be a distinguisher with (potentially two-sided) oracle access to some permutation $P \in \mathsf{Perm}(\mathcal{M})$, whose goal is to distinguish whether it is interacting with $E_K(\cdot)$ for some random key $K \leftarrow_\$ \mathcal{K}$, or with a uniformly random permutation $P \leftarrow_\$ \mathsf{Perm}(\mathcal{M})$. We classify distinguishers according to the type of attacks they can perform:

- chosen-plaintext attacks (CPA), where $\mathcal{D}$ can only make direct (i.e., plaintext) queries to the permutation oracle,
- and chosen-plaintext and ciphertext attacks (CCA), where $\mathcal{D}$ can make both direct and inverse (i.e., ciphertext) queries to the permutation oracle.

Additionally, we also consider the non-adaptive variants of these two types of attacks, namely NCPA and NCCA, where the distinguisher must choose all its queries before receiving any answer from the permutation oracle. We consider computationally unbounded distinguishers, and we assume *wlog* that the distinguisher is deterministic and never makes redundant queries.

The distinguishing advantage of $\mathcal{D}$ is defined as

$$\mathbf{Adv}(\mathcal{D}) = \left| \Pr\left[K \leftarrow_\$ \mathcal{K} : \mathcal{D}^{E_K} = 1\right] - \Pr\left[P \leftarrow_\$ \mathsf{Perm}(\mathcal{M}) : \mathcal{D}^P = 1\right]\right|,$$

where, depending on the type of the distinguisher, $\mathcal{D}$ can make one-sided or two-sided queries to the permutation oracle. For q a non-negative integer, the insecurity (or advantage) of E against ATK-attacks, where ATK $\in \{(\mathrm{N})\mathrm{CPA}, (\mathrm{N})\mathrm{CCA}\}$ is defined as

$$\mathbf{Adv}_E^{\mathrm{atk}}(q) = \max_{\mathcal{D}} \mathbf{Adv}(\mathcal{D}),$$

where the maximum is taken over all distinguishers $\mathcal{D}$ of type ATK making at most q oracle queries. We say that E is (q, ε)-ATK secure if $\mathbf{Adv}_E^{\mathrm{atk}}(q) \leq \varepsilon$.

Our analysis will rely on the H-coefficient method, first introduced by Patarin to prove the strong pseudorandomness of the 4-round Feistel scheme [Pat90, Pat91, Pat08]. We recall the two fundamental results of the H-coefficient method, regarding NCPA and CCA distinguishers respectively. For completeness, we give a proof of these results in Appendix A.

Lemma 1 (NCPA security). *Let E be a block cipher with message space $\mathcal{M}$. Then*

$$\mathbf{Adv}_E^{\mathrm{ncpa}}(q) = \max_{x \in (\mathcal{M})_q} \| \mathsf{p}_{E,x} - \mathsf{p}^* \|.$$

Lemma 2 (CCA security). *Let E be a block cipher with message space $\mathcal{M}$. Assume that there exists ε such that for any q-tuples $x, y \in (\mathcal{M})_q$, one has*

$$\mathsf{p}_E(x, y) \geq (1 - \varepsilon)\mathsf{p}^*.$$

Then

$$\mathbf{Adv}_E^{\mathrm{cca}}(q) \leq \varepsilon.$$

3   A Simple Proof of the "Two Weak Make One Strong" Theorem

In this section, we derive in a straightforward manner the "two weak make one strong" theorem [MP04, MPR07]. We start by giving a handful expression for the quantity $\mathsf{p}_{F \circ E}(x, y)$.

Lemma 3. *Let E and F be two block ciphers with the same message space $\mathcal{M}$ and respective key spaces $\mathcal{K}_E$ and $\mathcal{K}_F$. Then for any q-tuples x and y of pairwise distinct elements of $\mathcal{M}$, one has*

$$\mathsf{p}_{F \circ E}(x, y) = \mathsf{p}^* + \sum_{z \in (\mathcal{M})_q} (\mathsf{p}_E(x, z) - \mathsf{p}^*)(\mathsf{p}_F(z, y) - \mathsf{p}^*). \tag{2}$$

Proof. One has

$$\mathsf{p}_{F \circ E}(x, y) = \sum_{z \in (\mathcal{M})_q} \mathsf{p}_E(x, z)\mathsf{p}_F(z, y)$$

$$= \sum_z (\mathsf{p}_E(x, z) - \mathsf{p}^* + \mathsf{p}^*)(\mathsf{p}_F(z, y) - \mathsf{p}^* + \mathsf{p}^*)$$

$$= \sum_z (\mathsf{p}_E(x, z) - \mathsf{p}^*)(\mathsf{p}_F(z, y) - \mathsf{p}^*)$$

$$+ \mathsf{p}^* \underbrace{\sum_z (\mathsf{p}_E(x, z) - \mathsf{p}^*)}_{=0 \text{ by (1)}} + \mathsf{p}^* \underbrace{\sum_z (\mathsf{p}_F(z, y) - \mathsf{p}^*)}_{=0 \text{ by (1)}} + \underbrace{\sum_z (\mathsf{p}^*)^2}_{=\mathsf{p}^*}$$

$$= \mathsf{p}^* + \sum_z (\mathsf{p}_E(x, z) - \mathsf{p}^*)(\mathsf{p}_F(z, y) - \mathsf{p}^*),$$

from which the result follows. $\square$

The next step is to lower bound the sum appearing in the right hand-side of (2). Note that this term is exactly a covariance term. In particular, one could use the Cauchy-Schwarz inequality to get

$$\left| \sum_{z \in (\mathcal{M})_q} (\mathsf{p}_E(x, z) - \mathsf{p}^*)(\mathsf{p}_F(z, y) - \mathsf{p}^*) \right|$$

$$\leq \sqrt{\sum_{z \in (\mathcal{M})_q} (\mathsf{p}_E(x, z) - \mathsf{p}^*)^2} \sqrt{\sum_{z \in (\mathcal{M})_q} (\mathsf{p}_F(z, y) - \mathsf{p}^*)^2}.$$

However, the quantities appearing in the right hand-side involve the Euclidean distance between $\mathsf{p}_{E,x}$ (resp. $\mathsf{p}_{F^{-1},y}$) and p^*, which to the best of our knowledge is not related to any standard attack. Hence we prove in the next lemma a different bound which involves the statistical distance instead, which, as recalled in Lemma 1, is related to NCPA attacks.

Lemma 4. *Let E and F be two block ciphers with the same message space $\mathcal{M}$ and respective key spaces $\mathcal{K}_E$ and $\mathcal{K}_F$. Then for any q-tuples x and y of pairwise distinct elements of $\mathcal{M}$, one has*

$$\sum_{z \in (\mathcal{M})_q} (\mathsf{p}_E(x, z) - \mathsf{p}^*)(\mathsf{p}_F(z, y) - \mathsf{p}^*) \geq -\mathsf{p}^* \left(\|\mathsf{p}_{E,x} - \mathsf{p}^*\| + \|\mathsf{p}_{F^{-1},y} - \mathsf{p}^*\| \right).$$

Proof. Let

$$S \stackrel{\text{def}}{=} \sum_{z \in (\mathcal{M})_q} (\mathsf{p}_E(x, z) - \mathsf{p}^*)(\mathsf{p}_F(z, y) - \mathsf{p}^*) = \sum_{z \in (\mathcal{M})_q} (\mathsf{p}_{E,x}(z) - \mathsf{p}^*)(\mathsf{p}_{F^{-1},y}(z) - \mathsf{p}^*).$$

To simplify notation, we rename the probability distributions as $\mu := \mathsf{p}_{E,x}$ and $\nu := \mathsf{p}_{F^{-1},y}$. Then, keeping only the negative terms in the sum, we have

$$S \geq \sum_{z \in (\mathcal{M})_q : \left\{ \begin{smallmatrix} \mu(z) > \mathsf{p}^* \\ \nu(z) < \mathsf{p}^* \end{smallmatrix} \right.} (\mu(z) - \mathsf{p}^*)(\nu(z) - \mathsf{p}^*)$$

$$+ \sum_{z \in (\mathcal{M})_q : \left\{ \begin{smallmatrix} \mu(z) < \mathsf{p}^* \\ \nu(z) > \mathsf{p}^* \end{smallmatrix} \right.} (\mu(z) - \mathsf{p}^*)(\nu(z) - \mathsf{p}^*)$$

$$\geq \sum_{z \in (\mathcal{M})_q : \left\{ \begin{smallmatrix} \mu(z) > \mathsf{p}^* \\ \nu(z) < \mathsf{p}^* \end{smallmatrix} \right.} (\mu(z) - \mathsf{p}^*)(-\mathsf{p}^*) + \sum_{z \in (\mathcal{M})_q : \left\{ \begin{smallmatrix} \mu(z) < \mathsf{p}^* \\ \nu(z) > \mathsf{p}^* \end{smallmatrix} \right.} (-\mathsf{p}^*)(\nu(z) - \mathsf{p}^*)$$

$$= -\mathsf{p}^* \left(\sum_{z \in (\mathcal{M})_q : \left\{ \begin{smallmatrix} \mu(z) > \mathsf{p}^* \\ \nu(z) < \mathsf{p}^* \end{smallmatrix} \right.} (\mu(z) - \mathsf{p}^*) + \sum_{z \in (\mathcal{M})_q : \left\{ \begin{smallmatrix} \mu(z) < \mathsf{p}^* \\ \nu(z) > \mathsf{p}^* \end{smallmatrix} \right.} (\nu(z) - \mathsf{p}^*) \right)$$

$$\geq -\mathsf{p}^* (\|\mu - \mathsf{p}^*\| + \|\nu - \mathsf{p}^*\|),$$

where for the last inequality we used that

$$\|\mu - \mathsf{p}^*\| = \max_{S \subseteq (\mathcal{M})_q} \sum_{z \in S} (\mu(z) - \mathsf{p}^*)$$

(and the analogue equality for ν). This proves the result.   □

We can finally prove the "two weak make one strong" composition theorem.

Theorem 1. *Let E and F be two block ciphers with the same message space $\mathcal{M}$. For any integer q, one has*

$$\mathbf{Adv}_{F \circ E}^{\mathrm{cca}}(q) \leq \mathbf{Adv}_{E}^{\mathrm{ncpa}}(q) + \mathbf{Adv}_{F^{-1}}^{\mathrm{ncpa}}(q).$$

Proof. Fix any q-tuples $x, y \in (\mathcal{M})_q$. Then

$$\mathsf{p}_{F \circ E}(x, y) = \mathsf{p}^* + \sum_{z \in (\mathcal{M})_q} (\mathsf{p}_E(x, z) - \mathsf{p}^*)(\mathsf{p}_F(z, y) - \mathsf{p}^*) \qquad \text{(Lemma 3)}$$

$$\geq \mathsf{p}^* - \mathsf{p}^* \left(\|\mathsf{p}_{E,x} - \mathsf{p}^*\| + \|\mathsf{p}_{F^{-1}, y} - \mathsf{p}^*\| \right) \qquad \text{·(Lemma 4)}$$

$$\geq \mathsf{p}^*(1 - \mathbf{Adv}_{E}^{\mathrm{ncpa}}(q) - \mathbf{Adv}_{F^{-1}}^{\mathrm{ncpa}}(q)). \qquad \text{(Lemma 1)}$$

The result follows by Lemma 2.   □

To illustrate the usefulness of Eq. (2), we give a simple proof of the ε-amplification theorem for NCPA-secure ciphers [Vau98], as well as an amplification theorem for security against known-plaintext attacks (KPA), in the full version of this paper [CPS14].

4   Many Weak Make One Even Stronger

Let $n \geq 1$ be an integer. In this section, we extend Theorem 1 to the composition of n block ciphers (the special case $n = 3$ is treated in details in the full version of this paper [CPS14]).

We start by generalizing Lemma 3.

Lemma 5. *Let $E_1, \ldots, E_n$ be n block ciphers with the same message space $\mathcal{M}$. Then for any q-tuples x and y of pairwise distinct elements of $\mathcal{M}$, one has*

$$\mathsf{p}_{E_n \circ \cdots \circ E_1}(x, y) = \mathsf{p}^* + \sum_{x_1, \ldots, x_{n-1} \in (\mathcal{M})_q} \left(\prod_{i=1}^{n} (\mathsf{p}_{E_i}(x_{i-1}, x_i) - \mathsf{p}^*) \right) \qquad (3)$$

where $x_0 := x$ and $x_n := y$.

Proof. This result can be shown by induction. For $i \geq 1$, let (H_i) be the following proposition: for any $j \in \{1, \ldots, i\}$, for any block ciphers $E_1, \ldots, E_j$ with the same message space $\mathcal{M}$ and for any q-tuples x_0 and x_j of pairwise distinct elements of $\mathcal{M}$, one has

$$\mathsf{p}_{E_j \circ \cdots \circ E_1}(x_0, x_j) = \mathsf{p}^* + \sum_{x_1, \ldots, x_{j-1} \in (\mathcal{M})_q} \left(\prod_{i=1}^{j} (\mathsf{p}_{E_i}(x_{i-1}, x_i) - \mathsf{p}^*) \right).$$

Lemma 3 corresponds to (H_2).

Assume that (H_k) holds for an integer $k \geq 2$. Let $E_1, \ldots, E_{k+1}$ be block ciphers with the same message space $\mathcal{M}$ and $x_0, x_{k+1} \in (\mathcal{M})_q$. Then

$$\mathsf{p}_{E_{k+1} \circ \cdots \circ E_1}(x_0, x_{k+1})$$

$$= \mathsf{p}^* + \sum_{x_1 \in (\mathcal{M})_q} (\mathsf{p}_{E_1}(x_0, x_1) - \mathsf{p}^*)(\mathsf{p}_{E_{k+1} \circ \cdots \circ E_2}(x_1, x_{k+1}) - \mathsf{p}^*) \qquad (H_2)$$

$$= \mathsf{p}^* + \sum_{x_1 \in (\mathcal{M})_q} (\mathsf{p}_{E_1}(x_0, x_1) - \mathsf{p}^*) \sum_{\substack{x_2, \ldots, x_k \\ \in (\mathcal{M})_q}} \prod_{i=2}^{k+1} (\mathsf{p}_{E_i}(x_{i-1}, x_i) - \mathsf{p}^*) \qquad (H_k)$$

from which the result follows. □

We now have to study the sum appearing in the right hand-side of (3) in the same way as in the proof of Lemma 4, i.e., by splitting the sum according to the sign of each term of the product. In order to have a more compact notation, for a tuple $(t_0, \ldots, t_n) \in ((\mathcal{M})_q)^{n+1}$ and for each $i \in \{1, \ldots, n\}$ we denote:

- $C_{0,i}$ the inequality $\mathsf{p}_{E_i}(t_{i-1}, t_i) - \mathsf{p}* > 0$ and
- $C_{1,i}$ the inequality $\mathsf{p}_{E_i}(t_{i-1}, t_i) - \mathsf{p}* < 0$.

Then every part of the sum can be parametrized with a n-tuple $k = (k_1, \ldots, k_n)$ of integers in $\{0, 1\}$, the product being positive if and only if $k_1 + \ldots + k_n \equiv 0 \bmod 2$. Of course, the cases which have to be dealt carefully with are the ones where the product is negative (i.e., $k_1 + \ldots + k_n \equiv 1 \bmod 2$). This is what is done in the following lemma.

Lemma 6. *Let $E_1, \ldots, E_n$ be n block ciphers with the same message space $\mathcal{M}$ and $k = (k_1, \ldots, k_n) \in \{0, 1\}^n$ such that $k_1 + \ldots + k_n \equiv 1 \bmod 2$. For any fixed q-tuples t_0, t_n in $(\mathcal{M})_q$, denote*

$$A_k(t_0, t_n) := \{(t_1, \ldots, t_{n-1}) \in ((\mathcal{M})_q)^{n-1} \mid \forall i \in \{1, \ldots, n\}, C_{k_i, i} \text{ holds}\}.$$

Then

$$\sum_{t \in A_k(t_0, t_n)} \prod_{1 \leq i \leq n} (\mathsf{p}_{E_i}(t_{i-1}, t_i) - \mathsf{p}^*)$$

$$\geq -\mathsf{p}^* \max_{1 \leq i \leq n} \left(\prod_{1 \leq j \leq i-1} \mathbf{Adv}_{E_j}^{\mathrm{ncpa}}(q) \times \prod_{i+1 \leq j \leq n} \mathbf{Adv}_{E_j^{-1}}^{\mathrm{ncpa}}(q) \right).$$

Proof. Since $k_1 + \ldots + k_n \equiv 1 \bmod 2$, one can find an index j such that $k_j = 1$, i.e., $\mathsf{p}_{E_j}(t_{j-1}, t_j) - \mathsf{p}^* < 0$. Then, one has

$$\sum_{t \in A_k(t_0, t_n)} \prod_{1 \leq i \leq n} (\mathsf{p}_{E_i}(t_{i-1}, t_i) - \mathsf{p}^*) \geq -\mathsf{p}^* \sum_{t \in A_k(t_0, t_n)} \prod_{\substack{1 \leq i \leq n \\ i \neq j}} (\mathsf{p}_{E_i}(t_{i-1}, t_i) - \mathsf{p}^*).$$

In the sum appearing in the right hand-side, every term is positive since there is an even number of negative terms in each product. Hence,

$$\sum_{t \in A_k(t_0, t_n)} \prod_{1 \leq i \leq n} (\mathsf{p}_{E_i}(t_{i-1}, t_i) - \mathsf{p}^*) \geq -\mathsf{p}^* \sum_{t \in A_k(t_0, t_n)} \prod_{\substack{1 \leq i \leq n \\ i \neq j}} |\mathsf{p}_{E_i}(t_{i-1}, t_i) - \mathsf{p}^*|.$$

Let

$$B := \{(t_1, \ldots, t_{j-1}) \in ((\mathcal{M})_q)^{j-1} \,|\, \forall i \in \{1, \ldots, j-1\}, C_{k_i, i} \text{ holds}\} \text{ and}$$
$$C := \{(t_j, \ldots, t_{n-1}) \in ((\mathcal{M})_q)^{n-j} \,|\, \forall i \in \{j+1, \ldots, n\}, C_{k_i, i} \text{ holds}\}.$$

One has $A_k(t_0, t_n) \subseteq B \times C$ since the only difference between the sets is that in $B \times C$ we dropped the requirement that $C_{k_j, j}$ (i.e., inequality $\mathsf{p}_{E_j}(t_{j-1}, t_j) < \mathsf{p}^*$) holds. Hence,

$$\sum_{t \in A_k(t_0, t_n)} \prod_{1 \leq i \leq n} (\mathsf{p}_{E_i}(t_{i-1}, t_i) - \mathsf{p}^*) \geq -\mathsf{p}^* \sum_{t \in B \times C} \prod_{\substack{1 \leq i \leq n \\ i \neq j}} |\mathsf{p}_{E_i}(t_{i-1}, t_i) - \mathsf{p}^*|$$

$$\geq -\mathsf{p}^* \underbrace{\left(\sum_{(t_1, \ldots, t_{j-1}) \in B} \prod_{1 \leq i \leq j-1} |\mathsf{p}_{E_i}(t_{i-1}, t_i) - \mathsf{p}^*| \right)}_{S_1}$$

$$\times \underbrace{\left(\sum_{(t_j, \ldots, t_{n-1}) \in C} \prod_{j+1 \leq i \leq n} |\mathsf{p}_{E_i}(t_{i-1}, t_i) - \mathsf{p}^*| \right)}_{S_2}.$$

These sums S_1 and S_2 should be studied independently. For S_1, we have

$$S_1 = \sum_{\substack{t_1 \in (\mathcal{M})_q: \\ C_{k_1, 1}}} |\mathsf{p}_{E_1}(t_0, t_1) - \mathsf{p}^*| \sum_{\substack{t_2 \in (\mathcal{M})_q: \\ C_{k_2, 2}}} |\mathsf{p}_{E_2}(t_1, t_2) - \mathsf{p}^*| \ldots$$

$$\times \sum_{\substack{t_{j-1} \in (\mathcal{M})_q: \\ C_{k_{j-1}, j-1}}} |\mathsf{p}_{E_{j-1}}(t_{j-2}, t_{j-1}) - \mathsf{p}^*|$$

$$\leq \sum_{\substack{t_1 \in (\mathcal{M})_q: \\ C_{k_1, 1}}} |\mathsf{p}_{E_1}(t_0, t_1) - \mathsf{p}^*| \ldots$$

$$\times \sum_{\substack{t_{j-2}\in(\mathcal{M})_q:\\ C_{k_{j-2},j-2}}} |\mathsf{p}_{E_{j-2}}(t_{j-3},t_{j-2}) - \mathsf{p}^*| \times \|\mathsf{p}_{E_{j-1},t_{j-2}} - \mathsf{p}^*\|$$

$$\leq \mathbf{Adv}_{E_{j-1}}^{\mathrm{ncpa}}(q) \sum_{\substack{t_1\in(\mathcal{M})_q:\\ C_{k_1,1}}} |\mathsf{p}_{E_1}(t_0,t_1) - \mathsf{p}^*| \ldots \sum_{\substack{t_{j-2}\in(\mathcal{M})_q:\\ C_{k_{j-2},j-2}}} |\mathsf{p}_{E_{j-2}}(t_{j-3},t_{j-2}) - \mathsf{p}^*|$$

$$\vdots$$

$$\leq \prod_{2\leq i\leq j-1} \mathbf{Adv}_{E_i}^{\mathrm{ncpa}}(q) \sum_{\substack{t_1\in(\mathcal{M})_q:\\ C_{k_1,1}}} |\mathsf{p}_{E_1}(t_0,t_1) - \mathsf{p}^*|$$

$$\leq \prod_{2\leq i\leq j-1} \mathbf{Adv}_{E_i}^{\mathrm{ncpa}}(q) \times \|\mathsf{p}_{E_1,t_0} - \mathsf{p}^*\|$$

$$\leq \prod_{1\leq i\leq j-1} \mathbf{Adv}_{E_i}^{\mathrm{ncpa}}(q).$$

Similarly one has:

$$S_2 = \sum_{\substack{t_{n-1}\in(\mathcal{M})_q:\\ C_{k_n,n}}} |\mathsf{p}_{E_n}(t_{n-1},t_n) - \mathsf{p}^*| \ldots \sum_{\substack{t_j\in(\mathcal{M})_q:\\ C_{k_{j+1},j+1}}} |\mathsf{p}_{E_{j+1}}(t_j,t_{j+1}) - \mathsf{p}^*|$$

$$\leq \sum_{\substack{t_{n-1}\in(\mathcal{M})_q:\\ C_{k_n,n}}} |\mathsf{p}_{E_n}(t_{n-1},t_n) - \mathsf{p}^*| \ldots$$

$$\times \sum_{\substack{t_{j+1}\in(\mathcal{M})_q:\\ C_{k_{j+2},j+2}}} |\mathsf{p}_{E_{j+2}}(t_{j+1},t_{j+2}) - \mathsf{p}^*| \times \|\mathsf{p}_{E_{j+1}^{-1},t_{j+1}} - \mathsf{p}^*\|$$

$$\leq \mathbf{Adv}_{E_{j+1}^{-1}}^{\mathrm{ncpa}}(q)$$

$$\times \sum_{\substack{t_{n-1}\in(\mathcal{M})_q:\\ C_{k_n,n}}} |\mathsf{p}_{E_n}(t_{n-1},t_n) - \mathsf{p}^*| \ldots \sum_{\substack{t_{j+1}\in(\mathcal{M})_q:\\ C_{k_{j+2},j+2}}} |\mathsf{p}_{E_{j+2}}(t_{j+1},t_{j+2}) - \mathsf{p}^*|$$

$$\vdots$$

$$\leq \prod_{j+1\leq i\leq n} \mathbf{Adv}_{E_i^{-1}}^{\mathrm{ncpa}}(q),$$

from which the result follows. □

We can now prove the extension of Theorem 1.

Theorem 2. *Let $E_1,\ldots,E_n$ be n block ciphers with the same message space $\mathcal{M}$. For any integer q, one has*

$$\mathbf{Adv}_{E_n\circ\cdots\circ E_1}^{\mathrm{cca}}(q) \leq 2^{n-1} \max_{1\leq i\leq n} \left(\prod_{1\leq j\leq i-1} \mathbf{Adv}_{E_j}^{\mathrm{ncpa}}(q) \times \prod_{i+1\leq j\leq n} \mathbf{Adv}_{E_j^{-1}}^{\mathrm{ncpa}}(q) \right).$$

Proof. Fix any q-tuples $x_0, x_n \in (\mathcal{M})_q$. Then

$$\mathsf{p}_{E_n \circ \cdots \circ E_1}(x, y)$$

$$= \mathsf{p}^* + \sum_{(x_1, \ldots, x_{n-1}) \in ((\mathcal{M})_q)^{n-1}} \left(\prod_{1 \leq i \leq n} (\mathsf{p}_{E_i}(x_{i-1}, x_i) - \mathsf{p}^*) \right) \qquad \text{(Lemma 5)}$$

$$= \mathsf{p}^* + \sum_{\substack{k \in \{0,1\}^n}} \sum_{\substack{(x_1, \ldots, x_{n-1}) \in \\ A_k(x_0, x_n)}} \left(\prod_{1 \leq i \leq n} (\mathsf{p}_{E_i}(x_{i-1}, x_i) - \mathsf{p}^*) \right)$$

$$\geq \mathsf{p}^* + \sum_{\substack{k \in \{0,1\}^n: \\ k_1 + \ldots + k_n \equiv 1 \bmod 2}} \sum_{\substack{(x_1, \ldots, x_{n-1}) \in \\ A_k(x_0, x_n)}} \left(\prod_{1 \leq i \leq n} (\mathsf{p}_{E_i}(x_{i-1}, x_i) - \mathsf{p}^*) \right)$$

$$\geq \mathsf{p}^* - 2^{n-1} \mathsf{p}^* \max_{1 \leq i \leq n} \left(\prod_{1 \leq j \leq i-1} \mathbf{Adv}_{E_j}^{\mathrm{ncpa}}(q) \prod_{i+1 \leq j \leq n} \mathbf{Adv}_{E_j^{-1}}^{\mathrm{ncpa}}(q) \right). \qquad \text{(Lemma 6)}$$

The result follows by Lemma 2.   □

Remark 1. The upper bound of Theorem 2 is not tight in general already for $n = 2$. Indeed it is not hard to verify that Theorem 1 yields a better bound (at least when E_1 and E_2^{-1} have different levels of NCPA-security).

Corollary 1. *Let $E_1, \ldots, E_n$ be n block ciphers with the same message space $\mathcal{M}$. Fix $q \geq 1$. For $i = 1, \ldots, n$, let $\varepsilon_i = \max\{\mathbf{Adv}_{E_i}^{\mathrm{ncpa}}(q), \mathbf{Adv}_{E_i^{-1}}^{\mathrm{ncpa}}(q)\}$. Then one has*

$$\mathbf{Adv}_{E_n \circ \cdots \circ E_1}^{\mathrm{cca}}(q) \leq 2^{n-1} \max_{1 \leq i \leq n} \prod_{\substack{1 \leq j \leq n \\ j \neq i}} \varepsilon_i.$$

Remark 2. It is actually not hard to see that Corollary 1 also holds with $\varepsilon_1 = \mathbf{Adv}_{E_1}^{\mathrm{ncpa}}(q)$ and $\varepsilon_n = \mathbf{Adv}_{E_n}^{\mathrm{ncpa}}$, i.e., E_1 and E_n need only be secure in one direction. Only the "internal" components $E_2, \ldots, E_{n-1}$ are required to be secure in both directions.

In the case of self-composition, we obtain the following corollary.

Corollary 2. *Let E be a block cipher and $q \geq 1$. Denote*

$$\varepsilon = \max\{\mathbf{Adv}_E^{\mathrm{ncpa}}(q), \mathbf{Adv}_{E^{-1}}^{\mathrm{ncpa}}(q)\}.$$

Then, for any integer $n \geq 1$,

$$\mathbf{Adv}_{E^n}^{\mathrm{cca}}(q) \leq (2\varepsilon)^{n-1}.$$

Remark 3. The assumption required for Corollary 2, namely that both E and E^{-1} are (q, ε)-NCPA secure, might seem much stronger than simply assuming that E is (q, ε)-NCPA secure. However, the schemes used in block ciphers are often involutions or close to involutions (for example balanced Feistel schemes). Then one needs to determine only *one* of these upper bounds. We stress that there exists block cipher designs such that the NCPA-security of E^{-1} is much worse than the NCPA-security of E, the prominent example being type-1 generalized Feistel schemes [ZMI89, MV00], which is the basis for example of CAST-256.

5   On the Tightness of the Bound

The 2W1S theorem was shown to be tight in [MPR07] (see Appendix A of the full version of [MPR07]). In this section, we generalize the proof of tightness of [MPR07] to show that the bound of Theorem 2 is tight up to some constant.

As in [MPR07], denote G the family of all permutations of $\mathcal{M}$ such that 0 lies on a cycle of length 2 (i.e., $\forall g \in G, g(g(0)) = 0$). Seeing G as a block cipher[3], it can be shown that $\mathbf{Adv}_G^{\mathrm{ncpa}}(q) \leq \frac{2q}{|\mathcal{M}|}$ and $\mathbf{Adv}_G^{\mathrm{cca}}(2) \geq 1 - \frac{2}{|\mathcal{M}|}$. Then let us define the block cipher F such that:

- with probability ϵ, F is the identity function $\mathcal{I}$,
- with probability $1 - \epsilon$, F is uniformly randomly chosen in G.

Fix any constants $\delta, \delta', \delta'' > 0$. Then

$$\mathbf{Adv}_F^{\mathrm{ncpa}}(q) = \varepsilon \mathbf{Adv}_{\mathcal{I}}^{\mathrm{ncpa}}(q) + (1 - \varepsilon)\mathbf{Adv}_G^{\mathrm{ncpa}}(q) \leq \varepsilon + \frac{2q}{|\mathcal{M}|} \leq (1 + \delta)\varepsilon, \quad (4)$$

where for the last inequality we assumed $|\mathcal{M}|$ sufficiently large.

Now consider the block cipher F^n for a fixed integer $n \geq 2$. Consider the adaptive distinguisher $\mathcal{D}$ making two queries to its permutation oracle P, $P(0)$ and then $P(P(0))$, and outputs 1 *iff* $P(P(0)) = 0$. When interacting with a random permutation, $\mathcal{D}$ outputs 1 with probability exactly[4] $2/|\mathcal{M}|$, while when it is interacting with F^n, it outputs 1 (at least) whenever $n - 1$ among the n instances of F are the identity function, which happens with probability $n(1 - \varepsilon)\varepsilon^{n-1}$. Hence, for any $q \geq 2$, one has

$$\mathbf{Adv}_{F^n}^{\mathrm{cca}}(q) \geq n(1 - \varepsilon)\varepsilon^{n-1} - \frac{2}{|\mathcal{M}|} \geq \frac{n}{(1 + \delta')(1 + \delta'')}\varepsilon^{n-1},$$

where for the last inequality we assumed ε sufficiently small and $|\mathcal{M}|$ sufficiently large. Using (4), we finally obtain

$$\mathbf{Adv}_{F^n}^{\mathrm{cca}}(q) \geq \frac{n}{(1 + \delta)^{n-1}(1 + \delta')(1 + \delta'')}(\mathbf{Adv}_F^{\mathrm{ncpa}})^{n-1}.$$

---

[3] Ignoring efficiency considerations, this simply means that one defines the set of keys as $\mathcal{K} = G$.

[4] This can be seen as follows: with probability $1/|\mathcal{M}|$, 0 is a fixed point of P, and with probability $(|\mathcal{M}| - 1)/(\mathcal{M}|(|\mathcal{M}| - 1))$, one has $P(0) = y$ and $P(y) = 0$ for some $y \neq 0$.

Since δ, δ', and δ'' can be made arbitrarily close to zero, this essentially shows that the best upper bound one can hope for in Corollary 2 is $n\varepsilon^{n-1}$. Closing the gap between the proven upper bound $2^{n-1}\varepsilon^{n-1}$ and $n\varepsilon^{n-1}$ remains as an interesting open problem.

A   Omitted Proofs

Proof. (of Lemma 1). Fix some NCPA-distinguisher $\mathcal{D}$. Since we consider deterministic distinguishers, $\mathcal{D}$ is completely characterized by its q-tuple of queries $x = (x_1 \ldots, x_q)$ and its decision function $\phi_{\mathcal{D}} : (\mathcal{M})_q \to \{0, 1\}$, where $\phi_{\mathcal{D}}(y)$ is the output of $\mathcal{D}$ when receiving $y = (y_1, \ldots, y_q)$ as answers to its queries. By definition of the advantage,

$$\mathbf{Adv}(\mathcal{D}) = \Bigg| \sum_{y \in (\mathcal{M})_q : \phi_{\mathcal{D}}(y)=1} \Pr\left[K \leftarrow_{\$} \mathcal{K} : E_K(x) = y\right]$$

$$- \sum_{y \in (\mathcal{M})_q : \phi_{\mathcal{D}}(y)=1} \Pr\left[P \leftarrow_{\$} \mathsf{Perm}(\mathcal{M}) : P(x) = y\right] \Bigg|$$

$$= \Bigg| \sum_{y \in (\mathcal{M})_q : \phi_{\mathcal{D}}(y)=1} \left(\mathsf{p}_{E,x}(y) - \mathsf{p}^*\right) \Bigg|$$

$$\leq \|\mathsf{p}_{E,x} - \mathsf{p}^*\|.$$

By maximizing over $x \in (\mathcal{M})_q$, we obtain

$$\mathbf{Adv}_E^{\mathrm{ncpa}}(q) \leq \max_{x \in (\mathcal{M})_q} \|\mathsf{p}_{E,x} - \mathsf{p}^*\|.$$

To prove the equality of the two quantities, consider the distinguisher which queries the q-tuple x which maximizes $\|\mathsf{p}_{E,x} - \mathsf{p}^*\|$, and outputs 1 *iff* the answer y satisfies $\mathsf{p}_E(x, y) \geq \mathsf{p}^*$. Then the advantage of this distinguisher is exactly $\|\mathsf{p}_{E,x} - \mathsf{p}^*\|$, which concludes the proof.   □

Proof. (of Lemma 2). Fix some CCA-distinguisher $\mathcal{D}$. Let τ be the transcript of the interaction of $\mathcal{D}$ with its permutation oracle, i.e., the ordered q-tuple of queries and answers (b_i, z_i, z_i') where b_i is a bit indicating whether the i-th query is direct or inverse, z_i is the value queried to the oracle and z_i' the answer. From this transcript, we define the *directionless* transcript $\tau' = (x, y)$, with $x = (x_1, \ldots, x_q)$ and $y = (y_1, \ldots, y_q)$ as follows: if the i-th query was a direct query, we let $x_i = z_i$ and $y_i = z_i'$, and if it was an inverse query we let $x_i = z_i'$ and $y_i = z_i$. We say that a transcript τ is *attainable* if there exists a permutation $P \in \mathsf{Perm}(\mathcal{M})$ such that the interaction of $\mathcal{D}$ with P produces τ (in other words, the probability to obtain τ when $\mathcal{D}$ interacts with a random permutation is non-zero). Since the distinguisher is deterministic, there is a one-to-one mapping between attainable transcripts and attainable directionless transcripts. Let $\mathcal{T}$ denote the set of attainable directionless transcripts. Note that the interaction of $\mathcal{D}$ with some permutation $P \in \mathsf{Perm}(\mathcal{M})$ produces the directionless transcript $\tau' = (x, y)$ iff $P(x) = y$. Note also that

$$\sum_{(x,y)\in\mathcal{T}} \mathsf{p}_E(x,y) = \sum_{(x,y)\in\mathcal{T}} \mathsf{p}^* = 1.$$

The output of the distinguisher is a function of the transcript τ, or equivalently of the directionless transcript τ'. Let $\mathcal{T}_0$ (resp. $\mathcal{T}_1$) be the set of attainable directionless transcripts τ' such that $\mathcal{D}$ outputs 0 (resp. 1) when obtaining $\tau' = (x, y)$. Then, by definition of the advantage,

$$\mathbf{Adv}(\mathcal{D}) = \Bigg| \sum_{(x,y)\in\mathcal{T}_1} \Pr\left[P \leftarrow_\$ \mathsf{Perm}(\mathcal{M}) : P(x) = y\right]$$

$$- \sum_{(x,y)\in\mathcal{T}_1} \Pr\left[K \leftarrow_\$ \mathcal{K} : E_K(x) = y\right] \Bigg|$$

$$= \Bigg| \sum_{(x,y)\in\mathcal{T}_1} \mathsf{p}^* - \mathsf{p}_E(x,y) \Bigg|$$

Using the assumption of the lemma, we have

$$\sum_{(x,y)\in\mathcal{T}_1} (\mathsf{p}^* - \mathsf{p}_E(x,y)) \le \sum_{(x,y)\in\mathcal{T}_1} \varepsilon \mathsf{p}^* \le \varepsilon \sum_{(x,y)\in\mathcal{T}_1} \mathsf{p}^* \le \varepsilon,$$

and similarly

$$-\sum_{(x,y)\in\mathcal{T}_1} (\mathsf{p}^* - \mathsf{p}_E(x,y)) = \sum_{(x,y)\in\mathcal{T}_0} (\mathsf{p}^* - \mathsf{p}_E(x,y))$$

$$\le \sum_{(x,y)\in\mathcal{T}_0} \varepsilon \mathsf{p}^* \le \varepsilon \sum_{(x,y)\in\mathcal{T}_0} \mathsf{p}^* \le \varepsilon,$$

from which the result follows. $\square$

References

[ABCV98] Aiello, W., Bellare, M., Di Crescenzo, G., Venkatesan, R.: Security amplification by composition: the case of doubly-iterated, ideal ciphers. In: Krawczyk, H. (ed.) CRYPTO 1998. LNCS, vol. 1462, pp. 390–407. Springer, Heidelberg (1998)

[BR06] Bellare, M., Rogaway, P.: The security of triple encryption and a framework for code-based game-playing proofs. In: Vaudenay, S. (ed.) EUROCRYPT 2006. LNCS, vol. 4004, pp. 409–426. Springer, Heidelberg (2006)

[CPS14] Cogliati, B., Patarin, J., Seurin, Y.: Security amplification for the composition of block ciphers: simpler proofs and new results. Full version of this paper. Available from the authors of at http://eprint.iacr.org/

[GM09] Gaži, P., Maurer, U.: Cascade encryption revisited. In: Matsui, M. (ed.) ASIACRYPT 2009. LNCS, vol. 5912, pp. 37–51. Springer, Heidelberg (2009)

[HR10] Hoang, V.T., Rogaway, P.: On generalized Feistel networks. In: Rabin, T. (ed.) CRYPTO 2010. LNCS, vol. 6223, pp. 613–630. Springer, Heidelberg (2010)

[JÖS12] Jetchev, D., Özen, O., Stam, M.: Understanding adaptivity: random systems revisited. In: Wang, X., Sako, K. (eds.) ASIACRYPT 2012. LNCS, vol. 7658, pp. 313–330. Springer, Heidelberg (2012)

[KNR09] Kaplan, E., Naor, M., Reingold, O.: Derandomized constructions of k-wise (almost) independent permutations. Algorithmica 55(1), 113–133 (2009)

[Lee13] Lee, J.: Towards key-length extension with optimal security: cascade encryption and xor-cascade encryption. In: Johansson, T., Nguyen, P.Q. (eds.) EUROCRYPT 2013. LNCS, vol. 7881, pp. 405–425. Springer, Heidelberg (2013)

[LPS12] Lampe, R., Patarin, J., Seurin, Y.: An asymptotically tight security analysis of the iterated even-mansour cipher. In: Wang, X., Sako, K. (eds.) ASIACRYPT 2012. LNCS, vol. 7658, pp. 278–295. Springer, Heidelberg (2012)

[LR86] Luby, M., Rackoff, C.: Pseudo-random permutation generators and cryptographic composition. In: Symposium on Theory of Computing - STOC '86, pp. 356–363. ACM (1986)

[LS14] Lampe, R., Seurin, Y.: Security analysis of key-alternating Feistel ciphers. In: Fast Software Encryption - FSE 2014 (2014, to appear)

[Mau02] Maurer, U.M.: Indistinguishability of random systems. In: Knudsen, L.R. (ed.) EUROCRYPT 2002. LNCS, vol. 2332, pp. 110–132. Springer, Heidelberg (2002)

[MP04] Maurer, U.M., Pietrzak, K.: Composition of random systems: when two weak make one strong. In: Naor, M. (ed.) TCC 2004. LNCS, vol. 2951, pp. 410–427. Springer, Heidelberg (2004)

[MPR07] Maurer, U.M., Pietrzak, K., Renner, R.S.: Indistinguishability amplification. In: Menezes, A. (ed.) CRYPTO 2007. LNCS, vol. 4622, pp. 130–149. Springer, Heidelberg (2007). Full version available at http://eprint.iacr.org/2006/456

[MRS09] Morris, B., Rogaway, P., Stegers, T.: How to encipher messages on a small domain. In: Halevi, S. (ed.) CRYPTO 2009. LNCS, vol. 5677, pp. 286–302. Springer, Heidelberg (2009)

[MT09] Maurer, U., Tessaro, S.: Computational indistinguishability amplification: tight product theorems for system composition. In: Halevi, S. (ed.) CRYPTO 2009. LNCS, vol. 5677, pp. 355–373. Springer, Heidelberg (2009)

[MV00] Moriai, S., Vaudenay, S.: On the pseudorandomness of top-level schemes of block ciphers. In: Okamoto, T. (ed.) ASIACRYPT 2000. LNCS, vol. 1976, pp. 289–302. Springer, Heidelberg (2000)

[Mye04] Myers, S.: Black-box composition does not imply adaptive security. In: Cachin, C., Camenisch, J.L. (eds.) EUROCRYPT 2004. LNCS, vol. 3027, pp. 189–206. Springer, Heidelberg (2004)

[Pat90] Patarin, J.: Pseudorandom permutations based on the D.E.S. scheme. In: Cohen, G., Charpin, P. (eds.) EUROCODE 1990. LNCS, vol. 514, pp. 193–204. Springer, Heidelberg (1991)

[Pat91] Patarin, J.: New results on pseudorandom permutation generators based on the DES scheme. In: Feigenbaum, J. (ed.) CRYPTO 1991. LNCS, vol. 576, pp. 301–312. Springer, Heidelberg (1992)

[Pat08] Patarin, J.: The "coefficients H" technique. In: Avanzi, R.M., Keliher, L., Sica, F. (eds.) SAC 2008. LNCS, vol. 5381, pp. 328–345. Springer, Heidelberg (2009)

[Pie05a] Pietrzak, K.: Composition does not imply adaptive security. In: Shoup, V. (ed.) CRYPTO 2005. LNCS, vol. 3621, pp. 55–65. Springer, Heidelberg (2005)

[Pie05b] Pietrzak, K.: Indistinguishability and composition of random systems. Ph.D. thesis, ETH Zurich, Switzerland (2005)

[Pie06] Pietrzak, K.: Composition implies adaptive security in minicrypt. In: Vaudenay, S. (ed.) EUROCRYPT 2006. LNCS, vol. 4004, pp. 328–338. Springer, Heidelberg (2006)

[Vau98] Vaudenay, S.: Provable security for block ciphers by decorrelation. In: Morvan, M., Meinel, C., Krob, D. (eds.) STACS 1998. LNCS, vol. 1373, pp. 249–275. Springer, Heidelberg (1998)

[Vau99] Vaudenay, S.: Adaptive-attack norm for decorrelation and super-pseudorandomness. In: Heys, H.M., Adams, C.M. (eds.) SAC 1999. LNCS, vol. 1758, pp. 49–61. Springer, Heidelberg (2000)

[Vau03] Vaudenay, S.: Decorrelation: a theory for block cipher security. J. Cryptol. **16**(4), 249–286 (2003)

[ZMI89] Zheng, Y., Matsumoto, T., Imai, H.: On the construction of block ciphers provably secure and not relying on any unproved hypotheses. In: Brassard, G. (ed.) CRYPTO 1989. LNCS, vol. 435, pp. 461–480. Springer, Heidelberg (1990)

Improved Differential Cryptanalysis
of Round-Reduced Speck

Itai Dinur[(✉)]

Département d'Informatique, École Normale Supérieure, Paris, France
`dinur@di.ens.fr`

Abstract. Simon and Speck are families of lightweight block ciphers designed by the U.S. National Security Agency and published in 2013. Each of the families contains 10 variants, supporting a wide range of block and key sizes. Since the publication of Simon and Speck, several research papers analyzed their security using various cryptanalytic techniques. The best previously published attacks on all the 20 round-reduced ciphers are differential attacks, and are described in two papers (presented at FSE 2014) by Abed et al. and Biryukov et al.

In this paper, we focus on the software-optimized block cipher family Speck, and describe significantly improved attacks on all of its 10 variants. In particular, we increase the number of rounds which can be attacked by 1, 2, or 3, for 9 out of 10 round-reduced members of the family, while significantly improving the complexity of the previous best attack on the remaining round-reduced member. Our attacks use an untraditional key recovery technique for differential attacks, whose main ideas were published by Albrecht and Cid at FSE 2009 in the cryptanalysis of the block cipher PRESENT.

Despite our improved attacks, they do not seem to threaten the security of any member of Speck.

Keywords: Lightweight block cipher · Speck · Cryptanalysis · Differential attack · Key recovery

1 Introduction

In 2013 the U.S. National Security Agency published the Simon and Speck families of lightweight block ciphers [7]. Each block cipher family contains 10 variants and supports block sizes ranging from 32 to 128 and key sizes ranging from 64 to 256 bits. Both families of block ciphers have a simple and compact Feistel-like[1] design, but are optimized for different applications, where Simon is optimized for hardware and Speck is optimized for software implementations. Thus, Simon uses the basic hardware-friendly arithmetic operations of XOR, bitwise AND and

---

[1] Simon is a Feistel structure, while Speck can be represented as a composition of two Feistel maps [7].

© Springer International Publishing Switzerland 2014
A. Joux and A. Youssef (Eds.): SAC 2014, LNCS 8781, pp. 147–164, 2014.
DOI: 10.1007/978-3-319-13051-4_9

bit rotation, whereas Speck is a pure ARX cipher (i.e., it uses modular addition, bit rotation and XOR operations).

Since their publication, Simon and Speck received significant media attention, and were also subjects of extensive research in the cryptographic community, as several papers analyzed their security and performance [1,4,5,9,21]. In general, the best published attacks on all 20 round-reduced ciphers are differential attacks, described in the two papers [1,9]. However, despite the extensive analysis, all 20 variants seem to have a sufficiently large security margin, and the current attacks do not threaten their security.

In this paper, we present improved attacks on all 10 members of the Speck family of block ciphers. In particular, we increase the number of rounds which can be attacked by 1, 2, or 3, for 9 out of 10 members of the family, while significantly improving the complexity of the previous attack on the remaining member. More specifically, we increase the number of rounds which can be attacked by 1 for 4 members, by 2 for 2 members, and by 3 for 3 members. In 3 of these cases, not only do we attack more rounds, but we also improve the complexity of the best previous attacks, which were applied to a weaker cipher. Moreover, in all of these cases, our attacks use less data than the previous attacks, and all of them require only a few megabytes of memory (typically improving the previous attack with respect to this parameter as well).

Surprisingly, our attacks do not exploit any newly found differential characteristic of Speck. In fact, our attacks completely reuse the characteristics presented in [1,9], but are based on a significantly improved key recovery framework. As the basic idea behind this framework in very simple, at first, it seems quite strange that it was missed by the previous analysis. However, a closer look reveals that our key recovery technique is quite different from traditional techniques used in differential cryptanalysis. These key recovery techniques (called *counting* techniques) were published with the introduction of differential cryptanalysis [8]. Counting techniques remain, by far, the most common techniques to recover the key in differential attacks, and were thus naturally applied in the previous differential attacks on Speck [1,9].

One of the main features of counting techniques in differential attacks is that the key material is typically recovered in chunks (i.e., in a divide-and-conquer manner) using statistical analysis. In order to recover a chunk of key material (e.g., some bits of the first and last round-keys), we analyze encrypted input pairs, each pair suggesting a value (or a few values) for this chunk. Right pairs (conforming to the characteristic) always suggest the correct value for the chunk, while wrong pairs suggest an arbitrary value. In order to be able to distinguish the correct suggestions from the incorrect ones, we require strong filtering which eliminates a large fraction of the wrong pairs (and the arbitrary key suggestions). Such a strong filtering requirement places a restriction on the number of rounds of the iterated block cipher which we can attack with a given differential characteristic. Namely, in order to attack an $(r + a)$-round cipher with an r-round characteristic, a needs to be sufficiently small such that the characteristic can be extended to deduce some linear constraints on the output of the cipher (e.g., some bits of the output difference are zero), allowing us to filter many of the wrong pairs.

In contrast to standard key recovery techniques (in particular, the ones used in previous attacks on Speck), in this paper we extend a differential character-istic by a (relatively) large number of rounds, and thus simple linear filtering can eliminate only a small fraction of the data. Consequently, we remain with too many suggestions for the key to mount an efficient attack using counting. On the other hand, this situation resembles some self-similarity attacks (such as reflection-based attacks [12–14]), in which the attacker does not have any char-acteristic that allows simple filtering. In such self-similarity attacks, the attacker encrypts multiple plaintexts and awaits a special event to occur (such as a reflec-tion). The internal properties of the cipher assure that once this event occurs, the problem of attacking the full cipher is reduced to a simpler problem of attacking a sub-cipher with fewer rounds. The sub-cipher attack calculates suggestions for *the full secret key*, which the attacker tests using trial encryptions on the full cipher.[2] Since the attacker cannot detect the occurrence of the special event, the sub-cipher attack is executed for each plaintext (or plaintext structure). Thus, the complexity of the full attack is determined by the probability of the awaited event (which determines how many sub-cipher attacks we need to execute), and the average complexity of the sub-cipher attack.

In the scenario presented above for self-similarity attacks, the key is recovered in one chunk by a sub-cipher attack. However, there is nothing that prevents us from applying similar techniques in differential attacks. In fact, the last a rounds of the cipher in differential attacks can be viewed as a sub-cipher, and assuming the event that an encrypted pair conforms to the r-round characteristic (i.e., it is a right pair), we can mount a sub-cipher attack to obtain key suggestions for each encrypted pair, and enumerate each one of them, testing it using trial encryptions. As a right pair will always suggest the correct key value, the attack succeeds as soon as we finish executing the sub-cipher attack on this pair.

This generic key recovery framework for differential cryptanalysis was first proposed by Albrecht and Cid in [2], where it was applied to the block cipher PRESENT (and was further used in followup publications such as [3,22]). Albrecht and Cid used algebraic techniques to enhance differential cryptanalysis, and specifically, devised Attack-C which formulates the sub-cipher as a system of non-linear equations, and solves it using algebraic tools (e.g., SAGE [20]). On the other hand, the sub-cipher attack can use various methods which do not neces-sarily exploit algebraic tools. Indeed, while we use the same framework as [2], our sub-cipher attack on Speck applies guess-and-determine techniques, and does not directly solve any system of non-linear equations. Furthermore, in [6] the generic framework was (implicitly) applied to the block cipher Zorro using a complex two-phase sub-cipher attack (in which the only equation systems directly solved are linear).

We stress that the *only* difference between our approach and the algebraic approach of Attack-C [2], is in the details of the sub-cipher attack. While this

---

[2] Examples of sub-cipher attacks include the meet-in-the-middle and guess-and-determine attacks on round-reduced GOST, described in [12,13].

is a subtle difference, we believe that part of the reason that Attack-C was not considered in the previous attacks on Speck [1,9], is that it promoted the use of black-box algebraic tools to perform the sub-cipher attack. As such black-box algorithms are often highly heuristic, and their running time is not very well understood, they have not become mainstream analysis tools. In this paper, we show that the sub-cipher attack can sometimes be performed by a simple algorithm with a better understood running time, and we hope that cryptanalysts will consider similar attacks in the future.

In order to generalize Attack-C of [2] to a broader key recovery framework for differential attacks, we call it an *enumeration framework*, as it enumerates suggestions for the full key proposed by a sub-cipher attack. This should be contrasted with *counting* techniques which extract partial key material from a few rounds of the cipher using statistical analysis (e.g., the 1, 2 and 3-round attacks of [8]).

In most cases, counting techniques for differential attacks seem to give the best results. This is perhaps due to the reason that when we extend the characteristic beyond the reach of these techniques, the sub-cipher attack becomes too expensive (as it needs to analyze dependent round-keys according to the key schedule), making the full differential attack inefficient.[3] However, as we show in this paper, in the case of Speck, the sub-cipher attack can be performed very efficiently, and results in significantly improved differential attacks (as in the case of Zorro [6]).

As previously mentioned, our sub-cipher attack on Speck is a guess-and-determine attack, and it is related to the similar attack of [12]. Furthermore, since Speck is an ARX cipher, we use techniques that were developed in the analysis of ARX cryptosystems and similar designs. In particular, our tools are related to several search algorithms for differential characteristics on these designs, such as [10,15,16,18,23].

The rest of this paper is organized as follows. We introduce our notation in Sect. 2, and provide a brief description of Speck in Sect. 3. The previous and our new results on the 10 Speck variants are summarized in Sect. 4. In Sect. 5, we describe the auxiliary algorithms used in our attacks (and in particular, overview the specific sub-cipher attack on Speck), while our full differential attacks in the enumeration framework are described in Sect. 6. Finally, we give the details of the sub-cipher attack in Sect. 7, and conclude the paper in Sect. 8.

2   Notations and Conventions

In this section, we describe the notations and conventions used in the rest of the paper.

Given a positive integer r, we denote by $x \ggg i$ the n-bit word obtained by rotating x by i bits to the right, and by $x \lll i$ the word obtained by rotating x by

---

[3] We note that analysis of dependent round-keys can sometimes be performed efficiently using algebraic tools, as claimed in [2].

i bits to the left. Similarly, $x \gg i$ and $x \ll i$ denote a bitwise shift of x by i bits to the right and left, respectively. We denote by $\neg x$ the bitwise negation of x.

Given two n-bit words x and y, we denote by $x \oplus y$ their n-bit XOR, by $x \boxplus y$ their n-bit addition over $GF(2^n)$, and by $x \boxminus y$ their difference over $GF(2^n)$. We further denote by $x \wedge y$ the bitwise AND of x and y.

Given an n-bit word x, we denote its i'th bit for $i \in \{0, 1, \ldots n - 1\}$ by $x^{[i]}$. We note that operations on the bit indexes are performed modulo n, e.g. $x^{[n+5]} \equiv x^{[5]}$.

Conventions. Throughout this paper, we use the standard conventions and calculate the time complexity of our attacks in terms of evaluations of the full cipher. The memory complexity of the attacks is calculated in terms of bytes.

3   Description of Speck

In this section, we give a short description of Speck. More details can be found in [7].

Speck is a family of block ciphers containing 10 variants. The variants are characterized by a block size of $2n$ bits (where n is the internal word size), and a key size of mn bits. The 10 variants are identified with a $2n/mn$ label, and defined with rotation constants α and β and a number of rounds T, as shown in Table 1.

The key schedule of Speck expands the initial m-word master key $(\ell_{m-2}, \ldots, \ell_0, k_0)$ into T round-key words $k_0, k_1, \ldots, k_{T-1}$ according to the following algorithm:

> **for** $i = 0 \ldots T - 2$ **do**
> $\quad \ell_{i+m-1} \leftarrow (k_i \boxplus (\ell_i \ggg \alpha)) \oplus i$
> $\quad k_{i+1} \leftarrow (k_i \lll \beta) \oplus \ell_{i+m-1}$
> **end for**

The encryption function of Speck encrypts a plaintext of two n-bit words $P = (x_0, y_0)$, into a ciphertext $C = (x_T, y_T)$, using a sequence of T rounds according to the following algorithm (see Fig. 1 for the round function):

> **for** $i = 0 \ldots T - 1$ **do**
> $\quad x_{i+1} \leftarrow ((x_i \ggg \alpha) \boxplus y_i) \oplus k_i$
> $\quad y_{i+1} \leftarrow (y_i \lll \beta) \oplus x_{i+1}$
> **end for**

4   Summary of Previous and New Attacks on Speck

In this section, we summarize the previous and our new attacks on Speck, referring to Table 2. As the Speck family contains 10 variants, and each variant was analyzed by several papers, exhaustively listing all the dozens of previous attacks is too tedious. Instead, for each Speck variant, we first choose the attacks which

Table 1. The Speck family of block ciphers

Variant $2n/mn$	Word size n	Key words M	Rounds T	α	β
32/64	16	4	22	7	2
48/72	24	3	22	8	3
48/96	24	4	23	8	3
64/96	32	3	26	8	3
64/128	32	4	27	8	3
96/96	48	2	28	8	3
96/144	48	3	29	8	3
128/128	64	2	32	8	3
128/192	64	3	33	8	3
128/256	64	4	34	8	3

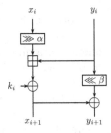

Fig. 1. The round-function of Speck

break the most number of rounds, and among these, we only refer to the attack with the best time complexity. As shown in Table 2, all the best previous attacks were described in the two papers [1,9], and we additionally note that all of them are based on differential cryptanalysis and related techniques (such as rectangle attacks).

For each variant of Speck, Table 2 summarizes our attack which breaks the most number of rounds. We note that for each variant, we can also use our techniques to attack fewer rounds (using a shorter differential characteristic), but once again, we do not explicitly refer to these numerous attacks in this paper (with the exception of the 32/64 variant). As our attacks reuse the differential characteristics of [1,9], we refer to these characteristics in the table, while describing them in more detail in Table 3. We note that since the internal differential transitions in each characteristic are not relevant for our attacks, Table 3 only gives the input and output differences for each characteristic, while their complete specification is described in [1,9].

We now highlight some interesting features of the attacks summarized in Table 2. We first look at the 32/64 variant, on which the best previous attack could break 11 out of its 22 rounds, with data complexity of about a quarter

Table 2. Previous attacks and our new attacks on Speck

Variant $2n/mn$	Rounds attacked/ Total rounds	Time	Data (CP)	Memory	Reference	Characteristic ID
32/64	11/22	$2^{46.7}$	$2^{30.1}$	$2^{37.1}$	[1]	-
32/64	**11/22**	2^{46}	2^{14}	2^{22}	This paper	1
32/64	**12/22**	2^{51}	2^{19}	2^{22}	This paper	2
32/64	**13/22**	2^{57}	2^{25}	2^{22}	This paper	3
32/64	**14/22**	2^{63}	2^{31}	2^{22}	This paper	4
48/72	12/22	2^{43}	2^{43}	NA	[9]	-
48/72	**14/22**	2^{65}	2^{41}	2^{22}	This paper	5
48/96	12/23	2^{43}	2^{43}	NA	[9]	-
48/96	**15/23**	2^{89}	2^{41}	2^{22}	This paper	5
64/96	16/26	2^{63}	2^{63}	NA	[9]	-
64/96	**18/26**	2^{93}	2^{61}	2^{22}	This paper	6
64/128	16/27	2^{63}	2^{63}	NA	[9]	-
64/128	**19/27**	2^{125}	2^{61}	2^{22}	This paper	6
96/96	15/28	$2^{89.1}$	2^{89}	2^{48}	[1]	-
96/96	**16/28**	2^{85}	2^{85}	2^{22}	This paper	7
96/144	16/29	$2^{135.9}$	$2^{90.9}$	$2^{94.5}$	[1]	-
96/144	**17/29**	2^{133}	2^{85}	2^{22}	This paper	7
128/128	16/32	2^{116}	2^{116}	2^{64}	[1]	-
128/128	**17/32**	2^{113}	2^{113}	2^{22}	This paper	8
128/192	18/33	$2^{182.7}$	$2^{126.9}$	$2^{121.9}$	[1]	-
128/192	**18/33**	2^{177}	2^{113}	2^{22}	This paper	8
128/256	18/34	$2^{182.7}$	$2^{126.9}$	$2^{121.9}$	[1]	-
128/256	**19/34**	2^{241}	2^{113}	2^{22}	This paper	8

The "Characteristic ID" column refers to the IDs of the characteristics in Table 3, which are used in our attacks. The data is given in chosen plaintexts (CP).

of the entire code-book. Compared to this attack, our attack on 11 rounds uses less than a square root of the code-book (2^{14} plaintexts), requires less memory and has a slightly better time complexity. Furthermore, we can attack up to 13 rounds in time complexity which is significantly faster than exhaustive search, and up to 14 rounds with a marginal attack. For additional 2 variants (48/96 and 64/128), we can attack 3 more rounds than the best previous attack. For the 2 variants 48/72 and 48/96, we increase the number of rounds that can be attacked by 2. For 4 variants (96/96, 96/144, 128/128 and 128/256), we can attack 1 more round than the best previous attack. Note that for the 3 variants 96/96, 96/144 and 128/128, our attacks are also more efficient than the previous attacks in all complexity parameters (and particularly use much less memory). Finally, for the 128/192 variant, we attack the same number of rounds as the best previous attack, but improve it in all complexity parameters, and in particular use much less memory.

Table 3. Differential characteristics used in our attacks

ID	Variant $2n$	Rounds	Probability	Reference	Input/Output differences
1	32	7	2^{-13}	[1]	211 a04/850a 9520
2	32	8	2^{-18}	[1]	a60 a205/850a 9520
3	32	9	2^{-24}	[1]	a60 a205/802a d4a8
4	32	10	2^{-30}	[9]	8054 a900/40 542
5	48	11	2^{-40}	[9]	202040 82921/80a0 2085a4
6	64	15	2^{-60}	[9]	9 1000000/40024 4200d01
7	96	14	2^{-84}	[1]	2a20200800a2 322320680801/ 1008004c804 c0180228c61
8	128	15	2^{-112}	[1]	144304280c010420 6402400040024/ 180208402886884 80248012c96c80

The n-bit halves in each input/ouput difference are separated by a space.

5 Auxiliary Algorithms Used by Our Attacks

In this section, we describe the two auxiliary algorithms that are used by our attacks on Speck.

5.1 Key-Schedule Inversion

Given a sequence of m key words $k_{j-m}, \ldots, k_{j-1}$ for any $j \in \{m, m+1, \ldots, T\}$, we can efficiently invert the key schedule and calculate the master key: first, we determine k_{j-m-1} using the following key schedule equalities

$$\ell_{j+m-3} = k_{j-1} \oplus (k_{j-2} \lll \beta)$$

$$\ell_{j-2} = ((\ell_{j+m-3} \oplus (j-2)) \boxminus k_{j-2}) \lll \alpha$$

$$k_{j-m-1} = (k_{j-m} \oplus \ell_{j-2}) \ggg \beta.$$

Next, given $k_{j-m-1}, \ldots, k_{j-2}$, we iteratively continue the inversion of the key schedule and derive the master key.

5.2 Overview of the 2-Round Attack on Speck

In our basic attacks on Speck, we use an r-round differential characteristic with an initial difference, denoted by $(\Delta x_0, \Delta y_0)$, and a final difference, denoted by $(\Delta x_r, \Delta y_r)$. We devise an attack on $r + 2$ rounds using a 2-round attack.

The enumeration framework poses the following problem: the $2n$-bit input difference $(\Delta x_r, \Delta y_r)$ to the last 2 rounds is fixed by the final difference of the differential characteristic, and the output difference $(\Delta x_{r+2}, \Delta y_{r+2})$ is known from the output. Furthermore, we are given actual output values (x_{r+2}, y_{r+2}) and $(x_{r+2} \oplus \Delta x_{r+2}, y_{r+2} \oplus \Delta y_{r+2})$. Our objective is to find all the possible independent round keys k_r and k_{r+1}, under which the difference of the 2-round partial decryptions of the pair (x_{r+2}, y_{r+2}) and $(x_{r+2} \oplus \Delta x_{r+2}, y_{r+2} \oplus \Delta y_{r+2})$ is equal to $(\Delta x_r, \Delta y_r)$. In general, we have $2n$ bits of variables and $2n$ bits of constraints (derived from the difference $(\Delta x_r, \Delta y_r)$). Thus, the problem can be formulated using an equation system, which has an average of one solution for an arbitrary pair of outputs (x_{r+2}, y_{r+2}) and $(x_{r+2} \oplus \Delta x_{r+2}, y_{r+2} \oplus \Delta y_{r+2})$. The goal of the 2-round attack is to enumerate all the possible solutions for each given output pair as efficiently as possible.[4]

We note that it is not trivial that the equation system has an average of one solution, as the pairs of outputs are ciphertexts, whose corresponding plaintexts have the fixed initial difference $(\Delta x_0, \Delta y_0)$ to the characteristic. If such a plaintext pair diverges from the characteristic at its later rounds, then the difference after r rounds can potentially be close to $(\Delta x_r, \Delta y_r)$, which may result in non-random behavior. In fact, our experiments show that the average number of solutions is about 4 for characteristic 1 in Table 3, which has a relatively high probability of 2^{-13}. However, for the lower probability characteristics which we could test experimentally, the average number of solutions was only slightly higher than 1 (and lower than 2).

Our 2-round attack is given in Sect. 7. This attack exploits the (relative) simplicity of the Speck round function in order to recover the 2 final round keys of Speck with very low average time complexity. Indeed, our experiments show that for an output pair (x_{r+2}, y_{r+2}) and $(x_{r+2} \oplus \Delta x_{r+2}, y_{r+2} \oplus \Delta y_{r+2})$ (generated by plaintexts with the fixed initial difference $(\Delta x_0, \Delta y_0)$), the 2-round attack requires an average time which is smaller than 2 time units (i.e., 2 full encryptions of round-reduced Speck) for any characteristic that we use in the full differential attacks on Speck.

6 Details of the Full Differential Attacks

In this section, we describe the details of our full differential attacks on Speck in the enumeration framework. In all the attacks, we assume that we have a differential characteristic that covers r rounds of the cipher with probability $p > 2 \cdot 2^{-2n}$. The attacks recover the mn-bit secret key of a variant with $r+m$ rounds using $2 \cdot p^{-1}$ chosen plaintexts, in expected time complexity of $2 \cdot p^{-1} \cdot 2^{(m-2)n}$. In other words, our attacks are faster than exhaustive search by a factor[5] of $p \cdot 2^{2n-1}$. For example, our attack on 11-round Speck 32/64 (with $m = 4$) uses a characteristic for $11 - 4 = 7$ rounds with $p = 2^{-13}$. Thus, its time complexity is $2 \cdot 2^{13} \cdot 2^{(4-2)16} = 2^{46}$, i.e. it is faster than exhaustive search for the 64-bit key by a factor of $p \cdot 2^{2n-1} = 2^{-13} \cdot 2^{31} = 2^{18}$.

---

[4] Recall that we have essentially no (linear) filtering conditions, and thus we must execute the sub-cipher attack for each encrypted input pair. Consequently, we are interested in the average time complexity of the algorithm.

[5] Note that information theoretically, without considering the internal transitions of the differential characteristic, $p \cdot 2^{2n-1}$ is the best improvement factor that one can hope for, given $2 \cdot p^{-1}$ data.

The Full Differential Attack for $m = 2$. We first present the details of our attack for the Speck instances with $m = 2$ key words (and $m + 2$ rounds), and then extend the attack to the remaining instances, in which $m = 3$ or $m = 4$. We denote the initial difference of the characteristic (at the input of the cipher) by $(\Delta x_0, \Delta y_0)$, and its final difference (after r rounds) by $(\Delta x_r, \Delta y_r)$.

1. Request the encryptions of p^{-1} plaintext pairs P and $P' = P \oplus (\Delta x_0, \Delta y_0)$, and denote the ciphertexts by C and C', respectively. For each plaintext pair P and P':
   (a) Execute the 2-round attack of Sect. 7 using $(\Delta x_r, \Delta y_r)$, C and C'.
   (b) For each returned value of k_r and k_{r+1}, iteratively calculate $k_{r-1}, \ldots, k_0$ (as described in Sect. 5.1), and finally recover the master key. Test the master key using trial encryptions, and return it if the trial encryptions succeed.[6]

The attack requires $2 \cdot p^{-1}$ chosen plaintexts, and given that the r-round characteristic has probability p, we expect one plaintext pair P, P' to be a right pair (i.e., to follow the characteristic, and have a difference of $(\Delta x_r, \Delta y_r)$ after r rounds). Given the ciphertexts C, C', corresponding to the right pair, the 2-round attack will find the correct key, which will be returned by the full attack.

According to the analysis of Sect. 7, the 2-round attack has an average time complexity which is smaller than 2 time units, and thus the average processing time for each analyzed plaintext pair remains about 2. This implies that the total time complexity of the attack is about $2 \cdot p^{-1}$.

The Full Differential Attack for $m = 3$ **and** $m = 4$. For $m = 3$ and $m = 4$, we attack variants with $r + 3$ and $r + 4$ rounds, respectively. The attacks on $m = 3$ and $m = 4$ are trivial extensions of the attack on the $m = 2$ variants, and work by guessing the last 1 and 2 round keys, respectively. Then, for each guess we apply a similar attack to the one applied for $m = 2$.

The data complexity of the attacks remain $2 \cdot p^{-1}$, while the time complexity increases to $2 \cdot p^{-1} \cdot 2^n$ and $2 \cdot p^{-1} \cdot 2^{2n}$ for $m = 3$ and $m = 4$, respectively.

7   The 2-Round Attack

In this section, we present the details of our 2-round attack on Speck. As described in Sect. 5.2, we have an input difference $(\Delta x_r, \Delta y_r)$ to the two rounds (which is fixed by a differential characteristic), and we are given the actual output values (x_{r+2}, y_{r+2}) and $(x_{r+2} \oplus \Delta x_{r+2}, y_{r+2} \oplus \Delta y_{r+2})$. Our goal is to enumerate all the possible independent round keys k_r and k_{r+1}, under which the difference of the

---

[6] This step can be slightly optimized to replace many of the full trial encryptions by lighter Speck round evaluations, if we consider the internal transitions of the differential characteristic: while iteratively calculating $k_{r-1}, \ldots, k_0$, we partially decrypt C and C', and verify that they satisfy the differential characteristic for each round. If the verification fails for some round, we discard the key and continue.

2-round partial decryptions of the pair (x_{r+2}, y_{r+2}) and $(x_{r+2} \oplus \Delta x_{r+2}, y_{r+2} \oplus \Delta y_{r+2})$ is equal to $(\Delta x_r, \Delta y_r)$.

The notation we use in our analysis is given in Fig. 2, where the XOR differential notation is given on the left, and the notation of the intermediate encryption values for (x_{r+2}, y_{r+2}) is given on the right. We further define $(x'_i, y'_i) = (x_i \oplus \Delta x_i, y_i \oplus \Delta y_i)$.

Note that $\Delta y_{r+1} = (\Delta x_{r+2} \oplus \Delta y_{r+2}) \ggg \beta$ and $\Delta x_{r+1} = \Delta y_{r+1} \oplus (\Delta y_r \lll \beta)$ are independent of the keys and can be calculated immediately. Thus, all the XOR differences in the scheme are completely determined. Similarly, the value $y_{r+1} = (x_{r+2} \oplus y_{r+2}) \ggg \beta$ can be calculated from the known (x_{r+2}, y_{r+2}), whereas (x_r, y_r) and x_{r+1} remain unknown. We further note that deriving the two round-keys is equivalent to deriving x_r and x_{r+1}, as their values allow us to calculate $k_{r+1} = (y_{r+1} \boxplus (x_{r+1} \ggg \alpha)) \oplus x_{r+2}$, and as $y_r = (x_{r+1} \oplus y_{r+1}) \ggg \beta$, then $k_r = (y_r \boxplus (x_r \ggg \alpha)) \oplus x_{r+1}$ can be calculated as well. Thus, in the following, we concentrate on deriving the intermediate values x_r and x_{r+1}.

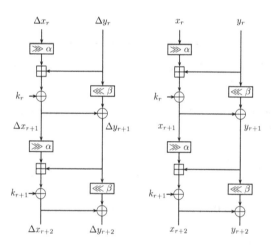

Our notation of differences is given on the left, whereas our notation of values is given on the right.

Fig. 2. Two rounds of speck

7.1   A Basic 2-Round Attack

The problem of solving differential equations of addition (DEA) of the form $(x \oplus \delta_1) \boxplus (y \oplus \delta_2) = (x \boxplus y) \oplus \delta_3$ (where $\delta_1, \delta_2, \delta_3$ are given and x, y are unknown variables) is a basic problem in the analysis of ARX cryptosystems, and was extensively studied in several papers. In particular, [19] described an algorithm for solving such equations in time complexity which is linear in the total number of solutions. However, the previous algorithm is not directly applicable in our case, as we actually have two dependant equation systems (generated by two addition operations), and we want to efficiently solve them simultaneously. Moreover, the value of y in the DEA is fixed to y_{r+1} for one of the addition operations, and since the solutions vary according to this fixed value, the second equation system is of a different type than the one analyzed in [19]. Note that a standard DEA has an average of 2^n solutions, whereas our equation system has an average of only 1 solution.

Given the complications above, it seems difficult to construct a generic algorithm that efficiently solves our type of equation systems for an arbitrarily large word size n. Thus, we concentrate on the word sizes $n \in \{16, 24, 32, 48, 64\}$ defined by the Speck family, and describe an algorithm whose complexity is estimated by experiments (rather than by a rigorous theoretical proof). As the full key recovery attack of Sect. 6 calls the 2-round attack with a fixed value of $(\Delta x_r, \Delta y_r)$ and many values of $(x_{r+2}, y_{r+2}), (x'_{r+2}, y'_{r+2})$, we are interested in the average complexity of the 2-round attack, where $(x_{r+2}, y_{r+2}), (x'_{r+2}, y'_{r+2})$ are chosen at random according to the procedure of the full key recovery attack.

In Appendix A, we devise a basic guess-and-determine algorithm which exploits the limited carry propagation of the addition operation in order to compute x_r and x_{r+1} bit by bit. It is related to several previous guess-and-determine algorithms such as the one of [12]. The analysis described in Appendix A shows that given some randomness assumptions on the problem, its average execution time is comparable to the execution time of a full Speck encryption.

The problem with this analysis is that the randomness assumptions do not hold in our case (as in many cases of DEA). In fact (as we show next), although we expect one solution on average, for almost any value of $(\Delta x_r, \Delta y_r)$, the distribution of solutions across the various instances is very far from uniform, and greatly depends on the values of the output pairs (x_{r+2}, y_{r+2}) and (x'_{r+2}, y'_{r+2}). More specifically, the solutions are distributed among a small fraction of the output pairs, whereas for the remaining output pairs, there are no solutions at all. Such non-randomness properties have a negative effect of the performance of our basic guess-and-determine algorithm, as it can potentially make a large number of guesses for some bits of x_{r+1} and x_r (i.e., guess partial solutions), while discarding all (or a large fraction) of them at a later stage.[7] Nevertheless, the theoretical analysis based on randomness assumptions strongly indicates that an optimized variant of the attack can perform very efficiently.

7.2   Optimizing the Basic 2-Round Attack Using Filters

In order to optimize the basic algorithm, we notice that we can filter out very quickly a large fraction of the non-useful instances (with no solutions). The idea is to use efficient "look-ahead" (non-linear) filters that try to find a contradiction in the equation constraints before actually computing the solutions. The techniques

---

[7] For example, assume that we want to solve the standard DEA over 16-bit words (given in hexadecimal) $(x \oplus 0000) \boxplus (y \oplus 0000) = (x \boxplus y) \oplus 8000$. If we solve the system from the LSB to the MSB, then we consider all 2^{30} partial solutions to the 15 LSBs of x and y, and then discard all of them at the MSB.

we use to implement the filters are closely related to various search algorithms for differential characteristics for ARX-based and related cryptosystems (e.g., [10,15,16,18,23]).

These filtering techniques allow us to concentrate our efforts on a small fraction of "interesting" instances, and obtain an algorithm whose average time complexity is estimated (according to our simulations) to be smaller than 2 encryptions of Speck.[8]

One-Bit Filter. This filter can be applied to any standard DEA $(x \oplus \delta_1) \boxplus (y \oplus \delta_2) = (x \boxplus y) \oplus \delta_3$. It was first described in [17], and it checks whether

$$eq(\delta_1 \ll 1, \delta_2 \ll 1, \delta_3 \ll 1) \wedge (\delta_1 \oplus \delta_2 \oplus \delta_3 \oplus (\delta_1 \ll 1)) = 0,$$

where $eq(a, b, c) = (\neg a \oplus b) \wedge (\neg a \oplus c)$ equals one at position i if and only if $a^{[i]} = b^{[i]} = c^{[i]}$. As shown in [17], a DEA for which the n-bit value of the filter is non-zero has no solutions and can be filtered out immediately.

This filter is called a 1-bit filter since for each bit position $i + 1$, it only depends the single bit position $i + 1$ of the input words (in addition to a 1-bit XOR difference $\delta_1 \oplus \delta_2 \oplus \delta_3$ at the previous position i). As applying the 1-bit filter involves only a few simple word operations, it requires much less time than a full Speck encryption, and given that it immediately filters out a large fraction of instances with no solutions, it can significantly reduce the running time of the algorithm.

In order to get an estimation of how the filter performs, we assume that all the values of $\delta_1, \delta_2, \delta_3$ are chosen at random. In this case, using the formula of [17], an instance of a DEA will have a solution with probability of $q = 1/2 \cdot (7/8)^{n-1}$. Specifically, for $n = 16$, we have $q \approx 2^{-4}$, and for $n = 24, 32, 48, 64$ we have $q \approx 2^{-5.5}, 2^{-7}, 2^{-10}, 2^{-13}$, respectively.

In the case of Speck, we can immediately apply the filter once for round r and once for round $r+1$ (since all the XOR differences at the inputs and outputs of the addition operations are known). However, as there are clear dependencies between the various input and output XOR differences in 2 rounds of Speck (in fact, the input differences $\Delta x_r, \Delta y_r$ are fixed by the characteristic), the formula does not apply. Nevertheless, our experiments show that for the values of $\Delta x_r, \Delta y_r$ used in our attacks, the filters actually give slightly better results than expected from a random instance. This can be partially explained since $\Delta x_r, \Delta y_r$ have a relatively low hamming weight, as they are outputs of a high probability differential characteristic. As a result, the equality predicate in the filter of round r, $eq(\delta_1 \ll 1, \delta_2 \ll 1, \delta_3 \ll 1) = 0$, holds with a probability which is lower than the expected $1/4$.

As described above, our simulations show that we remain with less than a $2^{-8}, 2^{-11}, 2^{-14}, 2^{-20}, 2^{-26}$ fraction of the instances for $n = 16, 24, 32, 38, 64$,

---

[8] We note that after applying the filters, one can try to apply standard counting techniques to recover some key bits in few first rounds of Speck. However, as we can solve the full equation system and test each suggested key efficiently, the counting techniques are not likely to significantly improve the complexity of the attacks.

respectively, after applying the two 1-bit filters on rounds r and $r+1$. Although the fraction of remaining instances is small, our experiments indicate that when executing the basic algorithm on this small fraction, there are still instances on which we waste a lot of time computing partial solutions that are later discarded. For the smaller values of $n = 16$ and $n = 24$, the effect of these wasteful instances on the average complexity of the algorithm seems to be limited. However, for $n \in \{32, 48, 64\}$, their effect seems to be more significant, and is also more difficult to predict, as we can sample only a small fraction of the possible output pairs after $r+2$ rounds. Consequently, we use additional filters in order to further reduce the number of wasted partial solutions.

Multiple-Bit Filters. These filters are generalizations of the 1-bit filters to larger blocks. They are built by breaking a system of n constraints into b-bit blocks of constraints with a relatively small number of parameters (e.g., a few bits of $\delta_1, \delta_2, \delta_3$ in a standard DEA), and analyzing each block independently. Obviously, if we encounter an equation system instance for which a block (with a certain value of the parameters) has no solutions for any possible values of its variables (e.g., a few bits of x and y in a standard DEA), then the full system has no solutions, and we can stop analyzing it. Given that the number of parameters that appear in a block is sufficiently small, we can exhaustively precompute and store for each of their possible values, a bit that specifies whether the block can potentially have a solution of not (by checking if the equations are satisfied for all possible values of its variables).

For a general system of equations, each block will contain many parameters and variables even for a small value of b, and thus the approach is useless. However, in our case, we can efficiently implement the b-bit filters, as it is easy to break simple equation systems based on ARX into blocks which are almost independent (the only dependency between the blocks are a few bits of carry).

The details of the b-bit filters that we use for our attacks are given in the full version of this paper [11]. As described in the full version, we select $b = 6$, and constructing the filters requires negligible precomputation time (compared to the time complexity of the full differential attacks), while their storage requires only a few megabytes of memory.

7.3   The Optimized 2-Round Attack

The details of our optimized 2-round attack are given in the full version of this paper [11]. We implemented the optimized 2-round attack and estimated its average complexity by running the differential attack of Sect. 5.2 for all the analyzed 10 Speck variants. For Speck variants with $m > 2$, we ran 2 types of tests: one type in which we arbitrarily guessed the values of round keys k_{r+3}, k_{r+2} (or only k_{r+2} for $m = 3$), and one type in which we assumed that their correct value is known. In each test, we executed the optimized 2-round attack for a few randomly chosen keys with 2^{30} arbitrary input pairs that have the fixed input difference of the corresponding characteristic $(\Delta x_0, \Delta y_0)$. In all the tests, the average number of discarded partial solutions for an analyzed pair (which determines the average running time of the algorithm, as the expected number of solutions is close to 1) was smaller than 4, and we estimate that the average time complexity of the 2-round attack is smaller than 2 full encryptions of Speck. We note that we are somewhat less confident in the results for Speck instances with larger word sizes of $n \in \{32, 48, 64\}$, as we can only sample a small fraction of the possible input pairs. However, given the very good results obtained for $n \in \{16, 24\}$, it seems reasonable to believe that the quality of our approximations does not degrade significantly, and the performance of the algorithm is close to what is claimed.

8 Conclusions

In this paper, we presented significantly improved attacks on all 10 variants of the lightweight block cipher Speck, based on an enumeration framework for differential cryptanalysis. This framework tests suggestions for the key that are calculated by a sub-cipher attack, generalizing the algebraic-based framework of Albrecht and Cid [2]. The type of attacks presented in this paper can potentially break a cipher with many more rounds than the number covered by a differential characteristic, especially when the cipher uses a secret key which is larger than the block size (e.g., in our attack on 14-round Speck 32/64, we used a 10-round characteristic to attack 14 rounds of the cipher). Consequently, such sub-cipher attacks should be considered by designers when proposing new cryptosystems.

Since the framework is generic, finding any improved differential characteristic for a Speck variant would (almost) immediately give an improved attack on the full cipher (without the need to perform the low-level statistical analysis, typically required in key recovery attacks based on counting techniques). Furthermore, designing efficient sub-cipher attacks on more rounds of a Speck variant, could also lead to improved attacks. However, such an attack would need to analyze dependencies in the round keys due to the key schedule.

Additional future work items include applying the enumeration framework to improve the best known attacks on more ciphers, and perhaps extending it to other types of attacks, which are different from attacks based on self-similarity and differential attacks.

A Details of the Basic 2-Round Attack

In this section, we give the details of our basic 2-round attack, which computes x_r and x_{r+1} bit by bit given $(x_{r+2}, y_{r+2}), (x'_{r+2}, y'_{r+2})$, assuming the knowledge of $(\Delta x_r, \Delta y_r)$.

In general, we have (for any round $\hat{r}$)

$$((x_{\hat{r}} \ggg \alpha) \boxplus y_{\hat{r}}) \oplus ((x'_{\hat{r}} \ggg \alpha) \boxplus y'_{\hat{r}}) = \Delta x_{\hat{r}+1},$$

and we denote by $z_{\hat{r}}$ the n-bit carry word generated by the addition operation $(x_{\hat{r}} \ggg \alpha) \boxplus y_{\hat{r}}$, and by $z'_{\hat{r}}$ the carry word generated by $(x'_{\hat{r}} \ggg \alpha) \boxplus y'_{\hat{r}}$.

The basic procedure $1RProcedure(\hat{r}, i)$ analyzes round $\hat{r}$ of Speck at bit index $i \in \{0, 1, 2, \ldots, n-1\}$, and is given below. The procedure assumes that we know the XOR input/output differences of round $\hat{r}$ (we actually need only a few bits of these differences), and requires the additional 1-bit value $y_{\hat{r}}^{[i]}$, and the values of the 2 carry bits $z_{\hat{r}}^{[i]}, z_{\hat{r}}^{'[i]}$. The procedure guesses the value of the bit $x_{\hat{r}}^{[i+\alpha]}$, and computes the next 2 carry bits $z_{\hat{r}}^{[i+1]}, z_{\hat{r}}^{'[i+1]}$.

1. For of the 2 possible values of $x_{\hat{r}}^{[i+\alpha]}$:
    (a) Compute $x_{\hat{r}}^{[i+\alpha]} \boxplus y_{\hat{r}}^{[i]} \boxplus z_{\hat{r}}^{[i]}$ and determine the next carry bit $z_{\hat{r}}^{[i+1]}$.
    (b) Compute $x_{\hat{r}}^{'[i+\alpha]} = x_{\hat{r}}^{[i+\alpha]} \oplus \Delta x_{\hat{r}}^{[i+\alpha]}$.
    (c) Compute $x_{\hat{r}}^{'[i+\alpha]} \boxplus y_{\hat{r}}^{'[i]} \boxplus z_{\hat{r}}^{'[i]}$ and determine the next carry bit $z_{\hat{r}}^{'[i+1]}$.
    (d) Check whether $z_{\hat{r}}^{[i+1]} \oplus z_{\hat{r}}^{'[i+1]} \oplus \Delta x_{\hat{r}}^{[i+1+\alpha]} \oplus \Delta y_{\hat{r}}^{[i+1]} = \Delta x_{\hat{r}+1}^{[i+1]}$, and if equality does not hold, discard the guess. Otherwise, output the current value of $x_{\hat{r}}^{[i+\alpha]}$ and the computed carry bits $z_{\hat{r}}^{[i+1]}, z_{\hat{r}}^{'[i+1]}$.

In order to analyze 2-rounds, we assume the we know the values of the XOR input/output differences of rounds $\hat{r}, \hat{r}-1$, and require the 2 additional 1-bit values $y_{\hat{r}}^{[i]}, y_{\hat{r}}^{[i+\alpha]}$, and the values of the 4 carry bits $z_{\hat{r}}^{[i]}, z_{\hat{r}}^{'[i]}, z_{\hat{r}-1}^{[i+\alpha-\beta]}, z_{\hat{r}-1}^{'[i+\alpha-\beta]}$. The procedure guesses the value of the bits $x_{\hat{r}}^{[i+\alpha]}, x_{\hat{r}-1}^{[i+2\alpha-\beta]}$, and computes the next 4 carry bits $z_{\hat{r}}^{[i+1]}, z_{\hat{r}}^{'[i+1]}, z_{\hat{r}-1}^{[i+1+\alpha-\beta]}, z_{\hat{r}-1}^{'[i+1+\alpha-\beta]}$. The $2RProcedure(\hat{r}, i)$ algorithm is given below.

1. Run $1RProcedure(\hat{r}, i)$ and for each returned solution:
    (a) Compute $y_{\hat{r}-1}^{[i+\alpha-\beta]} = x_{\hat{r}}^{[i+\alpha]} \oplus y_{\hat{r}}^{[i+\alpha]}$.
    (b) Run $1RProcedure(\hat{r}-1, i+\alpha-\beta)$

The guess-and-determine algorithm calls $2RProcedure(\hat{r}, i)$ with $\hat{r} = r+1$ for various indexes i in order to recover the full values of x_{r+1} and x_r. Note that all the input/output XOR differences to the last 2 rounds of Speck are known, and the full $y_{\hat{r}} = y_{r+1}$ is known as well, and thus we are only missing the value of the carry bits.

As the carries computed by index j are input to procedure $j+1$, we perform calls to $2RProcedure(r+1, i)$ with sequential labels $i, i+1, \ldots, n-1, 0, \ldots, i-1$ (from LSB to MSB) in order to recover x_{r+1} and x_r. As a result, we only have to guess the carries required by the initial procedure. Since there is no carry into the LSB (i.e., $z_{r+1}^{[0]} = z_{r+1}^{'[0]} = 0$), then we start with procedure $i = 0$ to minimize the number of carry guesses. Furthermore, the value of $z_r \oplus z_r'$ can be computed from $(\Delta x_r \ggg a) \oplus \Delta y_r \oplus \Delta x_{r+1}$, and thus we actually need to guess only one carry bit before executing the first procedure. Finally, after the execution of the last procedure (with index $n-1$), we can derive the actual value of this guessed carry bit and obtain an additional filtering condition.

As the time complexity of the guess-and-determine algorithm is proportional to the number of guesses it makes, we need to carefully analyze the ratio between the number of guessed bits, and the number of filtering conditions used to filter the guesses. The algorithm $2RProcedure(\hat{r}, i)$ guesses the values of the two bits $x_{\hat{r}}^{[i+\alpha]}, x_{\hat{r}-1}^{[i+2\alpha-\beta]}$ and uses two filtering conditions (one in each call to $1RProcedure(\hat{r}, i)$ at Step 1.(d)). Thus, assuming that the analyzed instance behaves randomly, we expect the number of guesses at each stage of the execution of the algorithm to remain constant. Such randomness assumptions lead to the conclusion that the average execution time of the algorithm is comparable to the execution time of a full Speck encryption (perhaps even smaller, as we analyze only 2 rounds). However, as noted in Sect. 7.1, these randomness assumptions do not hold in our case.

References

1. Abed, F., List, E., Wenzel, J., Lucks, S.: Differential Cryptanalysis of round-reduced Simon and Speck. Presented at FSE 2014. To Appear in Lecture Notes in Computer Science (2014)
2. Albrecht, M., Cid, C.: Algebraic techniques in differential cryptanalysis. In: Dunkelman, O. (ed.) FSE 2009. LNCS, vol. 5665, pp. 193–208. Springer, Heidelberg (2009)
3. Albrecht, M., Cid, C., Dullien, T., Faugère, J.-C., Perret, L.: Algebraic precomputations in differential and integral cryptanalysis. In: Lai, X., Yung, M., Lin, D. (eds.) Inscrypt 2010. LNCS, vol. 6584, pp. 387–403. Springer, Heidelberg (2011)
4. Alizadeh, J., Bagheri, N., Gauravaram, P., Kumar, A., Sanadhya, S.K.: Linear cryptanalysis of round reduced SIMON. Cryptology ePrint Archive, Report 2013/663 (2013). http://eprint.iacr.org/
5. Alkhzaimi, H.A., Lauridsen, M.M.: Cryptanalysis of the SIMON family of block ciphers. Cryptology ePrint Archive, Report 2013/543 (2013). http://eprint.iacr.org/
6. Bar-On, A., Dinur, I., Dunkelman, O., Lallemand, V., Tsaban, B.: Improved analysis of Zorro-like ciphers. IACR Cryptology ePrint Archive (2014)
7. Beaulieu, R., Shors, D., Smith, J., Treatman-Clark, S., Weeks, B., Wingers, L.: The SIMON and SPECK families of lightweight block ciphers. Cryptology ePrint Archive, Report 2013/404 (2013). http://eprint.iacr.org/
8. Biham, E., Shamir, A.: Differential cryptanalysis of des-like cryptosystems. J. Cryptol. **4**(1), 3–72 (1991)
9. Biryukov, A., Roy, A., Velichkov, V.: Differential analysis of block ciphers SIMON and SPECK. Presented at FSE 2014. To Appear in Lecture Notes in Computer Science (2014)
10. De Cannière, C., Rechberger, C.: Finding SHA-1 characteristics: general results and applications. In: Lai, X., Chen, K. (eds.) ASIACRYPT 2006. LNCS, vol. 4284, pp. 1–20. Springer, Heidelberg (2006)
11. Dinur, I.: Improved differential cryptanalysis of round-reduced Speck. IACR Cryptology ePrint Archive (2014)
12. Dinur, I., Dunkelman, O., Shamir, A.: Improved attacks on full GOST. In: Canteaut, A. (ed.) FSE 2012. LNCS, vol. 7549, pp. 9–28. Springer, Heidelberg (2012)
13. Isobe, T.: A single-key attack on the full GOST block cipher. J. Cryptol. **26**(1), 172–189 (2013)

14. Kara, O.: Reflection cryptanalysis of some ciphers. In: Chowdhury, D.R., Rijmen, V., Das, A. (eds.) INDOCRYPT 2008. LNCS, vol. 5365, pp. 294–307. Springer, Heidelberg (2008)

15. Leurent, G.: Analysis of differential attacks in ARX constructions. In: Wang, X., Sako, K. (eds.) ASIACRYPT 2012. LNCS, vol. 7658, pp. 226–243. Springer, Heidelberg (2012)

16. Leurent, G.: Construction of differential characteristics in ARX designs application to skein. In: Canetti, R., Garay, J.A. (eds.) CRYPTO 2013, Part I. LNCS, vol. 8042, pp. 241–258. Springer, Heidelberg (2013)

17. Lipmaa, H., Moriai, S.: Efficient algorithms for computing differential properties of addition. In: Matsui, M. (ed.) FSE 2001. LNCS, vol. 2355, pp. 336–350. Springer, Heidelberg (2002)

18. Mendel, F., Nad, T., Schläffer, M.: Finding SHA-2 characteristics: searching through a minefield of contradictions. In: Lee, D.H., Wang, X. (eds.) ASIACRYPT 2011. LNCS, vol. 7073, pp. 288–307. Springer, Heidelberg (2011)

19. Paul, S., Preneel, B.: Solving systems of differential equations of addition. In: Boyd, C., González Nieto, J.M. (eds.) ACISP 2005. LNCS, vol. 3574, pp. 75–88. Springer, Heidelberg (2005)

20. Stein, W.A., et al.: Sage Mathematics Software. The Sage Development Team. http://www.sagemath.org

21. Tupsamudre, H., Bisht, S., Mukhopadhyay, D.: Differential fault analysis on the families of SIMON and SPECK ciphers. Cryptology ePrint Archive, Report 2014/267 (2014). http://eprint.iacr.org/

22. Wang, M., Sun, Y., Mouha, N., Preneel, B.: Algebraic techniques in differential cryptanalysis revisited. In: Parampalli, U., Hawkes, P. (eds.) ACISP 2011. LNCS, vol. 6812, pp. 120–141. Springer, Heidelberg (2011)

23. Wang, X., Yu, H.: How to break MD5 and other hash functions. In: Cramer, R. (ed.) EUROCRYPT 2005. LNCS, vol. 3494, pp. 19–35. Springer, Heidelberg (2005)

Differential Cryptanalysis of SipHash

Christoph Dobraunig$^{(\boxtimes)}$, Florian Mendel, and Martin Schläffer

IAIK, Graz University of Technology, Graz, Austria
christoph.dobraunig@iaik.tugraz.at

Abstract. SipHash is an ARX based message authentication code developed by Aumasson and Bernstein. SipHash was designed to be fast on short messages. Already, a lot of implementations and applications for SipHash exist, whereas the cryptanalysis of SipHash lacks behind. In this paper, we provide the first published third-party cryptanalysis of SipHash regarding differential cryptanalysis. We use existing automatic tools to find differential characteristics for SipHash. To improve the quality of the results, we propose several extensions for these tools to find differential characteristics. For instance, to get a good probability estimation for differential characteristics in SipHash, we generalize the concepts presented by Mouha et al. and Velichkov et al. to calculate the probability of ARX functions. Our results are a characteristic for SipHash-2-4 with a probability of $2^{-236.3}$ and a distinguisher for the Finalization of SipHash-2-4 with practical complexity. Even though our results do not pose any threat to the security of SipHash-2-4, they significantly improve the results of the designers and give new insights in the security of SipHash-2-4.

Keywords: Message authentication code · MAC · Cryptanalysis · Differential cryptanalysis · SipHash · S-functions · Cyclic S-functions

1 Introduction

A message authentication code (MAC) is a cryptographic primitive, which is used to ensure the integrity and the origin of messages. Normally, a MAC takes a secret key K and a message M as input and produces a fixed size tag T. A receiver of such a message-tag-pair verifies the authenticity of the message by simply recalculating the tag T for the message and compare it with the received one. If the two tags are the same, the origin of the message and its integrity are ensured.

SipHash [1] was proposed by Aumasson and Bernstein due to the lack of MACs, which are fast on short inputs. Aumasson and Bernstein suggest two main fields of application for SipHash. The first application is as replacement for non-cryptographic hash functions used in hash-tables and the second application is to authenticate network traffic. The need for a fast MAC used in hash-tables arises from the existence of a denial-of-service attack called "hash flooding" [1]. This attack uses the fact, that is easy to find collisions for non-cryptographic

© Springer International Publishing Switzerland 2014
A. Joux and A. Youssef (Eds.): SAC 2014, LNCS 8781, pp. 165–182, 2014.
DOI: 10.1007/978-3-319-13051-4_10

hash functions. With the help of these collision producing inputs, an attacker is able to degenerate hash-tables to e.g. linked lists. Such a degeneration increases the time to perform operations like searching and inserting elements drastically and can lead to denial of service attacks.

So far, SipHash is already implemented in many applications. For example, SipHash is used as `hash()` in Python on all major platforms, in the dnschache instances of OpenDNS resolvers and in the hash-table implementation of Ruby. Besides these mentioned applications, other applications and dozens of third-party implementations of SipHash can be found on the SipHash website[1].

In this paper, we provide the first external security analysis regarding differential cryptanalysis. To find differential characteristics, we adapt techniques originally developed for the analysis of hash functions to SipHash. Using differential cryptanalysis to find collisions for hash functions has become very popular since the attacks on MD5 and SHA-1 by Wang et al. [15,16]. As a result, a number of automated tools have been developed to aid cryptographers in their search for valid characteristics [6–8]. For hash functions, the probability of a characteristic does not play an important role since message modification can be used to improve the probability and create collisions. However, this is not possible for keyed primitives like MACs. Therefore, we have to modify existing search tools to take the probability of a characteristic into account. With the help of these modified tools, we are able to improve the quality of the results for SipHash.

In cryptographic primitives consisting solely out of modular additions, rotations and xors (like SipHash), only the modular addition might contribute to the probability of a differential characteristic, if xor differences are considered for representation. A method to calculate the exact differential probability of modular additions is presented by Mouha et al. [9]. In constructions like SipHash, modular additions, rotations and xors interact together. Hence, the characteristic uses many intermediate values of the single rounds and is therefore divided into many small sections. To get a more exact prediction of the probability for the characteristic, it would be nice to calculate the probability of subfunctions combining modular additions, rotations and xors. Therefore, we introduce the concept of cyclic S-functions. This concept is a generalization of the work done by Mouha et al. [9] and Velichkov et al. [13] for generalized conditions [3]. Although all the basic concepts needed to create cyclic S-functions are already included in the work of Velichkov et al. [13], we do not think that the generalization to generalized conditions is trivial, since we have not seen a single use of it. Cyclic S-functions will help analysts and designers of ARX based cryptographic primitives to provide closer bounds for the probability of differential characteristics.

With the help of the extended search strategies and the probability calculation, we are able to find the first published collision producing characteristics for SipHash-1-x and SipHash-2-x (see Table 1). The characteristic for SipHash-2-x is also the best known characteristic for SipHash-2-4. Moreover, we are able to present a distinguisher for the Finalization (4 SipRounds) of SipHash-2-4.

---

[1] https://131002.net/siphash/

Table 1. Best found characteristics.

Instance	Type	Probability	Reference
SipHash-2-4	High probability	2^{-498}	[1]
SipHash-2-4	High probability	$2^{-236.3}$	Sect. 5.1
SipHash-2-x	Internal collision	$2^{-236.3}$	Sect. 5.1
SipHash-1-x	Internal collision	2^{-167}	Sect. 5.1
4 Round Finalization	High probability	2^{-35}	Sect. 5.2

The paper starts with a description of SipHash in Sect. 2. The following Sect. 3 explains the basic concepts and strategies used by us to search for differential characteristics. Section 4 deals with improvements of automatic search techniques to find suitable characteristics for SipHash. Finally, the most significant differential characteristics for SipHash found by us are presented in Sect. 5. Further results on SipHash are given in Appendix A.

2 Description of SipHash

SipHash is a cryptographic MAC consisting solely of modular additions, rotations and xors (ARX). SipHash has an internal state size of 256 bits, uses a 128-bit key and produces a 64-bit tag. The process of authenticating a single message can be split into three stages: Initialization, Compression and Finalization.

- **Initialization.** The internal state V of SipHash consists of the four 64-bit words V_a, V_b, V_c and V_d. The initial value consists of the ASCII representation of the string "somepseudorandomlygeneratedbytes" and is written to the internal state first. Then, the 128-bit key $K = K_1 \parallel K_0$ is xored to the state words $V_a \parallel V_b \parallel V_c \parallel V_d = V_a \parallel V_b \parallel V_c \parallel V_d \oplus K_0 \parallel K_1 \parallel K_0 \parallel K_1$.
- **Compression.** The message M is padded with as many zeros as needed to reach multiple block length minus 1 byte. Then, one byte, which encodes the length of the message modulo 256 is added to get a multiple of the block length. Afterwards, the message is split into t 8-byte blocks M_1 to M_t. The blocks M_i are in little-endian encoding. For each block M_i, starting with block M_1, the following is performed. The block M_i is xored to V_d. After that the SipRound function is performed c times on the internal state. Then the block M_i is xored to V_a.
- **Finalization.** After all message blocks have been processed, the constant ff_{16} is xored to V_c. Subsequently, d iterations of SipRound are performed. Finally, $V_a \oplus V_b \oplus V_c \oplus V_d$ is used as the MAC value $h_K = \text{SipHash-c-d}(K, M)$.

As shown above, SipHash is parameterizable using c SipRounds in the Compression and d SipRounds in the Finalization. Such a specific instantiation of SipHash is called SipHash-c-d. Aumasson and Bernstein propose two specific versions for use, which are SipHash-2-4 and SipHash-4-8.

Next, we describe one SipRound. As SipHash is an ARX based MAC, the SipRound network shown in Fig. 1 consists only of modular additions, xors and rotations. Every operation is an operation on 64-bit.

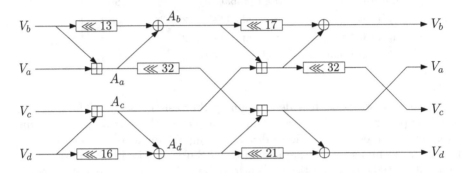

Fig. 1. One SipRound [1].

Now, we will discuss our naming scheme for the different variables involved in SipHash. In Fig. 1 one SipRound is shown. We will indicate a specific bit of a word by $V_{a,m,r}[i]$, where $i = \{0, ..., 63\}$ denotes the specific bit position of a word, m denotes the message block index, and r denotes the specific SipRound. Hence, to process the first message block, the input to the first SipRound is denoted by $V_{a,1,1}$, the intermediate variables by $A_{a,1,1}$, and the output by $V_{a,1,2}$. Words, which take part in the Finalization, are indicated with $m = f$.

3 Automatic Search for Differential Characteristics

For the search for differential characteristics, we have used an automatic search tool. To make use of such a tool, several key aspects have to be considered.

- The representation of the differential characteristic (Sect. 3.1).
- The description of the cryptographic primitive to perform propagation (Sect. 3.2).
- The used search strategy to search for characteristics (Sect. 3.3).

3.1 Generalized Conditions

We use generalized (one-bit) conditions introduced by De Cannière and Rechberger [3] to represent the differential characteristics within the automatic search tool. With the help of the 16 generalized conditions, we are able to express every possible condition on a pair of bits. For instance the generalized condition x denotes unequal values, - denotes equal values and ? denotes that every value for a pair of bits is possible.

In addition, we also use generalized two-bit conditions [6]. Using these conditions, every possible combination of a pair of two bits $|\Delta x, \Delta y|$ can be represented. In the most general form, these two bits can be any two bits of a characteristic. In our case, such two-bit conditions are used to describe differential information on carries, when computing the probability using cyclic S-functions.

3.2   Propagation of Conditions

Single conditions of a differential characteristic are connected via functions (additions, rotations, xors). Thus, the concrete value of a single condition affects other conditions. To be more precise, the information that certain values on a condition are allowed may lead to the effect, that certain values on other conditions are impossible. Therefore, we are able to remove impossible values on those conditions to refine their values. We can say that information propagates.

Within the automatic search tool, we do this propagation in a bitsliced manner like it is shown in [7]. This means, that we split the functions into single bitslices and brute force them by trying all possible combinations allowed by the generalized conditions. In this way, we are able to remove impossible combinations.

For the performance of the whole search for characteristics, it is crucial to find a suitable "size" of the subfunctions of a specific cryptographic primitive, which are used to perform propagation. The "size" of such a subfunction determines how many different conditions are involved during brute forcing a single bitslice. On one hand, "big" functions (many conditions involved in one bitslice) make the propagation slower. On the other hand, the amount of information that propagates is usually enhanced by using a few "big" subfunctions instead of many "small" ones. Generalized conditions are not able to represent every information that is gathered during propagation(mainly due to effects regarding the carry of the modular addition) [6]. So we loose information between single subfunctions. Usually, less information is lost if the subfunctions are "bigger". In fact, finding a good trade-off between speed and quality of propagation is not trivial.

3.3   Basic Search Strategy

For analyzing unkeyed primitives, especially when trying to find collisions for hash functions, the following search method, as used by Mendel et al. [7] has turned out to be a viable strategy.

– Find a good starting point for the search.
– Search for a good characteristic.
– Use message modification to find a colliding message pair.

The starting point describes the target problem to be solved and a good starting point can greatly reduce the complexity of a search. In the start characteristic, bits where no differences are allowed are represented by the condition – and bits where the characteristic may contain differences are represented by ?.

The search algorithm refines the conditions represented by a ? to − and x to get a valid differential characteristic in the end. An example of a search strategy can be found in [7]. Here, the search algorithm is split in three main parts:

- **Decision (Guessing).** In the guessing phase, a bit is selected, which condition is refined. This bit can be selected randomly or according to a heuristic.
- **Deduction (Propagation).** In this stage, the effects of the previous guess on other conditions is determined (see Sect. 3.2).
- **Backtracking (Correction).** If a contradiction is determined during the deduction stage the contradiction is tried to be resolved in this stage. A way to do this is to jump back to earlier stages of the search until the contradiction can be resolved.

There exist several message modification techniques [5,12,14]. The one used by us (in Appendix A) is refining a valid characteristic further until the colliding message pair is fixed [7]. For keyed primitives like SipHash, message modification to enhance the probability is usually out of reach, since the key is unknown. Hence, we need to stop the search at the characteristic and need an adapted search algorithm to find characteristics, which have a high probability.

4   Improvements in the Automatic Search for SipHash

We have used existing automatic search tools to analyze SipHash. Those tools use the search strategy describe in Sect. 3.3. This strategy has been developed to find collisions for hash functions. It turns out that this strategy is unsuitable for keyed primitives like MACs. Therefore, we extend the search strategy to the greedy strategy described in Sect. 4.1. This greedy strategy uses information on the probability of characteristics, or on the impact of one guess during the search for characteristics. We have created all the results of Sect. 5 with the help of this greedy strategy. To get closer bounds on the probability of the characteristic, we generalize the concepts presented by Mouha et al. [9] and Velichkov et al. [13] to cyclic S-functions (Sect. 4.2).

Another important point in the automatic search for differential characteristics is the representation of the cryptographic primitives within the search tool. We have evaluated dozens of different descriptions and present the most suitable in Sect. 4.3.

4.1   Extended Search Strategy

Our search strategy extends the strategy used in [7]. The search algorithm of Sect. 3.3 is split in three main parts decision (guessing), deduction (propagation), and backtracking (correction). We have extended this strategy to perform greedy searches using quality criteria like the probability. In short, we perform the guessing and propagation phase several times on the same characteristic. After that, we evaluate the resulting characteristics and take the characteristic with the best probability. Then, the next iteration of the search starts. The upcoming algorithm describes the search in more detail:

Let U be a set of bits with condition ? in the current characteristic A. In H we store all characteristics, which have been visited during the search. L is a set of candidate characteristics for A. n is the number of guesses. B_- and B_x are characteristics.

Preparation

1. Generate U from A. Clear L. Set i to 0.

Decision (Guessing)

2. Pick a bit from U.
3. Restrict this bit in A to – to get B_- and to x to get B_x.

Deduction (Propagation)

4. Perform propagation on B_- and B_x.
5. If B_- and B_x are inconsistent, mark bit as critical and go to Step 13, else continue.
6. If B_- is not inconsistent and not in H, add it to L. Do the same for B_x.
7. Increment i.
8. If i equals n, continue with Evaluation. Else go to Step 2.

Evaluation

9. Set A to the characteristic with the highest probability in L.
10. Add A to H.
11. If there are no ? in A, output A. Then set A to a characteristic of H.
12. Continue with Step 1.

Backtracking (Correction)

13. Jump back until the critical bit can be resolved.
14. Continue with Step 1.

To generate the set U, we use the following variables of SipHash A_a, A_c, V_a, and V_c. Except for $V_{a,m,1}$, $V_{c,m,1}$, $V_{a,f,1}$, and $V_{c,f,1}$, since they are only connected via an xor to their predecessors, or are even the same variable. We guess solely on those variables, since a guess on them always affects the probability of the differential characteristic. Experiments have shown that setting the number of guesses n to 25 leads to the best results.

Due to performance reasons, we store hash values of characteristics in H. In addition, we maintain a second list H^*. In this list, we store the next best characteristics of L according to a certain heuristic. The characteristics of H^* are also used for backtracking. If a characteristic is found and U is empty, we take a characteristic out of H^* instead of H in Step 11. After a while, we perform a soft restart, where everything is set to the initial values (also H^* is cleared) except for H.

This search strategy turns out to be good if we search for high probability characteristics, which do not lead to collisions. When searching for colliding characteristics, we have to adapt the given algorithm and perform a best impact strategy similar to Eichlseder et al. [4].

The best impact strategy differs in the following points from the strategy described above. In the best impact strategy, we do not calculate the characteristic B_x. Instead of taking the probability of B_- as a quality criterion for the selection, we use the variant of B_- of the candidate list L, where the most information propagates. As a figure of merit for the amount of information that propagates, we take the number of conditions with value ?, which have changed their value due to the propagation.

This best impact strategy has several advantages. The first one lies in the fact that mostly guesses will be made, which have a big impact on the characteristic. This ensures that no guesses are made, where nothing propagates. Such guesses often imply additional restrictions on the characteristic, which are not necessary. In addition, the big impact criterion also leads to rather sparse characteristics, which usually have a better probability than dense ones.

4.2   Calculating the Probability Using Cyclic S-Functions

In this section, we show a method to extend the use of S-functions [9] by introducing state mapping functions m_i and making the relationship between the states cyclic. For instance such cyclic states occur if rotations work together with modular additions. Velichkov et al. showed in [13] how to calculate the additive differential probability of ARX based functions. The method of cyclic S-functions is closely related to the methods shown in [13].

Concept of Cyclic S-Functions. According to Mouha et al. [9], a state function (S-function) is a function, where the output $s[i]$ can be computed using only the input bits $a_1[i]$, $a_2[i]$,... $a_k[i]$ and the finite state $S[i-1]$. This computation also leads to a next state $S[i]$. An example of such an S-function is the modular addition $a + b = s$. In ARX systems, we can discover the same behavior as for additions, except that the first state $S[0]$ depends on the outcome of the last operation and is therefore related to $S[n]$. We picture this relation by introducing so called state mapping functions m_i and making the state cyclic as it is shown in Fig. 2. The state mapping function m_i is a function, which maps distinct state values of $S_o[i]$ to $S_i[i]$. It is possible and often the case that more values of $S_o[i]$ map to the same value of $S_i[i]$. If m_i is the identity function, then the states $S_o[i]$ and $S_i[i]$ are the same state and we only write $S[i]$ in this case.

Note that every classic S-function can be transformed into a cyclic S-function by defining every m_i as the identity function except for m_n. The function m_n maps every value of $S_o[n]$ to the state $S_i[0] = 0$.

To give an example, we describe the function $((a + b) \lll 1) + c = s_b$ with the help of S-functions. In this example, we use 4-bit words. We picture the system as it is shown in Fig. 3. The carry c_a and c_b serve as state S together.

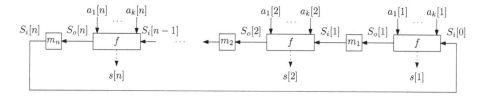

Fig. 2. Concept of cyclic S-functions.

They can be considered as a two-bit condition $|c_a, c_b|$. The black vertical lines in Fig. 3 mark transitions, where the state mapping function m_i is **not** the identity function. As $c_a[0]$ and $c_b[0]$ can only be 0, the state mapping functions perform the following mapping for any value v_a of c_a and v_b of c_b:

- $S_o[1] \Rightarrow S_i[1] : |v_a, v_b| \Rightarrow |0, v_b|$
- $S_o[4] \Rightarrow S_i[0] : |v_a, v_b| \Rightarrow |v_a, 0|$.

So the states in case of the system in Fig. 3 are:

- $S_i[0] = |c_a[3], 0|$
- $S_i[1] = |0, c_b[1]|$
- $S[3] = |c_a[2], c_b[3]|$
- $S_o[1] = |c_a[4], c_b[1]|$
- $S[2] = |c_a[1], c_b[2]|$
- $S_o[4] = |c_a[3], c_b[4]|$

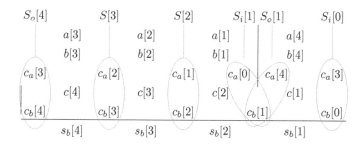

Fig. 3. Rewritten system to do $((a + b) \lll 1) + c = s_b$ in one step.

For a word length of n and a general rotation to the left by r, the state is $S[i] = |c_a[(i - r) \mod n], c_b[i]|$, except for states, where m_i is not the identity function. These are the states $S_o[r] = |c_a[n], c_b[r]|$, $S_i[r] = |c_a[0], c_b[r]|$, $S_o[n] = |c_a[n - r], c_b[n]|$ and $S_i[0] = |c_a[n - r], c_b[0]|$. The realization of additions with multiple rotations in between leads to more mapping functions m_i, which are not the identity function. Using additions with more inputs leads to bigger carries and bigger states.

Using Graphs for Description. Similar to S-functions [9], we can build a graph representing the respective cyclic S-function. The vertices in the graph stand for the single distinct states and circles in the graph represent valid solutions. Such a graph can be used to either propagate conditions, or to calculate the differential probability. An illustrative example for propagation and the probability calculation can be found in Appendix B.

The whole cyclic graph consists of subgraphs i. Each subgraph i consists of vertices representing $S_i[i-1]$, and $S_o[i]$ and single edges connecting them. So each subgraph represents a single bitslice of the whole function. For the system in Fig. 3, the edges of each subgraph are calculated by trying every possible pair of input bits for $a[((i-r-1) \mod n)+1]$, $b[((i-r-1) \mod n)+1]$ and $c[i]$, which is given by their generalized conditions and using every possible carry of the set of $S_i[i-1]$ to get an output $s_b[i]$ and a carry which belongs to $S_o[i]$. If the output is valid (with respect to the generalized conditions, which describe the possible values for s_b), an edge can be drawn from the respective value of the input vertex of $S_i[i-1]$ to the output vertex belonging to $S_o[i]$. Such a subgraph can be created for every bitslice.

Now, we have to form a graph out of these subgraphs. Subgraphs connected over a state mapping function m_i, which is the identity, stay the same. There exist two ways for connecting subgraphs i and $i+1$, which are separated by a state mapping function. Either the edges of graph i can be redrawn, so that they follow the mapping from $S_o[i]$ to $S_i[i]$, or the edges of graph $i+1$ can be redrawn so that they follow the inverse mapping from $S_i[i]$ to $S_o[i]$. We call the so gathered set of subgraphs "transformed subgraphs". After all subgraphs are connected, we can read out the valid input output combinations. These combinations are minimal circles in the directed graph. Since we are aware of the size of the minimal circles and of the shape of the graph, we can transform the search for those circle in a search for paths.

Probability Calculation Using Matrix Multiplication. To calculate the differential probability, we only need to divide the number of valid minimal circles of the graph by the number of total possible combinations of the input. The number of valid minimal circles can be calculated with the help of matrix multiplications. Similar to S-functions [9], we have to calculate the biadjacency matrix $A[i] = [x_{kj}]$ for each "transformed subgraph". x_{kj} stands for the number of edges, which connect vertex j of the group $S_i[i-1]$ with vertex k of the group $S_i[i]$. We define the $1 \times N$ matrices L_i and the $N \times 1$ matrices C_i.

$$L_1 = \begin{bmatrix} 1\,0\,0\,...\,0 \end{bmatrix} \qquad C_1 = \begin{bmatrix} 1\,0\,0\,...\,0 \end{bmatrix}^T$$
$$L_2 = \begin{bmatrix} 0\,1\,0\,...\,0 \end{bmatrix} \qquad C_2 = \begin{bmatrix} 0\,1\,0\,...\,0 \end{bmatrix}^T$$
$$...$$
$$L_n = \begin{bmatrix} 0\,0\,0\,...\,1 \end{bmatrix} \qquad C_n = \begin{bmatrix} 0\,0\,0\,...\,1 \end{bmatrix}^T$$

Here, N is the number of distinct states of $S[i]$. As $S_i[n]$ (this is $S_o[n]$ after applying the mapping) and $S_i[0]$ are in fact the same states, we can calculate the number of circles by summing up all paths which lead from a vertex in $S_i[0]$ to the same vertex in $S_i[n]$:

$$\#Circles = \sum_{i=1}^{n} (L_i \cdot A[n] \cdots A[2] \cdot A[1] \cdot C_i) \tag{1}$$

The formula shown above basically sums the numbers in the diagonal of the resulting matrix when all $A[i]$ are multiplied together. This sum divided by all possible input combinations gives us the exact differential probability of one step.

We consider the presented method based on cyclic S-functions to be equivalent to brute force and therefore to be optimal. The equivalence is only given if the words of the input and the output are independent of each other. For example, if the same input is used twice in the same function f, we do not have the required independence. Such a case is the calculation of $s = a + (a \lll 10)$.

Probability Calculation of SipHash. To calculate the probability of SipHash we group two subsequent additions together, considering also the intermediate outputs. In contrast to the propagation (Sect. 4.3), we do not do this overlapping, since we would calculate the probability twice. So we get two subfunctions per SipRound to calculate the probability (2). In the case of SipHash, we also consider the intermediate values of the additions, but they are omitted in the formulas.

$$\begin{aligned} V_{a,m,r+1} &= ((V_{a,m,r} + V_{b,m,r}) \lll 32) + A_{d,m,r} \\ V_{c,m,r+1} &= (V_{c,m,r} + V_{d,m,r} + A_{b,m,r}) \lll 32 \end{aligned} \tag{2}$$

In addition, we also consider the differential probability introduced by xors. Since we use generalized conditions, xors might contribute to the probability as well. To calculate the differential probability connected with xors, a simple bitsliced approach is used.

4.3   Bitsliced Description of SipHash

For SipHash, we have evaluated dozens of different descriptions. Because of several searches and other evaluations, we have chosen the following description. We combine every two subsequent two input modular additions to one subfunction, regardless if they are separated by a rotation. The resulting subfunctions overlap each other. This means, that every two input addition as shown in Fig. 1 takes part in two subfunctions. This results in the following subfunctions (the intermediate output is also considered, but not given in the formulas):

$$\begin{aligned} V_{a,m,r+1} &= ((V_{a,m,r} + V_{b,m,r}) \lll 32) + A_{d,m,r} \\ V_{c,m,r+1} &= (V_{c,m,r} + V_{d,m,r} + A_{b,m,r}) \lll 32 \\ A_{a,m,r+1} &= (A_{a,m,r} \lll 32) + A_{d,m,r} + V_{b,m,r+1} \\ A_{c,m,r+1} &= ((A_{c,m,r} + A_{b,m,r}) \lll 32) + V_{d,m,r+1} \end{aligned}$$

The xor operations are represented using only three variables (two inputs). We do this, since there is no information loss due to the representation capability of the generalized conditions.

5 Results

In this section, we give some results using the presented search strategies and the new probability calculation. At first, we start with characteristics, which lead to internal collisions. This type of characteristic can be used to create forgeries as described in [10,11]. To improve this attack, characteristics are needed, which have a probability higher than 2^{-128} (in the case of SipHash). Otherwise, a birthday attack should be preferred to find collisions. We are able to present characteristics that lead to an internal collision for SipHash-1-x (2^{-167}) and SipHash-2-x ($2^{-236.3}$). The characteristic for SipHash-2-x is also the best published characteristic for full SipHash-2-4.

The last part of this section deals with a characteristic for the Finalization of SipHash-2-4. This characteristic has a considerable high probability of 2^{-35}. Due to this high probability, this characteristic can be used as a distinguisher for the Finalization.

5.1 Colliding Characteristics for SipHash-1-x and SipHash-2-x

First, we want to start with an internal collision producing characteristic for SipHash-1-x. We have achieved the best result with the biggest impact strategy by using a starting point consisting of 7 message blocks. The bits of the first message block, the key, and the last state values are set to -. The rest of the characteristic is set to ?. We introduce one difference in a random way by picking a bit out of all A_a, A_c, V_a, and V_c. This strategy results in a characteristic with an estimated probability of 2^{-169}. The characteristic leads to an internal collision within 3 message blocks.

In the second stage of the search, we fix the values of the 3 message blocks to the values found before and set the internal state variables to ?. By using the so gotten starting point, we perform a high probability greedy search. This high probability greedy search results in the characteristic given in Table 2. This characteristic has an estimated probability of 2^{-167}.

Next, we handle the search for a characteristic, which results in an internal collision for SipHash-2-x. The collision with the best probability has been found by setting only the bits of one Compression iteration (including the corresponding message block) to ? and everything else to -. The difference is introduced in the most significant bit of the message block. The best characteristic we have found using this starting point and the best impact strategy has an estimated probability of $2^{-238.9}$. In a second stage, we use the value for the message block of this found characteristic to perform a high probability greedy search. With this method, we are able to get the characteristic of Table 3. This characteristic has an estimated probability of $2^{-236.3}$.

Table 2. Characteristic for SipHash-1-x, which leads to an internal collision (probability 2^{-167}).

M_1	-------------------------------------	K_0	-------------------------------------
M_2	--------------------------x----------	K_1	-------------------------------------
M_3	-xx-xxxx--------x----xxxx-----xxxxx-----x--x-x-		
M_4	xxx-----------xxx--------xxxxx-----xxxxx		
$V_{a,1,1}$	-------------------------------------	$V_{b,1,1}$	-------------------------------------
$V_{c,1,1}$	-------------------------------------	$V_{d,1,1}$	-------------------------------------
$V_{a,1,2}$	-------------------------------------	$V_{b,1,2}$	-------------------------------------
$V_{c,1,2}$	-------------------------------------	$V_{d,1,2}$	-------------------------------------
$V_{a,2,1}$	-------------------------------------	$V_{b,2,1}$	--------------------------x----------
$V_{c,2,1}$	----x-------------x-----------x-------	$V_{d,2,1}$	--------------------------x----------
$V_{c,2,2}$	----x-------------x-----------x-------	$V_{d,2,2}$	x-------------x---x-----------x-------
$V_{a,3,1}$	-------------------------------------	$V_{b,3,1}$	xxx-xxxx--------------xxx-----xxxxx---x--x-x-
$V_{a,3,2}$	-x-x-x---x-xxxxxx--xx--x--x-x-xxxx-x---xxx--xxxxx-x---x-xxx---	$V_{b,3,2}$	-x-xx--------------xxx-x----1-----xxxxx--x-x-
$V_{c,3,2}$	-----xxxxx--x-x------------x-x---x-----	$V_{d,3,2}$	xxx-------xxxxxxx------------xxxxx---xxx-------xx--xxxxx
$V_{a,4,1}$	--xxx-xxx-xxxxxxxxx----xx--xxxx-xxxx--x--xx-xxxxx---x-xxx---	$V_{b,4,1}$	-----xxxxx-------------xxxxxxxx------xxx---
$V_{a,4,2}$	xxx--------------xxx-----------xxxxx------xxxxx	$V_{b,4,2}$	-------------------------------------
$V_{c,4,2}$	-------------------------------------	$V_{d,4,2}$	-------------------------------------
$V_{a,f,1}$	-------------------------------------	$V_{c,f,1}$	-------------------------------------
h_K	-------------------------------------		

Table 3. Characteristic for SipHash-2-x, which leads to an internal collision (probability of $2^{-236.3}$). This is also the best characteristic for SipHash-2-4.

M_1	x--xxx--x--xxx---xxxxxxx---xx-xxx-x--xx-xx-xxxx---xx--x---xx--	K_0	-------------------------------------
		K_1	-------------------------------------
$V_{a,1,1}$	-------------------------------------	$V_{b,1,1}$	-------------------------------------
$V_{c,1,1}$	-------------------------------------	$V_{d,1,1}$	x--xxx--x--xxx---xxxxxxx---xx-xxx-x--xx-xx-xxxx---xx--x---xx--
$V_{a,1,2}$	xxx--x-x-x-x-xx----x---x----x-x-x-xxxx--xx-xxxxxxxxx--xx----x---	$V_{b,1,2}$	x-x--x-x-x-x---x--xx--xxx-------x---xxx--x-x-x-x-xx---x----x--
$V_{c,1,2}$	xxx--x-x-x-x-xx----x---x----x-x-x-xxxx--xx-xxxxxxxxx--xx----x---	$V_{d,1,2}$	xx-xxx-x--xx-xx-xxxx---xx----x---xx--x-x-xxx--x--xxx----xxxxxxx
$V_{a,1,3}$	x--xxx--x--xxx---xxxxxxx---xx-xxx-x--xx-xx-xxxx---xx--x---xx--	$V_{b,1,3}$	-------------------------------------
$V_{a,1,3}$	-------------------------------------	$V_{d,1,3}$	-------------------------------------
$V_{a,f,1}$	-------------------------------------	$V_{c,f,1}$	-------------------------------------
h_K	-------------------------------------		

Although both characteristics do not have a probability higher than 2^{-128}, they are the best collision producing characteristics published for SipHash so far. Especially the characteristic for SipHash-1-x (Table 2) is not far away from the bound of 2^{-128}, where it gets useful in an attack. Moreover, the characteristic for SipHash-2-x (Table 3) is the best published characteristic for full SipHash-2-4, with a probability of $2^{-236.3}$. The previous best published characteristic for SipHash-2-4 has a probability of 2^{-498} [1]. Nevertheless, SipHash-2-4 still has a huge security margin.

5.2 Characteristic for Finalization of SipHash-2-4

Now, we want to present a distinguisher for the full four SipRound Finalization of SipHash-2-4. For this result, we have used the greedy search algorithm presented in Sect. 4 considering solely the probability. With the help of this algorithm, we have found the distinguisher shown in Table 4. By using this characteristic, four rounds of the Finalization can be distinguished from a pseudo-random function with a complexity of 2^{35}.

Considering this new result, we are able to distinguish both building blocks of SipHash-2-4, the Compression and the Finalization from idealized versions (It is already shown in [1] that two rounds of the Compression are distinguishable). However, these two results do not endanger the full SipHash-2-4 function, which is still indistinguishable from a pseudo-random function.

Table 4. Distinguisher for 4 finalization rounds (probability 2^{-35}).

$V_{a,f,1}$	x------------x---------x----x-------------x--------------x	$V_{b,f,1}$	-------------x-x----------------x-----x-------x--------
$V_{c,f,1}$	------x----x-----x--x------x--------------x--x-	$V_{d,f,1}$	----x-----x----x---x--x-------------------------x-------
$V_{a,f,2}$	----------------x-------x----------x-----------	$V_{b,f,2}$	--x--------
$V_{c,f,2}$	------------------------------x---------------x-	$V_{d,f,2}$	---
$V_{a,f,3}$	x--	$V_{b,f,3}$	---
$V_{c,f,3}$	---	$V_{d,f,3}$	x--
$V_{a,f,4}$	x-----------------x--------x----------------x--	$V_{b,f,4}$	x-------------------x------x---------x----x----
$V_{c,f,4}$	---	$V_{d,f,4}$	x-----------------x---x------------x--x-x----
$V_{a,f,5}$	--------x---------x---------x----------------	$V_{b,f,5}$	-----------x--x------x--x-x----------x----x---
$V_{c,f,5}$	--x---------------------------x--------x--x-	$V_{d,f,5}$	-x---------x----------x--------x------------x---x--

6   Conclusion

This work deals with the differential cryptanalysis of SipHash. To be able to find good results, we had to introduce new search strategies. Those search strategies extend previously published strategies, which have solely been used in the search for collisions for hash functions. With the new presented concepts, also attacks on other primitives like MACs, block-ciphers and stream-ciphers are within reach.

Furthermore, we generalized the concept of S-functions to the concept of cyclic S-functions. With the help of cyclic S-functions, cryptanalysts will be able to get more exact results regarding the probability of differential characteristics for ARX based primitives.

With these new methods, we were able to improve upon the existing results on SipHash. Our results include the first published characteristics resulting in internal collisions, the first published distinguisher for the Finalization of SipHash and the best published characteristic for full SipHash-2-4.

Future work includes to apply the greedy search strategies to other ARX based primitives. Such cryptographic primitives may be for instance block ciphers or authenticated encryption schemes. Also, the further improvement of the used automatic search tools is part of future work.

Acknowledgments. The work has been supported by the Austrian Government through the research program FIT-IT Trust in IT Systems (Project SePAG, Project Number 835919).

A   Results Without Secret Key

In this section, we present results for SipHash without considering the secret key. This allows us to create semi-free-start collisions for the Compression as well as an internal collision using chosen related keys. Despite the fact that these attacks do not lie within the specified use of SipHash, the following results give at least some insight in the strength of the MAC. In addition, the existence of the semi-free-start collisions are a strong indicator that the characteristics given in Tables 2 and 3 are valid and that the estimated probability is at least somewhat realistic.

With the help of the characteristic of Table 2, we can produce a semi-free-start collision. Furthermore, we are able to fix the value of $V_{a,i,1}$ and $V_{b,i,1}$ to

0 in advance of the search. The values for the semi-free-start collision are given in Table 5. The message pair can be created out of the characteristic in seconds. However, we cannot state a time it takes to create the characteristic, since the best characteristic out of many searches has been selected.

Table 5. Message pair and state values for semi-free-start collision for SipHash-1-x using three message blocks.

$V_{a,i,1}$:	0000000000000000	$V_{b,i,1}$:	0000000000000000	$V_{c,i,1}$:	0C42F127F5B7A160	$V_{d,i-1,2}$:	8D8FA9B18E275ED4
$V_{a,i+3,1}$:	00747ADAB4A268A8	$V_{b,i+2,2}$:	CD49C9A065B1F2AE	$V_{c,i+2,2}$:	6045CA3667F1A304	$V_{d,i+2,2}$:	94ED96D23F686622
M_i:	CAE3C8DF846F8D00	M_{i+1}:	18701B50E5EABA01	M_{i+2}:	21027F74580E0EE8		
M'_i:	CAE3C8DF806F8D00	M'_{i+1}:	77709CD01DCFBA01	M'_{i+2}:	C1029F745BEE0EF7		

The same can be done for SipHash-2-x by using the characteristic of Table 3. Here, we are able to produce the semi-free-start collision for SipHash-2-x shown in Table 6 within 10 s. Due to the rather low probability of $2^{-236.3}$ of the characteristic given in Table 3, we are not able to fix any values in advance of a search for a semi-free-start collision.

Table 6. Message pair and state values for semi-free-start collision for SipHash-2-x using one message block.

$V_{a,i,1}$:	992E9AA7D76CEF0E	$V_{b,i,1}$:	A17197FCAADF73D4	$V_{c,i,1}$:	33E9CBC3EB8E4E32	$V_{d,i-1,3}$:	3B5E30192818D15C
$V_{a,i+1,1}$:	8255CD3D3A2B4213	$V_{b,i,3}$:	783B1ADCD7BC413C	$V_{c,i,3}$:	FA5B40A895829C5B	$V_{d,i,3}$:	230701332727C0b0
M_i:	C7DCDE77723E8AD8						
M'_i:	5B40A16CD4E0BA54						

Now, we want to present internal collision using chosen related keys for SipHash-2-4. For the creation of the best found characteristic for a Compression

Table 7. Message pair, key, state values after internal collision, and MAC value (SipHash-2-4) for internal collision of SipHash-2-x using chosen related keys and a fixed IV.

Key 1:	7F166B32181D1FE4041FA4A0DBCD3927
Key 2:	7D1EEB2218055CEE041724415BA73CA7
Message 1:	0C40E5F8510C351DBA045A72064A83
Message 2:	0C40E5F8510CF198BA045A72064A83
MAC value:	20A26EAD9B9855BE
V_a:	6B2FCACBF912BB2B
V_b:	4CB34F2A06657837
V_c:	6260226FF75DCB88
V_d:	45F20251CF5EC6CD

consisting of 2 SipRounds per iteration, we use a starting point, where we place the difference in the most significant bit of the first message. The rest of the message, as well as the key can be chosen completely free. The best characteristic we have found by using the impact oriented strategy has an estimated probability of 2^{-169}. By using this characteristic, we are able to create the pairs in Table 7. To be easier to verify, we have included the MAC value for SipHash-2-4. In addition, we give the state after the collision happens in the first message block. Again, we cannot state the time it takes to create the characteristic, because it is the best characteristic out of many searches. The collision producing message and key pair can be created within seconds out of the characteristic.

B   An Example for Cyclic S-Functions

In this section, we want to clarify the use of cyclic S-functions with the help of an example. We use $((a + b) \lll 1) + c = s_b$ (Fig. 3) as cyclic S-function. Throughout this example, we use the following values for inputs and outputs.

$$a = \texttt{1x11} \quad b = \texttt{AEn5} \quad c = \texttt{15nx} \quad s_b = \texttt{11Ax} \tag{3}$$

Before we start an explanation how to calculate the probability, we first show how the propagation is done by using the graph shown in Fig. 4. To represent the graph in Fig. 4 clearly, we use information, which can be gathered using a bitslice propagation to narrow the value for the carries and therefore decrease the amount of edges and vertices in the graph. This precomputation is only done to produce a graph with few edges. The concept also works if this bitslice precomputation is omitted.

Now, we want to describe the concept of propagation by looking at first at the state mapping. Performing a state mapping function for doing addition with rotations in between is in principle merging a set of vertices together. After this merging, some edges, which have led to separated vertices, may lead to the same vertex. In case of the example in Fig. 4 this means that the vertices $|u, n|$ and $|1, n|$ of $S_o[1]$ are mapped both on vertex $|0, n|$ of $S_i[1]$. Vertex $|1, 0|$ of $S_o[1]$ is mapped on vertex $|0, 0|$ of $S_i[1]$. The vertex $|u, n|$ of $S_o[4]$ is mapped on vertex $|u, 0|$ of $S_i[0]$, vertex $|1, 0|$ of $S_o[4]$ is mapped on vertex $|1, 0|$ of $S_i[0]$ and vertex $|n, n|$ of $S_o[4]$ is mapped on vertex $|n, 0|$ of $S_i[0]$. The vertices of $S_i[0]$ and $S_o[4]$ are in fact the same. Therefore, we can reduce the problem of finding circles to the problem of finding paths from a vertex $|v, 0|$ of state $S_i[0]$ to the in fact same vertex $|v, 0|$ of state $S_i[4]$ (this is $S_o[4]$ after applying the mapping), where v stands for any value of a state. Edges, which do not belong to a path can be deleted. These are the dashed edges in Fig. 4. So we get following values after propagation:

$$a = \texttt{1u11} \quad b = \texttt{AAn5} \quad c = \texttt{15nn} \quad s_b = \texttt{11Au} \tag{4}$$

Note that that this result is optimal with respect to the limited representation capability of generalized conditions (The graph of Fig. 4 shows that in fact only two valid solutions exist).

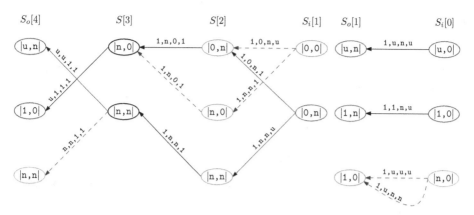

Fig. 4. Graph for $((a + b) \lll 1) + c = s_b$. States of the same color can be considered as equivalent (except for the black ones). The values on the edges represent $a[i]$, $b[i]$, $c[i]$ and $s_b[i]$.

Now we show, how the probability calculation is done. Because of space constraints, we only give the biadjacency matrix $A[1]$ (5) out of the set of transformed subgraphs. The matrix $A[1]$ corresponds to the rightmost subgraph shown in Fig. 4.

$$A[1] = \begin{bmatrix} 0&0&0&0&0&0&0&0&2&0&0&0&0&0&0&0 \\ 0&0&0&0&0&0&0&0&0&0&0&0&0&0&0&0 \\ 0&0&0&0&1&0&0&0&0&0&0&1&0&0&0 \\ 0&0&0&0&0&0&0&0&0&0&0&0&0&0&0&0 \\ 0&0&0&0&0&0&0&0&0&0&0&0&0&0&0&0 \\ 0&0&0&0&0&0&0&0&0&0&0&0&0&0&0&0 \\ 0&0&0&0&0&0&0&0&0&0&0&0&0&0&0&0 \\ 0&0&0&0&0&0&0&0&0&0&0&0&0&0&0&0 \\ 0&0&0&0&0&0&0&0&0&0&0&0&0&0&0&0 \\ 0&0&0&0&0&0&0&0&0&0&0&0&0&0&0&0 \\ 0&0&0&0&0&0&0&0&0&0&0&0&0&0&0&0 \\ 0&0&0&0&0&0&0&0&0&0&0&0&0&0&0&0 \\ 0&0&0&0&0&0&0&0&0&0&0&0&0&0&0&0 \\ 0&0&0&0&0&0&0&0&0&0&0&0&0&0&0&0 \\ 0&0&0&0&0&0&0&0&0&0&0&0&0&0&0&0 \\ 0&0&0&0&0&0&0&0&0&0&0&0&0&0&0&0 \end{bmatrix} \qquad (5)$$

After doing matrix multiplication and adding the elements of the diagonal together, we get two valid circles in the graph (1). This result can also be verified by looking at the graph of Fig. 4. The number of valid solutions divided by all possible input combinations gives us the exact differential probability of this subfunction. Here, we have to distinguish if we consider the input values given before (3) or after propagation (4). In the case of (3), we have 96 possible input pairs resulting in a probability of $2^{-5.58}$. For (4), we only have 16 possible input pairs and get a probability of 2^{-3}. Both results are exact.

References

1. Aumasson, J.-P., Bernstein, D.J.: SipHash: a fast short-input PRF. In: Galbraith, S., Nandi, M. (eds.) INDOCRYPT 2012. LNCS, vol. 7668, pp. 489–508. Springer, Heidelberg (2012)
2. Cramer, R. (ed.): EUROCRYPT 2005. LNCS, vol. 3494. Springer, Heidelberg (2005)
3. De Cannière, C., Rechberger, C.: Finding SHA-1 characteristics: general results and applications. In: Lai, X., Chen, K. (eds.) ASIACRYPT 2006. LNCS, vol. 4284, pp. 1–20. Springer, Heidelberg (2006)
4. Eichlseder, M., Mendel, F., Schläffer, M.: Branching heuristics in differential collision search with applications to SHA-512. IACR Cryptology ePrint Archive 2014:302 (2014)
5. Klima, V.: Tunnels in hash functions: MD5 collisions within a minute. IACR Cryptology ePrint Archive 2006:105 (2006)
6. Leurent, G.: Construction of differential characteristics in ARX designs application to Skein. In: Canetti, R., Garay, J.A. (eds.) CRYPTO 2013, Part I. LNCS, vol. 8042, pp. 241–258. Springer, Heidelberg (2013)
7. Mendel, F., Nad, T., Schläffer, M.: Finding SHA-2 characteristics: searching through a minefield of contradictions. In: Lee, D.H., Wang, X. (eds.) ASIACRYPT 2011. LNCS, vol. 7073, pp. 288–307. Springer, Heidelberg (2011)
8. Mendel, F., Nad, T., Schläffer, M.: Improving local collisions: new attacks on reduced SHA-256. In: Johansson, T., Nguyen, P.Q. (eds.) EUROCRYPT 2013. LNCS, vol. 7881, pp. 262–278. Springer, Heidelberg (2013)
9. Mouha, N., Velichkov, V., De Cannière, C., Preneel, B.: The differential analysis of S-functions. In: Biryukov, A., Gong, G., Stinson, D.R. (eds.) SAC 2010. LNCS, vol. 6544, pp. 36–56. Springer, Heidelberg (2011)
10. Preneel, B., van Oorschot, P.C.: MDx-MAC and building fast MACs from hash functions. In: Coppersmith, D. (ed.) CRYPTO 1995. LNCS, vol. 963, pp. 1–14. Springer, Heidelberg (1995)
11. Preneel, B., van Oorschot, P.C.: On the Security of Iterated Message Authentication Codes. IEEE Trans. Inf. Theory 45(1), 188–199 (1999)
12. Sugita, M., Kawazoe, M., Perret, L., Imai, H.: Algebraic cryptanalysis of 58-Round SHA-1. In: Biryukov, A. (ed.) FSE 2007. LNCS, vol. 4593, pp. 349–365. Springer, Heidelberg (2007)
13. Velichkov, V., Mouha, N., De Cannière, C., Preneel, B.: The additive differential probability of ARX. In: Joux, A. (ed.) FSE 2011. LNCS, vol. 6733, pp. 342–358. Springer, Heidelberg (2011)
14. Wang, X., Lai, X., Feng, D., Chen, H., Yu, X.: Cryptanalysis of the Hash Functions MD4 and RIPEMD. In: Cramer, R. [2], pp. 1–18
15. Wang, X., Yin, Y.L., Yu, H.: Finding collisions in the full SHA-1. In: Shoup, V. (ed.) CRYPTO 2005. LNCS, vol. 3621, pp. 17–36. Springer, Heidelberg (2005)
16. Wang, X., Yu, H.: How to break MD5 and other hash functions. In: Cramer, R. [2], pp. 19–35

Weak Instances of PLWE

Kirsten Eisenträger[1,2]([✉]), Sean Hallgren[3], and Kristin Lauter[4]

[1] Department of Mathematics, The Pennsylvania State University,
University Park, State College, PA 16802, USA
eisentra@math.psu.edu
[2] Department of Mathematics, Harvard University, Cambridge, MA 02138, USA
[3] Department of Computer Science and Engineering,
The Pennsylvania State University, University Park, State College, PA 16802, USA
hallgren@cse.psu.edu
[4] Microsoft Research, One Microsoft Way, Redmond, WA 98052, USA
klauter@microsoft.com

Abstract. In this paper we present a new attack on the polynomial version of the Ring-LWE assumption, for certain carefully chosen number fields. This variant of RLWE, introduced in [BV11] and called the PLWE assumption, is known to be as hard as the RLWE assumption for 2-power cyclotomic number fields, and for cyclotomic number fields in general with a small cost in terms of error growth. For general number fields, we articulate the relevant properties and prove security reductions for number fields with those properties. We then present an attack on PLWE for number fields satisfying certain properties.

1 Introduction

Lattice-based cryptography has been an active area of study for at least two decades. The Ajtai-Dwork [AD99] public-key cryptosystem was based on the worst-case hardness of a variant of the Shortest Vector Problem (SVP). The NTRU family of cryptosystems [HPS98] were defined in particularly efficient lattices connected to number theory and were standardized in the IEEE P1363.1 Lattice-Based Public Key Cryptography standard [IEEE]. Recently, a new assumption has been introduced, Learning-With-Errors (LWE) [Reg09] and the Ring-Learning-With-Errors (RLWE) variant [LPR10], which is related via various security reductions from hard lattice problems such as (Gap-)SVP and Bounded Distance Decoding (BDD) [LPR10,Reg09,BLP+13]. NTRUEncrypt

Kirsten Eisenträger - Partially supported by National Science Foundation grant DMS-1056703 and by the National Security Agency (NSA) under Army Research Office (ARO) contract number W911NF-12-1-0522. Part of this work was done while the first author was visiting Microsoft, Harvard University and MIT.

Sean Hallgren - Partially supported by National Science Foundation awards CCF-0747274 and CCF- 1218721, and by the National Security Agency (NSA) under Army Research Office (ARO) contract number W911NF-12-1-0522. Part of this work was done while visiting Microsoft and MIT.

© Springer International Publishing Switzerland 2014
A. Joux and A. Youssef (Eds.): SAC 2014, LNCS 8781, pp. 183–194, 2014.
DOI: 10.1007/978-3-319-13051-4_11

and NTRUSign can be slightly modified so that their security also based on the hardness of a variant of the RLWE problem [SS11], and applications to Homomorphic Encryption were proposed in [BV11] and extended in [BGV11] and [GHS12].

The advantage of RLWE over LWE is better efficiency and functionality for cryptosystems based on this hardness assumption, but with extra structure in the ring variant of the assumption comes the possibility of special attacks which take advantage of this structure. So far, we have not seen special attacks which take advantage of the extra structure.

The PLWE decisional hardness assumption was proposed in [BV11] as the basis for a fully homomorphic encryption scheme, and was introduced as "a variant of the RLWE assumption". But the worst-case to average-case reduction from the shortest vector problem on ideal lattices to the PLWE problem was only proved for the special case of 2-power cyclotomic fields and the proof of the reduction was cited from [LPR10]. A clear explanation of why this reduction works in the 2-power cyclotomic case was given in [DD12], which identified the necessary properties of the 2-power cyclotomic ring, and extended the proof to work for general cyclotomic fields, with minimal loss in the growth of the error bounds.

On the other hand, we can ask about the hardness of the PLWE assumption for general number rings, and its relationship to the hardness of the RLWE assumption. The key point is the distortion in the error distribution which occurs when passing between the Gaussian error distribution in the continuous complex space, where the ring is embedded via the Minkowski (or "canonical") embedding, and the error distribution when sampling error vectors coefficient-wise.

In this paper, we investigate the extent to which the hardness of these problems holds in more general number rings, that is, when the number field is not necessarily a cyclotomic field generated by roots of unity. We present an attack on the PLWE problem in certain carefully constructed examples of number fields. We also give a sequence of reductions between the Search and Decision versions of RLWE and the PLWE assumptions, under various conditions on the number field. An intuitive way to explain our results is that, for number fields satisfying our conditions, our attack on PLWE works by "guessing" one of q possibilities for the value of the secret polynomial evaluated at 1, and distinguishing PLWE samples with non-negligible probability when the error vectors are sampled from Gaussian distributions coefficient-wise in the polynomial ring.

Practical encryption schemes based on PLWE all work based on the assumption that the error distribution is sampled coefficient-wise in the polynomial ring. For 2-power cyclotomic fields, this is equivalent to sampling from the usual Gaussian error distribution for the ring embedded in a real vector space, but our attack does not work for these fields. On the other hand, our attack shows that it is not safe to work directly with the PLWE assumption in arbitrary number fields. So for the purpose of constructing secure and efficient cryptosystems, it is a reasonable conclusion that one should stick to cyclotomic number fields, until the class of fields for which there exists a reduction to RLWE is enlarged.

More specifically, for a degree n number field $K = \mathbb{Q}[x]/(f(x))$ and an integer modulus q, if $f(1)$ is congruent to zero modulo q, then our attack runs in time $\tilde{O}(q)$. For all current recommendations on parameter selection for RLWE, our attack runs much faster than the known distinguishing attacks based on solutions to the shortest vector problem, or decoding attacks based on computing a reduced basis for a lattice, which run in time exponential in n. For example, recommended high security parameters for LWE and RLWE-based cryptosystems given in [LP11, Figure 4] specify $n = 320$ and $q = 4093$. While the distinguishing and decoding attacks, estimated to run in time 2^{122} and 2^{119} seconds, are impractical, an attack which runs in time $\tilde{O}(q)$ is certainly feasible.

We emphasize that this does not constitute a practical attack on existing PLWE/RLWE-based cryptosystems. First of all, all practical systems known to us are based on the RLWE problem in a cyclotomic ring, normally a 2-power cyclotomic ring, $R = \mathbb{Z}[x]/(\Phi_m(x))$ where m is a power of 2. Our attack does not apply to $f = \Phi_m$ because $f(1) = \Phi_m(1)$ cannot be zero modulo q when q is much larger than m. Secondly, our attack runs in time proportional to q. While this is an improvement over algorithms which need to find a short vector or compute a reduced basis and run in time $O(2^n)$, it is still far from a practical attack when q is taken to be of size 2^{128}, (with $n = 2^{12}$), which is the minimum size required for homomorphic computations in [GLN12] and [BLN14], for example.

2 Background

2.1 Distances and Distributions

For the definition of the RLWE and PLWE hardness assumptions and for the implementation of related cryptosystems, it is necessary to define certain distributions, which will be used in particular for error distributions.

Adopting the notation from [LPR10], we will let $H \subseteq \mathbb{R}^{s_1} \times \mathbb{C}^{2s_2}$ denote the space

$$H = \{(x_1, \ldots, x_n) \in \mathbb{R}^{s_1} \times \mathbb{C}^{2s_2} \mid x_{s_1+s_2+j} = \overline{x_{s_1+j}}, j = 1, \ldots, s_2\},$$

where s_1 and s_2 are non-negative integers and $n = s_1 + 2s_2$. Since the last s_2 complex coordinates depend on the previous s_2 coordinates, as they are just the complex conjugates of them, H is isomorphic to $\mathbb{R}^n$. It inherits the usual inner product $\langle (x_i), (y_i) \rangle := \sum_{i=1}^{n} x_i \cdot y_i$.

There are several natural notions of distance on an inner product space, and we will primarily need the ℓ_2-norm, given by $\|x\|_2 := (\sum_{i=1}^{n} x_i^2)^{1/2} = \sqrt{\langle x, x \rangle}$ for $x \in H$, and the ℓ_∞-norm, given by $\|x\|_\infty := \max |x_i|$.

For a real number $\sigma > 0$, the Gaussian function $\rho_\sigma : H \to (0, 1]$ is given by

$$\rho_\sigma(x) := \exp(-\pi \langle x, x \rangle / \sigma^2).$$

The *continuous Gaussian probability distribution* D_σ is given by $D_\sigma(x) = \frac{\rho_\sigma(x)}{\sigma^n}$.

As in [LPR10, Definition 5] we now define the family of LWE error distributions to which the results apply.

Definition 1. *For a positive real $\alpha > 0$, the family $\Psi_{\leq \alpha}$ is the set of all elliptical Gaussian distributions $D_{\mathfrak{r}}$ (over $K \otimes_{\mathbb{Q}} \mathbb{R}$) where each parameter $r_i \leq \alpha$.*

2.2 Lattices

A lattice is a discrete subgroup of a continuous space. For example, in $\mathbb{R}^n$, the real vector space of dimension n, a lattice can be specified by a set of n linearly independent vectors, and the lattice is the integral span of those vectors. An orthogonal basis for a lattice, if one exists, is a basis such that the basis vectors are pairwise orthogonal, with respect to the given inner product.

For a lattice $\Lambda \subset H$, define the dual lattice as

$$\Lambda^{\vee} = \{y \in H \mid \forall x \in \Lambda, \langle x, y \rangle = \sum_{i=1}^{n} x_i \overline{y_i} \in \mathbb{Z}\}.$$

We will also need to refer to the *smoothing parameter* of a lattice, $\eta_\epsilon(\Lambda)$ introduced by Micciancio and Regev [MR07], which for a lattice Λ and a positive real number ϵ is defined to be the smallest s such that $\rho_{1/s}(\Lambda^{\vee} \setminus \{0\}) \leq \epsilon$.

2.3 Number Fields

A *number field* is a finite algebraic extension of the field of rational numbers $\mathbb{Q}$. It is a field which contains $\mathbb{Q}$ and is a finite dimensional vector space over $\mathbb{Q}$. The *degree* of the number field is its dimension as a vector space. A number field K is *Galois* if $K/\mathbb{Q}$ is a Galois extension, which means it is both separable and normal. An extension is *separable* if every element in the extension is separable, which means that its minimal polynomial has distinct roots. An extension $K/\mathbb{Q}$ is *normal* if every irreducible polynomial with rational coefficients which has one root in K has all of its roots in K. In particular, this means that every isomorphism of the field K into its algebraic closure which fixes $\mathbb{Q}$ actually maps into K, and thus is an automorphism of K. For a Galois extension $K/\mathbb{Q}$, the set of automorphisms of K which fix $\mathbb{Q}$ forms a group, and is called the Galois group, $\mathrm{Gal}(K/\mathbb{Q})$, of $K/\mathbb{Q}$.

The *ring of integers* in a number field K is the set of all algebraic integers in the number field, which means the elements which satisfy a *monic* irreducible polynomial with integer coefficients. This set is a ring, called a number ring, and is usually denoted by $\mathcal{O}_K$. If the ring $R = \mathcal{O}_K$ is generated over $\mathbb{Z}$ by (sums and powers of multiples of) a single element, $R = \mathbb{Z}[\beta]$, then we say that R is *monogenic*.

The mth cyclotomic field is the number field generated by the mth roots of unity. Let ζ_m be a primitive m^{th} root of 1, i.e. $\zeta_m^m = 1$ but no smaller power is 1. Then $K = \mathbb{Q}(\zeta_m) = \mathbb{Q}[x]/(\Phi_m(x))$, where $\Phi_m(x)$ is the m-th cyclotomic polynomial $\Phi_m(x) = \prod_{k \in (\mathbb{Z}/m\mathbb{Z})^{\times}} (x - \zeta_m^k)$ with degree equal to $n = \varphi(m)$. When m is an odd prime we have $\Phi_m(x) = 1 + x + x^2 + \cdots + x^{m-1}$, and when m is a power of 2, $\Phi_m(x) = x^{m/2} + 1$.

For a finite algebraic extension $K/\mathbb{Q}$, the Trace and Norm maps from K to $\mathbb{Q}$ are defined as the sum (*resp.* the product) of all the algebraic conjugates of an element. The Trace map induces a non-degenerate bilinear form $\text{Tr}(xy)$ on K. The *dual* or the *codifferent* of the ring of integers, R, with respect to this bilinear form is the collection of elements

$$R^{\vee} = \{y \in K \mid \text{Tr}(xy) \in \mathbb{Z}, \forall x \in R\}.$$

This is often denoted $\mathcal{D}_K^{-1}$ in algebraic number theory.

It is known that if $R = \mathcal{O}_K = \mathbb{Z}[\beta] = \mathbb{Z}[x]/(f(x))$ is monogenic, then the codifferent is generated by the single element $(1/f'(\beta))$ [Ser79, p. 56, Cor 2]. In that case, there is a simple isomorphism between $R^{\vee}$ and R which scales elements by multiplication by $f'(\beta)$. For $K = \mathbb{Q}(\zeta_m)$, we have $\mathcal{O}_K = \mathbb{Z}[\zeta_m]$.

The ring of integers R is embedded in H via the Minkowski embedding (called the *canonical embedding* in [LPR10]) which sends any $x \in K$ to $(\sigma_1(x), \dots \sigma_n(x))$, where σ_i are the real and complex embeddings of K, ordered to coincide with the definition of H. Under this embedding, the notions of duality and codifferent coincide. In particular, we can define an *ideal lattice* to be the image under this embedding of any fractional ideal of R by taking the lattice generated by the image of the n basis elements of the $\mathbb{Z}$-basis for the ideal.

An *ideal I* in a commutative ring R is an additive subgroup which is closed under multiplication by elements of R. A *prime ideal $I \neq R$* is an ideal with the property that if the product of two elements $a, b \in R$ is such that $ab \in I$, then either a or b is in I. Ideals in the ring of integers of number fields have unique factorization into products of prime ideals. We say that a prime ideal $(p) = p\mathbb{Z}$ *splits completely* in an extension of number fields $K/\mathbb{Q}$ if the ideal $p\mathcal{O}_K$ factors into the product of n distinct ideals of degree 1, where $n = [K : \mathbb{Q}]$ is the degree of the extension $K/\mathbb{Q}$.

2.4   Definition of the Ring-LWE Distribution and Problem

The Ring-LWE distribution and hardness assumptions were introduced in [LPR10, Section 3] using the notation $K_{\mathbb{R}} = K \otimes \mathbb{R}$ and $\mathbb{T} = K_{\mathbb{R}}/R^{\vee}$. For an integer q, let R_q denote R/qR.

Definition 2. *(Ring-LWE Distribution) For $s \in R_q^{\vee}$ a secret, and an error distribution ψ over $K_{\mathbb{R}}$, the Ring-LWE distribution $A_{s,\psi}$ over $R_q \times \mathbb{T}$ consists of samples generated as follows: choose a uniformly at random from R_q and choose the error vector e from the error distribution ψ, then the samples are pairs of the form $(a, (a \cdot s)/q + e)$.*

Definition 3. *(Ring-LWE Search Problem) Let Ψ be a family of distributions over $K_{\mathbb{R}}$. The Ring-LWE Search problem $(\text{R} - \text{LWE}_{q,\Psi})$, for some $s \in R_q^{\vee}$ and $\psi \in \Psi$, is to find s, given arbitrarily many independent samples from $A_{s,\psi}$.*

Definition 4. *(Ring-LWE Average-Case Decision Problem) Let Υ be a family of error distributions over $K_{\mathbb{R}}$. The Ring-LWE Average-Case Decision problem*

$(R - DLWE_{q,\Upsilon})$ *is to distinguish with non-negligible advantage between arbitrarily many independent samples from* $A_{s,\psi}$, *for a random choice of* $s \in R_q^{\vee}$ *and* $\psi \in \Upsilon$, *and the same number of samples chosen independently and uniformly at random from* $R_q \times \mathbb{T}$.

2.5 Worst-Case Hardness of Search Version of Ring-LWE

Theorem 1. *([LPR10]) Let* K *be an arbitrary number field of degree* n, *with* $R = \mathcal{O}_K$, $\alpha = \sigma/q \in (0,1)$, *and* $q \geq 2 \in N$ *such that* $\alpha \cdot q \geq \omega(\sqrt{\log n})$. *Then there is a probabilistic polynomial-time quantum reduction from the* $\tilde{O}(\sqrt{n}/\alpha)$-*approximate SIVP problem on ideal lattices in* K *to* $R - LWE_{q,\Psi_{\leq\alpha}}$. *For* K *a cyclotomic number field, this gives a reduction from the* $\tilde{O}(\sqrt{n}/\alpha)$-*approximate SVP problem.*

2.6 Known Attacks

When selecting secure parameters for cryptographic applications of the hardness of RLWE, the following known attacks are currently taken into account. The *distinguishing attack* considered in [MR09, RS10] for LWE requires the adversary to find a short vector in the scaled dual of the LWE lattice. The distinguishing advantage is then given in terms of the length of the vector found. According to [LP11], the vector should be of length less than $q/(2\sigma)$. Writing q in terms of n, this amounts to solving a short-vector problem in an n-dimensional lattice, and if q is too large with respect to n, this problem will be easy. This gives some insight as to why q cannot be too large with respect to n.

Concrete security estimates given in [LP11, Figure 4] against this attack lead to suggested parameters, for example at the "high security" level, of $n = 320$, $q \approx 2^{12}$, and $\sigma = 8$ (however recall that for 2-power cyclotomic fields, n should be a power of 2). For those parameter choices, the distinguishing attack is estimated to run in time 2^{122} (seconds) to obtain a distinguishing advantage of 2^{-64}.

The *decoding attack* presented in [LP11] is an attack which actually recovers the secret error vector in the ciphertext. To run the attack requires a reduced basis, and the estimated time to compute the reduced basis when $n = 320$ and $q \approx 2^{12}$ is 2^{119} seconds for decoding probability 2^{-64}.

3 Overview of Results

We work with the ring of integers $R = \mathcal{O}_K$ in a number field K of degree n and a prime number q and consider the following properties:

1. (q) splits completely in K, and $q \nmid [R : \mathbb{Z}[\beta]]$;
2. K is Galois over $\mathbb{Q}$;
3. the ring of integers of K is generated over $\mathbb{Z}$ by β, $\mathcal{O}_K = \mathbb{Z}[\beta] = \mathbb{Z}[x]/(f(x))$ with $f'(\beta) \mod q$ "small";
4. the transformation between the Minkowski embedding of K and the power basis representation of K is given by a scaled orthogonal matrix.

5. let $f \in \mathbb{Z}[x]$ be the minimal polynomial for β. Then $f(1) \equiv 0 \pmod{q}$;
6. q can be chosen suitably large.

Note that by the Chebotarev Density Theorem, there are infinitely many choices for q satisfying this and only finitely many exclusions.

In Sect. 4 we will show that for pairs (K, q) satisfying conditions (1) and (2) we have a search-to-decision reduction from $\mathrm{R} - \mathrm{LWE}_q$ to $\mathrm{R} - \mathrm{DLWE}_q$.

In Sect. 5 we will consider a second reduction, from $\mathrm{R} - \mathrm{DLWE}_q$ to PLWE, which is essentially a slightly more general version of the reduction given in [LPR10] and [DD12]. For that step we require that K satisfies conditions (3) and (4).

In Sect. 6 we will then give an attack which breaks instances of the PLWE decision problem whenever (K, q) satisfy conditions (5) and (6), and we will consider possible extensions of our attack.

4   Search to Decision Reduction for the Ring-LWE Problem

In this section we will prove the following theorem.

Theorem 2. *Let K be a number field such that $K/\mathbb{Q}$ is Galois of degree n and let $R = \mathcal{O}_K$ be its ring of integers. Let $R^\vee$ be the dual (the codifferent ideal) of R. Let β be an algebraic integer such that $K = \mathbb{Q}(\beta)$, and let $f(x) \in \mathbb{Z}[x]$ be the minimal polynomial of β over $\mathbb{Q}$. Let q be a prime such that (q) splits completely in K and such that $q \nmid [R : \mathbb{Z}[\beta]]$. Let α be such that $\alpha \cdot q \geq \eta_\varepsilon(R^\vee)$ for some negligible $\varepsilon = \varepsilon(n)$. Then there is a randomized polynomial-time reduction from $\mathrm{R} - \mathrm{LWE}_{q, \Psi_{\leq \alpha}}$ to $\mathrm{R} - \mathrm{DLWE}_{q, \Upsilon_\alpha}$.*

Proof. Let n denote the degree of K over $\mathbb{Q}$. Since $K/\mathbb{Q}$ is Galois, f factors completely in K as $f(x) = (x - \beta) \cdot (x - \beta_2) \cdots (x - \beta_m)$. Let q be a prime as in the theorem statement, which factors as $(q) = \mathfrak{q}_1 \ldots \mathfrak{q}_n$.

The field K has n embeddings, and since $K/\mathbb{Q}$ is Galois either all of these embeddings are real or they are all complex.

Let $\beta_1 := \beta$, and for $i = 1, \ldots, n$, let $\sigma_i : K \hookrightarrow \mathbb{C}$ $(i = 1, \ldots, n)$ be the embedding which sends β_1 to β_i. Let $\sigma : K \hookrightarrow \mathbb{C} \times \cdots \times \mathbb{C}$ be the Minkowski embedding sending $x \in K$ to $(\sigma_1(x) = x, \sigma_2(x), \ldots, \sigma_n(x))$.

Before we can finish the proof we need two lemmas:

Lemma 1. *Let $K = \mathbb{Q}(\beta_1)$ be as above. For any $\alpha > 0$, the family $\Psi_{\leq \alpha}$ is closed under every automorphism τ of K, i.e. $\psi \in \Psi_{\leq \alpha}$ implies that $\tau(\psi) \in \Psi_{\leq \alpha}$.*

Proof. Let τ be an automorphism of K. Then $\tau(\beta_1) = \beta_j$ for some $1 \leq j \leq n$. Let $x \in K$. Then $x = \sum_{i=0}^{n-1} k_i \beta_1^i$ for $k_i \in \mathbb{Q}$ and

$$\sigma(x) = \left(\sum_{i=0}^{n-1} k_i \beta_1^i, \sum_{i=0}^{n-1} k_i \beta_2^i, \ldots, \sum_{i=0}^{n-1} k_i \beta_n^i \right).$$

On the other hand, $\sigma(\tau(\beta_1))$ is a vector whose entries are simply a permutation of $\beta_1, \ldots, \beta_n$ and whose first entry is β_j, and so for any $x \in K$, the coordinates of $\sigma(x)$ and $\sigma(\tau(x))$ are simply a rearrangement of each other.

Hence for any $\psi = D_{\mathfrak{r}} \in \Psi_{\leq \alpha}$, we have $\tau(D_{\mathfrak{r}}) = D_{\mathfrak{r}'} \in \Psi_{\leq \alpha}$, where the entries of $\mathfrak{r}'$ are simply a rearrangement of the entries of $\mathfrak{r}$ and hence are all at most α. $\qquad\Box$

Worst-Case Search to Worst-Case Decision

Definition 5. *The $\mathfrak{q}_i$-$\mathrm{LWE}_{q,\Psi}$ problem is: given access to $A_{s,\psi}$ for some arbitrary $s \in R_q^{\vee}$ and $\psi \in \Psi$, find $s \bmod \mathfrak{q}_i R^{\vee}$.*

Lemma 2. *(LWE to $\mathfrak{q}_i$-LWE) Suppose that the family Ψ is closed under all automorphisms of K. Then for every $1 \leq i \leq n$ there is a deterministic polynomial-time reduction from $\mathrm{LWE}_{q,\psi}$ to $\mathfrak{q}_i$-$\mathrm{LWE}_{q,\psi}$.*

Proof. The proof proceeds almost word for word as the proof in [LPR10]. Given two prime ideals $\mathfrak{q}_i$ and $\mathfrak{q}_j$ above q, [LPR10] uses the explicit automorphism τ_k with $\tau_k(\zeta) = \zeta^k$ where $k = j/i \in \mathbb{Z}_m^*$ that maps $\mathfrak{q}_j$ to $\mathfrak{q}_i$. Instead we use the fact that the Galois group of K over $\mathbb{Q}$ acts transitively on the prime ideals above $\mathfrak{q}$. Hence in our situation, given $\mathfrak{q}_i, \mathfrak{q}_j$ there is also an automorphism τ of K such that $\tau(\mathfrak{q}_i) = \mathfrak{q}_j$. The rest of the argument is identical to the argument in [LPR10]. $\qquad\Box$

Conclusion of the proof of Theorem 2: To finish the proof based on these Lemmas, we argue as in [LPR13a, Lemma 5.9] and the proof given there goes through for Galois fields exactly as stated. $\qquad\Box$

5   Reduction from $\mathbf{R - DLWE}_q$ to PLWE

This section essentially summarizes and slightly generalizes one of the main results from [DD12]. The reduction for general cyclotomic fields is also covered in [LPR13b].

5.1   The PLWE Problem

The PLWE problem was first defined in [LPR10] and [BV11].

Definition 6. *(The PLWE assumption). For all $\kappa \in \mathbb{N}$, let $f(x) = f_\kappa(x)$ be a polynomial of degree $n = n(\kappa)$, and let $q = q(\kappa)$ be a prime integer. Let $R = \mathbb{Z}[x]/(f)$, let $R_q = R/qR$ and let χ denote a distribution over R.*

The PLWE assumption $\mathrm{PLWE}_{f,q,\chi}$ states that for any $\ell = poly(\kappa)$ it holds that

$$\{(a_i, a_i \cdot s + e_i)\}_{i \in [\ell]} \text{ is computationally indistinguishable from } \{a_i, u_i\}_{i \in [\ell]},$$

where s is sampled from the noise distribution χ, the a_i are uniform in R_q, the error polynomials e_i are sampled from the error distribution χ and the ring elements u_i are uniformly random over R_q.

The PLWE assumption is a decisional assumption.

5.2   Reduction

In [DD12, p. 39], the authors explain the reduction in the 2-power cyclotomic case in terms of the two key properties of the ring $R = \mathcal{O}_K$ which are used:

1. When $R = \mathbb{Z}[\zeta_m]$ with m a power of 2, then $nR^\vee = R$, for $n = m/2$.
2. The transformation between the embedding of R in the continuous real vector space H and the representation of R as a $\mathbb{Z}$-vector space with the power basis consisting of powers of ζ_m is an orthogonal linear map.

Their argument shows that one can slightly generalize those conditions to our Properties (3) and (4) and obtain the reduction for general number fields with those properties. Note that the claim is that these conditions are sufficient to obtain the reduction, not that they are necessary. There may be a reduction which works for an even more general class of number fields.

Step 1 of the reduction uses the property that $R = \mathbb{Z}[\beta] = \mathbb{Z}[x]/(f(x))$ is monogenic to transform the ring-LWE samples between distributions on $R^\vee$ and R, at the cost of a scaling by $f'(\beta)$, where $f'(\beta)$ is "small" modulo q.

When reducing RLWE to PLWE we take samples from the Minkowski embedding and consider them in the coefficient embedding. The main point is whether vectors that are short in the Minkowski embedding have small coefficients in the coefficient embedding.

Step 2 uses the fact that the matrix which transforms between the embedding of R in H and the power basis representation of R is a scaled orthogonal matrix, so it transforms the spherical Gaussian distribution in H into a spherical Gaussian distribution in the power basis representation. Thus the error distribuition can be sampled directly from small values coefficient-wise in the polynomial ring.

Note that a different reduction is given in [DD12] for general cyclotomic fields because of the fact that ζ_m potentially does not satisfy the requirement that $\Phi'_m(\zeta_m)$ is small modulo q compared to n. As noted there, according to a result of Erdős [Erd46], the coefficients of Φ_m can be superpolynomial in size. In any case, even for m prime, Φ'_m has coefficients of size up to roughly m.

6   Breaking Certain Instances of PLWE

6.1   The Attack

Let K be a number field such that $f(1) \equiv 0 \pmod{q}$, and such that q can be chosen large enough. Let $R := \mathcal{O}_K$, and let $R_q := R/qR$.

Now, given samples, $(a_i, b_i) \in R_q \times R_q$, we have to decide whether the samples are uniform or come from a PLWE distribution. To do this we take the representatives of a_i and b_i in R, call them a_i and b_i again, and evaluate them at 1. This gives us elements $a_i(1), b_i(1) \in \mathbb{F}_q$. If (a_i, b_i) are PLWE samples, then by fdefinition,

$$b_i = a_i \cdot s + e_i,$$

and so
$$b_i(1) \equiv (a_i \cdot s)(1) + e_i(1) \pmod{q}.$$

Since $f(1) \equiv 0 \pmod{q}$, the Chinese Remainder Theorem gives us that

$$b_i(1) \equiv a_i(1) \cdot s(1) + e_i(1) \pmod{q}.$$

Now we can guess $s(1)$, and we have q choices. For each of our guesses we compute $b_i(1) - a_i(1) \cdot s(1)$. If (a_i, b_i) are PLWE samples and our guess for $s(1)$ is correct, then $b_i(1) - a_i(1) \cdot s(1) = e_i(1)$, and we will detect that it is non-uniform, because e_i is taken from χ. (For example, if e_i is taken from a Gaussian with small radius, then $e_i(1)$ will be "small" for all i and hence not uniform.) If (a_i, b_i) are uniform samples, then $b_i(1) - a_i(1) \cdot s(1)$ for any fixed choice of $s(1)$ will still be uniform, since $a_i(1)$, $b_i(1)$ are both uniform modulo q.

6.2   A Family of Examples

Let $f(x) = X^n + (k-1)pX + p$, where p is a prime less than n, and k is chosen such that $1 + kp = q$ with q prime and $q > n$. This polynomial is Eisenstein at p and hence irreducible. By Dirichlet's theorem about primes in arithmetic progressions, there are infinitely many values of k that give a prime q.

Also, by construction

$$f(1) = 1 + (k-1)p + p = 1 + kp = 1 \equiv 0 \pmod{q}.$$

Moreover, $f'(1)$ is not zero modulo q since

$$f'(1) = n + (k-1)p = (1 + kp) + (n - 1 - p) = q + a,$$

with a a number which, by construction is $< n$. Hence f has 1 as a simple root modulo q, and by the Chinese Remainder Theorem

$$\mathbb{Z}[X]/(f(X)) \cong Z[X]/(X - 1) \times Z[X]/(h(X))$$

with $h(X)$ coprime to $(X-1)$. As explained in the previous section, this allows us to guess $s(1)$, since $(a_i \cdot s)(1) = a_i(1) \cdot s(1)$. Hence for this choice of polynomials and choice of q, we can distinguish uniform samples from PLWE samples and break PLWE.

6.3   Extension of the Attack on PLWE

The attack we presented in Sect. 6.1 above on PLWE for number fields satisfying property (5) can be extended to a more general class of number fields as follows:

Suppose that $f(x)$ has a root β modulo q which has small order in $(\mathbb{Z}/q\mathbb{Z})^*$. If q is a prime, then this is equivalent to β having small order modulo $q - 1$. If $f(\beta) \equiv 0 \mod q$, then the same attack above will work by evaluating samples at β, instead of at 1. Now unfortunately, the value of the error polynomials $e_i(\beta)$ are harder to distinguish from random ones than in the case $\beta = 1$: although the $e_i(x)$ have small coefficients modulo q, the powers of β may grow large and also may wrap around modulo q. However, if β has small order in $(\mathbb{Z}/q\mathbb{Z})^*$, then the set $\{\beta^i\}_{i=0,\ldots,n-1}$ takes on only a small number values, and this can be used to distinguish samples arising from $e_i(\beta)$ from random ones with non-negligible advantage.

6.4 Security Implications for RLWE and PLWE-based Cryptosystems

Putting all the results of this paper together, if there exist number fields satisfying all 6 properties, then for those number fields we would also have an attack on RLWE. A toy example of a field satisfying the first five conditions listed above is $K = \mathbb{Q}(\sqrt{11})$, $\beta = \sqrt{11}$, $f(x) = x^2 - 11$, and $q = 5$.

In general, for a given degree n, it is not hard to generate irreducible polynomials $f(x)$ of degree n, such that, letting $q = f(1)$, q is sufficiently large. Each such polynomial $f(x)$ gives rise to a weak instance of PLWE, according to our attack. However, to obtain an attack on RLWE, we would need to check that the first 4 properties are also satisfied. The first two properties are easy to check, but not necessarily easy to assure by construction. Properties (3) and (4) are not as easy to check, and harder to assure by construction.

The security of RLWE in general and its reduction to hard lattice problems is an interesting theoretical question and thus the construction of a number field satisfying all 6 properties would be a significant result. But from the point of view of practical applications to cryptography and homomorphic encryption, the security of the proposed cryptosystems is based on the hardness of the PLWE assumption. Thus the attack presented here and the results of this section are of interest in themselves.

References

[IEEE]   P1363.1: Standard specifications for public-key cryptographic techniques based on hard problems over lattices, December 2008. http://grouper.ieee. org/groups/1363/

[AD99]   Ajtai, M., Dwork, C.: A public-key cryptosystem with worst-case/average-case equivalence. In: STOC '97: Proceedings of the Twenty-ninth Annual ACM Symposium on Theory of Computing, pp. 284–293. ACM, New York (1999)

[BLN14] Bos, J.W., Lauter, K., Naehrig, M.: Private predictive analysis on encrypted medical data. J. Biomed. Inform. (2014). doi:10.1016/j.jbi.2014.04.003

[BLP+13] Brakerski, Z., Langlois, A., Peikert, C., Regev, O., Stehlé, D.: Classical hardness of learning with errors. In: STOC'13: Proceedings of the 45th Annual ACM Symposium on Theory of Computing, pp. 575–584. ACM, New York (2013)

[BV11]   Brakerski, Z., Vaikuntanathan, V.: Fully homomorphic encryption from ring-LWE and security for key dependent messages. In: Rogaway, P. (ed.) CRYPTO 2011. LNCS, vol. 6841, pp. 505–524. Springer, Heidelberg (2011)

[BGV11] Brakerski, Z., Gentry, C., Vaikuntanathan, V.: Fully homomorphic encryption without bootstrapping. In: Innovations in Theoretical Computer Science–ITCS 2012, pp. 309–325. ACM (2012)

[DD12] Ducas, L., Durmus, A.: Ring-LWE in polynomial rings. In: Fischlin, M., Buchmann, J., Manulis, M. (eds.) PKC 2012. LNCS, vol. 7293, pp. 34–51. Springer, Heidelberg (2012)

[Erd46] Erdős, P.: On the coefficients of the cyclotomic polynomial. Bull. Am. Math. Soc. **52**, 179–184 (1946)

[GHS12] Gentry, C., Halevi, S., Smart, N.P.: Fully homomorphic encryption with polylog overhead. In: Pointcheval, D., Johansson, T. (eds.) EUROCRYPT 2012. LNCS, vol. 7237, pp. 465–482. Springer, Heidelberg (2012)

[GLN12] Graepel, T., Lauter, K., Naehrig, M.: ML confidential: machine learning on encrypted data. In: Kwon, T., Lee, M.-K., Kwon, D. (eds.) ICISC 2012. LNCS, vol. 7839, pp. 1–21. Springer, Heidelberg (2013)

[HPS98] Hoffstein, J., Pipher, J., Silverman, J.H.: NTRU: a ring-based public key cryptosystem. In: Buhler, J.P. (ed.) ANTS 1998. LNCS, vol. 1423, pp. 267–288. Springer, Heidelberg (1998)

[LP11] Lindner, R., Peikert, C.: Better key sizes (and attacks) for LWE-based encryption. In: Kiayias, A. (ed.) CT-RSA 2011. LNCS, vol. 6558, pp. 319–339. Springer, Heidelberg (2011)

[LPR10] Lyubashevsky, V., Peikert, C., Regev, O.: On ideal lattices and learning with errors over rings. In: Gilbert, H. (ed.) EUROCRYPT 2010. LNCS, vol. 6110, pp. 1–23. Springer, Heidelberg (2010)

[LPR13a] Lyubashevsky, V., Peikert, C., Regev, O.: On ideal lattices and learning with errors over rings. J. ACM **60**(6(Art. 43)), 35 (2013)

[LPR13b] Lyubashevsky, V., Peikert, C., Regev, O.: A toolkit for ring-LWE cryptography. In: Johansson, T., Nguyen, P.Q. (eds.) EUROCRYPT 2013. LNCS, vol. 7881, pp. 35–54. Springer, Heidelberg (2013)

[MR07] Micciancio, D., Regev, O.: Worst-case to average-case reductions based on Gaussian measures. SIAM J. Comput. **37**(1), 267–302 (2007). (electronic)

[MR09] Micciancio, D., Regev, O.: Lattice-based cryptography. In: Bernstein, D.J., Buchmann, J., Dahmen, E. (eds.) Post-Quantum Cryptography, pp. 147–191. Springer, Heidelberg (2009)

[Reg09] Regev, O.: On lattices, learning with errors, random linear codes, and cryptography. J. ACM **56**(6), 1–40 (2009)

[RS10] Rückert, M., Schneider, M.: Selecting secure parameters for lattice-based cryptography. Cryptology ePrint Archive, Report 2010/137 (2010)

[Ser79] Serre, J.-P.: Local Fields. Graduate Texts in Mathematics, vol. 67. Springer, New York (1979)

[SS11] Stehlé, D., Steinfeld, R.: Making NTRU as secure as worst-case problems over ideal lattices. In: Paterson, K.G. (ed.) EUROCRYPT 2011. LNCS, vol. 6632, pp. 27–47. Springer, Heidelberg (2011)

The Usage of Counter Revisited: Second-Preimage Attack on New Russian Standardized Hash Function

Jian Guo[1], Jérémy Jean[1(✉)], Gaëtan Leurent[2], Thomas Peyrin[1], and Lei Wang[1]

[1] Division of Mathematical Sciences, School of Physical and Mathematical Sciences, Nanyang Technological University, Singapore, Singapore
{GuoJian,JJean,Thomas.Peyrin,Wang.Lei}@ntu.edu.sg
[2] INRIA, Paris, France
Gaetan.Leurent@inria.fr

Abstract. `Streebog` is a new Russian hash function standard. It follows the HAIFA framework as domain extension algorithm and claims to resist recent generic second-preimage attacks with long messages. However, we demonstrate in this article that the specific instantiation of the HAIFA framework used in `Streebog` makes it weak against such attacks. More precisely, we observe that `Streebog` makes a rather poor usage of the HAIFA counter input in the compression function, which allows to construct second-preimages on the full `Streebog-512` with a complexity as low as $n \times 2^{n/2}$ (namely 2^{266}) compression function evaluations for long messages. This complexity has to be compared with the expected 2^{512} computations bound that an ideal hash function should provide. Our work is a good example that one must be careful when using a design framework for which not all instances are secure. HAIFA helps designers to build a secure hash function, but one should pay attention to the way the counter is handled inside the compression function.

Keywords: `Streebog` · Cryptanalysis · Second-preimage attack · Diamond structure · Expandable message · HAIFA

1 Introduction

Hash functions are among the most fundamental primitives in modern cryptography. Informally, a cryptographic hash function maps an arbitrarily long message into a short random looking digest, which acts as the fingerprint of the original message. As for any cryptographic primitive, one expects some security properties to be fulfilled and in the case of hash functions we can point to three classical notions:

- **Collision Resistance:** it should be computationally infeasible for an adversary to find a pair of distinct messages that have the same hash digest.

© Springer International Publishing Switzerland 2014
A. Joux and A. Youssef (Eds.): SAC 2014, LNCS 8781, pp. 195–211, 2014.
DOI: 10.1007/978-3-319-13051-4_12

- **Second-Preimage Resistance:** for any given message M, it should be computationally infeasible for an adversary to find a distinct message M' that leads to the same hash digest than M.
- **Preimage Resistance:** for any given hash digest h, it should be computationally infeasible for an adversary to find a message M that leads to the hash digest h.

By "computationally infeasible", we mean that an attacker should not be able to break that property with less than a certain number of computations that depends on n, the bit length of the hash digest. More precisely, we expect that the best attacks on a cryptographic hash function are generic attacks. In the case of an ideal hash function, one expects to find a (second)-preimage only after trying about 2^n distinct messages, and to find a collision only after trying about $2^{n/2}$ distinct messages (due to the birthday paradox).

A cryptographic hash function is commonly built by iterating a fixed input-length function called *compression function* in order to handle arbitrarily long messages, and the iteration algorithm is referred to as *domain extension*. In this article, we mainly discuss the domain extension schemes for cryptographic hash functions, and consider the compression function as an ideal component.

Generic Attacks. The well-known Merkle-Damgård scheme [13,26] has been the most popular domain extension scheme in order to build a hash function, e.g., MD5, SHA-1 and SHA-2 are built upon such design strategy. However, since 2004, several weaknesses of Merkle-Damgård scheme have been discovered. In particular, Kelsey and Schneier published a generic second-preimage attack for long messages against the Merkle-Damgård scheme [23] in 2005. The attack complexity is roughly 2^{n-k} compression function calls if the original given message is 2^k-block long, with $k \leq n/2$. Later, Andreeva et al. gave an alternative attack using a diamond structure [3]. Their attack also require 2^{n-k} compression function calls if the original given message is 2^k-block long, but only for $k \leq n/3$. On the other hand, it is applicable to a wider range of designs; in particular it can accommodate a small dithering input in the compression function. It also gives some more freedom to the adversary: as mentioned in [3], this variant allows "the attacker to leave most of the target message intact in the second preimage, or to arbitrarily choose the contents of roughly the first half of the second preimage, while leaving the remainder identical to the target message."

Therefore, regardless of how the compression function is designed, a Merkle-Damgård hash function can simply not achieve the security of 2^n with respect to second-preimage resistance. Consequently, the research community designed new domain extension schemes in order to overcome the inherent weaknesses of the original Merkle-Damgård construction. In their original second-preimage attack, Kelsey and Schneier already suggest this approach, and mention that "XORing in a monotomic counter as part of the round function would resist the attacks". Later, Biham and Dunkelman proposed the HAIFA domain extension scheme [7], which became quite popular. The main feature of HAIFA is that it adds a counter (which corresponds to the number of previously hashed message bits) as an extra input parameter to the compression function during the iteration process, in order to make each compression function call different. On the one hand, this is widely believed to provide resistance against second-preimage attacks, and this can be proved under strong randomness assumptions for the compression function [10]. On the other hand, this means the compression function must accept an extra input, which must be processed securely to avoid security issues. In particular, compression function attacks can take advantage of this input [4, 9, 16, 19], even though the effect on the iterated function is not obvious. Recently, many new dedicated hash functions have been designed following the HAIFA framework, including some SHA-3 candidates (BLAKE [5], ECHO [6], Shavite-3 [8], Shabal [12], Skein [14]), as well as Streebog, which has been standardized by the Russian government as GOST R 34.11-2012 [27] and by IETF as RFC 6896 [20].

Our Contributions. In this article, we focus on the security of Streebog hash function with respect to the second-preimage resistance. According to the designers, Streebog is based on the HAIFA framework, and is explicitly claimed to resist second-preimage attacks with long message [17, 30][1].

While we are not aware of any generic second-preimage attack on the HAIFA framework, we emphasize that HAIFA acts as a *generic* framework, without explicitly specifying how the counter should be involved in the compression function computation. On the other hand, Streebog, as an instantiation of the HAIFA framework, has fully specified the way how the counter is used inside the compression function. This instantiation is quite provocative as the counter is simply XORed to the internal state variable of the compression function. Thus, it is necessary to evaluate whether this simple approach is sound or not (at least with respect to the second-preimage resistance). This analysis will also shed some light on the statement of Kelsey and Schneier that "XORing in a monotomic counter" is sufficient to avoid those attacks.

Unfortunately, we show in this article that Streebog's method to incorporate the counter *does not strengthen its security with respect to second-preimage resistance*. More precisely, we observe that during the sequential iteration of the compression function, the counter injection at block i interacts with the counter injection at next block $i + 1$. The iteration of the compression function in Streebog can then be transformed into an equivalent form, for which a counter-independent function is used multiple times during the hashing process. This behavior reduces to almost zero the extra security brought by the HAIFA framework over the regular Merkle-Damgård construction. Thanks to our findings, we describe two second-preimage attacks on the full Streebog-512. In Sect. 4, we give an attack using a diamond structure, similar to the attack of [3]. It requires about 2^{342} compression function evaluations for long messages with at least 2^{179} blocks. In Sect. 5, we give attack using an expandable message, similar to the attack of [23]. It requires only 2^{266} compression function evaluations

---

[1] These documents also claim that Streebog is resistant to the herding attack from Kerlsey and Kohno [22], but it is well known that this attack is applicable to HAIFA if no salt is used [7].

for long messages with at least 2^{259} blocks. For short messages of 2^x blocks, the first attack gives a complexity of about $2x \cdot 2^{512-x}$ when $x < 179$, while the second attack gives a complexity of about 2^{523-x} when $x < 259$. Note that this increases *linearly* with the decrease of the message block length (ignoring the logarithmic factor).

The rest of the article is organized as follows. In Sect. 2, we provide a description of the Streebog hash function, and then discuss our main observation on the usage of the counter value in Sect. 3. We detail how this observation can be used in order to mount second-preimage attacks of the full Streebog-512 hash function in Sect. 4 (using a diamond structure), and in Sect. 5 (using an expandable message). Finally, we draw conclusions in Sect. 6.

2 Specifications of Streebog

2.1 Domain Extension of Streebog

Streebog is a family of two hash functions, Streebog-256 and Streebog-512 that has hash output sizes 256 and 512 bits respectively [20,27]. In this article, we only consider the large version Streebog-512 and we simply refer to it as Streebog.

During the computation process, Streebog updates the internal state h as well as two other internal variables: Σ that denotes the checksum of the message blocks already processed, and the counter N that refers to the number of already hashed bits. Both the message block size and the intermediate hash variable size are 512 bits. The dedicated domain extension consists of three stages that we describe below (see also Fig. 1). Let M be the input message, and we denote $|M|$ its bit length. In the rest of the article, we also denote h_i the internal state variable h after the i-th application of the compression function g, which is defined in more details in Sect. 2.2.

Stage 1. This phase initializes the hash state. The three variables Σ, N and h are assigned to 0, 0 and IV respectively, where IV refers to the initialization vector of Streebog, and has been publicly defined by the designers.

Stage 2. The input message M is divided into 512-bit blocks $m_1||m_2||\cdots||m_t$, where $t = \left\lceil \frac{|M|}{512} \right\rceil$. The block m_i, $1 \le i \le t$, is processed according to the following operations:

$$h_i \longleftarrow g(N, h_{i-1}, m_i); \qquad N \longleftarrow N + 512; \qquad \Sigma \longleftarrow \Sigma + m_i.$$

Stage 3. Pad the last block with $10\cdots0$ so that it becomes full, and we denote this padded block m. Then, process this padded last block with:

$$h_{t+1} \longleftarrow g(N, h_t, m); \qquad N \longleftarrow N + (|M| \mod 512); \qquad \Sigma \longleftarrow \Sigma + m.$$

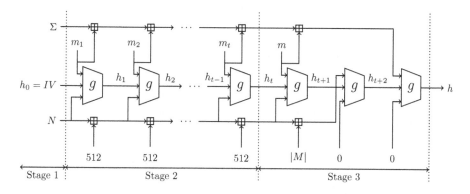

Fig. 1. The domain extension algorithm of `Streebog`.

After all the message blocks have been processed, two extra compression function calls are applied:

$$h_{t+2} \longleftarrow g(0, h_{t+1}, |M|); \qquad\qquad h_{t+3} \longleftarrow g(0, h_{t+2}, \Sigma).$$

Finally, h_{t+3} is the hash digest for `Streebog-512`. In the case of `Streebog-256`, the 256 MSBs of h_{t+3} are outputted as hash digest.

2.2 The Compression Function of `Streebog`

As described in the introduction, the designers of `Streebog` have chosen to adopt the HAIFA model in the design of the compression function g. This framework has been initially introduced to differentiate the successive applications of the compression function calls by adding a counter as additional input parameter. Here, we mainly focus on how the counter N is used in the compression function $g(N, h_{i-1}, m_i)$, which is described in Fig. 2. Particularly, we emphasize that f is a deterministic function independent of the counter N. Since the detailed algorithm of f is not related to our attack, we omit its description in this paper, and refer the interested reader to the original document [20,27]. Yet we would like to point out that f shares high similarity with the compression function of Whirlpool hash function [28], which leads to the analysis results on `Streebog` [1,2,31] that share similarity with the attacks on Whirlpool [25,29].

For the sake of simplicity, we consider that the counter value equals the number of compression calls rather than the number of processed bits. Practically, this only consists in performing a right-shift operations of 9 bit positions on the counter value. This simplification does not change any of the results described in this article, while easing the reading of the technical contents.

3   Our Observation

In this section, we propose an equivalent representation of the domain extension algorithm of `Streebog`, which we use in the next section to launch a second-preimage attack on the full hash function.

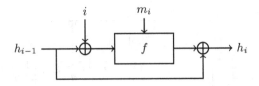

Fig. 2. The compression function $g(N, h_{i-1}, m_i)$ of Streebog produces the new chaining variable h_i.

First of all, we describe this equivalent description of the compression function, which is depicted in Fig. 3. The counter variable N coming from the HAIFA design is simply XORed to the internal state h_{i-1} prior to the application of the function f (but after the feed-forward branching, see Fig. 2), which makes it possible to linearly shift the addition before and after the feed-forward in the original compression function. Formally, we have the following equivalence:

$$h_i = h_{i-1} \oplus f(h_{i-1} \oplus i, m_i) \qquad \Longleftrightarrow \qquad \begin{cases} h_i = F(h_{i-1} \oplus i, m_i) \oplus i, \\ F(x, m_i) = f(x, m_i) \oplus x. \end{cases}$$

Note that the counter value i is now XORed to both the input hash variable and the output hash variable of F (see Fig. 3), while F itself is a deterministic function which is independent of the counter parameter i.

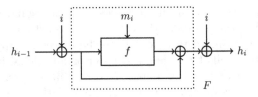

Fig. 3. An equivalent representation of Streebog's compression function: the internal function F has been made independent of the counter value.

We now pay attention to the sequential iteration of the above equivalent compression function in Stage 2 of the domain extension. For the sake of simplicity, we detail here the case of two consecutive blocks (see Fig. 4).

As we can see, during the end of the i-th message block computation until the beginning of the $(i + 1)$-th message block computation, the output of F is updated twice by XORing consecutively the counter values i and $i+1$. We define

$$\Delta(i) \stackrel{\text{def}}{=} i \oplus (i + 1),$$

$$F_{\Delta(i)}(X, Y) \stackrel{\text{def}}{=} F(X, Y) \oplus \Delta(i).$$

From this observation, we get yet another equivalent representation of the consecutive compression function iterations during Stage 2 of Streebog, as shown in Fig. 5.

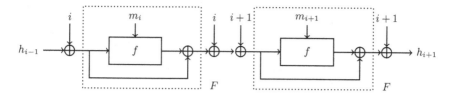

Fig. 4. Two consecutive compression function calls in the equivalent representation: the counter addition in between the two calls can be combined and controlled.

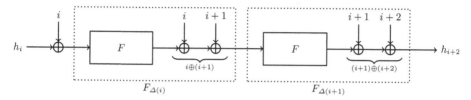

Fig. 5. Two consecutive compression function blocks in the equivalent representation.

Next, we investigate the relation between the functions $F_{\Delta(i)}$, $1 \leq i \leq t$. In the most simple case, we can easily see that $\Delta(i) = i \oplus (i+1) = 1$ always holds as long as i is an even integer. Consequently, the very same function F_1 is used every even integer index during the iterations in Fig. 5. We list the first values of $\Delta(i)$ in Table 1, and one can see that there is a lot of structure: sequences of length $2^s - 1$ seem to repeat every 2^s steps. More formally, we compare the functions $F_{\Delta(i)}$ and $F_{\Delta(i+2^s)}$ for any $0 \leq i < 2^s - 1$, where s can be any positive integer smaller than 512. Let $\langle i \rangle$ denote the s-bit binary representation of that integer i. We have:

$$\Delta(i) = \langle \mathtt{i} \rangle \oplus \langle \mathtt{i} + \mathtt{1} \rangle$$
$$\Delta(i + 2^s) = (\mathtt{1} || \langle \mathtt{i} \rangle) \oplus (\mathtt{1} || \langle \mathtt{i} + \mathtt{1} \rangle) = \langle \mathtt{i} \rangle \oplus \langle \mathtt{i} + \mathtt{1} \rangle.$$

Thus, we conclude that $F_{\Delta(i)}$ and $F_{\Delta(i+2^s)}$ are the same function for any $0 \leq i < 2^s - 1$. By extending this simple reasoning, we can generalize and demonstrate that $F_{\Delta(i)}$ and $F_{\Delta(i+j \times 2^s)}$ are the same function for any $0 \leq i < 2^s - 1$ and any integer j. This is illustrated in Fig. 6.

Table 1. First values of $\Delta(i)$.

i:	0	1	2	3	4	5	6	7	8	9	10	11	12	13	14	15	16	17	18	19	20	21	22	23
$\Delta(i)$:	1	3	1	7	1	3	1	15	1	3	1	7	1	3	1	31	1	3	1	7	1	3	1	15

Finally, we present an equivalent representation of the sequential iteration in Stage 2 of the domain extension of **Streebog** in Fig. 7, where F_i denotes the function for $F_{\Delta(j \times 2^s + i)}$ with $0 \leq i \leq 2^s - 2$, and G_j denotes the functions $F_{\Delta(j \times 2^s - 1)}$, where j is any integer. Let l be $\lfloor \frac{t}{2^s} \rfloor$ and p be the reminder of $t \bmod 2^s$.

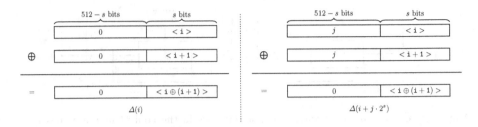

Fig. 6. Functions $F_{\Delta(i)}$ and $F_{\Delta(j \times 2^s + i)}$ are the same

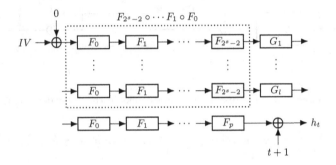

Fig. 7. The equivalent representation of Stage 2

4   Second-Preimage Attack on Full Streebog with a Diamond

Based on the equivalent description of the Stage 2 computation of Streebog presented in the previous section, we now describe a second-preimage attack on the full Streebog-512 hash function with time complexity equivalent to 2^{342} compression function evaluations for an original message of at least 2^{179} blocks.

Our main observation provides a way to remove the security benefits brought by the counter of the HAIFA design in the Streebog hash function. This is due to a poor usage of this counter, which allows an adversary to reuse previously known second-preimage techniques on the classical Merkle-Darmgård construction. In particular, we can use the *diamond structure* introduced by Kelsey and Kohno [22] on the function $F_{2^s-2} \circ \cdots \circ F_1 \circ F_0$, which is reused several times. Indeed, this technique allows to construct a large multicollision set of 2^d d-block messages, all hashing to a single chaining variable $h_\diamond$. This is similar to the second-preimage attack on dithered hash functions by Andreeva et al. [3].

We first give in Sect. 4.1 a detailed explanation concerning the construction of this structure with $2^{(n+d)/2}$ computations, and we later describe in Sect. 4.2 how to use it inside a second-preimage attack for the full Streebog-512.

4.1   The Diamond Structure

As depicted in Fig. 8, a 2^d-diamond construction refers to a complete binary tree of depth d, i.e., the distance from the leaves to the root is d. There are exactly 2^{d-l} nodes at level l, for $0 \leq l \leq d$, where $l = 0$ refers to the leaf level and $l = d$ to the root level. All nodes except the leaves have two children from lower level. In [22], Kelsey and Kohno introduced this structure to launch herding attacks. In this diamond, a node refers to a chaining value, and an edge represents a message connecting one chaining value to another.

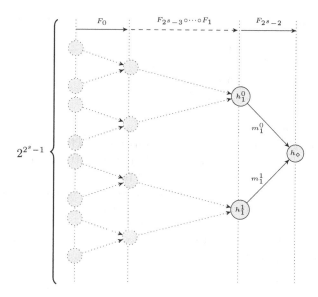

Fig. 8. The diamond structure of depth $2^s - 1$ used in our second-preimage attack.

Given the leaves, i.e., 2^d chaining values at level 0, one can construct the diamond in $2^{(n+d)/2}$ compression function evaluations. The construction algorithm was initially proposed by Kelsey and Kohno [22] and later refined by Kortelainen and Kortelainen [24] and verified in [18]. The algorithm works level by level recursively and independently. Below in Algorithm 1, we show how the next level of 2^{d-1} chaining values are computed given the current level of 2^d nodes and compression function $f = F_0$ as input. The output L_{out} of the current level is then fed into the algorithm as input L_{in} for next level, until root is reached. The overall complexity has been estimated as $2^{(n+d)/2}$ in [24].

4.2   Details of the Attack

At this point, we are able to build a diamond structure, and we would like to use it for a second-preimage attack. An overview of our attack is given in Fig. 9,

---

Algorithm 1. Construction of one level of a diamond

Input: input chaining value list L_{in} of size 2^d
Input: compression function f
Output: next layer chaining values list L_{out} of size 2^{d-1}
1: initialize an empty hash table T, a message list L_M.
2: **while** L_{in} is not empty **do**
3:     pick random message block M, add to L_M.
4:     **for all** $h_{in} \in L_{in}$ **do**
5:         evaluate $h_{out} = f(h_{in}, M)$
6:         **if** $T[h_{out}]$ is not empty **then**
7:             fetch the pair (h'_{in}, M') in entry $T[h_{out}]$
8:             add h_{out} to L_{out}, along with (h_{in}, M), (h'_{in}, M') as the connecting edges.
9:             remove $f(h_{in}, m)$ and $f(h'_{in}, m)$ from T for all $m \in L_M$.
10:            remove h_{in} and h'_{in} from L_{in}.
11:        **else**
12:            add (h_{in}, M) to $T[h_{out}]$.

---

where one can see that we use $d = 2^s - 1$ in order to fully control the effect of the counter. The diamond structure is constructed with the function $F_{2^s-1} \circ \cdots F_1 \circ F_0$. Then, as in the original second-preimage attack using the diamond structure, we use a single message block $m_\diamond^\searrow$ to connect the root chaining value $h_\diamond$ to the known message we are attacking. The connection is done after the next F function, but before the addition of the counter, i.e. we match of set of values $\{F(h_\diamond, m) \mid m \leftarrow \$\}$, and $\{h_i \oplus i \mid i \equiv 0 \bmod 2^s\}$. If the original message m consists of t 2^s-bit blocks, we have $l = \lfloor \frac{t}{2^s} \rfloor$ possible connecting points, meaning that we expect to pick about $2^n/l$ random message blocks $m_\diamond^\searrow$ before hitting a known point $h'_\diamond$.

This point of connection gives the value $l' \times 2^s - 1$ of the counter N used in Streebog at that position. Once we have found the 1-block connecting message $m_\diamond^\searrow$ at the end of the diamond structure, we need to connect one of the 2^d leaves of the diamond structure to the IV of the hash function.

Before finding a valid second-preimage, there are two additional points that we need to consider. First, the second-preimage needs to have the exact same length $|M|$ as the first message since Streebog processes the length of the message at the end of the hashing process. Second, the additive checksum computed over the new blocks of the second-preimage needs to match the targeted one Σ of the original message.

In order to overcome both of these two points, we first construct a 2^{512}-multicollision (with a technique similar to the one from Joux [21]) over the first $2 \times 512 = 1024$ message blocks so as to handle the checksum issue. This step can be performed efficiently with $512 \times 2^{n/2}$ computations using the technique described in [15]. The idea is that, at each step i of the multicollision search, we compute two sets of 2-block messages: $\{(A_i)||(-A_i)\}$, for $2^{n/2}$ random choices of A_i, and $\{(B_i + 2^i)||(-B_i)\}$, for $2^{n/2}$ random choices of B_i, in order to find a collision between the two sets.

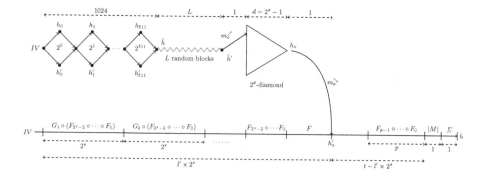

Fig. 9. Overview of the second-preimage attack.

Then, starting from the IV, we reach a chaining value $\tilde{h}$, such that we can find a 1024-block message that verifies any given additive checksum value σ. Indeed, the binary decomposition of σ gives precisely the path to follow (and incidentally the message blocks to use) in the multicollision graph we just built in order to reach σ.

We would like now to match the correct message length $|M|$. For that task, we first evaluate the number of blocks already fixed by the attack. The diamond uses $d = 2^s - 1$ blocks, the multicollision uses 1024 blocks, and we use one block for $m_\diamond'$ to connect to $h_\diamond'$ in the original message chain. After the collision on $h_\diamond'$, we use the same values as in the original message, such that we want to use exactly $l' \times 2^s$ blocks between the IV and $h_\diamond'$. We use an additional message block $m_\diamond'$ to connect to one leaf of the diamond, so that in total there are $L = l' \times 2^s - 1024 - 1 - 2^s - 1$ blocks left between $\tilde{h}$ and $\tilde{h}'$. We pick random values for all those blocks, obtain the value of $\tilde{h}'$, and then pick about 2^{n-d} random blocks $m_\diamond'$ to hit one of the 2^d leaves of the diamond.

Finally, we compute the reduced checksum value σ of all the message blocks except the 1024 first ones, and we choose the correct 1024 message blocks in the graph so as to match the local checksum $\Sigma - \sigma$. At this point, the attack is over: all the message blocks are fixed, and the second-preimage is constructed.

Overall, the total complexity of this attack requires $2^{(n+d)/2}$ computations to construct the diamond, $2^n/l$ computations to connect the root of the diamond to the original message chain, and $512 \times 2^{n/2}$ computations for the Joux's multicollision. The time complexity

$$512 \times 2^{n/2} + 2^{n-d} + 2^{(n+d)/2} + 2^{n-\log_2(l)}$$

can be minimized by fixing $d = n/3$ and $l \geq 2^{n/3}$, which reaches an overall time complexity of about $2^{2n/3}$ computations for the second-preimage attack. With the parameters of Streebog-512, $n = 512$ gives the integer value $s = 8$ and $d = n/3$, and a total time complexity equivalent to about 2^{342} compression function evaluations. We note that our attack imposes a certain length on the original message as $n - \log_2(l) \leq 341$ imposes $l \geq 2^{171}$, which constraints M to have at least $2^{171+8} = 2^{179}$ message blocks.

For shorter messages with 2^x blocks and $x < 179$, the complexity is mainly dominated by the complexity of linking IV to one leaf node of the diamond structure, which is 2^{n-d}, and the complexity of linking $h_\diamond$ to $h'_\diamond$, which is $2^{n-x+\lceil \log_2(d) \rceil}$. Let $x = d$, and we get the complexity is upper bounded by $2x \cdot 2^{n-x}$. Thus the complexity increases *linearly* with the decrease of the message block length (ignoring logarithmic factors).

5   Second-Preimage Attack on Full Streebog with an Expandable Message

The equivalent description of Streebog given in the previous sections can also be used to mount a variant of the attack of Kelsey and Schneier using an *expandable message* [23]. This gives a second-preimage attack on the full Streebog-512 hash function with time complexity equivalent to 2^{266} compression function calls for an original message of at least 2^{259} blocks.

We first give in Sect. 5.1 a detailed explanation concerning the construction of this structure with $n/2 \times 2^{n/2}$ computations, and we later describe in Sect. 5.2 how to use it inside a second-preimage attack for the full Streebog-512.

5.1   The Expandable Message

In order to build an expandable message, we use the technique of [23], i.e. we build a multicollision where the messages in each colliding pair have a different length, as shown by Algorithm 2. If we have colliding pairs with length $(1, 2^k + 1)$, for $0 \le k < t$, this implicitly defines a set of 2^t messages with length in the range $[t, 2^t + t - 1]$, that all reach the same final chaining value x_*. More precisely, one can build a message of length $t + L$ using the binary expression of L to select a message in each pair.

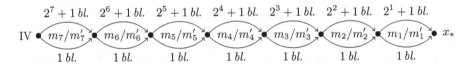

In a second-preimage attack, we hash random blocks starting from x_* until we find a link to one of the intermediate values reached when hashing the challenge message. This gives the required length for the expandable message, and we build the second preimage using the expandable message, the linking block, and the end of the challenge message.

However, this does not work for a HAIFA compression function: depending on which message is selected in the pair k (m_k or m'_k), the message length before the following block will be different, and the counter will have a different value. Therefore, the collision (m_{k-1}, m'_{k-1}) will only be valid in one case.

In the case, of Streebog, the weak use of the counter makes this attack still possible thanks to the equivalent representation of Sect. 3. Indeed, the sequence

---

Algorithm 2. Construction of an expandable message (Merkle-Damgård)

Input: Initial chaining value x
Input: Compression function f
Output: Message pairs (m_i, m_i'), final chaining value x
1: **for** $0 \leq i < n/2$ **do**
2:      Initialize an empty hash table T
3:      **for** $0 \leq r < 2^{n/2}$ **do**
4:          $T[f(x, r)] \leftarrow r$
5:      $y \leftarrow x$
6:      **for** $0 \leq j < 2^i$ **do**
7:          $y \leftarrow f(y, 0)$
8:      **repeat** $r \leftarrow \$$
9:      **until** $T[f(y, r)]$ not empty
10:      $m_i \leftarrow [0]^{2^i} \| r$
11:      $m_i' \leftarrow T[f(y, r)]$
12:      $x \leftarrow f(y, r)$

$\Delta(i)$ has a lot of regularity and repetitions (as seen in Table 1), and with a careful construction, we can ensure that the message pairs (m_i, m_i') are only used at positions with same sequences of $\Delta(i)$. More precisely, we must build pairs with large difference first, and use differences that are powers of two, while more general constructions can be used for plain Merkle-Damgård. We must also stop the construction a few steps before reaching a difference of 1 (as explained later, the smallest difference is $O(n)$). This means that we can only use a fraction of the intermediate states reached by the challenge message.

In the following, we call an expandable message that can reach lengths between a and b by increment of c an (a, b, c)-expandable message. Let us assume we have built an $(l, l + L, 2^i)$-expandable message for Streebog, with $l < 2^{i-1} - 1$. Since $l < 2^i - 1$, we have $\Delta(l + x) = \Delta(l + x + j \cdot 2^i)$, for all $0 \leq x < 2^i - l - 1$ and $j \geq 0$. In particular, if we append a new message pair (m, m') with $|m| - 2^{i-1} + 1, |m'| - 1$ to the expandable message, the sequence of $\Delta(i)$ used for the messages will be same for every choice of the $(l, l + L, 2^i)$-expandable message. This allows to extend the $(l, l + L, 2^i)$-expandable message into a $(l + 1, l + L + 1 + 2^{i-1}, 2^{i-1})$-expandable message. If we iterate this construction, starting from a single message of length l and a maximal increment of 2^t, we can build a $(l + t - s, l + t - s + 2^{t+1} - 2^s, 2^s)$-expandable message for Streebog, assuming that $l + t - s < 2^s - 1$ (Fig. 10).

5.2   Details of the Attack

The second preimage attack on full Streebog-512 uses an initial multicollision with 1024 blocks in order to adjust the checksum, like the attack of Sect. 4. Then, we build the expandable message starting for the final value of the multicollision. With the parameters of Streebog-512, we use $l = 1024$, $s = 11$, $t = 258$, i.e. we build a $(1271, 2^{259} - 777, 2048)$-expandable message. After building the

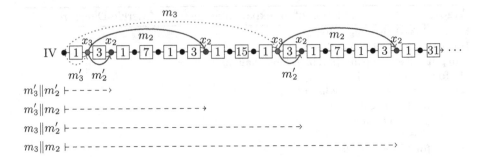

Fig. 10. Construction of a $(2, 14, 4)$-expandable message for `Streebog`. Note that m_2 and m_2' have the same Δ indices in both positions, and the Δ for the block after $m_3'\|m_2'$, $m_3'\|m_2$, $m_3\|m_2'$, or $m_3\|m_2$ is the same (here, $\Delta = 1$).

expandable message, the attack mostly follows the procedure given by Kelsey and Schenier. An overview of our attack is given in Fig. 11.

We first use a message block m_* to connect the final chaining value h_* to the known message we are attacking. More precisely, if the original message m consists of t 2^s-bit blocks, we have $l = \lfloor \frac{t}{2^s} \rfloor$ possible connecting points, meaning that we expect to pick about $2^n/l$ random message blocks m_* before hitting a known point h_*'. With the parameters used for `Streebog-512`, we use connecting points[2] with $i \equiv 1272 \bmod 2048$. This point of connection gives the value of the counter N used in `Streebog` at that position, and the length $L = N - 1024 - 1$ required for the expandable message. In order to build the second preimage, we select the message with the correct length L in the expandable message, and we select a message in the initial multicollision to adjust the checksum.

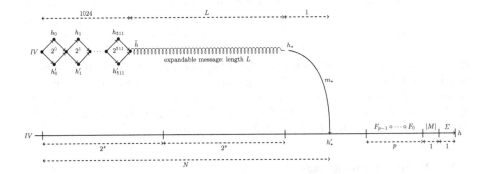

Fig. 11. Overview of the second-preimage attack.

---

[2] This correspond to the set of positions such that $i+1$ can be reached by a $(1271, 2^{259} - 777, 2048)$-expandable message.

Overall, the attack requires about $512 \times 2^{n/2}$ computations for the Joux's multicollision, $256 \times 2^{n/2}$ for the expandable message, and $2^n/l$ computations to connect the end of the expandable message to the original message chain. The time complexity

$$768 \times 2^{n/2} + 2^n/l$$

can be minimized with $l > 2^{n/2}/n$, and reaches an overall time complexity in the order of $n \cdot 2^{n/2}$ computations for the second-preimage attack. With the parameters of Streebog-512, we have $n = 512$ and $s = 11$, and a total time complexity equivalent to about 2^{266} compression function evaluations, if the message has more than 2^{259} blocks (so that $2^n/l \leq 256 \times 2^{n/2}$).

6   Open Discussion and Conclusion

In this article, we have studied the security of the Russian hash function standard Streebog. We showed that an attacker can find second-preimages much faster than what is expected from an ideal hash function, even though Streebog uses HAIFA as the domain extension algorithm. Our main observation is that the counter is not very well handled in Streebog and this enables the attacker to apply a more complex variation of the now classical generic second-preimage attacks. As a result, Streebog is only marginally stronger than a plain Merkle-Damgåd iteration.

This analysis also contradicts the remark by Kelsey and Schneier that "XOR-ing in a monotomic counter" would be sufficient to avoid the second-preimage attacks with long messages: there is at least one way to XOR the counter that do not provide any extra security.

Our work is a good example why one should be careful when using a design framework: problems might arise if bad instances in that framework exist. In the particular case of HAIFA, it is crucial to make sure the counter is properly handled. We have the intuition that the security property that a compression function in HAIFA has to follow with regards to the counter input is quite strong (even if the counter might controlled by the adversary, he must not be able to distinguish the output). Clearly, Streebog would not meet that criteria (inserting a difference δ in both the counter and the chaining variable input, one always get δ on the output). It would be interesting to study what is exactly the minimal security assumption that is required on the counter input for HAIFA in order to ensure only secure instances.

Acknowledgment. We would like to thank the anonymous reviewers for their detailed feedback and comments. Jian Guo, Jérémy Jean, Thomas Peyrin and Lei Wang were supported by the Singapore National Research Foundation Fellowship 2012 (NRF-NRFF2012-06).

References

1. AlTawy, R., Kircanski, A., Youssef, A.M.: Rebound attacks on Stribog. IACR Cryptology ePrint Archive 2013, 539 (2013)
2. AlTawy, R., Youssef, A.M.: Preimage attacks on reduced-round stribog. In: Pointcheval, D., Vergnaud, D. (eds.) AFRICACRYPT. LNCS, vol. 8469, pp. 109–125. Springer, Heidelberg (2014)
3. Andreeva, E., Bouillaguet, C., Fouque, P.-A., Hoch, J.J., Kelsey, J., Shamir, A., Zimmer, S.: Second preimage attacks on dithered hash functions. In: Smart, N.P. (ed.) EUROCRYPT 2008. LNCS, vol. 4965, pp. 270–288. Springer, Heidelberg (2008)
4. Aumasson, J.-P., Guo, J., Knellwolf, S., Matusiewicz, K., Meier, W.: Differential and invertibility properties of BLAKE. In: Hong, S., Iwata, T. (eds.) FSE 2010. LNCS, vol. 6147, pp. 318–332. Springer, Heidelberg (2010)
5. Aumasson, J.P., Henzen, L., Meier, W., Phan, R.C.W.: SHA-3 proposal BLAKE. Submission to NIST (Round 3) (2010)
6. Benadjila, R., Billet, O., Gilbert, H., Macario-Rat, G., Peyrin, T., Robshaw, M., Seurin, Y.: SHA-3 Proposal: ECHO. Submission to NIST (updated) (2009)
7. Biham, E., Dunkelman, O.: A framework for iterative hash functions - HAIFA. Cryptology ePrint Archive, Report 2007/278 (2007)
8. Biham, E., Dunkelman, O.: The SHAvite-3 hash function. Submission to NIST (Round 2) (2009)
9. Biryukov, A., Gauravaram, P., Guo, J., Khovratovich, D., Ling, S., Matusiewicz, K., Nikolić, I., Pieprzyk, J., Wang, H.: Cryptanalysis of the LAKE hash family. In: Dunkelman, O. (ed.) FSE 2009. LNCS, vol. 5665, pp. 156–179. Springer, Heidelberg (2009)
10. Bouillaguet, C., Fouque, P.A.: Practical hash functions constructions resistant to generic second preimage attacks beyond the birthday bound. Submitted to Information Processing Letters (2010)
11. Brassard, G. (ed.): CRYPTO 1989. LNCS, vol. 435. Springer, Heidelberg (1990)
12. Bresson, E., Canteaut, A., Chevallier-Mames, B., Clavier, C., Fuhr, T., Gouget, A., Icart, T., Misarsky, J.F., Naya-Plasencia, M., Paillier, P., Pornin, T., Reinhard, J.R., Thuillet, C., Videau, M.: Shabal, a submission to NIST's cryptographic hash algorithm competition. Submission to NIST (2008)
13. Damgård, I.: A design principle for hash functions. In: [11], pp. 416–427
14. Ferguson, N., Lucks, S., Schneier, B., Whiting, D., Bellare, M., Kohno, T., Callas, J., Walker, J.: The skein hash function family. Submission to NIST (Round 3) (2010)
15. Gauravaram, P., Kelsey, J.: Linear-XOR and additive checksums don't protect Damgård-Merkle hashes from generic attacks. In: Malkin, T. (ed.) CT-RSA 2008. LNCS, vol. 4964, pp. 36–51. Springer, Heidelberg (2008)
16. Gauravaram, P., Leurent, G., Mendel, F., Naya-Plasencia, M., Peyrin, T., Rechberger, C., Schläffer, M.: Cryptanalysis of the 10-round hash and full compression function of SHAvite-3-512. In: Bernstein, D.J., Lange, T. (eds.) AFRICACRYPT 2010. LNCS, vol. 6055, pp. 419–436. Springer, Heidelberg (2010)
17. Grebnev, S., Dmukh, A., Dygin, D., Matyukhin, D., Rudskoy, V., Shishkin, V.: Asymmetrical reply to SHA-3: Russian hash function draft standard. CTCrypt 2012, abstract available from http://agora.guru.ru/csr2012/files/6.pdf
18. Guo, J.: A program confirmation of the diamond construction by Kortelainen and Kortelainen (Feburary 2014). http://guo.crypto.sg/diamond.zip

19. Guo, J., Karpman, P., Nikolic, I., Wang, L., Wu, S.: Analysis of BLAKE2. In: Benaloh, J. (ed.) CT-RSA 2014. LNCS, vol. 8366, pp. 402–423. Springer, Heidelberg (2014)
20. IETF: GOST R 34.11-2012: Hash Function. RFC6896 (2013)
21. Joux, A.: Multicollisions in iterated hash functions. Application to cascaded constructions. In: Franklin, M. (ed.) CRYPTO 2004. LNCS, vol. 3152, pp. 306–316. Springer, Heidelberg (2004)
22. Kelsey, J., Kohno, T.: Herding hash functions and the nostradamus attack. In: Vaudenay, S. (ed.) EUROCRYPT 2006. LNCS, vol. 4004, pp. 183–200. Springer, Heidelberg (2006)
23. Kelsey, J., Schneier, B.: Second preimages on n-bit hash functions for much less than 2^n work. In: Cramer, R. (ed.) EUROCRYPT 2005. LNCS, vol. 3494, pp. 474–490. Springer, Heidelberg (2005)
24. Kortelainen, T., Kortelainen, J.: On diamond structures and trojan message attacks. In: Sako, K., Sarkar, P. (eds.) ASIACRYPT 2013, Part II. LNCS, vol. 8270, pp. 524–539. Springer, Heidelberg (2013)
25. Lamberger, M., Mendel, F., Rechberger, C., Rijmen, V., Schläffer, M.: Rebound distinguishers: results on the full whirlpool compression function. In: Matsui, M. (ed.) ASIACRYPT 2009. LNCS, vol. 5912, pp. 126–143. Springer, Heidelberg (2009)
26. Merkle, R.C.: One way hash functions and DES. In: [11], pp. 428–446
27. REGULATION, F.A.O.T., METROLOGY: Information technology - CRYPTO-GRAPHIC DATA SECURITY - Hash-function. GOST R 34.11-2012 (2012)
28. Rijmen, V., Barreto, P.S.L.M.: The WHIRLPOOL hashing function. Submitted to NISSIE, September 2000
29. Sasaki, Y., Wang, L., Wu, S., Wu, W.: Investigating fundamental security requirements on whirlpool: improved preimage and collision attacks. In: Wang, X., Sako, K. (eds.) ASIACRYPT 2012. LNCS, vol. 7658, pp. 562–579. Springer, Heidelberg (2012)
30. GOST R 34.11-2012: Streebog Hash Function. https://www.streebog.net/
31. Wang, Z., Yu, H., Wang, X.: Cryptanalysis of GOST R hash function. Cryptology ePrint Archive, Report 2013/584 (2013). http://eprint.iacr.org/

Side-Channel Analysis of Montgomery's Representation Randomization

Eliane Jaulmes[✉], Emmanuel Prouff, and Justine Wild

ANSSI, 51, Bd de la Tour-Maubourg, 75700 Paris 07 SP, France
{eliane.jaulmes,emmanuel.prouff,justine.wild}@ssi.gouv.fr

Abstract. Elliptic curve cryptography is today widely spread in embedded systems and the protection of their implementation against side-channel attacks has been largely investigated. At CHES 2012, a countermeasure has been proposed which adapts Montgomery's arithmetic to randomize the intermediate results during scalar point multiplications. The approach turned out to be a valuable alternative to the previous strategies based on hiding and/or masking techniques. It was argued to be specifically dedicated to hardware implementations and it aimed to defeat first-order side-channel attacks involving Pearson's correlation as distinguisher. In this paper however, we exhibit an important flaw in the countermeasure and we show, through various simulations, that it leads to efficient first-order correlation-based attacks.

1 Introduction

Elliptic Curves Cryptosystems (ECC) have been introduced by N. Koblitz [22] and V. Miller [29]. Their security relies on the hardness of the *discrete logarithm problem*. Elliptic curve based algorithms usually require keys far smaller than those involved in other public-key cryptosystems like RSA. This explains the current popularity of ECC and their involvement in a large variety of applications implemented over all kinds of devices: smart-cards, micro-controllers, and so on. Since such devices are widespread and in the hands of end-users, they are confronted to a wide range of threats. In particular, physical attacks need to be taken into account when assessing the overall security of the implementation. Thus, countermeasures are often conceived and implemented alongside the algorithms.

Physical attacks are traditionally divided into the two following families: *perturbation analysis* and *observation analysis*. The first one aims to modify the cryptosystem processing with any physical mean such as laser beams, clock jitter or voltage perturbation (*e.g.* fault injection attacks [12,13]). The attacker then learns information on the secret parameter by observing the response of the cryptosystem to this perturbation. Such attacks can be prevented by monitoring the device environment with captors and by verifying the result of the computation before output. The second family of attacks consists in measuring physical data during the algorithm execution and then in exploiting this information to recover the secret. Such leakage sources can be the power consumption or

© Springer International Publishing Switzerland 2014
A. Joux and A. Youssef (Eds.): SAC 2014, LNCS 8781, pp. 212–227, 2014.
DOI: 10.1007/978-3-319-13051-4_13

the electro-magnetic emanation. Among observation attacks, we can distinguish several categories. *Simple Power Analysis* [24] directly deduces the value of the secret from a single input processing (possibly averaged), while *Advanced Power Analysis* exploits observations for several algorithm inputs. The latter kind of attacks requires the choice of a *leakage model* to compare predictions based on key-hypotheses with the measured traces. The comparison is done with a statistical tool, also called *distinguisher*. For example, the well-known *Differential Power Analysis* (DPA) [25] uses the *difference of means* whereas the *Correlation Power Analysis* (CPA) [10] involves *Pearson's correlation coefficient*[1].

Countermeasures against observation analysis fall into two categories: *hiding* techniques aim at reducing the *Signal to Noise Ratio* (SNR) by increasing the noise or by equalizing the current in the circuit [35], while *masking* techniques consists in randomizing the sensitive computations. In practice, masking is always applied (possibly combined with hiding) since it provides strong security guaranty. Recently, Lee *et al.* [26] proposed a new efficient countermeasure to overcome first-order CPA[2] attacks. It assumes that the field operations are performed in the Montgomery domain [30] and consists in randomizing the Montgomery representation of the internal results (thus defining a so-called *Randomized Montgomery Domain*). This countermeasure has been considered as a very valuable alternative to the previous techniques because it avoids the need for scalar blinding (see *e.g.* [5]) and is much more efficient from a hardware implementation point of view.

Our work. In this paper, we show that the countermeasure proposed in [26] is flawed and can be efficiently broken by first-order CPA, even in presence of a large amount of noise in the measurements. After a presentation of the techniques proposed by Lee *et al.* in Sect. 2, the flaw is exhibited in Sect. 3.1. The attack is afterwards detailed in Sect. 3.2 and its efficiency is demonstrated in Sect. 4 thanks to simulations in various contexts. Finally, Sect. 5 analyses the experiments given in [26] and concludes with a short discussion about possible countermeasures.

2   On Randomized Implementations of Modular Operations

In this section, we recall some mathematical background on elliptic curve and the associated scalar multiplication. The reader may also refer to [3] for a more complete overview. Here, we will focus on the efficient implementation of the scalar multiplication in embedded devices. In particular we recall the use of the Montgomery domain for efficient modular operations [31].

---

[1] The need for a leakage model may be relaxed when using the so-called *collision attacks* [40] which look for colliding values during a computation. Such attacks compare only real traces to each other.

[2] A side channel attack is said to be *of first order* if it exploits the dependency between the mean of an instantaneous leakage and a function of the secret parameter. The original CPA attack in [10] is of first-order.

2.1   Background on Elliptic Curves and Montgomery Multiplication

Elliptic Curves. In this paper, we focus on *elliptic curves* E defined over a prime field $\mathbb{F}_p$ according to the following *short Weierstrass's Equation*:

$$E : y^2 = x^3 + ax + b, \tag{1}$$

where a and b are elements in $\mathbb{F}_p$ satisfying $4a^3 + 27b^2 \neq 0$. The set of rational points of E is denoted by $E(\mathbb{F}_p)$. It contains all the points whose coordinates $(x, y) \in \mathbb{F}_p^2$ satisfy (1). This set, augmented with a neutral element $\mathcal{O}$ called *point at infinity*, has an Abelian group structure for the following addition law: let $P = (x_1, y_1)$ and $Q = (x_2, y_2)$ then the coordinates (x_3, y_3) of $P + Q$ satisfy;

$$x_3 = \lambda^2 - x_1 - x_2 \text{ and } y_3 = \lambda(x_1 - x_3) - y_1, \tag{2}$$

where λ equals $(y_2 - y_1)/(x_2 - x_1)$ if $P \neq Q$ and $(3x_1^2 + a)/(2y_1)$ otherwise. The *scalar multiplication* (ECSM for short) of a point $P \in E(\mathbb{F}_p)$ by a natural integer k is denoted by kP. It is the core operation of many cryptographic protocols such as ECDSA [22] and defining an efficient scalar multiplication arithmetic is hence a central issue (the interested reader is referred to [16] for a good overview). The point addition formula (2) is also central in elliptic curve implementations and several papers have been published on this subject [1,4,14,15,18,27,28, 41]. They aim at proposing sequences of operations over $\mathbb{F}_p$ which are optimal according to some relations between the cost of a field inversion, the cost of an addition/subtraction and that of a multiplication. To defeat SPA attacks [5] at the arithmetic level, some works also try to propose sequences that stay unchanged whether (2) defines an addition ($P \neq Q$) or a doubling ($P = Q$). In this paper, we make no particular assumption on the type of ECSM algorithm nor on the sequence of operations which is used to process a point addition or a doubling. We indeed present a side-channel attack that does not exploit some particularity of the operations sequence or modus operandi but exploits information leakage during the manipulation of an intermediate result when the masking proposed in [26] is used. The latter type of result is likely to appear whatever the implementation choice.

Montgomery Domain. In order to efficiently perform the field operations involved in the point addition/doubling, developers often use the well-known Montgomery arithmetic [31]. This technique allows to replace the divisions occurring during the modular reduction by very efficient binary shifts. The only costly operation is the transformation to and from the Montgomery domain, but it is only done twice: at the beginning and at the end of the whole point operation (*e.g.* the scalar multiplication).

Let x represent an element in the prime field $\mathbb{F}_p \simeq \mathbb{Z}_p$ and let $R = 2^m$ be defined such that $2^m < p < 2^{m+1}$ (*i.e.* $m = \lfloor \log_2(p) \rfloor$). The value $\bar{x} = x \cdot R \bmod p$ is called the *Montgomery representation* of x and R is called the *Montgomery constant*. There is an isomorphism between $(\mathbb{Z}_p, +, \cdot)$ and the *Montgomery domain* $\mathrm{MD}(p) = (\{\bar{x}\}, +, \otimes)$, where $\otimes$ represents the *Montgomery multiplication* $\bar{x} \otimes \bar{y} = (\bar{x} \cdot \bar{y}) \cdot R^{-1} \bmod p$. The result of the multiplication corresponds to $\overline{xy}$, *i.e.* the Montgomery representation of $xy \bmod p$.

2.2 Randomized Montgomery Domain

Instead of using always $R = 2^m$, the authors of [26] propose to use a randomized Montgomery constant $r = 2^\lambda$ where λ is the Hamming Weight of a random m-bit value. The randomized Montgomery representation (RMR for short) of $x \in \mathbb{Z}_p$ is $x \cdot 2^\lambda \bmod p$. It is denoted $\widetilde{x}_\lambda$ or just $\widetilde{x}$ when there is no ambiguity on λ. The randomized Montgomery domain is denoted by $\mathtt{RMD}_\lambda(p)$. The multiplication in $\mathtt{RMD}_\lambda(p)$ works as follows: $\widetilde{x} \otimes_\lambda \widetilde{y} = (\widetilde{x} \cdot \widetilde{y})2^{-\lambda} \bmod p$. The result of the multiplication corresponds to $\widetilde{xy}$, i.e. the RMR of $xy \bmod p$. To ensure that the number of final subtractions stays upper-bounded by 1 (as in the classical Montgomery multiplication), the random power λ must be chosen in $[0..m]$.

In [26], the idea of RMR is applied to secure the implementation of an elliptic curve scalar multiplication against first-order side channel attack (*e.g.* CPA). The principle of these attacks is to observe the device behaviour during the processing of several scalar multiplications kP where the secret scalar $k \in \mathbb{N}$ stays unchanged and the public point P varies. Such attacks can be applied, for example, against some implementations of semi-static Diffie-Hellman key exchange, as found in the IEEE P1363 standard [17]. The main steps of the secure algorithm proposed in [26] are recalled hereafter, where the point P is assumed to belong to the set of rational points $E(\mathbb{F}_p)$ defined as in (1).

1. **Inputs:** a (public) point $P = (x_1, y_1) \in E(\mathbb{F}_p)$, a (secret) scalar $k \in \mathbb{N}$, a random number $\alpha \in [0..2^m - 1]$ with $m = \lfloor \log_2(p) \rfloor$.
2. **Conversion to RMD:** for $\lambda = \mathrm{HW}(\alpha)$, process

$$\widetilde{x_1} = x_1 \cdot 2^\lambda \bmod p$$

and

$$\widetilde{y_1} = y_1 \cdot 2^\lambda \bmod p.$$

Also convert the curve parameters (a, b) into $\mathtt{RMD}_\lambda(p)$. The point with coordinates $(\widetilde{x_1}, \widetilde{y_1})$ is denoted by $\widetilde{P}$.
3. **Elliptic Curve Scalar Multiplication (ECSM):** in $\mathtt{RMD}_\lambda(p)$, process

$$\widetilde{Q} = (\widetilde{x_2}, \widetilde{y_2}) = k\widetilde{P}. \tag{3}$$

4. **Conversion to Integer Domain:** process

$$x_2 = \widetilde{x_2} \otimes_\lambda 1 = \widetilde{x_2} \cdot 2^{-\lambda} \bmod p$$

and

$$y_2 = \widetilde{y_2} \otimes_\lambda 1 = \widetilde{y_2} \cdot 2^{-\lambda} \bmod p$$

5. **Output:** $Q = (x_2, y_2) = kP$

The Elliptic Curve Scalar Multiplication (ECSM) may be done with any algorithm (*e.g.* [31]). The authors of [26] do not recommend any particular one, even if the resistance tests reported in their paper are applied against an ECSM based on Montgomery Ladder (see Algorithm 5 in Appendix A). Similarly, our attack described in the next section does not exploit any particular feature of the ECSM and thus applies independently from the choice of algorithm. To allow for comparisons with [26] we however chose to target an ECSM based on Montgomery Ladder in our attack simulations.

3   Our Attack

3.1   Core Idea

The goal of our attack is to recover the bits of k one after another during the processing of ECSM in the randomized Montgomery domain (*i.e.* the processing of (3)). In this section, we detail how the second MSB (respectively the LSB) of k is recovered[3]. Once this bit is obtained, the attack can be applied similarly on the other bits from left to right (respectively right to left).

We denote by $R(\widetilde{u}, \widetilde{v})$ the value of an intermediate point during the processing of ECSM in the randomized Montgomery domain. We assume that the coordinates of the point R depend on a small sub part of k (*e.g.* a bit). This part is denoted by s. It can for instance correspond to the result of the second point operation in ECSM with Montgomery Ladder (see (8)) or to the result of the third point operation in ECSM with double-and-add always method (see (6)–(7)). Our side channel attack targets the manipulation of the first coordinate $\widetilde{u}$ of this point by the device: by assumption, it satisfies $\widetilde{u} = f(\widetilde{x}, \widetilde{y}, s)$ where we recall that $(\widetilde{x}, \widetilde{y})$ denotes the RMR coordinates of the input point P and where f is a known function that depends on the curve parameters and the algorithm used to process ECSM (see Appendix A for examples). To simplify the presentation, we assume that s is reduced to a single bit of k (the MSB or the LSB) but our attack straightforwardly applies to higher values (the only restriction is that the upper bound must be small enough to allow for an exhaustive test of all the elements). Our attack is based on the following statement which essentially means that a RMR representation leaks information on the un-blinded coordinate and that this information can be exploited by a *first-order* side-channel attack:

Statement: the function $f : x \mapsto \mathbb{E}\big[\mathrm{HW}(\widetilde{X}_\lambda) \mid X = x\big]$ is not constant.

Remark 1. Taking into account the specificities of the randomized Montgomery representation recalled in Sect. 2.2, the mean is computed over the random variable λ defined such that $\lambda = \mathrm{HW}(\alpha)$ with α having a uniform distribution on $[0..2^m - 1]$.

To argue on the statement above, we evaluated f on 1000 different values x for $m = 256$ and $m = 384$ (these parameters are recommended for instance by the American National Security Agency when ECSM is used to process an ECDSA). The results are plotted in Fig. 1(a) and (b). For comparison, we also

---

[3] To easy the explanation of the attack against left-to-right implementations of ECSM, we make the classical assumption that the MSB of k equals 1.

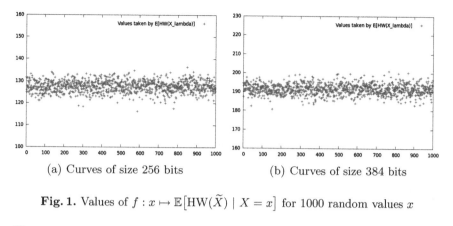

Fig. 1. Values of $f : x \mapsto \mathbb{E}\big[\mathrm{HW}(\widetilde{X}) \mid X = x\big]$ for 1000 random values x

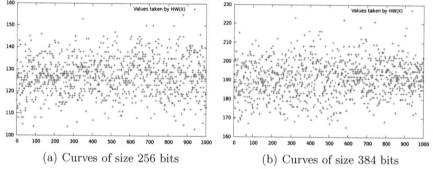

Fig. 2. Values of $f : x \mapsto \mathrm{HW}(x)$ for 1000 random values x

plotted in Fig. 2(a) and (b) the value $\mathrm{HW}(x)$ for 1000 values x in $[0..p)$, which corresponds to the case of non blinded values. Finally, in Fig. 3(a) and (b) we plotted $\mathbb{E}\big[\mathrm{HW}(\lambda \cdot X \bmod p) \mid X = x\big]$ where λ is a random uniformly distributed 16-bit value. The latter corresponds to the case where ECSM is protected by blinding the projective coordinates of the point P.

As expected, we can see in Fig. 2(a) and (b) that the Hamming weight of x varies with x when no blinding is involved; this implies that a first-order CPA is possible if x is sensitive. On the contrary, Fig. 3(a) and (b) show that coordinate blinding cancels any leakage on x since the average Hamming weight of the blinded coordinate is almost constant (even in our case where we limited λ to 16-bits words which can be considered as too limited in practice). Between the two previous extreme cases, Fig. 1(a) and (b) show that $f(x)$ varies with x (not with a high variance as in Fig. 2(a) and (b) but visibly much more than in Fig. 3(a) and (b)): this implies that the countermeasure in [26] can be attacked with a first-order CPA involving f to process the predictions on $\widetilde{x}$. In the following section, we detail the latter attack.

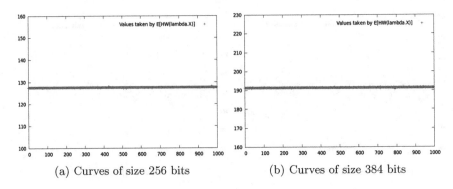

(a) Curves of size 256 bits   (b) Curves of size 384 bits

Fig. 3. Values of $f : x \mapsto \mathbb{E}\big[\mathrm{HW}(\lambda \cdot X \bmod p) \mid X = x\big]$ for 1000 random values x

3.2 Attack Description

As explained in the previous section, our attack recovers the secret piece by piece. Here, we detail how it allows for the recovery of the second MSB or the LSB s of the secret scalar k (the principle can then be repeated to recover the other bits one after another). The intermediate result/point $R(\widetilde{u}, \widetilde{v})$ exploited by the attack differs with the algorithm used to process the scalar multiplication. We give some examples below but our first-order side-channel attack also applies in other contexts (*e.g.* against atomic implementations or implementations applying the *window* principle [21]) as long as the point is blinded with the RMR representation proposed in [26] and the scalar is not blinded.

Several scalar multiplication algorithms are recalled in Appendix A. We recall that, for left-to-right versions, we make the (classical) assumption that the MSB equals 1. Our attack targets the manipulation of the first coordinate $\widetilde{u}$ of the intermediate result/point $R(\widetilde{u}, \widetilde{v})$ corresponding to:

– [**left-to-right double-and-add** ECSM] the second point operation

$$R = 3P + (1 - s)P. \tag{4}$$

– [**right-to-left double-and-add** ECSM] the first point operation

$$R = P + (1 - s)P. \tag{5}$$

– [**left-to-right double-and-add-always** ECSM] the third point operation

$$R = 2(P + sP). \tag{6}$$

– [**right-to-left double-and-add-always** ECSM] the third point operation

$$R = 2P + sP. \tag{7}$$

– [**Montgomery Ladder** ECSM] the second point operation

$$R = 2(P + sP). \tag{8}$$

– [**Joye ECSM**] the second point operation

$$R = P + (1 - s)P. \tag{9}$$

Following the classical outlines of a first-order side-channel attack, our attack starts by the observation of the device behaviour for several executions of the algorithm ECSM parametrized by different public inputs P but a same secret scalar k. Note that only the part of the observation corresponding to the manipulation of the coordinate $\tilde{u}$ of R is used in the attack described hereafter. Since the ith observation (*e.g.* the power consumption or the electromagnetic emanation) is algorithmically related to the ith input point P_i and the secret bit s, it is denoted by $\mathcal{L}(s, P_i)$ in the following. At the end of this measurement step of the attack, the adversary is assumed to be provided with a sample of pairs $(P_i, \mathcal{L}(s, P_i))_i$. The size of this sample is denoted by N.

To underline the functional dependency between $\tilde{u}$ and (s, P_i), we use the notation $\tilde{u}(s, P_i)$ in the following. For testing an hypothesis $\hat{s}$ on s, the attack continues with the computation of the values $u(\hat{s}, P_i)$ for $i \in [1..N]$. These values correspond to the *unmasked* version of the u-coordinates under the hypothesis $\hat{s} = s$. Then applying the strategy already used in [6,36] and argued in [38], we choose a *device model* m (usually the Hamming weight) and we compute the sample of predictions $(h_i)_{i \leqslant N}$ defined such that:

$$h_i(\hat{s}) = \mathbb{E}_\lambda \big[\mathsf{m}(u(\hat{s}, P_i) \cdot 2^\lambda \bmod p) \big] = \sum_{i=1}^{m} \mathsf{m}(u(\hat{s}, P_i) \cdot 2^i \bmod p) \times \mathrm{p}[\lambda = i].$$

By construction λ is defined as the Hamming weight of an m-bit random element. Its distribution is therefore binomial with parameters m and $1/2$ and the equation above can be developed as follows:

$$h_i(\hat{s}) = \frac{1}{2^m} \sum_{i=1}^{m} \mathsf{m}(u(\hat{s}, P_i) \cdot 2^i \bmod p) \times \binom{m}{i}. \tag{10}$$

Eventually, the absolute value of the correlation $\rho_{\hat{s}}$ between the samples $(h_i(\hat{s}))_i$ and $(\mathcal{L}(s, P_i))_i$ is computed for $\hat{s} \in \{0,1\}$ and the attack returns the hypothesis with the greatest correlation.

4   Simulations

Setting. In this section, we assume that the ECSM is implemented on a 32-bit architecture and according to the Montgomery ladder (Algorithm 5 in Appendix A). In such environment which corresponds to a classical context, data will only be manipulated through 32-bit registers. Without loss of generality, we hence assume that the leakage exploited in our attacks is the Hamming weight of the 32 least significant bits of $\tilde{u}$ (namely $\mathrm{HW}(\tilde{u} \bmod 2^{32})$) instead of the Hamming weight of the whole 256 or 384 bit value[4]. The described attack aims at

---

[4] The attack applies similarly whether the attacker chose any 32-bit part of the targeted value.

recovering the most significant bit s of the secret k and, according to (8), the observations are hence assumed to be related to the manipulation of the coordinate $\widetilde{u}$ of the point $R(\widetilde{u}, \widetilde{v}) = 2(P + sP)$. The latter observations are simulated in the classical Hamming weight model with Gaussian Noise. Namely, the leakage observation $\mathcal{L}(s, P_i)$ corresponding to the ith scalar multiplication is simulated such that:

$$\mathcal{L}(s, P_i) = \text{HW}(\widetilde{u} \bmod 2^{32}) + \mathcal{N}, \tag{11}$$

where $\widetilde{u} = u \cdot 2^{\lambda_i} \bmod p$ is the first coordinate of $2(P_i + sP_i)$ represented in the **randomized** Montgomery domain $\text{RMD}_{\lambda_i}(p)$ (with λ_i generated at random in its definition set) and where $\mathcal{N}$ is a Gaussian random variable with mean 0 and standard deviation σ.

Since the leakage is assumed to satisfy (11), the model function m in Eq. (10) has been simply chosen to be the Hamming weight of the 32 least significant bits of the input. The predictions $h_i(\hat{s})$ associated to the binary hypothesis $\hat{s}$ on s hence satisfy:

$$h_i(\hat{s}) = \frac{1}{2^m} \sum_{i=1}^{m} \text{HW}((u \cdot 2^i \bmod p) \bmod 2^{32}) \times \binom{m}{i}, \tag{12}$$

where u is the first coordinate of $2(P_i + \hat{s}P_i)$ in the standard integer domain and where m equals 256 or 384.

To sum-up, the simulations reported in this section are split in two parts. The first part aims to show that the CPA (*i.e.* the correlation coefficient) succeeds in distinguishing the correct hypothesis from the wrong one. The second part serves to estimate the success rate in recovering the first secret bit in a Montgomery ladder implementation of ECSM. The latter success rate is estimated for different noise levels σ and different number of observations.

Pearson Correlation Coefficient. First, we are comparing the correlation values for the correct and the wrong predictions. Following the attack description in Sect. 3.2, we compute the set of hypotheses $(h_i(\hat{s} = 1))_{i \leqslant N}$ according to (12), and the two sets of leakages $(\mathcal{L}(1, P_i))_{i \leqslant N}$ and $(\mathcal{L}(0, P_i))_{i \leqslant N}$, according to (11), where we recall that N represents the total number of randomly chosen points P_i. The first set corresponds to leakages that match the chosen hypotheses and the second one to leakages that do not match the hypotheses. These experiments have been repeated for different values of $N \in [500..10000]$ and different noise standard deviation $\sigma \in [0..30]$. Then, we compute the two correlation coefficients $\rho_{\text{correct}} = \rho((h_i(\hat{s} = 1))_i, (\mathcal{L}(1, P_i))_i)$ and $\rho_{\text{wrong}} = \rho((h_i(\hat{s} = 1))_i, (\mathcal{L}(0, P_i))_i)$. The obtained correlation values are then averaged over the execution of 100 such attacks and the standard deviation of the correlation values is also computed to measure how much the mean is informative.

Results of this first set of experiments are presented in Fig. 4. It represents the average and the standard deviation cone for the correlation coefficients obtained with the correct and wrong predictions. It may be seen that the correlation

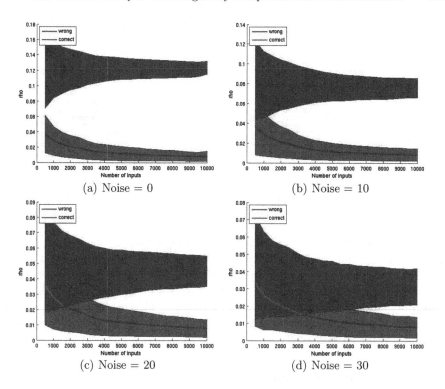

Fig. 4. Correlation for 256-bit curve attacking with known messages

coefficient allows us to distinguish a correct key bit hypothesis from a wrong one with high probability as long as the number of observations satisfies:

$$\begin{cases} N \geq 500 & \text{if } \sigma = 0 \\ N \geq 1500 & \text{if } \sigma = 10 \\ N \geq 3000 & \text{if } \sigma = 20 \\ N \geq 6000 & \text{if } \sigma = 30 \end{cases}.$$

Success Rate. We then proceed to test the attack efficiency. For such a purpose, we randomly choose the secret s we are trying to recover. We then simulate the leakage observations according to (11) for N pairs (P_i, λ_i) of values generated at random in their respective definition set. This step provides us with a set $(\mathcal{L}(s, P_i))_{i \leq N}$ playing the role of the registered traces in a real attack scenario. We then compute the two sets of predictions $(h_i(\hat{s} = 0))_{i \leq N}$ and $(h_i(\hat{s} = 1))_{i \leq N}$, and we process the absolute value of the two corresponding correlation coefficients $\rho_0 = \rho((\mathcal{L}(s, P_i)_i, (h_i(\hat{s} = 0))_i)$ and $\rho_1 = \rho((\mathcal{L}(s, P_i)_i, (h_i(\hat{s} = 1))_i)$. The prediction $\hat{s}$ such that $|\rho_{\hat{s}}|$ has the highest value is set as the most likely one. If $\hat{s} = s$ the attack succeeds, else it fails. In our experiments we repeated each attack 1000 times with different random values to build a success rate.

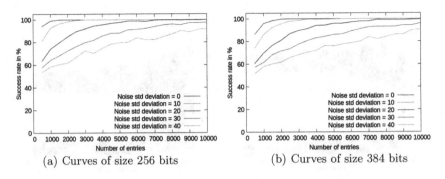

(a) Curves of size 256 bits   (b) Curves of size 384 bits

Fig. 5. Success rate for several noise values

Figure 5 represents the success rate in percentage when trying to guess the bit s. In abscissa is the number of different inputs used for the computation of the correlation coefficient. Several noise standard deviation values have been used. It may be checked that even for $\sigma = 40$ (*i.e.* SNR$= 0,005$) the attack succeeds with probability greater than 80 % if the number N of observations used by the attacker is greater than 5500 for curves of size 256 bits, respectively greater than 7000 for 384-bit curves.

5   Analysis and Conclusion

In this paper, we have argued that the countermeasure proposed in [26] is flawed and does not defeat first-order CPA. Actually, through attack simulations conducted under reasonable and classical assumptions on the execution environment we have shown that a CPA is likely to be very efficient even when the noise is huge. The problem identified in the countermeasure under study is that the distribution of the masking values λ is binomial (and not uniform). This choice, which has been done for efficiency reason, is in fact dramatic from a security point of view. It must be mentioned that the authors of [26] tested a school-book CPA against their implementation and used the failure of this attack to argue on the resistance of their countermeasure. As we have shown here, the failure of the CPA performed in [26] is not a consequence of the countermeasure quality but of a wrong attack parametrization. By adapting to a first-order context the argumentation given in [38] for second-order attacks, we were indeed able to use a much better parametrization for the CPA. Our work demonstrates once again that countermeasures must not be only validated by performing some *ad-hoc* attacks (even classical) but must be formally analysed, for instance by following the approaches proposed in [7,9,37].

Concerning the proposed countermeasure, a possible patching could consist in generating λ uniformly in $[0..m]$ but a careful security analysis is needed to assess on the pertinence of this patch. The proposal of Dupaquis and Venelli in [8] may offer an interesting alternative in the case where efficiency is required. Otherwise, well-studied approaches such as classical exponent and message blinding seem to offer better security garanties.

A   Examples of Algorithms for Elliptic Curve Scalar Multiplication

We recall hereafter two basic algorithms (see Algorithms 1 and 2) to calculate the scalar multiplication either from left-to-right or from right-to-left[5].

Algorithm 1. Left-to-Right Binary ECSM

> **Input** : a point P on E and a secret scalar $k = (1, k_{n-2}, \cdots, k_0)_2$
> **Output**: the point $Q = kP$

1   $R_0 \leftarrow P$
2   $R_1 \leftarrow P$
3   **for** $i = n - 2$ **to** 0 **do**
4   $\quad$ $R_0 \leftarrow 2R_0$
5   $\quad$ **if** $k_i = 1$ **then**
6   $\quad\quad$ $R_0 \leftarrow R_0 + R_1$

7   **return** R_0

Algorithm 2. Right-to-Left Binary ECSM

> **Input** : a point P on $E(\mathbb{F}_p)$ and a secret scalar $k = (k_{n-1}, \cdots, k_0)_2$
> **Output**: the point $Q = kP$

1   $R_0 \leftarrow \mathcal{O}$
2   $R_1 \leftarrow P$
3   **for** $i = 0$ **to** $n - 1$ **do**
4   $\quad$ **if** $k_i = 1$ **then**
5   $\quad\quad$ $R_0 \leftarrow R_0 + R_1$
6   $\quad$ $R_1 \leftarrow 2R_1$

7   **return** R_1

The previous algorithms are simple and relatively efficient (n additions and $n/2$ doublings in average). However, they are not regular and can hence induce an information leakage exploitable by SPA. As for instance explained in [39], regularity can be achieved by using unified formulae for point addition and point doubling [11] or by the mean of side-channel *atomicity* whose principle is to build point addition and point doubling algorithms from the same atomic pattern of field operations [2]. Another possibility is to render the scalar multiplication algorithm itself regular, independently of the field operation flows in each point operation. Namely, one designs a scalar multiplication with a constant flow of point operations. This approach was first followed by Coron in [5] who proposed to perform a dummy addition in the binary algorithm loop whenever the scalar

---

[5] To simplify the attacks description, and because it has no impact on their feasibility, it is assumed for the left-to-right versions that the most significant bit of k always equals 1, *i.e.* that the bit-length is exactly n.

bit equals 0 or not. The obtained *double-and-add-always* algorithm (see Algorithms 3 and 4) performs a point doubling and a point addition at every loop iteration and the scalar bits are no more distinguishable by SPA.

Algorithm 3. Left-to-Right double-and-add always ECSM [5]

Input : a point P on $E(\mathbb{F}_p)$ and a secret scalar $k = (1, k_{n-2}, \cdots, k_0)_2$
Output: the point $Q = kP$

1 $R_0 \leftarrow P$
2 **for** $i = n - 2$ **to** 0 **do**
3 $\quad$ $R_0 \leftarrow 2R_0$
4 $\quad$ $R_1 \leftarrow R_0 + P$
5 $\quad$ $R_0 \leftarrow R_{k_i}$
6 **return** R_0

Algorithm 4. Right-to-Left double-and-add always ECSM [19]

Input : a point P on $E(\mathbb{F}_p)$ and a secret scalar $k = (k_{n-1}, \cdots, k_0)_2$
Output: the point $Q = kP$

1 $R_0 \leftarrow P$
2 $R_1 \leftarrow \mathcal{O}$
3 **for** $i = 0$ **to** $n - 1$ **do**
4 $\quad$ $R_2 \leftarrow R_0 + R_1$
5 $\quad$ $R_0 \leftarrow 2R_0$
6 $\quad$ $R_1 \leftarrow R_{1+k_i}$
7 **return** R_1

Other regular binary algorithms exist such as the *Montgomery ladder* [32] (see Algorithm 5) and the double-and-add algorithm proposed by Joye in [20] (see Algorithm 6). These algorithms which are recalled hereafter not only counteract SPA but also some fault attacks (such that the *safe-error* ones).

Algorithm 5. Montgomery Ladder ECSM [32]

Input : a point P on $E(\mathbb{F}_p)$ and a secret scalar $k = (1, k_{n-2}, \cdots, k_0)_2$
Output: the point $Q = kP$

1 $R_0 \leftarrow P$
2 $R_1 \leftarrow 2P$
3 **for** $i = n - 2$ **to** 0 **do**
4 $\quad$ $b \leftarrow k_i$
5 $\quad$ $R_{1-b} \leftarrow R_0 + R_1$
6 $\quad$ $R_b \leftarrow 2R_b$
7 **return** R_0

Algorithm 6. Joye double-and-add always ECSM [20]

Input : a point P on $E(\mathbb{F}_p)$ and a secret scalar $k = (k_{n-1}, \cdots, k_0)_2$
Output: the point $Q = kP$

1   $R_0 \leftarrow \mathcal{O}$
2   $R_1 \leftarrow P$
3   **for** $i = 0$ **to** $n - 1$ **do**
4     $b \leftarrow k_i$
5     $R_{1-b} \leftarrow 2R_{1-b} + R_b$
6   **return** R_0

References

1. Baldwin, B., Goundar, R.R., Hamilton, M., Marnane, W.P.: Co-z ecc scalar multiplications for hardware, software and hardware-software co-design on embedded systems. J. Crypt. Eng. **2**(4), 221–240 (2012)
2. Chevallier-Mames, B., Ciet, M., Joye, M.: Low-cost solutions for preventing simple side-channel analysis: side-channel atomicity. IEEE Trans. Comput. **53**(6), 760–768 (2004)
3. Cohen, H., Frey, G. (eds.): Handbook of Elliptic and Hyperelliptic Curve Cryptography. CRC Press, Boca Raton (2005)
4. Cohen, H., Miyaji, A., Ono, T.: Efficient elliptic curve exponentiation using mixed coordinates. In: Ohta, K., Pei, D. (eds.) ASIACRYPT 1998. LNCS, vol. 1514, pp. 51–65. Springer, Heidelberg (1998)
5. Coron, J.-S.: Resistance against differential power analysis for elliptic curve cryptosystems. In: Koç, Ç.K., Paar, C. [23], pp. 292–302
6. Coron, J.-S., Prouff, E., Roche, T.: On the use of shamir's secret sharing against side-channel analysis. In: Mangard, S. (ed.) CARDIS 2012. LNCS, vol. 7771, pp. 77–90. Springer, Heidelberg (2013)
7. Duc, A., Dziembowski, S., Faust, S.: Unifying leakage models: from probing attacks to noisy leakage. In: Nguyen, P.Q., Oswald, E. [34], pp. 423–440
8. Dupaquis, V., Venelli, A.: Redundant modular reduction algorithms. In: Prouff, E. (ed.) CARDIS 2011. LNCS, vol. 7079, pp. 102–114. Springer, Heidelberg (2011)
9. Durvaux, F., Standaert, F.-X., Veyrat-Charvillon, N.: How to certify the leakage of a chip? In: Nguyen, P.Q., Oswald, E. [34], pp. 459–476
10. Brier, E., Clavier, C., Olivier, F.: Correlation power analysis with a leakage model. In: Joye, M., Quisquater, J.-J. (eds.) CHES 2004. LNCS, vol. 3156, pp. 16–29. Springer, Heidelberg (2004)
11. Brier, É., Joye, M.: Weierstraß elliptic curves and side-channel attacks. In: Naccache, D., Paillier, P. [33], pp. 335–345
12. Giraud, C., Thiebeauld, H.: A survey on fault attacks. In: Quisquater, J.-J., Paradinas, P., Deswarte, Y., El Kalam, A.A. (eds.) CARDIS 2004. LNCS, vol. 153, pp. 159–176. Springer, Heidelberg (2004)
13. Giraud, C., Thiebeauld, H.: Basics of fault attacks. In: Breveglieri, L., Koren, I. (eds.) Workshop on Fault Diagnosis and Tolerance in Cryptography - FDTC'04, pp. 343–347. IEEE Computer Society (2004)
14. Goundar, R.R., Joye, M., Miyaji, A.: Co-Z addition formulæ and binary ladders on elliptic curves. In: Mangard, S., Standaert, F.-X. (eds.) CHES 2010. LNCS, vol. 6225, pp. 65–79. Springer, Heidelberg (2010)

15. Goundar, R.R., Joye, M., Miyaji, A., Rivain, M., Venelli, A.: Scalar multiplication on weierstraß elliptic curves from co-z arithmetic. J. Crypt. Eng. **1**(2), 161–176 (2011)

16. Hankerson, D., Menezes, A.J., Vanstone, S.: Guide to Elliptic Curve Cryptography. Springer Professional Computing Series. Springer, New York (2003)

17. IEEE Std 1363–2000. IEEE Standard Specifications for Public Key Cryptography. IEEE Computer Society, January 2000

18. Itoh, K., Takenaka, M., Torii, N., Temma, S., Kurihara, Y.: Fast implementation of public-key cryptography ona DSP TMS320C6201. In: Koç, Ç.K., Paar, C. [23], pp. 61–72

19. Izu, T., Takagi, T.: A fast parallel elliptic curve multiplication resistant against side channel attacks. In: Naccache, D., Paillier, P. [33], pp. 280–296

20. Joye, M.: Highly regular right-to-left algorithms for scalar multiplication. In: Paillier, P., Verbauwhede, I. (eds.) CHES 2007. LNCS, vol. 4727, pp. 135–147. Springer, Heidelberg (2007)

21. Knuth, D.E.: The Art of Computer Programming, vol. 2, 3rd edn. Addison Wesley, Reading (1988)

22. Koblitz, N.: Elliptic curve cryptosystems. Math. Comput. **48**(177), 203–209 (1987)

23. Koç, Ç.K., Paar, C. (eds.): CHES 1999. LNCS, vol. 1717. Springer, Heidelberg (1999)

24. Kocher, P.C.: Timing attacks on implementations of Diffie-Hellman, RSA, DSS, and other systems. In: Koblitz, N. (ed.) CRYPTO 1996. LNCS, vol. 1109, pp. 104–113. Springer, Heidelberg (1996)

25. Kocher, P.C., Jaffe, J., Jun, B.: Differential power analysis. In: Wiener, M. (ed.) CRYPTO 1999. LNCS, vol. 1666, pp. 388–397. Springer, Heidelberg (1999)

26. Lee, J.-W., Chung, S.-C., Chang, H.-C., Lee, C.-Y.: An efficient countermeasure against correlation power-analysis attacks with randomized montgomery operations for DF-ECC processor. In: Prouff, E., Schaumont, P. (eds.) CHES 2012. LNCS, vol. 7428, pp. 548–564. Springer, Heidelberg (2012)

27. Longa, P., Miri, A.: New composite operations and precomputation scheme for elliptic curve cryptosystems over prime fields. In: Cramer, R. (ed.) PKC 2008. LNCS, vol. 4939, pp. 229–247. Springer, Heidelberg (2008)

28. Meloni, N.: New point addition formulae for ECC applications. In: Carlet, C., Sunar, B. (eds.) WAIFI 2007. LNCS, vol. 4547, pp. 189–201. Springer, Heidelberg (2007)

29. Miller, V.S.: Use of elliptic curves in cryptography. In: Williams, H.C. (ed.) CRYPTO 1985. LNCS, vol. 218, pp. 417–426. Springer, Heidelberg (1986)

30. Montgomery, P.L.: Evaluating recurrences of form $X_{m+n} = f(X_m, X_n, X_{m-n})$ via Lucas chains 1983, Revised (1992). ftp.cwi.nl:/pub/pmontgom/Lucas.ps.gz

31. Montgomery, P.L.: Modular multiplication without trial division. Math. Comput. **44**(170), 519–521 (1985)

32. Montgomery, P.L.: Speeding the pollard and elliptic curve methods of factorization. Math. Comput. **48**, 243–264 (1987)

33. Naccache, D., Paillier, P. (eds.): PKC 2002. LNCS, vol. 2274. Springer, Heidelberg (2002)

34. Nguyen, P.Q., Oswald, E. (eds.): EUROCRYPT 2014. LNCS, vol. 8441. Springer, Heidelberg (2014)

35. Popp, T., Mangard, S.: Masked dual-rail pre-charge logic: DPA-resistance without routing constraints. In: Rao, J.R., Sunar, B. (eds.) CHES 2005. LNCS, vol. 3659, pp. 172–186. Springer, Heidelberg (2005)

36. Prouff, E., McEvoy, R.: First-order side-channel attacks on the permutation tables countermeasure. In: Clavier, C., Gaj, K. (eds.) CHES 2009. LNCS, vol. 5747, pp. 81–96. Springer, Heidelberg (2009)

37. Prouff, E., Rivain, M.: Masking against side-channel attacks: a formal security proof. In: Johansson, T., Nguyen, P.Q. (eds.) EUROCRYPT 2013. LNCS, vol. 7881, pp. 142–159. Springer, Heidelberg (2013)

38. Prouff, E., Rivain, M., Bévan, R.: Statistical analysis of second order differential power analysis. IEEE Trans. Comput. **58**(6), 799–811 (2009)

39. Rivain, M.: Fast and regular algorithms for scalar multiplication over elliptic curves. IACR Cryptology ePrint Arch. **2011**, 338 (2011)

40. Schramm, K., Wollinger, T., Paar, C.: A new class of collision attacks and its application to DES. In: Johansson, T. (ed.) FSE 2003. LNCS, vol. 2887, pp. 206–222. Springer, Heidelberg (2003)

41. Venelli, A., Dassance, F.: Faster side-channel resistant elliptic curve scalar multiplication. In: Kohel, D., Rolland, R. (eds.) Arithmetic, Geometry, Cryptography and Coding Theory 2009, Contemporary Mathematics, vol. 521, pp. 29–40. American Mathematical Society (2010)

Practical Cryptanalysis of PAES

Jérémy Jean[1], Ivica Nikolić[1]($\boxtimes$), Yu Sasaki[2], and Lei Wang[1]

[1] Nanyang Technological University, Singapore, Singapore
{JJean,INikolic,Wang.Lei}@ntu.edu.sg
[2] NTT Secure Platform Laboratories, Tokyo, Japan
sasaki.yu@lab.ntt.co.jp

Abstract. We present two practical attacks on the CAESAR candidate PAES. The first attack is a universal forgery for any plaintext with at least 240 bytes. It works for the nonce-repeating variant of PAES and in a nutshell it is a state recovery based on solving differential equations for the S-box leaked throught the ciphertext that arise when the plaintext has a certain difference. We show that to produce the forgery based on this method the attacker needs only 2^{11} time and data. The second attack is a distinguisher for 2^{64} out of 2^{128} keys that requires negligible complexity and only one pair of known plaintext-ciphertext. The attack is based on the lack of constants in the initialization of the PAES which allows to exploit the symmetric properties of the keyless AES round. Both of our attacks contradict the security goals of PAES.

Keywords: PAES · Universal forgery · Distinguisher · Symmetric property · Authenticated encryption

1 Introduction

The CAESAR competition [2] (Competition for Authenticated Encryption: Security, Applicability, and Robustness) has started in March 2014, and its goal is to improve the understanding of the crypto community in the area of authenticated ciphers through a public competition for submitting authenticated encryption schemes that offer advantages over the widely used AES-GCM [8]. In total, 58 ciphers were submitted to the open call, and in the following three years, through security analysis and investigation of the implementations advantages, it is expected that among these ciphers, a few to be selected in a portfolio of recommended authenticated schemes that are suitable for widespread adoption.

A number of the proposed CAESAR candidates (as well as the benchmark AES-GCM) are based on the current encryption standard: the AES family of block ciphers. The reason for this is twofold. First, the AES has undergone an extensive analysis and is assumed that its security is well understood (or at least better understood compared to all of the remaining unbroken ciphers). Second,

J. Jean, I. Nikolić and L. Wang are supported by the Singapore National Research Foundation Fellowship 2012 NRF-NRFF2012-06.

© Springer International Publishing Switzerland 2014
A. Joux and A. Youssef (Eds.): SAC 2014, LNCS 8781, pp. 228–242, 2014.
DOI: 10.1007/978-3-319-13051-4_14

AES offers a large software implementation advantage on the latest processor through the so-called AES-NI instruction set, i.e., modern processors have dedicated instructions that allow to reduce the execution time of the AES cipher calls.

In general, the CAESAR candidates based on the AES use the block cipher in two ways: either as a whole (or a variant consisting of at least a certain number of rounds), or only its round function. The first type of candidates (OCB [6], AES-COPA [1], etc., and AES-GCM) are constructions that require calls to the full 10-round AES-128 (or at least 4-round variants with independent round keys, e.g., SHELL [11]). Usually, they are provable modes based on security reduction to the security of AES, and thus benefit from the current state-of-the-art cryptanalysis of AES-128 [4,5]. The second type uses only the AES round function and has no strict security proof, i.e., the mode is not provably secure, however, the resistance against common attacks is provided through ad-hoc techniques. Such candidates (see AEGIS [12], PAES [13], Tiaoxin-346 [9]) benefit from the good security properties and the software performance of the AES round function. They tend to use less than 10 AES round calls per message blocks, and as such are extremely fast.

Our Contributions. We provide a cryptanalysis of the CAESAR candidate PAES [13] and show two attacks that contradict the security claims given by the designers. Common for both of the attacks are the low complexity requirements and misuse of the AES round function in PAES.

The first attack targets the nonce-repeating mode of PAES (called PAES-8) and is a universal forgery attack of any plaintext with at least 240 bytes. It requires 2^{11} time and data complexity to fully recover the internal state and produce forgery. To launch the attack, we use a special differential trail that can take two different paths. By analyzing the ciphertext difference, the path is uniquely determined and allows state recovery based on the differential property of the AES S-Box. Our attack shows that a mere differential analysis (often given by providing the best differential characteristic of a construction) is insufficient for proving security in the nonce-repeating mode, even when the candidates guarantees multiple applications of AES round function.

The second attack comes in a form of a distinguisher for a class of 2^{64} weak keys among the total 2^{128} keys of PAES. We show that if the attacker can control the nonce, then a single pair of known plaintext and corresponding ciphertext is sufficient to distinguish PAES from an ideal authenticated encryption scheme. The attack relies on the initialization phase of PAES that does not use constants, while the AES round function preserves certain symmetric properties when constants are absent. The results of this paper are summarized in Table 1.

Organization of the Paper. We recall the design details of the PAES submissions in Sect. 2 and present the universal forgery attack on PAES-8 in Sect. 3. Then, in Sect. 4 we introduce the distinguisher for PAES in the context of weak keys, and we conclude the paper in Sect. 5.

Table 1. Attacks on PAES.

Design	Supported nonce modes	Attack	Attack mode	Key class size (out of 2^{128})	Time complexity
PAES-4	respecting	distinguisher	respecting	2^{64}	1
PAES-8	respecting+repeating	universal forgery	repeating	2^{128}	2^{11}
PAES-8	respecting+repeating	distinguisher	respecting+repeating	2^{64}	1

2 Description of PAES

The family of authenticated encryption (AE) algorithms PAES has been submitted to the ongoing CAESAR competition and consists of two concrete proposals: PAES-4 and PAES-8. As the name suggests, they both use the AES design strategy [3], and take as input a variable-length plaintext, a 128-bit key, a 128-bit nonce and produce a variable-length ciphertext and a 128-bit authentication tag. The difference between PAES-4 and PAES-8 lies in the size of the internal state, which amounts to four 128-bit blocks for the former, and eight 128-bit blocks for the latter. A functional difference between these two variants is in the mode: PAES-4 has security claims only in the nonce-respecting mode, while PAES-8 in both, the nonce-respecting and nonce-repeating modes.

To simplify the presentation, we describe only PAES-8 in the sequel, and only as authenticated encryption. The design resembles a stream cipher: it has an initialization (where the key and the nonce are loaded into the state), then it processes the input message and produces the ciphertext, and finally in the finalization it produces the tag. The internal state S has eight words $S_1, S_2, \ldots, S_8$, each of 128 bits, i.e., $|S_i| = 128, i = 1, \ldots, 8$. The state update function $StateUpdate(S, M)$ is the round transformation and uses eight keyless[1] AES-round calls (denoted further as AES_0) to update the state as depicted in Fig. 1.

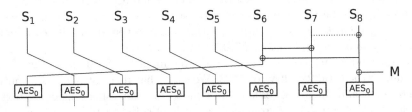

Fig. 1. The round function $StateUpdate(S, M)$. During the processing of the plaintext, the XOR from S_7 to S_8 is absent.

---

[1] We emphasize that all the AES calls are keyless, that is, composed of SubBytes, ShiftRows and MixColumns (but no AddRoundKey).

Initialization. The 128-bit master key K and the nonce N are loaded into the eight words of the state, the state goes through 10 rounds and at the end the key is XORed to all eight words of the state:

$$S_1 = K \oplus N, \qquad\qquad S_5 = L^4(K) \oplus L^7(N)$$
$$S_2 = L(K) \oplus L^3(N), \qquad\qquad S_6 = L^5(K) \oplus L^3(N)$$
$$S_3 = L^2(K) \oplus L(N), \qquad\qquad S_7 = L^6(K) \oplus L^5(N)$$
$$S_4 = L^3(K) \oplus L^2(N), \qquad\qquad S_8 = L^7(K) \oplus L^6(N)$$
$$for \ i = 1 \ to \ 10$$
$$\qquad S = StateUpdate(State, 0)$$
$$for \ i = 1 \ to \ 8$$
$$\qquad S_i = S_i \oplus K$$

where L is the linear transformation that operates on the four 32-bit columns a, b, c, d of a 128-bit word $a||b||c||d$, and is defined as $L(a, b, c, d) = (b, c, d \oplus a, a)$. We denote L^i the i-th functional power of the transformation L, e.g., $L^2 = L \circ L$.

Processing the Plaintext. In one round, from 16-byte plaintext P_i, 16-byte ciphertext C_i is obtained with one call to the *StateUpdate* function (see Fig. 2):

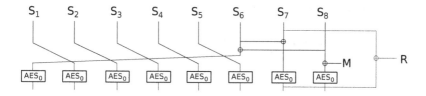

Fig. 2. One round of the encryption.

$$tmp = S_7$$
$$StateUpdate(S, P_i)$$
$$R_i = tmp \oplus S_7$$
$$C_i = P_i \oplus R_i$$

Finalization and the Tag Production. Let $|M|$ be the 128-bit encoding of the message length. Then, the tag T is produced after 10 rounds of the *StateUpdate* function where the message input is set to $|M|$:

$$for \ i = 1 \ to \ 10$$
$$\qquad StateUpdate(S, |M|)$$
$$T = S_7 \oplus S_8$$

Claimed Security of PAES. The claimed security of PAES is given in Table 2. We emphasize in particular that 128-bit security is claimed for the integrity of PAES in the nonce-repeating mode.

Table 2. Bits of security goals of PAES [13, Table 3.1].

Goal	Nonce-respecting	Nonce-repeating	
	PAES-4/PAES-8	PAES-4	PAES-8
Confidentiality for the plaintext	128	-	-
Integrity for the plaintext	128	-	128
Integrity for the associated data	128	-	128
Integrity for the public message number	128	-	128

3   Practical Universal Forgery Attack Against PAES-8

In this section, we show a universal forgery attack for PAES-8 in the nonce-repeating mode. The attack works for any plaintext with length of at least 240 bytes, and requires only a small time and data complexity. The steps of the attack can be summarized as follows:

1. Inject differences in two consecutive plaintext blocks such that they cancel in S_8 with a high probability.
2. The ciphertext difference after eight rounds will reveal if the cancellation in S_8 occurred and if so, it will leak information about the state bits.
3. Once the state is recovered, the tag is produced by going through the remaining of the transformations of the (now) public construction.

3.1   Differential Trail and Detection of Difference Cancellation

The differential trail used in the attack is given in Fig. 3. We inject difference $\Delta\alpha$ in the plaintext P_0, and try to cancel it with another difference $\Delta\beta$ in the plaintext P_1. Interestingly, this type of trail has been discussed by the designers of PAES (see [13, Figure 4.3]), however, they focused on the standard case of propagating the difference through eight rounds and tried to predict it. On the other hand, we use a different approach: our goal is not to predict the difference after eight rounds, but only to detect if the initial differences in $\Delta\alpha$ and $\Delta\beta$ have canceled. In Fig. 3, the trail can take two paths:

1. The differences $\Delta\alpha$ and $\Delta\beta$ cancel, thus only the words with bold lines are active,
2. The differences $\Delta\alpha$ and $\Delta\beta$ do not cancel and there are additional active words depicted with red lines.

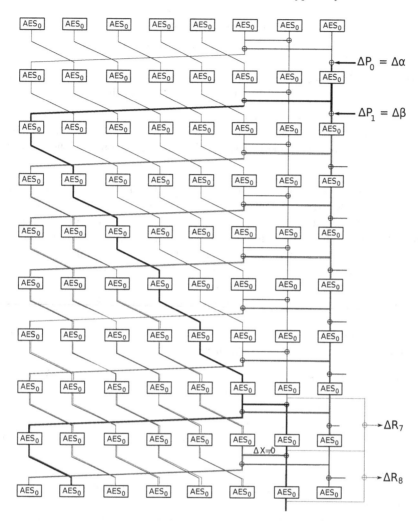

Fig. 3. Differential trail used in the attack. The black bold lines denote active state words. The red lines denote active words when $\Delta\alpha$ and $\Delta\beta$ do not cancel in S_8.

We further show how to choose optimal $\Delta\alpha$ and $\Delta\beta$ and how to detect the cancellation.

Choosing Plaintext Differences $\Delta\alpha$ and $\Delta\beta$. For an arbitrary difference $\Delta\alpha$ in the plaintext P_0, the difference $\Delta\beta$ in the plaintext P_1 should be chosen such that it will cancel $\Delta\alpha$ and thus will avoid activating the state S_8. Therefore, $\Delta\alpha$ and $\Delta\beta$ are chosen so that the cancellation can occur with a high probability – this happens when $\Delta\alpha$ has only one active byte. Let α and β be the input and output difference transition of the S-Box, i.e., α changes to β with a probability 2^{-6}. Then, $\Delta\alpha$ and $\Delta\beta$ are defined as

$$\Delta\alpha = (\alpha, 0, 0, 0,\ 0, 0, 0, 0,\ 0, 0, 0, 0,\ 0, 0, 0, 0),$$
$$\Delta\beta = \texttt{MixColumns} \circ \texttt{ShiftRows}(\beta, 0, 0, 0,\ 0, 0, 0, 0,\ 0, 0, 0, 0,\ 0, 0, 0, 0),$$

and thus $\Delta\alpha$ after $\texttt{AES}_0$ will change to $\Delta\beta$ with probability 2^{-6}. We note that the difference α can be located in any of the 16 bytes of the state.

Detecting the Cancellation Between $\Delta\alpha$ and $\Delta\beta$. We can detect if the cancellation occurred by observing the differences in the ciphertexts C_i (or equivalently, the difference in the key streams R_i) after eight rounds. There are two cases:

- **Cancellation occurred.** From the trail on Fig. 3, it follows that the difference $\Delta R_8 \oplus \Delta R_7$ is obtained when ΔR_7 goes through one $\texttt{AES}_0$ round. It means that the difference in each of the 16 bytes of ΔR_7 can be matched through the S-Box with the corresponding differences in the bytes of $\texttt{ShiftRows}^{-1} \circ \texttt{MixColumns}^{-1}(\Delta R_8 \oplus \Delta R_7)$. We note that the probability of matching is one.
- **Cancellation did not occurred.** If the cancellation did not occur, then there are additional state words with differences (depicted with red lines in Fig. 3). In this case, $\Delta R_8 \oplus \Delta R_7$ is obtained when $\Delta R_7 \oplus \Delta X$ (where ΔX is the non-zero difference in S_6) goes through $\texttt{AES}_0$. In contrast to the above case, now ΔR_7 and $\texttt{ShiftRows}^{-1} \circ \texttt{MixColumns}^{-1}(\Delta R_8 \oplus \Delta R_7)$ can be matched through the S-Box only with some probability lower than one.

Two randomly chosen differences can be matched through the S-Box with a probability $127/256 \approx 2^{-1}$. Without loss of generality, we can assume that ΔX is active in all 16 bytes[2]. Therefore, when $\Delta\alpha$ and $\Delta\beta$ cancel, the probability of a 16-byte match is 1, however, when they do not cancel, then the probability drops to 2^{-16}. As a result, we can easily distinguish the above two cases, by analyzing ΔR_7 and ΔR_8.

The same distinguishing method can be applied to 4 additional rounds (see Fig. 4). This way, we can increase the probability of distinguishing the two cases, and end up with a very low probability of matching differences through S-Boxes in the case when $\Delta\alpha$ and $\Delta\beta$ do not cancel. As we apply it to five rounds, the probability becomes $2^{-5\cdot16} = 2^{-80}$.

3.2   Recovery of State Words

Assume that $\Delta\alpha$ and $\Delta\beta$ have canceled (as demonstrated above, we can single out the case when they cancel). It means that we have the input difference ΔR_7 and the output difference $\Delta R_8 \oplus \Delta R_7$ of an active $\texttt{AES}_0$ for the word S_7, i.e., $\texttt{SubBytes}(\Delta R_7) = \texttt{ShiftRows}^{-1} \circ \texttt{MixColumns}^{-1}(\Delta R_8 \oplus \Delta R_7)$. As in S_7, all 16 bytes are active (with a probability very close to 1), we can easily find the values of the individual bytes by the well-known method of solving 16 differential

---

[2] The difference ΔX is produced after some initial difference goes through multiple $\texttt{AES}$ rounds, thus we can assume ΔX is a random 16-byte difference. As a result, the probability that in ΔX all 16 bytes are active is $(1 - 1/256)^{16} \approx 1$.

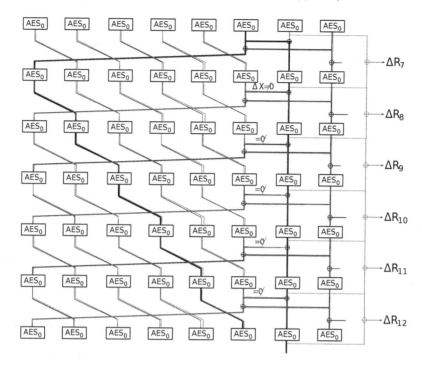

Fig. 4. Extending the previous trail for 4 additional rounds.

equations of the form $S(x \oplus \Delta_{input}) \oplus S(x) = \Delta_{output}$ that come from the system using S-Box S. Each such equation on average has two solutions, because if x is a solution, then $x \oplus \Delta_{input}$ is also a solution. To find a single solution for each byte, we repeat once the recovery for different $\Delta\alpha$ and $\Delta\beta$. As a result, we can recover the value of S_7 at round 8 of the encryption.

Using the very same method, we can recover S_7 at rounds 9, 10, 11 and 12. For instance, for round 9, the input (resp. output) difference of AES_0 is $\Delta R_7 \oplus \Delta R_8$ (resp. $\Delta R_7 \oplus \Delta R_8 \oplus \Delta R_9$). With the knowledge of the values of 5 consecutive S_7, we can uniquely recover the values of S_6, S_5, S_4, S_3 at round 8 by simple computation using those words. Let S_u^{vR} be the u-th variable of the state for round v. For instance, S_6^{8R} is computed by $S_7^{8R} \oplus AES_0^{-1}(S_7^{9R})$.

We can recover two more S_7 words (of additional 2 rounds) if we shift the round where we apply the difference $\Delta\alpha$ and instead to P_0 we introduce $\Delta\alpha$ at P_2 and $\Delta\beta$ at P_3. Hence, we will have the values of S_7 for 7 consecutive rounds.

The state word S_8 is different compared to the remaining seven words and it is not possible to recover it by using the above method. Nevertheless, we can still recover S_8 at round 0 of the encryption based on the differences $\Delta\alpha$ and $\Delta\beta$, i.e., we can recover the active byte where the difference $\Delta\alpha$ is non-zero. By repeating the recovery with 16 different positions of active bytes, we can deduce the whole state word S_8 at round 0. As S_8 does not take feedback from any other word (but the plaintext), we can easily find the value of S_8 at any round, including our target round 8. That is, with the knowledge of S_7 of seven consecutive rounds (8, 9,...14) which can be deduced as shown above, and S_8 at round 8, we can recover the full state at round 8.

3.3   The Attack

We now present the universal forgery attack. The goal of the attack is to produce a tag of an arbitrary plaintext. In our case, the attack works as long as the length of the plaintext is at least 16 blocks (240 bytes). Our forgery is based on a state recovery, i.e., if at some round the whole state is known, then the tag can easily be produced by performing the remaining operations of the finalization, and therefore it can be produced offline.

Let $P_0, P_1, \ldots, P_{14}$ be the first 15 blocks of the plaintext. Then, the forgery can be described with the following algorithm:

1. Query the first 15 plaintext blocks of the target $(P_0 \| P_1 \| \cdots \| P_{14})$, and obtain the key stream $R_0, R_1, \cdots, R_{14}$.
2. FOR *position* $= 1$ to 16 DO
3.   FOR $i = 1$ to 2^7 DO
4.   Choose 1-byte difference $\Delta\alpha^i$ with active byte at *position* and find the corresponding $\Delta\beta^i$.
5.   Query $(P_0 \oplus \Delta\alpha^i \| P_1 \oplus \Delta\beta^i \| P_2 \| \cdots \| P_{14})$ and obtain the key stream $R_0^i, \cdots, R_{14}^i$.
6.   Check if the difference $R_7 \oplus R_7^i$ can result in $R_7 \oplus R_7^i \oplus R_8 \oplus R_8^i$ by AES_0.
7.   Check the same property for additional 4 rounds.
8.   Save the pairs that pass all the above checks.
9.   END FOR
10.   Recover the byte at *position* of the state word S_8 at round 0
11. END FOR
12. Recover S_7 at rounds 8,9,10,11,12
13. FOR $i = 1$ to 2^7 DO
14.   Choose 1-byte difference $\Delta\alpha^i$ and find the corresponding $\Delta\beta^i$.
15.   Query $(P_0 \| P_1 \| P_2 \oplus \Delta\alpha^i \| P_3 \oplus \Delta\beta^i \| P_4 \| \cdots \| P_{14})$ and obtain the key stream $R_0^i, \cdots, R_{14}^i$.
16.   Check if the difference $R_9 \oplus R_9^i$ can result in $R_9 \oplus R_9^i \oplus R_{10} \oplus R_{10}^i$ by AES_0.
17.   Check the same property for next 4 additional rounds.
18.   Save the pairs that pass all the above checks.
19. END FOR
20. Recover S_7 at rounds 13 and 14.
21. Deduce all the state words at round 8.
22. Go through the remaining of the transformations and produce the tag.

The first loop is used to recover S_8, and to recover five S_7, and the second to recover the remaining two S_7. Note, each of the loops (the inner loop of the first loop) will produce two pairs, as the probability of the trail in the top ($\Delta\alpha$ will be canceled by $\Delta\beta$) is 2^{-6}. In case no good trails with probability 2^{-6} exist, the attacker can switch to ones with probability 2^{-7} and run the loops 2^8 times. Furthermore, as we have seen from the previous analysis, a probability of false positives is very low (around 2^{-80}).

From the algorithm, it follows that the time complexity of the attack is $16 \cdot 2^7 + 2^7 \approx 2^{11}$ computations. The data complexity is similar and comes in a form of chosen plaintexts. To solve efficiently the differential equations, the attack needs about 2^{16} bytes in memory.

4   Practical Distinguisher for a Weak-Key Class of PAES

We continue our analysis by presenting a distinguisher for a class of 2^{64} weak keys (out of 2^{128} keys) in PAES-8. The distinguisher requires negligible time complexity and only a single pair of known plaintext-ciphertext and a chosen nonce. It exploits the lack of constants in the design and the symmetric properties of the keyless AES round function. Although we give the distinguisher for PAES-8, we note that a similar attack is applicable to the nonce-respecting mode PAES-4.

4.1   Symmetric Properties of the AES Round Function

We first recall the known symmetric property of the AES round function [7]. Namely, if a state is symmetric in the sense that its two halves are equal, then the keyless round function AES_0 of the AES maintains this property. We recall the property of [7] using block matrices, and we introduce the following more general notations:

$$U(A,B) = \left(\begin{array}{c|c} A & A \\ \hline B & B \end{array}\right), \qquad V(A,B) = \left(\begin{array}{c|c} A & B \\ \hline B & A \end{array}\right), \qquad W(A,B) = \left(\begin{array}{c|c} A & B \\ \hline A & B \end{array}\right).$$

Additionally, we denote by $\mathcal{U}$, $\mathcal{V}$ and $\mathcal{W}$ the associated sets respectively for all possible values of the 2×2 block matrices A and B. Finally, we denote M the constant MDS matrix used in the AES round function, and observe that:

$$M = \begin{pmatrix} 2 & 3 & 1 & 1 \\ 1 & 2 & 3 & 1 \\ 1 & 1 & 2 & 3 \\ 3 & 1 & 1 & 2 \end{pmatrix} = \left(\begin{array}{c|c} M_1 & M_2 \\ \hline M_2 & M_1 \end{array}\right) = V(M_1, M_2) \in \mathcal{V}.$$

Property 1. Let $S \in \mathcal{U}$. Then, $\text{AES}_0(S) \in \mathcal{U}$.

Proof. Let $S = U(A,B) \in \mathcal{U}$, and write the bytes in S as:

$$\left(\begin{array}{c|c} A & A \\ \hline B & B \end{array}\right) = \begin{pmatrix} x_0 & x_4 & x_0 & x_4 \\ x_1 & x_5 & x_1 & x_5 \\ x_2 & x_6 & x_2 & x_6 \\ x_3 & x_7 & x_3 & x_7 \end{pmatrix}.$$

As the SubBytes operation applies the same bijection to all the bytes in the state, we ignore it here as it obviously preserves the structure. After the ShiftRows operation, the state becomes

$$
\begin{pmatrix}
x_0 \ x_4 \,|\, x_0 \ x_4 \\
x_5 \ x_1 \,|\, x_5 \ x_1 \\
\hline
x_2 \ x_6 \,|\, x_2 \ x_6 \\
x_7 \ x_3 \,|\, x_7 \ x_3
\end{pmatrix}
\overset{\text{def}}{=}
\left(\begin{array}{c|c} A' & A' \\ \hline B' & B' \end{array} \right),
$$

thus it still belongs to $\mathcal{U}$. Then, the MixColumns operation results in:

$$
\begin{pmatrix}
2\,3\,1\,1 \\
1\,2\,3\,1 \\
1\,1\,2\,3 \\
3\,1\,1\,2
\end{pmatrix}
\times
\begin{pmatrix}
x_0 \ x_4 \ x_0 \ x_4 \\
x_5 \ x_1 \ x_5 \ x_1 \\
x_2 \ x_6 \ x_2 \ x_6 \\
x_7 \ x_3 \ x_7 \ x_3
\end{pmatrix}
=
\left(\begin{array}{c|c} M_1 & M_2 \\ \hline M_2 & M_1 \end{array} \right)
\times
\left(\begin{array}{c|c} A' & A' \\ \hline B' & B' \end{array} \right)
$$

$$
=
\left(\begin{array}{c|c} M_1 A' \oplus M_2 B' & M_1 A' \oplus M_2 B' \\ \hline M_2 A' \oplus M_1 B' & M_2 A' \oplus M_1 B' \end{array} \right)
\overset{\text{def}}{=}
\left(\begin{array}{c|c} A'' & A'' \\ \hline B'' & B'' \end{array} \right) \in \mathcal{U}.
$$

$\square$

Property 2. Let $S \in \mathcal{W}$. Then, $\mathtt{AES}_0(S) \in \mathcal{V}$, and $\mathtt{AES}_0(\mathtt{AES}_0(S)) \in \mathcal{W}$.

Proof. Let $S = W(A, B) \in \mathcal{W}$, and write the bytes in S as:

$$
\left(\begin{array}{c|c} A & B \\ \hline A & B \end{array} \right)
=
\begin{pmatrix}
x_0 \ x_2 \,|\, x_4 \ x_6 \\
x_1 \ x_3 \,|\, x_5 \ x_7 \\
\hline
x_0 \ x_2 \,|\, x_4 \ x_6 \\
x_1 \ x_3 \,|\, x_5 \ x_7
\end{pmatrix}.
$$

Again, we ignore the SubBytes operation as the applied bijection preserves the structure of the internal states. However, after the ShiftRows operation the state becomes:

$$
\begin{pmatrix}
x_0 \ x_2 \,|\, x_4 \ x_6 \\
x_3 \ x_5 \,|\, x_7 \ x_1 \\
\hline
x_4 \ x_6 \,|\, x_0 \ x_2 \\
x_7 \ x_1 \,|\, x_3 \ x_5
\end{pmatrix}
\overset{\text{def}}{=}
\left(\begin{array}{c|c} A' & B' \\ \hline B' & A' \end{array} \right) \in \mathcal{V},
$$

which is transformed by the subsequent MixColumns transformation into the state:

$$
\left(\begin{array}{c|c} M_1 & M_2 \\ \hline M_2 & M_1 \end{array} \right)
\times
\left(\begin{array}{c|c} A' & B' \\ \hline B' & A' \end{array} \right)
=
\left(\begin{array}{c|c} M_1 A' \oplus M_2 B' & M_1 B' \oplus M_2 A' \\ \hline M_2 A' \oplus M_1 B' & M_2 B' \oplus M_1 A' \end{array} \right)
$$

$$
\overset{\text{def}}{=}
\left(\begin{array}{c|c} A'' & B'' \\ \hline B'' & A'' \end{array} \right) \in \mathcal{V}.
$$

After applying a second keyless AES round, we get:

$$
\left(\begin{array}{c|c} A'' & B'' \\ \hline B'' & A'' \end{array} \right)
=
\begin{pmatrix}
y_0 \ y_2 \,|\, y_4 \ y_6 \\
y_1 \ y_3 \,|\, y_5 \ y_7 \\
\hline
y_4 \ y_6 \,|\, y_0 \ y_2 \\
y_5 \ y_7 \,|\, y_1 \ y_3
\end{pmatrix}
\xrightarrow{\text{SR}}
\begin{pmatrix}
y_0 \ y_2 \,|\, y_4 \ y_6 \\
y_3 \ y_5 \,|\, y_7 \ y_1 \\
\hline
y_0 \ y_2 \,|\, y_4 \ y_6 \\
y_3 \ y_5 \,|\, y_7 \ y_1
\end{pmatrix}
\overset{\text{def}}{=}
\left(\begin{array}{c|c} A''' & B''' \\ \hline A''' & B''' \end{array} \right) \in \mathcal{W},
$$

and by the MixColumns:

$$\left(\frac{M_1\,|\,M_2}{M_2\,|\,M_1}\right) \times \left(\frac{A'''\,|\,B'''}{A'''\,|\,B'''}\right) = \left(\frac{M_1A''' \oplus M_2A'''\,|\,M_1B''' \oplus M_2B'''}{M_2A''' \oplus M_1A'''\,|\,M_2B''' \oplus M_1B'''}\right)$$
$$\stackrel{\text{def}}{=} \left(\frac{A''''\,|\,B''''}{A''''\,|\,B''''}\right) \in \mathcal{W},$$

which concludes the proof. □

Finally, we can represent the action of the keyless AES round function AES_0 on the three sets $\mathcal{U}$, $\mathcal{V}$ and $\mathcal{W}$ as follows on Fig. 5.

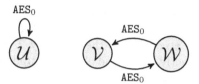

Fig. 5. Action of AES_0 of the symmetrical states from $\mathcal{U}$, $\mathcal{V}$ and $\mathcal{W}$.

4.2 Symmetric Properties of the PAES Transformations

Along with AES_0, PAES uses a few more transformations, in particular, the XOR and the linear transformation L. We investigate here how these two transformations preserve the class belongings.

Property 3. Let $\mathcal{X}$ be either $\mathcal{U}$, $\mathcal{V}$ or $\mathcal{W}$, and let $S_1, S_2 \in \mathcal{X}$. Then, $S_1 \oplus S_2 \in \mathcal{X}$.

Proof. Let $S_1 = U(A_1, B_1)$, $S_2 = U(A_2, B_2) \in \mathcal{U}$. Then:

$$S_1 \oplus S_2 = \left(\frac{A_1\,|\,A_1}{B_1\,|\,B_1}\right) \oplus \left(\frac{A_2\,|\,A_2}{B_2\,|\,B_2}\right) = \left(\frac{A_1 \oplus A_2\,|\,A_1 \oplus A_2}{B_1 \oplus B_2\,|\,B_1 \oplus B_2}\right) \in \mathcal{U}.$$

The cases for $\mathcal{V}$ and $\mathcal{W}$ can be proven similarly. □

Property 4. Let $S \in \mathcal{W}$. Then, $L(S) \in \mathcal{W}$.

Proof. Let $S = W(A, B) \in \mathcal{W}$, and write the bytes in S as:

$$S = \left(\frac{A\,|\,B}{A\,|\,B}\right) = \begin{pmatrix} x_0\ x_2\,|\,x_4\ x_6 \\ x_1\ x_3\,|\,x_5\ x_7 \\ x_0\ x_2\,|\,x_4\ x_6 \\ x_1\ x_3\,|\,x_5\ x_7 \end{pmatrix}.$$

Then:

$$L(S) = L\begin{pmatrix} x_0\ x_2\,|\,x_4\ x_6 \\ x_1\ x_3\,|\,x_5\ x_7 \\ x_0\ x_2\,|\,x_4\ x_6 \\ x_1\ x_3\,|\,x_5\ x_7 \end{pmatrix} = \begin{pmatrix} x_2 & x_4 & x_6 \oplus x_0 & x_0 \\ x_3 & x_5 & x_7 \oplus x_1 & x_1 \\ x_2 & x_4 & x_6 \oplus x_0 & x_0 \\ x_3 & x_5 & x_7 \oplus x_1 & x_1 \end{pmatrix} \in \mathcal{W}.$$

□

4.3   The Distinguisher

To distinguish PAES, we use the first ciphertext C_0 produced during the encryption of an arbitrary plaintext P_0 with a secret key $K \in \mathcal{W}$ and nonce $N \in \mathcal{W}$. The key K can be any of such 2^{64} keys (the first two rows equal to the second two rows), and the same structure holds for the nonce N.

We first inspect how the state words $S_1, S_2, \ldots, S_8$ change the class belongings (either $\mathcal{W}$ or $\mathcal{V}$) from the very first to the last steps of the initialization phase:

- $K, N \in \mathcal{W}$. By Properties 3 and 4 $S_1, S_2, \ldots, S_8 \in \mathcal{W}$ after the initial assignments in the initialization.
- After the first update. By Property 3, the XORs do not change the class belongings, thus each S_6, S_7, S_8 stay in $\mathcal{W}$ after the XORs at the top of the *StateUpdate*. Further, according to the Property 2, AES$_0$ changes the class from $\mathcal{W}$ to $\mathcal{V}$. Consequently, at the end of the first update, $S_i \in \mathcal{V}, i = 1, \ldots, 8$.
- The second update is similar to the previous one, but this time the class of S_i changes to $\mathcal{W}$.
- $\ldots$
- After the tenth update. The classes of all S_i are $\mathcal{W}$.
- After the XORs of the key. As each S_i is in $\mathcal{W}$ and the key is in $\mathcal{W}$, by Property 3, it follows that each S_i will be in $\mathcal{W}$.

We now focus on the production of the ciphertext C_0. Obviously, $tmp = S_7 = W(A_1, B_1) \in \mathcal{W}$ and after the application of the *StateUpdate*, $S_7 = V(A_2, B_2) \in \mathcal{V}$ by Property 2. Thus, from the definition of the ciphertext $C_0 = P_0 \oplus tmp \oplus S_7$, we get:

$$C_0 \oplus P_0 = \left(\frac{A_1 \,|\, B_1}{A_1 \,|\, B_1} \right) \oplus \left(\frac{A_2 \,|\, B_2}{B_2 \,|\, A_2} \right) = \left(\frac{A_1 \oplus A_2 \,|\, B_1 \oplus B_2}{A_1 \oplus B_2 \,|\, B_1 \oplus A_2} \right) = \left(\frac{X \,|\, Z}{Y \,|\, T} \right).$$

Obviously $X \oplus Y \oplus Z \oplus T = 0$, hence the xor of the four 32-bit blocks of the first ciphertext and plaintext must result in a zero block. Therefore, we have a distinguisher which requires negligible complexity and only a single block of plaintext/ciphertexts to distinguish PAES when instantiated with any of the 2^{64} keys and nonces from the class $\mathcal{W}$. We note that our computer simulation confirmed the correctness of the distinguisher.

5   Conclusion

We have shown two practical attacks on the CAESAR candidate PAES: a universal forgery attack and a distinguisher, which contradict the security claims of this authenticated encryption scheme.

Our analysis gives insights into possible misuses of the AES round function. Although this transformation per se provides excellent resistance against differential and linear attacks (once it has been iterated several times), by no means it is sufficient proof of security against all attacks. The designs based on the round function that does not apply any constants, as we have seen on the example of our distinguisher and the chosen-key rotational distinguisher [10] of PAES, are susceptible to attacks that exploit the symmetry of the AES transformations. Consequently, using random constants in such designs should be taken as a requirement to destroy those symmetric behaviors. Furthermore, as our forgery attack shows, evaluating the differential properties in a straightforward manner (providing the best in terms of probability differential characteristic), does not guarantee security against differential attacks in the nonce-repeating mode.

We would also like to emphasize the importance of the technique used in the forgery attack on the nonce-repeating mode. Due to the mode and the attack framework, there is no need to provide a valid tag at the beginning of the attack (forgery or state recovery). Hence the attacker can focus only on finding a differential characteristic that will leak differences in state words sufficient for recovery based on solving differential equations. The characteristic does not necessarily need to hold with a high probability, but for the forgery on PAES this was required in the first two rounds only because there was an alternative path that does not permit state recovery. In general, the probability of the characteristic is irrelevant, however, it is important for the characteristic to leak input and output differences of non-linear operations which subsequently will be used to recover the state bits. We believe that this technique (improved or modified variants) can be a valuable approach for cryptanalysis of other CAESAR submissions and authenticated encryption schemes.

Acknowledgment. We would like to thank the anonymous reviewers for their detailed feedback and comments.

References

1. Andreeva, E., Bogdanov, A., Luykx, A., Mennink, B., Tischhauser, E., Yasuda, K., Compute, D.: AES-COPA v1. Submitted to the CAESAR competition, March 2014
2. Bernstein, D.: CAESAR Competition. http://competitions.cr.yp.to/caesar.html
3. Daemen, J., Rijmen, V.: The Design of Rijndael: - The Advanced Encryption Standard. Springer, New York (2002)
4. Derbez, P., Fouque, P.A., Jean, J.: Improved key recovery attacks on reduced-round AES in the single-key setting. IACR Cryptology ePrint Archive 2012, 477 (2012)
5. Derbez, P., Fouque, P.-A., Jean, J.: Improved key recovery attacks on reduced-round AES in the single-key setting. In: Johansson, T., Nguyen, P.Q. (eds.) EURO-CRYPT 2013. LNCS, vol. 7881, pp. 371–387. Springer, Heidelberg (2013)
6. Krovetz, T., Rogaway, P.: OCB v1. Submitted to the CAESAR competition, March 2014
7. Van Le, T., Sparr, R., Wernsdorf, R., Desmedt, Y.G.: Complementation-like and cyclic properties of AES round functions. In: Dobbertin, H., Rijmen, V., Sowa, A. (eds.) AES 2005. LNCS, vol. 3373, pp. 128–141. Springer, Heidelberg (2005)
8. McGrew, D., Viega, J.: The Galois/Counter mode of operation (GCM). Submission to NIST (2004). http://csrc.nist.gov/CryptoToolkit/modes/proposedmodes/gcm/gcm-spec.pdf

9. Nikolić, I.: Tiaoxin-346 v1. Submitted to the CAESAR competition, March 2014
10. Saarinen, M.J.O.: PAES and rotations, March 2014. https://groups.google.com/forum/#!topic/crypto-competitions/vRmJdRQBzOo
11. Wang, L.: SHELL v1. Submitted to the CAESAR competition, March 2014
12. Wu, H., Preneel, B.: AEGIS v1. Submitted to the CAESAR competition, March 2014
13. Ye, D., Wang, P., Hu, L., Wang, L., Xie, Y., Sun, S., Wang, P.: PAES v1. Submitted to the CAESAR competition, March 2014

Diffusion Matrices from Algebraic-Geometry Codes with Efficient SIMD Implementation

Daniel Augot[1,2], Pierre-Alain Fouque[3,4], and Pierre Karpman[1,2,5](✉)

[1] Inria, Saclay, France
{daniel.augot,pierre.karpman}@inria.fr
[2] LIX — École Polytechnique, Palaiseau, France
[3] Université de Rennes 1, Rennes, France
pierre-alain.fouque@irisa.fr
[4] Institut Universitaire de France, Paris, France
[5] Nanyang Technological University, Singapore, Singapore

Abstract. This paper investigates large linear mappings with very good diffusion and efficient software implementations, that can be used as part of a block cipher design. The mappings are derived from linear codes over a small field (typically $\mathbf{F}_{2^4}$) with a high dimension (typically 16) and a high minimum distance. This results in diffusion matrices with equally high dimension and a large branch number. Because we aim for parameters for which no MDS code is known to exist, we propose to use more flexible *algebraic-geometry* codes.

We present two simple yet efficient algorithms for the software implementation of matrix-vector multiplication in this context, and derive conditions on the generator matrices of the codes to yield efficient encoders. We then specify an appropriate code and use its automorphisms as well as random sampling to find good such matrices.

We provide concrete examples of parameters and implementations, and the corresponding assembly code. We also give performance figures in an example of application which show the interest of our approach.

Keywords: Diffusion matrix · Algebraic-geometry codes · Algebraic curves · SIMD · Vector implementation · SHARK

1 Introduction

The use of *MDS* matrices over finite fields as a linear mapping in block cipher design is an old trend, followed by many prominent algorithms such as the AES/Rijndael family [7]. These matrices are called MDS as they are derived from *maximum distance separable* linear error-correcting codes, which achieve the highest minimum distance possible for a given length and dimension. This notion of minimum distance coincides with the one of *branch number* of a mapping [7], which is a measure of the effectiveness of a diffusion layer. MDS matrices thus have an optimal diffusion, in a cryptographic sense, which makes them attractive for cipher designs.

© Springer International Publishing Switzerland 2014
A. Joux and A. Youssef (Eds.): SAC 2014, LNCS 8781, pp. 243–260, 2014.
DOI: 10.1007/978-3-319-13051-4_15

The good security properties that can be derived from MDS matrices are often counter-balanced by the cost of their computation. The standard matrix-vector product is quadratic in the dimension of the vector, and finite field operations are not always efficient. For that reason, there is often a focus on finding matrices allowing efficient implementations. For instance, the AES matrix is circulant and has small coefficients. More recently, the PHOTON hash function [9] introduced the use of matrices that can be obtained as the power of a companion matrix, which sparsity may be useful in lightweight hardware implementations. The topic of finding such so-called recursive diffusion layers has been quite active in the past years, and led to a series of papers investigating some of their various aspects [2,18,23]. One of the most recent developments shows how to systematically construct some of these matrices from BCH codes [1]. This allows in particular to construct very large recursive MDS matrices, for instance of dimension 16 over $\mathbf{F}_{2^8}$. This defines a linear mapping over a full 128-bit block with excellent diffusion properties, at a moderate hardware implementation cost.

As interesting as it may be in hardware, the cost in software of a large linear mapping tends to make these designs rather less attractive than more balanced solutions. An early attempt to use a large matrix was the block cipher SHARK, a Rijndael predecessor [16]. It is a 64-bit cipher which uses an MDS matrix of dimension 8 over $\mathbf{F}_{2^8}$ for its linear diffusion. The usual technique for implementing such a mapping in software is to rely on a table of precomputed multiples of the matrix rows. However, table-based implementations now tend to be frown upon as they may lead to *timing attacks* [21], and this could leave ciphers with a structure similar to SHARK's without reasonable software implementations when resistance to these attacks is required. Yet, such designs also have advantages of their own; their diffusion acts on the whole state at every round, and therefore makes structural attacks harder, while also ensuring that many S-Boxes are kept active. Additionally, the simplicity of the structure makes it arguably easier to analyze than in the case of most ciphers.

Our Contributions. In this work, we revisit the use of a *SHARK structure* for block cipher design and endeavour to find good matrices and appropriate algorithms to achieve both a linear mapping with very good diffusion and efficient software implementations that are not prone to timing attacks. To be more specific on this latter point, we target software running on 32 or 64-bit CPUs featuring an SIMD vector unit.

An interesting way of trying to meet both of these goals is to decrease the size of the field from $\mathbf{F}_{2^8}$ to $\mathbf{F}_{2^4}$. However, according to the *MDS conjecture*, there is no MDS code over $\mathbf{F}_{2^4}$ of length greater than 17, and no such code is known [15]. Because a diffusion matrix of dimension n is typically obtained from a code of length $2n$, MDS matrices over $\mathbf{F}_{2^4}$ are therefore restricted to dimensions less than 8. Hence, the prospect of finding an MDS matrix over $\mathbf{F}_{2^4}$ diffusing on more than $8 \times 4 = 32$ bits is hopeless. Obviously, 32 bits is not enough for a large mapping *à la* SHARK. We must therefore search for codes with a slightly smaller minimum distance in the hope that they can be made longer.

Our proposed solution to this problem is to use *algebraic-geometry codes* [22], as they precisely offer this tradeoff. One way of defining these codes is as evaluation codes on algebraic curves; thus our proposal brings a nice connection between these objects and symmetric cryptography. Although elliptic and hyperelliptic curves are now commonplace in public-key cryptography, we show a rare application of an hyperelliptic curve to the design of block ciphers. We present a specific code of length 32 and dimension 16 over $\mathbf{F}_{2^4}$ with minimum distance 15, which is only 2 less than what an MDS code would achieve. This lets us deriving a very good diffusion matrix on $16 \times 4 = 64$ bits in a straightforward way. Interestingly, this matrix can also be applied to vectors over an extension of $\mathbf{F}_{2^4}$ such as $\mathbf{F}_{2^8}$, while keeping the same good diffusion properties. This allows for instance to increase the diffusion to $16 \times 8 = 128$ bits.

We also study two simple yet efficient algorithms for implementing the matrix-vector multiplication needed in a SHARK structure, when a *vector permute* instruction is available. From one of these, we derive conditions on the matrix to make the product faster to compute, in the form of a cost function; we then search for matrices with a low cost, both randomly, and by using automorphisms of the code and of the hyperelliptic curve on which it is based. The use of codes automorphisms to derive efficient encoders is not new [6,11], but it is not generally applied to the architecture and dimensions that we consider in our case.

We conclude this paper by presenting examples of performance figures of assembly implementations of our algorithms when used as the linear mapping of a block cipher.

Structure of the Paper. We start with a few background notions in Sect. 2. We then present our algorithms for matrix-vector multiplication and their context in Sect. 3, and derive a cost function for the implementation of matrices. This is followed by the definition of the algebraic-geometry code used in our proposed linear mapping, and a discussion of how to derive efficient encoders in Sect. 4. We conclude with insights into the performance of the mapping when used over both $\mathbf{F}_{2^4}$ and $\mathbf{F}_{2^8}$ in Sect. 5.

2   Preliminaries

We note $\mathbf{F}_{2^m}$ the finite field with 2^m elements. We often consider $\mathbf{F}_{2^4}$, and implicitly use this specific field if not mentioned otherwise. W.l.o.g. we use the representation $\mathbf{F}_{2^4} \cong \mathbf{F}_2[\alpha]/(\alpha^4 + \alpha + 1)$. We freely use "integer representation" for elements of $\mathbf{F}_{2^4}$ by writing $n \in \{0 \ldots 15\} = \sum_{i=0}^{3} a_i 2^i$ to represent the element $x \in \mathbf{F}_{2^4} = \sum_{i=0}^{3} a_i \alpha^i$.

Bold variables denote vectors (in the sense of elements of a vector space), and subscripts are used to denote their i^{th} coordinate, starting from zero. For instance, $\mathbf{x} = (1, 2, 7)$ and $\mathbf{x}_2 = 7$. If M is a matrix of n columns, we call

$\mathbf{m}^i = (M_{i,j}, \ j = 0 \ldots n - 1)$ the row vector formed from the coefficients of its ith row. We use angle brackets "$\langle$" and "$\rangle$" to write ordered sets.

Arrays, or tables, (in the sense of software data structures) are denoted by regular variables such as x or T, and their elements are accessed by using square brackets. For instance, $T[i]$ is the ith element of the table T, starting from zero.

We conclude with two definitions.

Definition 1 (Systematic form and dual of a code). *Let the code $\mathcal{C}$ be an $[n, k, d]_{\mathbf{F}_{2^m}}$ code of length n, dimension k and minimum distance d with symbols in $\mathbf{F}_{2^m}$. A generator matrix for $\mathcal{C}$ is in* systematic form *if it is of the form $(I_k \ A)$, with I_k the identity matrix of dimension k and A a matrix of k rows and $n - k$ columns. A systematic generator matrix for the dual of $\mathcal{C}$ is given by $(I_{n-k} \ A^t)$.*

Definition 2 (Branch number [7]). *Let A be the matrix of a linear mapping over $\mathbf{F}_{2^m}$, and $\mathrm{w}_m(\mathbf{x})$ be the number of non-zero positions of the vector $\mathbf{x}$ over $\mathbf{F}_{2^m}$. Then the* differential branch number *of A is equal to $\min_{\mathbf{x} \neq 0}(\mathrm{w}_m(x) + \mathrm{w}_m(A(x)))$, and the* linear branch number *of A is equal to $\min_{\mathbf{x} \neq 0}(\mathrm{w}_m(x) + \mathrm{w}_m(A^t(x)))$.*

Note that if A is such that $(I_k \ A)$ is a generator matrix of a code of minimum distance d which dual code has minimum distance d', then A has a differential (resp. linear) branch number of d (resp. d').

3 Efficient Algorithms for Matrix-Vector Multiplication

This section presents software algorithms for matrix-vector multiplication over $\mathbf{F}_{2^4}$. We focus on square matrices of dimension 16. This naturally defines linear operations on 64 bits, which can also be extended to 128 bits, as it will be made clear in Sect. 5. Both cases are a common block size for block ciphers.

Targeted Architecture. The algorithms in this section target CPUs featuring vector instructions, including in particular a *vector shuffle* instruction such as Intel's pshufb from the SSSE3 instruction set extension [12]. These instructions are now widespread and have already been used successfully in fast cryptographic implementations, see *e.g.* [4,10,20]. We mostly considered SSSE3 when designing the algorithms, but other processor architectures do feature vector instructions. This is for instance the case of ARM's NEON extensions, which may also yield efficient implementations, see *e.g.* [5]. We do not consider these explicitly in this paper, however.

Because it plays an important role in our algorithms, we briefly recall the semantics of pshufb. The pshufb instruction takes two 128-bit inputs[1]. The first (the destination operand) is an xmm SSE vector register which logically represents a vector of 16 bytes. The second (the source operand) is either a similar

---

[1] The instruction can actually also be used on 64-bit operands, but we do not consider this possibility here.

xmm register, or a 128-bit memory location. The result of calling pshufb x y is to overwrite the input x with the vector x' defined by:

$$x'[i] = \begin{cases} x[\lfloor y[i] \rfloor_4] & \text{if the most significant bit of } y[i] \text{ is not set} \\ 0 & \text{otherwise} \end{cases}$$

where $\lfloor \cdot \rfloor_4$ denotes truncation to the 4 least significant bits. This instruction allows to arbitrarily *shuffle* a vector according to a mask, with possible repetition and omission of some of the vector values[2]. Notice that this instruction can also be used to perform 16 parallel 4-to-8-bit table lookups: let us call T this table; take as first operand to pshufb the vector $x = (T[i], i = 0 \ldots 15)$, as second operand the vector $y = (a, b, c, d, \ldots)$ on which to perform the lookup; then we see that the first byte of the result is $x[y[0]] = T[a]$, the second is $x[y[1]] = T[b]$, etc.

Finally, there is a three-operand variant of this instruction in the more recent AVX instruction set and onward [12], which allows not to overwrite the first operand.

Targeted Properties. In this paper we focus solely on algorithms that can easily be implemented in a way that makes them immune to timing-attacks [21]. Specifically, we consider the matrix as a known constant but the vector as a secret, and we wish to perform the multiplication without secret-dependent branches or memory accesses. It might not always be important to be immune (or even partially resistant) to this type of attacks, but we consider that it should be important for any cryptographic primitive or structure to possibly be implemented in such a way. Hence we try to find efficient such implementations for the SHARK structure and therefore for dense matrix-vector multiplications.

We now go on to describe the algorithms. In all of the remainder of this section, $\mathbf{x}$ and $\mathbf{y}$ are two (column) vectors of $\mathbf{F}_{2^4}^{16}$, and M a matrix of $\mathcal{M}_{16}(\mathbf{F}_{2^4})$. We first briefly recall the principle of table implementations, which are unsatisfactory when timing attacks are taken into account.

3.1   Table Implementation

We wish to compute $\mathbf{y} = M \cdot \mathbf{x}$. The idea behind this algorithm is to use table lookups to perform the equivalent multiplication $\mathbf{y}^t = \mathbf{x}^t \cdot M^t$, *i.e.* $\mathbf{y}^t = \sum_{i=0}^{15} \mathbf{x}_i \cdot (\mathbf{m}^t)^i$ (where $(\mathbf{m}^t)^i$ is the ith row of M^t). This can be computed efficiently by tabulating beforehand the products $\lambda \cdot (\mathbf{m}^t)^i, \lambda \in \mathbf{F}_{2^4}$ (resulting in 16 tables, each of 16 entries of 64 bits), and then for each multiplication by accessing the table for $(\mathbf{m}^t)^i$ at the index $\mathbf{x}_i$ and summing all the retrieved table entries together. This only requires 16 table lookups per multiplication. However, the memory accesses depend on the value of $\mathbf{x}$, which makes this algorithm inherently vulnerable to timing attacks.

---

[2] We will use the word *shuffle* with this precise meaning in the remainder of this paper.

Note that there is a more memory-efficient alternative implementation of this algorithm which consists in computing each term $\lambda \cdot (\mathbf{m}^t)^i$ with a single pshufb instruction instead of using a table-lookup. In that case, only the 16 multiplication tables need to be stored, but their accesses still depend on the secret value $\mathbf{x}$.

3.2   A Generic Constant-Time Algorithm

We now describe our first algorithm, which can be seen as a variant of table multiplication that is immune to timing attacks. The idea consists again in computing the right multiplication $\mathbf{y}^t = \mathbf{x}^t \cdot M^t$, i.e. $\mathbf{y}^t = \sum_{i=0}^{15} \mathbf{x}_i \cdot (\mathbf{m}^t)^i$. However, instead of tabulating the results of the scalar multiplication of the matrix rows $(\mathbf{m}^t)^i$, those are always recomputed, in a way that does not explicitly depend on the value of the scalar.

Description of Algorithm 1. We give the complete description of Algorithm 1 in the full version of the paper [3], and focus here on the intuition. We want to perform the scalar multiplication $\lambda \cdot \mathbf{z}$ for an unknown scalar λ and a known, constant vector $\mathbf{z}$, over $\mathbf{F}_{2^4}$. Let us write λ as the polynomial $\lambda_3 \cdot \alpha^3 + \lambda_2 \cdot \alpha^2 + \lambda_1 \cdot \alpha + \lambda_0$ with coefficients in $\mathbf{F}_2$. Then, the result of $\lambda \cdot \mathbf{z}$ is simply $\lambda_3 \cdot (\alpha^3 \cdot \mathbf{z}) + \lambda_2 \cdot (\alpha^2 \cdot \mathbf{z}) + \lambda_1 \cdot (\alpha \cdot \mathbf{z}) + \lambda_0 \cdot \mathbf{z}$. Thus we just need to precompute the products $\alpha^i \cdot \mathbf{z}$, select the right ones with respect to the binary representation of λ, and add these together. This can easily be achieved thanks to a *broadcast* function defined as:

$$broadcast(x, i)_n = \begin{cases} \mathbf{1}_n \text{ if the } i^{\text{th}} \text{ bit of } x \text{ is set} \\ \mathbf{0}_n \text{ otherwise} \end{cases}$$

where $\mathbf{1}_n$ and $\mathbf{0}_n$ denote the n-bit binary string made all of one and all of zero respectively. The full algorithm then just consists in using this scalar-vector multiplication 16 times, one for each row of the matrix.

Implementation of Algorithm 1 with SSSE3 Instructions. We now consider how to efficiently implement Algorithm 1 in practice. The only non-trivial operation is the *broadcast* function, and we show that this can be performed with only one or two pshufb instructions.

To compute $broadcast(\lambda, i)_{64}$, with λ a 4-bit value, we can use a single pshufb with first operand x, such that $x[j] = 11111111_2$ if the ith bit of j is set and 0 otherwise, and with second operand $y = (\lambda, \lambda, \lambda, \ldots)$. The result of pshufb x y is indeed $(x[\lambda], x[\lambda], \ldots)$ which is $\mathbf{1}_{64}$ if the ith bit of λ is set, and $\mathbf{0}_{64}$ otherwise, that is $broadcast(\lambda, i)_{64}$.

In practice, the vector x can conveniently be constructed offline and stored in memory, but the vector y might not be readily available before performing this computation[3]. However, it can easily be computed thanks to an additional pshufb. Alternatively, if the above computation is done with a vector

---

[3] And because it depends on what we assume to be a secret value, it cannot either be fetched from memory.

$y = (\lambda, ?, ?, \ldots)$ instead (with ? denoting unknown values) and call z its result $(x[\lambda], ?, ?, \ldots)$, then we have $broadcast(\lambda, i)_n = \texttt{pshufb}\ z\ (0, 0, \ldots)$.

In the specific case of matrices of dimension 16 over $\mathbf{F}_{2^4}$, one can take advantage of the 128-bit wide $\texttt{xmm}$ registers by interleaving, say, $8 \cdot \mathbf{x}$ with $4 \cdot \mathbf{x}$, and $2 \cdot \mathbf{x}$ with $\mathbf{x}$, and by computing a slightly more complex version of the $broadcast$ function $broadcast(x, i, j)_{2n}$ which interleaves $broadcast(x, i)_n$ with $broadcast(x, j)_n$. In that case, an implementation of one step of algorithm 1 only requires two $broadcast$ calls, two logical and, folding back the interleaved vectors (which only needs a couple of logical shift and exclusive or), and adding the folded vectors together. We give a snippet of such an implementation in the full version [3].

3.3   A Faster Algorithm Exploiting Matrix Structure

The above algorithm is already reasonably efficient, and has the advantage of being completely generic w.r.t. the matrix. Yet, better solutions may exist in more specific cases. We present here an alternative that can be much faster when the matrix possesses a particular structure.

The idea behind this second algorithm is to take advantage of the fact that in a matrix-vector product, the same constant values may be used many times in finite-field multiplications. Hence, we try to take advantage of this fact by performing those in parallel. The fact that we now focus on multiplications by constants (*i.e.* matrix coefficients) allows us to compute these multiplications with a single $\texttt{pshufb}$ instead of using the process from Algorithm 1.

Description of Algorithm 2. We give the complete description of Algorithm 2 in the full version [3], and focus here on the intuition. Let us first consider a small example, and compute $M \cdot \mathbf{x}$ defined as:

$$\begin{pmatrix} 1 & 0 & 2 & 2 \\ 3 & 1 & 2 & 3 \\ 2 & 3 & 3 & 2 \\ 0 & 2 & 3 & 1 \end{pmatrix} \cdot \begin{pmatrix} x_0 \\ x_1 \\ x_2 \\ x_3 \end{pmatrix}. \tag{1}$$

It is obvious that this is equal to:

$$\begin{pmatrix} x_0 \\ x_1 \\ 0 \\ x_3 \end{pmatrix} + 2 \cdot \begin{pmatrix} x_2 \\ x_2 \\ x_0 \\ x_1 \end{pmatrix} + 2 \cdot \begin{pmatrix} x_3 \\ 0 \\ x_3 \\ 0 \end{pmatrix} + 3 \cdot \begin{pmatrix} 0 \\ x_0 \\ x_1 \\ x_2 \end{pmatrix} + 3 \cdot \begin{pmatrix} 0 \\ x_3 \\ x_2 \\ 0 \end{pmatrix}, \tag{2}$$

where both the constant multiplications of the vector $(x_0\ x_1\ x_2\ x_3)^t$ and the shuffles of its coefficients can be computed with a single $\texttt{pshufb}$ instruction each, while none of these operations directly depends on the value of the vector. This type of decomposition can be done for any matrix, but the number of operations depends on the value of its coefficients.

We now sketch one way of obtaining an optimal decomposition as above. We consider a matrix product $M \cdot \mathbf{x}$ with M constant and $\mathbf{x}$ unknown, where $\mathbf{x}$ is seen as the formal arrangement of variables $\mathbf{x}_i$. Let us define $\mathcal{S}(M, \gamma)$ as one of the minimal sets of shuffles of coefficients of $\mathbf{x}$, such that there exists a unique vector $\mathbf{z} \in \mathcal{S}(M, \gamma)$ with $\mathbf{z}_i = \mathbf{x}_j$ iff $M_{i,j} = \gamma$. For instance, in the above example, we have $\mathcal{S}(M, 2) = \{(x_2\ x_2\ x_0\ x_1)^t, (x_3\ 0\ x_3\ 0)^t\}$. Equivalently, we could have taken $\mathcal{S}(M, 2) = \{(x_3\ x_2\ x_3\ 0)^t, (x_2\ 0\ x_0\ x_1)^t\}$. These sets are straightforward to compute from this particular matrix, and so are they in the general case.

From the definition of $\mathcal{S}$, it is clear that we have:

$$M \cdot \mathbf{x} = \sum_{\gamma \in \mathbf{F}_{2^4}^*} \sum_{\mathbf{s} \in \mathcal{S}(M, \gamma)} \gamma \cdot \mathbf{s}. \tag{3}$$

Once the values of the sets $\mathcal{S}$ have been determined, it is clear that we only need to compute this sum to get our result, and this is precisely what this second algorithm does.

Cost of Algorithm 2. The cost of computing a matrix-vector product with Algorithm 2 depends on the coefficients of the matrix, since the size of the sets $\mathcal{S}(M, \gamma)$ depends both on the density of the matrix and of how its coefficients are arranged.

If we assume that a vector implementation of this algorithm is used, and if the dimension and the field of the matrix are well chosen, we can assume that both the scalar multiplication of $\mathbf{x}$ by a constant and its shuffles can be computed with a single `pshufb` and a few ancillary instructions. Hence, we can define a cost function for a matrix with respect to its implementation with Algorithm 2 to be $cost2(M) = \left(\sum_{\gamma \in \mathbf{F}_{2^4}^*} \mathbb{1}(\mathcal{S}(M, \gamma)) + \# \mathcal{S}(M, \gamma) \right) - \mathbb{1}(\mathcal{S}(M, 1))$, where $\mathbb{1}(\mathcal{E})$ with $\mathcal{E}$ a set is one if $\mathcal{E} \neq \emptyset$, and zero otherwise. We may notice that $\# \mathcal{S}(M, \gamma)$ is equal to the maximum number of occurrence of γ in a single row of M, and the $cost2$ function is therefore easy to compute. As an example the cost of the matrix M from Eq. 1 is 7.

In order to find matrices that minimize the $cost2$ function, we would like to minimize the sum of the maximum number of occurrence of γ for every $\gamma \in \mathbf{F}_{2^4}^*$. A simple observation is that for matrices with the same number of non-zero coefficients, this amount is minimal when every row can be deduced by permutation of a single one; an important particular case being the one of *circulant* matrices. More generally, we can heuristically hope that the cost of a matrix will be low if all of its rows can be deduced by permutation of a small subset thereof.

We can try to estimate the minimum cost for an arbitrary dense circulant matrix of dimension 16 over $\mathbf{F}_{2^4}$. It is fair to assume that nearly all of the values of $\mathbf{F}_{2^4}$ should appear as coefficients of such a matrix, 14 of them needing a multiplication. Additionally, 15–16 permutations are needed if all the rows are to be different. Hence we can assume that the $cost2$ function of such a matrix is about 30.

Finally, let us notice that special cases of this algorithm have already been used for circulant matrices, namely in the case of the AES MixColumn matrix [4,10].

Implementation of Algorithm 2 with SSSE3 Instructions. The implementation of Algorithm 2 is straightforward. We refer to the full version for a small code snippet [3].

3.4 Performance

In Table 2 of Sect. 5, we give a few performance figures for ciphers with a SHARK structure using assembly implementations of Algorithms 1 and 2 for their linear mapping. From there it can be seen without surprise that Algorithm 2 is more efficient if the matrix is well chosen. However, Algorithm 1 still performs reasonably well, without imposing any condition on the matrix.

4 Diffusion Matrices from Algebraic-Geometry Codes

In this section, we present so-called algebraic-geometry codes and show how they can give rise to diffusion matrices with interesting parameters. We also focus on implementation aspects, and investigate how to find matrices with efficient implementations with respect to the algorithms of Sect. 3, and in particular Algorithm 2.

4.1 A Short Introduction to Algebraic-Geometry Codes

We first briefly present the concept of algebraic-geometry codes (or AG codes for short), which are linear codes, and how to compute their generator matrices. Because the codes are linear, these encoders are matrices. We do not give a complete description of AG codes, and refer to *e.g.* [22] for a more thorough treatment. We present a class of AG codes as a generalization of Reed-Solomon (RS) codes.

We see AG codes as *evaluation codes*: to build the codeword for a message w, we consider w as a function, and the codeword as a vector of values of this function evaluated on some "elements". In our case, the elements are points of the two-dimensional affine space $\mathbf{A}^2(\mathbf{F}_{2^m})$, and the functions are polynomials in two variables, that is elements of $\mathbf{F}_{2^m}[x, y]$. The core idea of AG codes is to consider points of a (smooth) projective curve of the projective space $\mathbf{P}^2(\mathbf{F}_{2^m})$ and functions from the curve's function space. Points at infinity are never included in the (ordered) set of points. However, points of the curve at infinity *are* useful in defining the curve's function (sub)-space, which is why we do consider the curve in the projective space instead of the affine one.

We first give the definition of the Riemann-Roch space in the special case where it is defined from a divisor made of a single point at infinity. We refer to *e.g.* [19] or [22] for a more complete and rigorous definition.

Definition 3 (Riemann-Roch space). *Let $\mathcal{X}$ be a smooth projective curve of $\mathbf{P}^2(\mathbf{F}_{2^m})$ defined by the homogeneous polynomial $p(x, y, z)$, and let $p'(x, y)$ be the dehomogenized of p. We define $\mathbf{F}_{2^m}[\mathcal{X}] = \mathbf{F}_{2^m}[x, y]/p'$ as the* coordinate *ring of* $\mathcal{X}$, *and its corresponding quotient field* $\mathbf{F}_{2^m}(\mathcal{X})$ *as the* function field *of* $\mathcal{X}$. *Assume* Q *is the only point of* $\mathcal{X}$ *at infinity, and let* r *be a positive integer. The Riemann-Roch space* $\mathcal{L}(rQ)$ *is the set of all functions of* $\mathbf{F}_{2^m}(\mathcal{X})$ *with poles only at* Q *of order less than* r. *This is a finite-dimensional* $\mathbf{F}_{2^m}$-*vector-space. Furthermore, let* $o_Q(x)$ *and* $o_Q(y)$ *be the order of the poles of* x *and* y *in* Q^4, *then a basis of* $\mathcal{L}(rQ)$ *is formed by all the monomial functions* $x^i y^j$ *that are such that* $i \cdot o_Q(x) + j \cdot o_Q(y) \leq r$.

This space is particularly important because of the following theorem, which links its dimension with the genus of $\mathcal{X}$ [19].

Theorem 1 (Riemann and Roch). *Let* $\mathcal{L}(rQ)$ *be a Riemann-Roch space defined on* $\mathcal{X}$, *and* g *be the genus of* $\mathcal{X}$. *We have* $\dim(\mathcal{L}(rQ)) \geq r + 1 - g$, *with equality when* $r > 2g - 2$.

We have also mentioned earlier that a basis for a space $\mathcal{L}(rQ)$ can be computed as soon as the order of the poles of x and y in Q are known, and the dimension of the space can obviously be computed from the basis. In practice, computing $o_Q(x)$ and $o_Q(y)$ can be done from a local parameterization of x and y in Q. Both this parameterization and the values $o_Q(x)$ and $o_Q(y)$ can easily be obtained from a computational algebra software such as Magma. Again, we refer to [22] for more details.

We are now ready to define a simple class of AG codes.

Definition 4 (Algebraic-Geometry Codes). *Let* $\mathcal{X}$ *be a smooth projective curve of* $\mathbf{P}^2(\mathbf{F}_{2^m})$ *with a unique point* Q *at infinity, and call* $\#\mathcal{X}$ *its number of affine points (that is not counting* Q). *Assume that* $\#\mathcal{X} \geq n$ *and let* r *be s.t.* $\dim(\mathcal{L}(rQ)) = k$, *and call* $(f_0, \dots, f_{k-1})$ *one basis of this space. We define the codeword of the* $[n, k, d]_{\mathbf{F}_{2^m}}$ *algebraic-geometry code* $\mathcal{C}_{AG}$ *associated with the message* $\mathbf{m}$ *as the vector* $\sum_{i=0\dots k-1}(\mathbf{m}_i \cdot f_i(p_j), j = 0 \dots n - 1)$, *where* $P = \langle p_0, \dots, p_{n-1} \rangle$ *is an ordered set of points of* $\mathcal{X}/\{Q\}$. *The code* $\mathcal{C}_{AG}$ *is the set of all such codewords.*

These codes have the following properties: for fixed parameters n and k and a curve $\mathcal{X}$, there are $\binom{\#\mathcal{X}}{n} \cdot n!$ equivalent codes, which corresponds to the number of possible ordered sets P; it is also obvious that the maximal length of a code over $\mathcal{X}$ is $\#\mathcal{X}$. We also have the following proposition [22]:

Proposition 1. *Let* $\mathcal{C}_{AG}$ *be a code of length* n *and dimension* k, *and let* r *be an integer such that* $\dim(\mathcal{L}(rQ)) = k$. *Then the minimum distance of* $\mathcal{C}_{AG}$ *is at least* $n - r$. *If* $\mathcal{X}$ *is of genus* g *and* $r > 2g - 2$, *this is equal to* $n - ((k - 1) + g) = n - k - g + 1$. *Therefore, the "gap" between this code and an MDS code of the same length is* g. *The same holds for the dual code.*

---

[4] We slightly abuse the notations here and actually mean x/z and y/z. But we prefer manipulating their dehomogenized equivalents x and y. It is obvious that x/z and y/z indeed have poles in Q, which is at infinity and hence has a zero z coordinate.

The minimum distance of AG codes thus depends on the genus of the curves used to define them. Because the maximal number of points on a curve increases with its genus, there is a tradeoff between the length of a code and its minimum distance.

Construction of a Generator Matrix of an AG Code. Once the parameters of a code have been fixed, including the ordered set P, one just has to specify a basis of $\mathcal{L}(rQ)$, and to form the encoding matrix $M \in \mathcal{M}_{k,n}(\mathbf{F}_{2^m})$ obtained by evaluating this basis on P. A useful basis is one such that the encoding matrix is in systematic form, but it does not necessarily exist for any P. Note however that in the case of MDS codes (such as RS codes) this basis always exists whatever the parameters and the choice of P: this is because in this case every minor of M is of full rank [15]. When such a basis exists, it is easy to find as one just has to start from an arbitrary basis and to compute the reduced row echelon form of the matrix thus obtained.

Example 1: An AG Code from an Elliptic Curve. We give parameters for a code built from the curve defined on $\mathbf{P}^2(\mathbf{F}_{2^4})$ by the homogeneous polynomial $x^2z + xz^2 = y^3 + yz^2$, or equivalently defined on $\mathbf{A}^2(\mathbf{F}_{2^4})$ by $x^2 + x = y^3 + y$. It is of genus 1, and hence it is an elliptic curve. It has 25 points, including one point at infinity, the point $Q = [1 : 0 : 0]$; the order of the poles of x and y in Q are respectively 3 and 2. From this, a basis for the space $\mathcal{L}(12Q)$ can easily be obtained. This space has dimension $12 + 1 - g = 12$, and can be used to define a $[24, 12, 12]_{\mathbf{F}_{2^4}}$ code by evaluation over the affine points of the curve. This allows to define a matrix of dimension 12 over $\mathbf{F}_{2^4}$, which diffuses over $12 \times 4 = 48$ bits and has a differential and linear branch number of 12.

Example 2: An AG Code from an Hyperelliptic Curve. We increase the length of the code by using a curve with a larger genus. We give parameters for a rather well-known code, built from the curve defined on $\mathbf{P}^2(\mathbf{F}_{2^4})$ by the homogeneous polynomial $x^5 = y^2z + yz^4$. This curve has 33 points, including one point at infinity, the point $Q = [0 : 1 : 0]$; the order of the poles of x and y in Q are respectively 2 and 5. From this, a basis for the space $\mathcal{L}(17Q)$ can easily be defined. This space has dimension $17 + 1 - g = 16$, and can be used to define a $[32, 16, 15]_{\mathbf{F}_{2^4}}$ code by evaluation over the affine points of the curve. This code has convenient parameters for defining diffusion matrices: from a generator matrix in systematic form $(I_{16}\ A)$, we can extract the matrix A, which naturally diffuses over 64 bits and has a differential and linear branch number of 15. Furthermore, the code is *self-dual*, which means that A is orthogonal: $A \cdot A^t = I_{16}$. The inverse of A is therefore easy to compute.

We give the right matrix of two matrices of this code in systematic form in the full version [3], the latter further including an example of a basis of $\mathcal{L}(17Q)$ and the order of the points used to construct the matrix.

The problem for the rest of the section is now to find good point orders P for the hyperelliptic code of Exaqmple 2 such that efficient encoders can be constructed thanks to Algorithm 2 of Sect. 3.3. For convenience, we name $\mathcal{C}_{\mathcal{HE}}$ any of the codes equivalent to the one of Example 2.

4.2   Compact Encoders Using Code Automorphisms

We consider matrices in systematic form $(I_{16}\ A)$. For dense matrices, Algorithm 2 tends to be most efficient when all the rows of a matrix can be deduced by permutation of one of them, or more generally of a small subset of them. Our objective is thus to find matrices of this form.

The main tool we use to achieve this goal are *automorphisms* of $\mathcal{C}_{\mathcal{HE}}$. Let us first give a definition. (In the following, $\mathfrak{S}_n$ denotes the group of permutations of n elements.)

Definition 5. (Automorphisms of a code). *The automorphism group $Aut(\mathcal{C})$ of a code $\mathcal{C}$ of length n is a subgroup of $\mathfrak{S}_n$ such that $\pi \in Aut(\mathcal{C}) \Rightarrow (c \in \mathcal{C} \Rightarrow \pi(c) \in \mathcal{C})$.*

Because we consider here the code $\mathcal{C}_{\mathcal{HE}}$ which is an evaluation code, we can equivalently define its automorphisms as being permutations of the points on which the evaluation is performed. If π is an automorphism of $\mathcal{C}_{\mathcal{HE}}$, if $\{O_0, \ldots, O_l\}$ are its orbits, and if the code is defined with a point order P such that for each orbit all of its points are neighbours in the order P, then the effect of π on a codeword of $\mathcal{C}_{\mathcal{HE}}$ is to cyclically permute its coordinates along each orbit.

To see that this is useful, assume that there is an automorphism π with two orbits O_0 and O_1 of size $n/2$ each. Then, if $M = (I_{n/2}\ A)$ is obtained with point order $P = \langle O_0, O_1 \rangle$, each row of M can be obtained by the repeated action of π on, say, $\mathbf{m}^0$, and it follows that A is circulant (and therefore has a low cost w.r.t. Algorithm 2). More generally, if an automorphism can be found such that it has orbits of size summing up to $n/2$, the corresponding matrix M can be deduced from a small set of rows. We give two toy examples with Reed-Solomon codes, which can easily be verified.

$\pi : \mathbf{F}_{2^4} \to \mathbf{F}_{2^4}$, $x \mapsto 8x$. This automorphism has $O_0 = \langle 1, 8, 12, 10, 15 \rangle$ and $O_1 = \langle 2, 3, 11, 7, 13 \rangle$ for orbits, among others. The systematic matrix for the $[10, 5, 6]_{\mathbf{F}_{2^4}}$ code obtained with the points in that order is then such that A is circulant and obtained from the cyclic permutation of the row $(12, 10, 2, 6, 3)$.

$\pi : \mathbf{F}_{2^4} \to \mathbf{F}_{2^4}$, $x \mapsto 7x$. This automorphism has $O_0 = \langle 1, 7, 6 \rangle$, $O_1 = \langle 2, 14, 12 \rangle$, $O_2 = \langle 4, 15, 11 \rangle$, and $O_3 = \langle 8, 13, 5 \rangle$ for orbits, among others. The systematic matrix for the $[12, 6, 7]_{\mathbf{F}_{2^4}}$ code obtained with the points in that order is then of the form $\begin{pmatrix} I_3 & 0_3 & A & B \\ 0_3 & I_3 & C & D \end{pmatrix}$ with A, B, C and D circulant matrices. It can thus be obtained by cyclic permutation of only two rows.

Application to $\mathcal{C}_{\mathcal{HE}}$. Automorphisms of $\mathcal{C}_{\mathcal{HE}}$ are quite harder to find than ones of RS codes. They can however be found within automorphisms of the curve $\mathcal{X}$ on which it is based [19]. This is quite intuitive, as these will precisely permute points on the curve, which are the points on which the code is defined.

We mostly need to be careful to ensure that the point at infinity is fixed by these automorphisms. We considered the degree-one automorphisms of $\mathcal{X}$ described by Duursma [8]. They have two generators: $\pi_0 : \mathbf{F}_{2^4}^2 \to \mathbf{F}_{2^4}^2$, $(x,y) \mapsto (\zeta x, y)$ with $\zeta^5 = 1$, and $\pi_{1_{(a,b)}} : \mathbf{F}_{2^4}^2 \to \mathbf{F}_{2^4}^2$, $(x,y) \mapsto (x + a, y + a^8 x^2 + a^4 x + b^4)$, with (a,b) an affine point of $\mathcal{X}$. These generators span a group of order 160. When considering their orbit decomposition, the break-up of the size of the orbits can only be of one of five types, given in Table 1.

Table 1. Possible combination of orbit sizes of automorphisms of $\mathcal{C}_{\mathcal{HE}}$ spanned by π_0 and π_1. A number n in col. c means that an automorphism of this type has n orbits of size c.

Orbit size	1	2	4	5	10
Type 1	32	0	0	0	0
Type 2	0	16	0	0	0
Type 3	0	0	8	0	0
Type 4	2	0	0	6	0
Type 5	0	1	0	0	3

From these automorphisms, it is possible to define a partitions of P in two sets of size 16, which are union of orbits. We may therefore hope to obtain systematic matrices of the type we are looking for. Unfortunately, after an extensive search[5], it appears that ordering P in this fashion *never* results in obtaining a systematic matrix. We recall that indeed, because AG codes are not MDS, it is not always the case that computing the reduced row echelon form of an arbitrary encoding matrix yields a systematic matrix.

Extending the Automorphisms with the Frobenius Mapping. We extend the previous automorphisms with the Frobenius mapping $\theta : \mathbf{F}_{2^4}^2 \to \mathbf{F}_{2^4}^2$, $(x,y) \mapsto (x^2, y^2)$; this adds another 160 automorphisms for $\mathcal{X}$. However, these will not anymore be automorphisms for the *code* $\mathcal{C}_{\mathcal{HE}}$ in general, and we will therefore obtain matrices of a form slightly different from what we first hoped to achieve.

The global strategy is still the same, however, and consists in ordering the points along orbits of the curve automorphisms. By using the Frobenius, new combinations of orbits are possible, notably 4 of size 8. We study below the result of ordering P along the orbits of one such automorphism. We take the example of $\sigma = \theta \circ \sigma_2 \circ \sigma_1$, with $\sigma_1 : (x,y) \mapsto (x+1, y+x^2+x+7)$, $\sigma_2 : (x,y) \mapsto (12x, y)$, and θ the Frobenius mapping. The key observation is that in this case, only σ^0 and σ^4 are automorphisms of $\mathcal{C}_{\mathcal{HE}}$. Note that not all orbits orderings of σ for P yield a systematic matrix. However, unlike as above, we were able to find some

---

[5] Both on $\mathcal{C}_{\mathcal{HE}}$ and on the smaller elliptic code of Example 1. However, we are not as yet able to explain this fact.

orders that do. In these cases, the right matrix "A" of the full generator matrix $(I_{16}\ A)$ is of the form:

$$(\mathbf{a}^0,\ldots,\mathbf{a}^3,\sigma^4(\mathbf{a}^0),\ldots,\sigma^4(\mathbf{a}^3),\mathbf{a}^8,\ldots,\mathbf{a}^{11},\sigma^4(\mathbf{a}^8),\ldots,\sigma^4(\mathbf{a}^{11}))^t,$$

with $\mathbf{a}^0,\ldots,\mathbf{a}^3,\ \mathbf{a}^8,\ldots,\mathbf{a}^{11}$ row vectors of dimension 16. For instance, the first and fifth row of one such matrix are:

$$\mathbf{a}^0 = (5,2,1,3,8,5,1,5,12,10,14,6,7,11,4,11)$$
$$\mathbf{a}^4 = \sigma^4(\mathbf{a}^0) = (8,5,1,5,5,2,1,3,7,11,4,11,12,10,14,6).$$

We give the full matrix in the full version [3]. We have therefore partially reached our goal of being able to describe A from a permutation of a subset of its rows. However this subset is not small, as it is of size 8—half of the matrix dimension. Consequently, these matrices have a moderate cost according to the *cost2* function, when implemented with Algorithm 2, but it is not minimal. Interestingly, all the matrices of this form that we found have the same cost of 52.

4.3 Fast Random Encoders

We conclude this section by presenting the results of a very simple random search for efficient encoders of $\mathcal{C}_{\mathcal{HE}}$ with respect to Algorithm 2. Unlike the above study, this one does not exploit any kind of algebraic structure. Indeed, the search only consists in repeatedly generating a random permutation of the affine points of the curve, building a matrix for the code with the corresponding point order, tentatively putting it in systematic form $(I_{16}\ A)$, and if successful evaluating the *cost2* function from Sect. 3.3 on A. We then collect matrices with a minimum cost.

Because there are $32! \approx 2^{117,7}$ possible point orders, we can only explore a very small part of the search space. However, matrices of low cost can be found even after a moderate amount of computation, and we found many matrices of cost 43, though none of a lower cost. We give the number of matrices of cost strictly less than 60 that we found during a search of 2^{38} encoders in the full version [3]. We also give in the full paper an example of a matrix of cost 43, which is only about a factor 1.5 away from the estimate of the minimum cost of a circulant matrix given in Sect. 3.3. We observe that the transpose of this matrix also has a cost of 43.

5   Applications and Performance

This last section presents the performance of straightforward assembly implementations of both of our algorithms when applied to a matrix of the code $\mathcal{C}_{\mathcal{HE}}$ from Sect. 4, of cost 43. For convenience, we denote $M_{\mathcal{H}16}$ this matrix. It is of dimension 16 over $\mathbf{F}_{2^4}$ and has a differential and linear branch number of 15.

We do this study in the context of block ciphers, by assuming that $M_{\mathcal{H}16}$ is used as the linear mapping of two ciphers with a SHARK structure: one with

4-bit S-Boxes and a 64-bit block, and one with 8-bit S-Boxes and a 128-bit block. What we wish to measure in both cases is the speed in cycles per byte of such hypothetical ciphers, so as to be able to gauge the efficiency of this linear mapping and of the resulting ciphers. In order to do this, we need to estimate how many rounds would be needed for the ciphers to be secure.

Basic Statistical Properties of a 64-bit Block Cipher with 4-bit S-Boxes and $M_{\mathcal{H}16}$ as a Linear Mapping. We use standard *wide-trail* considerations to study differential and linear properties of this cipher [7]. This is very easy to do thanks to the simple structure of the cipher. The branch number of $M_{\mathcal{H}16}$ is 15, which means that at least 15 S-boxes are active in any two rounds of a differential path or linear characteristic. The best 4-bit S-boxes have a maximum differential probability and a maximal linear bias of 2^{-2} (see *e.g.* [14,17]). By using such S-boxes, one can upper-bound the probability of a single differential path or linear characteristic for $2n$ rounds by $2^{-2.15n}$. This is smaller than 2^{-64} as soon as $n > 2$. Hence we conjecture that 6 to 8 rounds are enough to make a cipher resistant to standard statistical attacks.

If one were to propose a concrete cipher, a more detailed analysis would of course be needed, especially w.r.t. more dedicated structural attacks. However, it seems reasonable to consider at a first glance that 8 rounds would indeed be enough to bring adequate security. It is only 2 rounds less than AES-128, which uses a round with comparatively weaker diffusion. Also, 6 rounds might be enough. Consequently, we present software performance figures for 6 and 8 rounds of such an hypothetical 64-bit cipher with 4-bit S-Boxes in Table 2, on the left[6]. We include data both for a strict SSSE3 implementation and for one using AVX extensions, which can be seen to bring a considerable benefit. Note that the last round is complete and includes the linear mapping, unlike *e.g.* AES. Also, note that the parallel application of the S-Boxes can be implemented very efficiently with a single `pshufb`, and thus has virtually no impact on the speed.

Basic Statistical Properties of a 128-Bit Block Cipher with 8-Bit S-Boxes and $M_{\mathcal{H}16}$ as a Linear Mapping. Although the code $\mathcal{C}_{\mathcal{HE}}$ from which the matrix $M_{\mathcal{H}16}$ is built was initially defined with $\mathbf{F}_{2^4}$ as an alphabet, this latter can be replaced by an algebraic extension of $\mathbf{F}_{2^4}$ such as $\mathbf{F}_{2^8}$, to yield a code $\mathcal{C}_{\mathcal{HE}}'$ with the same parameters, namely a $[32, 16, 15]_{\mathbf{F}_{2^8}}$ code. Indeed, by using a suitable representation such as $\mathbf{F}_{2^4} \cong \mathbf{F}_2[\alpha]/(\alpha^4 + \alpha^3 + \alpha^2 + \alpha + 1)$; $\mathbf{F}_{2^8} \cong \mathbf{F}_{2^4}[t]/(t^2 + t + \alpha)^7$, an element of $\mathbf{F}_{2^8}$ is represented as a degree-one polynomial $at + b$ over $\mathbf{F}_{2^4}$. It follows that the minimum weight of a codeword $w = (a_i t + b_i)$, $i \in \{0 \dots 31\}$ of $\mathcal{C}_{\mathcal{HE}}'$ is at least equal to the minimum weight of words (a_i), $i = 0 \dots 31$ and (b_i), $i \in \{0 \dots 31\}$. If those are taken among codewords of $\mathcal{C}_{\mathcal{HE}}$, their minimum weight is 15, and thus so is the one of w. It is also possible to efficiently compute the multiplication by $M_{\mathcal{H}16}$ over $\mathbf{F}_{2^8}$ from two computations over $\mathbf{F}_{2^4}$: because the coefficients of $M_{\mathcal{H}16}$ are in $\mathbf{F}_{2^4}$, we have

---

[6] The figures are given for computation of a single block; better performance can be achieved for longer messages.

[7] This representation is used by Hamburg in [10].

$M_{\mathcal{H}16} \cdot (a_i t + b_i) = (M_{\mathcal{H}16} \cdot (a_i) \cdot t) + (M_{\mathcal{H}16} \cdot (b_i))$. As a result, applying $M_{\mathcal{H}16}$ to $\mathbf{F}_{2^8}$ has only twice the cost of applying it to $\mathbf{F}_{2^4}$, while effectively doubling the size of the block.

From wide-trail considerations, the resistance of such a cipher to statistical attacks is comparatively even better than when using 4-bit S-Boxes, when an appropriate 8-bit S-Box is used. For instance, the AES S-box has a maximal differential probability and linear bias of 2^{-6} [7]. This implies that the probability of a single differential path or linear characteristic for $2n$ rounds is upper-bounded by $2^{-6 \cdot 15n}$, which is already much smaller than 2^{-128} as soon as $n > 1$. Again, 8 rounds of such a cipher should bring adequate security, and 6 rounds might be enough. We provide performance figures for both an SSSE3 and an AVX implementation in Table 2, on the right(see footnote 6). However, in the case of 8-bit S-Boxes, the S-Box application is rather more complex and expensive a step than with 4-bit S-Boxes. In these test programs, we decided to use the efficient vector implementation of the AES S-Box from Hamburg [10].

Table 2. Performance of software implementations of the hypothetical 64 and 128-bit cipher, in cycles per byte (cpb). Figures in parentheses are for an AVX implementation (when applicable).

Processor type	# rounds	64-bit Block		128-bit Block	
		cpb (Algorithm 1)	cpb (Algorithm 2)	cpb (Algorithm 1)	cpb (Algorithm 2)
Intel Xeon E5-2650 @ 2.00 GHz	6	50 (45.5)	33 (24.2)	58 (52.3)	32.7 (26.5)
	8	66.5 (60.2)	44.5 (31.9)	76.8 (69.6)	43.8 (35.7)
Intel Xeon E5-2609 @ 2.40 GHz	6	72.3 (63.7)	45.3 (33.2)	79.8 (75.6)	47.1 (36.8)
	8	95.3 (84.7)	63.3 (45.6)	106.6 (97.1)	62.1 (50.3)
Intel Xeon E5649 @ 2.53 GHz	6	84.7	46	84.5	47
	8	111.3	59.8	111	61.9

Discussion. The performance figures given in Table 2 are average for a block cipher. For instance, it compares favourably with the optimized vector implementations of 64-bit ciphers LED and Piccolo in sequential mode from [4], which run at speeds between 70 and 90 cpb., depending on the CPU. It is however slower than Hamburg's vector implementation of AES, with reported speeds of 6 to 22 cpb. (9 to 25 for the inverse cipher) [10,20].

6   Conclusion

We revisited the SHARK structure by replacing the MDS matrix of its linear diffusion layer by a matrix built from an algebraic-geometry code. Although this code is not MDS, it has a very high minimum distance, while being defined over $\mathbf{F}_{2^4}$ instead of $\mathbf{F}_{2^8}$. This allows to reduce the size of the coefficients of the matrix from 8 to 4 bits, and has important consequences for efficient implementations of this linear mapping. We studied algorithms suitable for a vector implementation of the multiplication by this matrix, and how to find matrices that are most efficiently implemented with those algorithms. Finally, we gave performance figures for assembly implementations of hypothetical SHARK-like ciphers using this matrix as a linear layer.

This work provided generalizations of SHARK that are not vulnerable to timing attacks as is the original cipher, and also a generalization to 128-bit blocks. It also showed that even if not the fastest, such potential design could be implemented efficiently in software.

As a future work, it would be interesting to investigate how to use the full automorphism group of the code to design matrices with a lower cost.

Acknowledgments. Pierre Karpman is partially supported by the Direction générale de l'armement and by the Singapore National Research Foundation Fellowship 2012 (NRF-NRFF2012-06).

References

1. Augot, D., Finiasz, M.: Direct construction of recursive MDS diffusion layers using shortened BCH codes. In: Cid, C., Rechberger, C. (eds.) FSE 2014. LNCS. Springer (2014). https://eprint.iacr.org/2014/566
2. Augot, D., Finiasz, M.: Exhaustive search for small dimension recursive MDS diffusion layers for block ciphers and hash functions. In: ISIT, IEEE, pp. 1551–1555 (2013)
3. Augot, D., Fouque, P.A., Karpman, P.: Diffusion Matrices from Algebraic-Geometry Codes with Efficient SIMD Implementation. IACR Cryptology ePrint Archive 2014, 551 (2014). https://eprint.iacr.org/2014/551
4. Benadjila, R., Guo, J., Lomné, V., Peyrin, T.: Implementing lightweight block ciphers on x86 architectures. In: Lange, T., Lauter, K., Lisoněk, P. (eds.) SAC 2013. LNCS, vol. 8282, pp. 324–352. Springer, Heidelberg (2014)
5. Bernstein, D.J., Schwabe, P.: NEON crypto. In: Prouff, E., Schaumont, P. (eds.) CHES 2012. LNCS, vol. 7428, pp. 320–339. Springer, Heidelberg (2012)
6. Chen, J.P., Lu, C.C.: A serial-in-serial-out hardware architecture for systematic encoding of Hermitian codes via Gröbner Bases. IEEE Trans. Commun. **52**(8), 1322–1332 (2004)
7. Daemen, J., Rijmen, V.: The Design of Rijndael: AES – The Advanced Encryption Standard. Information Security and Cryptography. Springer, New York (2002)
8. Duursma, I.: Weight distributions of geometric Goppa codes. Trans. Am. Math. Soc. **351**(9), 3609–3639 (1999)
9. Guo, J., Peyrin, T., Poschmann, A.: The PHOTON family of lightweight hash functions. In: Rogaway, P. (ed.) CRYPTO 2011. LNCS, vol. 6841, pp. 222–239. Springer, Heidelberg (2011)
10. Hamburg, M.: Accelerating AES with vector permute instructions. In: Clavier, C., Gaj, K. (eds.) CHES 2009. LNCS, vol. 5747, pp. 18–32. Springer, Heidelberg (2009)
11. Heegard, C., Little, J., Saints, K.: Systematic encoding via Gröbner bases for a class of algebraic-geometric goppa codes. IEEE Trans. Inf. Theor. **41**(6), 1752–1761 (1995)
12. Intel Corporation: Intel® 64 and IA-32 Architectures Software DeveloperG's Manual, March 2012

13. Knudsen, L.R., Wu, H. (eds.): SAC 2012. LNCS, vol. 7707. Springer, Heidelberg (2013)
14. Leander, G., Poschmann, A.: On the classification of 4 bit S-Boxes. In: Carlet, C., Sunar, B. (eds.) WAIFI 2007. LNCS, vol. 4547, pp. 159–176. Springer, Heidelberg (2007)
15. MacWilliams, F.J., Sloane, N.J.A.: The Theory of Error-Correcting Codes. North-Holland Mathematical Library. North-Holland, Amsterdam (1978)
16. Rijmen, V., Daemen, J., Preneel, B., Bosselaers, A., Win, E.D.: The cipher SHARK. In: Gollmann, D. (ed.) FSE 1996. LNCS, vol. 1039, pp. 99–111. Springer, Heidelberg (1996)
17. Saarinen, M.-J.O.: Cryptographic analysis of all 4 × 4-Bit S-Boxes. In: Miri, A., Vaudenay, S. (eds.) SAC 2011. LNCS, vol. 7118, pp. 118–133. Springer, Heidelberg (2012)
18. Sajadieh, M., Dakhilalian, M., Mala, H., Sepehrdad, P.: Recursive diffusion layers for block ciphers and hash functions. In: Canteaut, A. (ed.) FSE 2012. LNCS, vol. 7549, pp. 385–401. Springer, Heidelberg (2012)
19. Stichtenoth, H.: Algebraic Function Fields and Codes. Graduate Texts in Mathematics, vol. 254, 2nd edn. Springer, New York (2009)
20. Suzaki, T., Minematsu, K., Morioka, S., Kobayashi, E.: TWINE: A Lightweight Block Cipher for Multiple Platforms. [13], pp. 339–354
21. Tromer, E., Osvik, D.A., Shamir, A.: Efficient cache attacks on AES, and counter-measures. J. Cryptol. 23(1), 37–71 (2010)
22. Van Lint, J.H.: Introduction to Coding Theory. Graduate Texts in Mathematics, vol. 86, 3rd edn. Springer, Berlin (1999)
23. Wu, S., Wang, M., Wu, W.: Recursive Diffusion Layers for (Lightweight) Block Ciphers and Hash Functions. [13], pp. 355–371

Error-Tolerant Side-Channel Cube Attack Revisited

Zhenqi Li[1]([⊠]), Bin Zhang[2,3], Arnab Roy[4,5], and Junfeng Fan[6]

[1] Trusted Computing and Information Assurance Laboratory,
Institute of Software, Chinese Academy of Sciences,
Beijing, China
lizhenqi@tca.iscas.ac.cn
[2] Trusted Computing and Information Assurance Laboratory,
Institute of Software, Chinese Academy of Sciences,
Beijing, China
[3] State Key Laboratory of Computer Science, Institute of Software,
Chinese Academy of Sciences, Beijing, China
zhangbin@tca.iscas.ac.cn
[4] University of Luxembourg, Luxembourg, Luxembourg
[5] Technical University of Denmark,
Kongens Lyngby, Denmark
arroy@dtu.dk
[6] Nationz Technologies Inc, Shenzhen, China
fanjunfeng@gmail.com

Abstract. Error-tolerant side-channel cube attacks have been recently introduced as an efficient cryptanalytic technique against block ciphers. The known Dinur-Shamir model and its extensions require error-free data for at least part of the measurements. Then, a new model was proposed at CHES 2013, which can recover the key in the scenario that each measurement contains noise. The key recovery problem is converted to a decoding problem under a binary symmetric channel. In this paper, we propose a high error-tolerant side-channel cube attack. The error-tolerant rate is significantly improved by utilizing the polynomial approximation and a new variant of cube attack. The simulation results on PRESENT show that given about $2^{21.2}$ measurements, each with an error probability of 40.5%, the new model achieves a success probability of 50% for the key recovery. The error-tolerant level can be enhanced further if the attacker can obtain more measurements.

Keywords: Cube attack · Side-channel attack · PRESENT

This work was supported by the National Grand Fundamental Research 973 Program of China(Grant No. 2013CB338002), the programs of the National Natural Science Foundation of China (Grant No. 60833008, 60603018, 61173134, 91118006, 61272476).

Most of Arnab Roy's work was done when he was in the University of Luxembourg.

© Springer International Publishing Switzerland 2014
A. Joux and A. Youssef (Eds.): SAC 2014, LNCS 8781, pp. 261–277, 2014.
DOI: 10.1007/978-3-319-13051-4_16

1   Introduction

Cube attack was formally proposed by Dinur and Shamir at Eurocrypt 2009 [8] as an efficient cryptanalytic technique which can be applied to many types of well-designed cryptosystems by exploiting an low degree multivariate polynomial of a single output bit. It is an extension of high-order differential attacks [13] and algebraic IV differential attacks [17,18]. It shows superior performance on several stream ciphers [1,2,7,8,11], however, most block ciphers are immune to it, as they iterate a highly non-linear round function for a number of times and the degree of the polynomial for the ciphertext bits is much higher.

Since the master polynomials of some intermediate variables in the early rounds are of relatively low degree, cube attack becomes a convincing method to attack block ciphers by combining physical attacks, where the attackers can exploit some leaked information about the intermediate variables, i.e., state registers. The attacker only needs to learn the value of a single wire or register in each execution, it is thus ideal for probing attacks. The main challenge is to overcome the measurement noise, thus how to launch an efficient error-tolerant side-channel cube attack in a realistic setting is a highly interesting topic.

Dinur and Shamir initialized the first study on error-tolerant side-channel cube attack (ET-SCCA) [10]. They treat the uncertain bits as new erasure variables and it was further enhanced in [6,9] by utilizing more trivial equations of high dimensional cubes to correct the errors. The success of this model is based on an assumption that the adversary possesses the exact knowledge of error positions and partial measurements are error-free. Then, at CHES 2013 [19], Li *et. al.* proposed a new model, which can recover the key when each measurement contains noise. The key recovery problem is converted to decoding a $[L, n]$ linear code. However, the error-tolerant level is still very low.

This paper introduces a new ET-SCCA which can tolerate heavy noise interference. The error-tolerant rate can be improved significantly by utilizing the polynomial approximation technique and applying a new variant of cube attack. The main idea of polynomial approximation is to appropriately remove some key variables to reduce the code dimension n of a $[L, n]$ code. Moreover, a new variant of cube attack is proposed, inspired by the idea of dynamic cube attack [7]. The main idea is to increase the number of linear equations, i.e., code length L, by adaptively choosing the plaintext. Consequently, the bound of error probability has been refined. Compared with the simulation results on PRESENT in [19], the error probability for each measurement can be improved to 40.5 % given about $2^{21.2}$ measurements and $2^{27.6}$ time complexity. The error-tolerant rate can be enhanced further if the attacker can obtain more measurements. Table 1 summarized our simulation results on PRESENT.

This paper is organized as follows. We first introduce the basic idea of cube attack and ET-SCCA in Sect. 2. In Sect. 3, we present the new model. Error probability evaluation is developed and analyzed in Sect. 4. Section 5 presents the simulations on PRESENT. The comparison is given in Sect. 6, followed by some further discussions. Finally, we conclude the paper in Sect. 7.

Table 1. Simulation results on PRESENT

Time complexity	Data (measurements)	Error probability	Scenario
$2^{31.6}$	$2^{10.1}$	23.2 %	Lower measurements
$2^{27.6}$	$2^{16.2}$	29.5 %	Balanced
$2^{27.6}$	$2^{21.2}$	40.5 %	Higher error tolerance

The success probability is about 50 %.
The memory requirement is negligible.

2   Preliminaries

2.1   Cube Attack

Consider a block cipher $\mathbb{T}$ and its encryption function $(c_1, ..., c_m) = \mathrm{E}(k_1, ..., k_n, v_1, ..., v_m)$, where c_i, k_j and v_s are ciphertext, encryption key and plaintext bits, respectively. One can always represent c_i, $i \in [1, m]$, with a multivariate polynomial in the plaintext and key bits, namely, $c_i = p(k_1, ..., k_n, v_1, ..., v_m)$. Let $I \subseteq \{1, ..., m\}$ be an index subset, and $t_I = \prod_{i \in I} v_i$, the polynomial p is divided into two parts: $p(k_1, ..., k_n, v_1, ..., v_m) = t_I \cdot p_{S(I)} + q(k_1, ..., k_n, v_1, ..., v_m)$, where no item in q contains t_I. Here $p_{S(I)}$ is called the superpoly of I in p. A maxterm of p is a term t_I such that $\deg(p_{S(I)}) \equiv 1$ verified by the BLR test [4] and this $p_{S(I)}$ is called maxterm equation of t_I.

Example 1. Let $p(k_1, k_2, k_3, v_1, v_2, v_3) = v_2 v_3 k_1 + v_2 v_3 k_2 + v_1 v_2 v_3 + v_1 k_2 k_3 + k_2 k_3 + v_3 + k_1 + 1$ be a polynomial of degree 3 in 3 secret variables and 3 public variables. Let $I = \{2, 3\}$ be an index subset of the public variables. We can represent p as $p(k_1, k_2, k_3, v_1, v_2, v_3) = v_2 v_3(k_1 + k_2 + v_1) + (v_1 k_2 k_3 + k_2 k_3 + v_3 + k_1 + 1)$, where $t_I = v_2 v_3$, $p_{S(I)} = k_1 + k_2 + v_1$ and $q(k_1, k_2, k_3, v_1, v_2, v_3) = v_1 k_2 k_3 + k_2 k_3 + v_3 + k_1 + 1$.

Let d be the size of I, then a *cube* on I is defined as a set C_I of 2^d vectors that cover all the possible combinations of t_I and leave all the other variables undetermined. Any vector $\tau \in C_I$ defines a new derived polynomial $p_{|\tau}$ with $n - d$ variables. Summing these derived polynomials over all the 2^d possible vectors in C_I results in exactly $p_{S(I)}$ (cf. Theorem 1, [8]). For p and I defined in Example 1, we have $C_I = \{\tau_1, \tau_2, \tau_3, \tau_4\}$, where $\tau_1 = [k_1, k_2, k_3, v_1, 0, 0]$, $\tau_2 = [k_1, k_2, k_3, v_1, 0, 1]$, $\tau_3 = [k_1, k_2, k_3, v_1, 1, 0]$, and $\tau_4 = [k_1, k_2, k_3, v_1, 1, 1]$. It is easy to verify that $p_{|\tau_1} + p_{|\tau_2} + p_{|\tau_3} + p_{|\tau_4} = k_1 + k_2 + v_1 = p_{S(I)}$. Here $p_{S(I)}$ is called the maxterm equation of t_I. In the off-line phase, the attacker tries to find as many maxterms and their corresponding maxterm equations as possible.

In the on-line phase, the secret key is fixed. The attacker chooses plaintexts $\tau \in C_I$ and obtains the evaluation of p at τ. By summing up $p_{|\tau_i}$ for all the 2^d vectors in C_I, he obtains $p_{S(I)}$, a linear equation in k_i. The attacker repeats this process for all the maxterms found in the off-line phase, and obtains a group of equations, which he can solve to recover the key.

2.2 Error-Tolerant Side-Channel Cube Attack(ET-SCCA)

At CHES 2013, Li *et. al.* [19] proposed a new model for ET-SCCA, which can retrieve the key when *all* the leaked bits are noisy. The leaked data observed is regarded as the received channel output of some linear code transmitted through a binary symmetric channel (BSC). The problem of recovering the n secret key bits in L linear equations can be considered as the problem of decoding a binary linear $[L, n]$ code with L being the code length and n the dimension as follows.

$$\begin{cases} l_1 : a_1^1 k_1 + a_1^2 k_2 + ... + a_1^n k_n = b_1 \\ l_2 : a_2^1 k_1 + a_2^2 k_2 + ... + a_2^n k_n = b_2 \\ \quad \vdots \\ l_L : a_L^1 k_1 + a_L^2 k_2 + ... + a_L^n k_n = b_L \end{cases} \quad (1)$$

where $a_i^j \in \{0,1\}$ $(1 \le i \le L, 1 \le j \le n)$ denotes the coefficient. Note that $b_i \in \{0,1\}$ is obtained by summing up the evaluation of the maxterm equation over the ith cube C_i, namely, $b_i = \sum_{\tau \in C_i} p_{|\tau}$. The value of $p_{|\tau}$ is obtained via measurements. Ideally, the measurement is error-free and the attacker obtains the correct sequence $B = [b_1, b_2, ..., b_L]$. In reality, however, the attacker is likely to observe a different sequence $Z = z_1, z_2, ..., z_L$ due to the measurement errors.

Denote q as the probability that a bit may flip in each measurement and assume that $q < 1/2$, then $1 - q = 1/2 + \mu$ is the probability of an accurate measurement and $\mu = 0$ means a random guess. Since $b_i = \sum_{\tau \in C_i} p_{|\tau}$, and C_i has $t = 2^{\bar{d}}$ elements ($\bar{d}$ is the average size of cubes), and each measurement can be treated as an independent event, according to the piling-up lemma [14], $Pr\{b_i = z_i\} \triangleq 1 - p = \frac{1}{2} + 2^{t-1}\mu^t$. Thus, each z_i can be regarded as the output of a BSC with $p = 1/2 - \varepsilon$ ($\varepsilon = 2^{t-1}\mu^t$) being the crossover probability. Therefore, the key recovery problem is converted to decoding a $[L, n]$ linear code. Maximum likelihood decoding (ML-decoding, see Appendix C) is used and they derive the error-tolerant bound in Lemma 1.

Lemma 1. *To ensure 50 % success probability of decoding a $[L, n]$ code to retrieve the key, the error probability q of each measurement should satisfy $q \le \frac{1}{2}$.* $(1 - (\frac{0.35 \cdot n}{L})^{\frac{1}{2 \cdot t}} \cdot 2^{\frac{1}{t}})$, where $t = 2^{\bar{d}}$ denotes the number of summations to evaluate each linear equation.

The simulation results on PRESENT-80 show that given about $2^{10.2}$ measurements, each with an error probability $q = 19.4\%$, it achieves 50.1 % of success rate for the key recovery. However, the error-tolerant rate is still very low.

3 A New ET-SCCA with Higher Error-Tolerant Rate

3.1 Polynomial Approximation

The main target is to remove several secret variables while keeping the number of maxterm equations reduced as few as possible. In this way, the code dimension

n can be reduced while keeping the code length L reduced as little as possible. However, removing secret variables might be a challenging task as previous studies on Trivium [8], Serpent [9,10], KATAN [12], LBlock [20] and PRESENT [19] show that most of the maxterm equations have a low density. Removing secret variables will probably lead to the reduction of the maxterm equations. We propose two basic strategies of removing key variables as follows.

Lower Reduction Factor. The removed secret variables should not be those that solely exist in the maxterm equations and should be those that exist in the maxterm equations with multiple secret variables. (e.g., suppose we have derived 2 maxterm equations, one is $k_1 + k_2$ and the other is $k_3 + 1$, then the removed secret variables should contain k_1 or k_2, but not k_3, since removing k_3 will lead to the second maxterm equation become a trivial one.) Note that this selection process can be finished in the off-line phase, since all the maxterm equations are available. Suppose the number of removed key variables is n' and the number of maxterm equations reduced is $\gamma \cdot n'$, where γ is the reduction factor. $\gamma = 0$ means the removed secret variables will not influence the number of the maxterm equations and the value of γ depends on the choice of removed secret variables. The problem now convert to decoding a $[L - \gamma \cdot n', n - n']$ code.

Higher Approximation Rate. Suppose a polynomial p containing n secret variables and m public variables, the removed key set is $R = \{k_{i_1}, k_{i_2}, ..., k_{i_r}\}$, where $1 \leq i_q \leq n$, $1 \leq q \leq r$. The approximation rate between p and $\tilde{p}$ after removing variables in R is defined as $\Lambda(p, \tilde{p})|_R = e/2^{m+n} = 1/2 + \sigma$, where e is the number of the equal evaluations and σ $(0 < \sigma < 1/2)$[1] is the bias factor. In reality, there might be more than one leakage function. Suppose $P = \{p_1, p_2, ..., p_u\}$ are all the associated leakage functions and the corresponding removed key variable sets are $R_1, R_2, ..., R_u$ respectively[2], the average approximation rate is defined as

$$\bar{\Lambda} = \frac{\sum_{t=1}^{u} \Lambda(p_t, \tilde{p}_t)|_{R_t}}{u}. \tag{2}$$

The candidate key variables to remove should be those with maximum $\bar{\Lambda}$. Note that this process can also be finished in the off-line phase, i.e., all the removed key variables are set to 0 for the evaluation of $\tilde{p}$.

3.2 A New Variant of Cube Attack

The main idea is to increase the number of maxterm equations by choosing the static public variables, which are those variables that are not part of the cube variables. In the traditional applications of cube attacks and cube testers [1,2,8], these static variables will be set to constant values. We find that multiple

---

[1] For the case of $1/2 - \sigma$, convert it to $1/2 + \sigma$ by adding 1 to the evaluation of $\tilde{p}$
[2] $R_1 = \{k_{i_1}^{[1]}, k_{i_2}^{[1]}, ..., k_{i_{r_1}}^{[1]}\}$, $R_2 = \{k_{i_1}^{[2]}, k_{i_2}^{[2]}, ..., k_{i_{r_2}}^{[2]}\},...$

maxterm equations can be derived for each maxterm by choosing static variables. In Example 1 of Sect. 2, the maxterm equation for the maxterm $t_I = v_2 v_3$ is $p_{S(I)} = k_1 + k_2 + v_1$, where v_1 is a static variable. If we set $v_1 = 0$, then we can derive a maxterm equation $k_1 + k_2$. Similarly, if we set $v_1 = 1$, another variant maxterm equation $k_1 + k_2 + 1$ can be derived. Then, we have the following theorem (please refer to Appendix A for the details of the proof).

Theorem 1. *For the maxterm equation $p_{S(I)}$ of maxterm t_I, the number of variant maxterm equations which can be derived is at most 2^{m-d} and each can be classified into the following two types.*

1. *$p_{S(I)}^* + C$, where $C \in \{0,1\}$. (Type I)*
2. *$p_{S(I)}^* + C_0 + C_1 k_{n_1} + C_2 k_{n_2} + ... + C_r k_{n_r}$, where $C_i \in \{0,1\}$, $\bigvee_{i=1}^r C_i \neq 0$ and C_0 represents a constant term. (Type II)*

$p_{S(I)}^*$ is the equation of $p_{S(I)}$ when we set all static variables to 0.

The previous Example 1 of Sect. 2 describes the scenario of Type I. The following example shows the scenario of Type II.

Example 2. Suppose a polynomial $p = v_1 v_2 k_1 + v_1 v_2 v_3 k_2 + v_3 v_4 k_1 k_2 k_3 + v_1 v_2 = v_1 v_2 \cdot (k_1 + v_3 k_2 + 1) + v_3 v_4 k_1 k_2 k_3$, then $t_I = v_1 v_2$ with $I = \{1,2\}$ is a maxterm, $p_{S(I)} = k_1 + v_3 k_2 + 1$ is the maxterm equation and $p_{S(I)}^* = k_1 + 1$.

The static variables is thus $\{v_3, v_4\}$. If we choose $v_3 = 1$, then a variant of maxterm equation appears as $k_1 + k_2 + 1 = p_{S(I)}^* + k_2$, which fits into Type II.

In the traditional cube attack, most of these variant maxterm equations are trivial and make no contribution to the key recovery. However, in our model, these variants can be treated as redundant information, which are beneficial to the decoding algorithm. For a linear code considering polynomial approximation $[L - \gamma \cdot n', n - n']$, the total number of maxterm equations can be increased by a factor of E. Now the problem of key recovery is converted to decoding a $[(L - \gamma \cdot n') \cdot E, n - n']$ linear code, where $1 \leq E \leq 2^{m-d}$.

4   Error Probability Evaluation

By utilizing ML-decoding, we derive a new bound for error-tolerant rate in Corollary 1 (Please refer to Appendix B for the details of the proof).

Corollary 1. *To ensure 50 % success probability of decoding a $[L^*, n^*]$ code to retrieve the key, the error probability q of each measurement should satisfy*

$$q \leq \frac{1}{2} \cdot \left(1 - \left(\frac{0.35 \cdot n^*}{L^*}\right)^{\frac{1}{2 \cdot t}} \cdot 2^{\frac{1}{t}}\right), \tag{3}$$

where $n^* = n - n'$, $L^* = (L - \gamma \cdot n') \cdot E$, $1 \leq E \leq 2^{m-\bar{d}}$ and $t = 2^{\bar{d}}$.

If $n' = 0$ and $E = 1$, it reduces to the original ET-SCCA. The cost for the polynomial approximation is that the removed key variables will add more noise to those associated maxterm equations, but this kind of noise can be ignored if we only remove a few key variables and keep the number of maxterm equations influenced as little as possible. Moreover, the rest of the n' key variables removed can be exhaustively searched. The cost for choosing static public variables is that the number of measurements will increase accordingly.

Suppose $L = 1000$, then the error probabilities under different number of removed key variables $n' = 0, 10, 30$ with $\gamma = 1$ and $E = 1$ are depicted in Fig. 1.

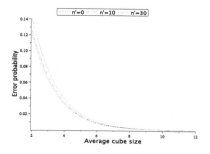
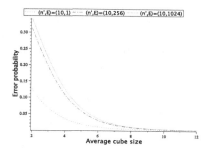

Fig. 1. Error probability q as a function of $\bar{d}$ (Given $n' = 0, 10, 30$, $\gamma = 1$, $E = 1$ and $L = 1000$)

Fig. 2. Error probability q as a function of $\bar{d}$ (Given $n' = 10$, $\gamma = 1$, $E = 1, 256, 1024$ and $L = 1000$)

Figure 1 shows that the error probability gradually increased with the growth of n'. By applying the new variant of cube attack with $E = 1, 256$ and 1024, shown in Fig. 2, which demonstrates that the error probability increased with the growth of E. Similar results can also be obtained if we choose other size of L. These results demonstrate that the error probability can be further improved under the same noise channel and utilizing the same decoding algorithm.

5   Simulations on PRESENT

To compare our model with the original ET-SCCA in [19], we will apply the model to PRESENT-80, a standardized round based lightweight block cipher [5]. We assume PRESENT cipher is implemented on a 8-bit processor. Under Hamming weight leakage model, the attacker exploits the Hamming weight leakage containing noise when the state variables are loaded from memory to ALU.

5.1   Off-Line Phase

We enumerate all the small candidate cubes, each size is at most 2. The time complexity is thus $P = \binom{64}{1} \cdot 2 + \binom{64}{2} \cdot 2^2 = 2^{13}$ encryptions. The leakage function is the LSB (least significant bit) of the Hamming weight of the state byte after the *first* round. There are altogether 8 state bytes $byte_1, byte_2, ..., byte_8$, corresponding to 8 leakage functions. We can derive 304 maxterm equations containing 64 key variables (Appendix D) and the average cube size is $\bar{d} = 1.9$. Compared with the special cube searching strategy in the original ET-SCCA [19], the process of off-line phase in our model requires no knowledge of the internal round function.

Now we need to figure out which key variables should be removed according to the distribution of key variables in all the maxterm equations. For the leakage function of $byte_1$ (or $byte_2$), there are only 16 maxterm equations (Tables 9, 10 Appendix D), each of which only contains a single key variable, it is thus hard to decide which key variables should be removed. For the leakage function of $byte_3$, considering those maxterm equations containing $\{k_{17}, k_{18}, k_{19}, k_{20}\}$, removing k_{17} will lead to 3 maxterm equations (corresponding to maxterm $\{2, 3\}$, $\{2, 4\}$ and $\{3, 4\}$ respectively) become trivial and removing k_{20} will lead to 2 maxterm equations (corresponding to maxterm $\{1, 2\}$ and $\{1, 3\}$ respectively) reduced for the leakage function of $byte_5$. Removing k_{18} or k_{19} is a good choice since it only lead to one maxterm equation (corresponding to maxterm $\{3\}$ or $\{2\}$) reduced for the leakage function of $byte_1$ and it will not affect other state bytes. We choose k_{18} as a representative variable. Similarly, we can also derive other representative variables $k_{22}, k_{26}, k_{30}, k_{34}, k_{38}, k_{42}, k_{46}$ from other ranges for the leakage function of $byte_3$. All the candidate key variables are summarized in Table 2.

Table 2. Candidate key variables for removing

State byte	Candidate key variables
$byte_3, byte_7$	$k_{18}, k_{22}, k_{26}, k_{30}, k_{34}, k_{38}, k_{42}, k_{46}$
$byte_4, byte_8$	$k_{50}, k_{54}, k_{58}, k_{62}, k_{66}, k_{70}, k_{74}, k_{78}$

These variables will not lead to any reduction of the maxterm equations for both $byte_5$ and $byte_6$, each only lead to one reduction of the maxterm equations for $byte_1$ or $byte_2$. Therefore, $n' \in [0, 16]$ and the reduction factor $\gamma = 1$.

5.2   Polynomial Approximation for PRESENT-80

Now we need to select the optimal combinations of these candidates that can maximize the approximation rate. We enumerate all the possible combinations and calculate the average approximation rate according to equation (2). The optimal combinations for each $|R| \in [1, 8]$ are listed in the following table.

From Table 3, we can see that the value of σ decreases with the growth of $|R|$. We will not consider those candidate key variables set with $|R| > 8$, since the bias σ becomes trivial and more removed key variables will add more noise to the evaluations of those associated maxterm equations.

Table 3. Optimal combinations

R	σ	R	σ
$\{k_{30}\}$	0.267	$\{k_{18}, k_{22}, k_{26}, k_{42}, k_{50}\}$	0.162
$\{k_{26}, k_{74}\}$	0.263	$\{k_{42}, k_{50}, k_{54}, k_{58}, k_{66}, k_{74}\}$	0.154
$\{k_{34}, k_{38}, k_{78}\}$	0.207	$\{k_{22}, k_{50}, k_{54}, k_{58}, k_{62}, k_{70}, k_{74}\}$	0.152
$\{k_{18}, k_{66}, k_{70}, k_{78}\}$	0.176	$\{k_{18}, k_{26}, k_{30}, k_{34}, k_{38}, k_{42}, k_{46}, k_{54}\}$	0.148

5.3   On-Line Phase

From the previous analysis, we know that $n' = |R|$ and $\gamma = 1$. The key recovery problem is now equivalent to decoding a $[L - n', n - n']$ linear code.

For the sake of comparison, grouping strategy and list decoding are also utilized in the model as in [19]. More precisely, all the key variables are divided into 4 groups G_1, G_2, G_3 and G_4 with several overlapping bits. ML-decoding is applied in each group as a direct application of the ML-decoding has a time complexity of $2^{64-n'}$. To increase the success probability, we save a candidate list of T closest solutions for each group. The configurations with $R = \{k_{18}, k_{66}, k_{70}, k_{78}\}$ are listed in the following Table 4.

Table 4. Groups with $R = \{k_{18}, k_{66}, k_{70}, k_{78}\}$

Group	$[L, n]$	Key bits	Overlapping bits
G_1	$[113, 23]$	$[k_{17}, k_{19}, ..., k_{40}]$	8 with G_2
G_2	$[114, 24]$	$[k_{33}, k_{34}, ..., k_{56}]$	8 with G_1, 8 with G_3
G_3	$[94, 19]$	$[k_{49}, k_{50}, ..., k_{68}]$	8 with G_2, 8 with G_4
G_4	$[92, 17]$	$[k_{61}, k_{62}, ..., k_{80}]$	8 with G_3

We have simulated the decoding algorithm for 100 runs with $T = 200$. For each run, we randomly generate a key and construct the linear code in each group. The noise was simulated by a random binary number generator according to the crossover probability p (e.g., suppose $k_0 = 1, k_1 = 0$ and there is a maxterm equation $1 + k_0 + k_1 = 0$, the value 0 will flip to 1 with probability p and remain unchanged with probability $1 - p$). We have conducted the simulation for 10 times and the average number of successful decoding out of a batch of 100 runs are recorded. The comparison results are shown in Fig. 3, which indicates that under the same success probability, decoding with removing 4 key variables can tolerate more noise. The comparisons with various size of R are summarized in Fig. 4, which demonstrates that under the same success probability, more noise can be tolerated with the growth of $|R|$.

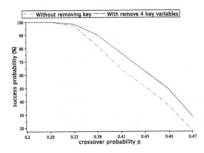
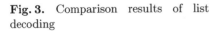

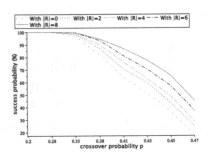

Fig. 3. Comparison results of list decoding

Fig. 4. Comparison results of list decoding with various size of R

p is the crossover probability for each evaluation of the maxterm equation. Since $p = 1/2 - 2^{t-1}\mu^t$, $t = 2^d$ and $1 - q = 1/2 + \mu$, the error probability q for each measurement are listed in Table 5.

Table 5. q with various R under the decoding success probability of 50 %

$|R|$	p	q	$|R|$	p	q
0	0.428	0.191	6	0.459	0.232
2	0.439	0.204	8	0.467	0.247
4	0.447	0.214			

The whole attack contains two phases, the first phase is the decoding in each group. The results are the candidate lists. Denote t_i as the time complexity of decoding in group G_i, η as the number of the groups and n_i as the code dimension in G_i, the time complexity is thus $\sum_{i=1}^{\eta} t_i$, where $t_i = 2^{n_i}$ key trials. The second is the verification phase. Since each candidate only contains $64 - n'$ master key variables, we need to verify it by combining the removed n' key variables and the rest of 16 master key bits, using the known plaintext/ciphertext pairs. The time complexity is thus $V(T) = T^{\eta} \cdot 2^{n'+16}/2^r$ encryptions[3], where 2^r is the reduction factor related to the number of overlapping bits. Therefore, the attack complexity is bounded by $max\{\ \sum_{i=1}^{\eta} t_i,\ T^{\eta} \cdot 2^{n'+16}/2^r\ \}$. The attack results with various R on PRESENT are given in Table 6.

From Table 6, we can see that the model can tolerate more errors if more candidate key variables are removed. The growth of the key variables removed will lead to the increase of the time complexity in the verification phase and will add more noise due to the polynomial approximation. Therefore, R should be carefully chosen so that the error-tolerant rate can be optimal while keeping the time complexity practical.

---

[3] The verification complexity in [19] is actually incorrect. When 16 bits of the key are missing, they can be exhaustively searched (in time complexity of 2^{16}).

Table 6. Simulations by utilizing polynomial approximation

Size of R	Time	Data (measurements)	Reduction factor r	Error probability
0	$2^{25.1}$	1152	24	19.1 %
2	$2^{24.7}$	1148	24	20.4 %
4	$2^{27.6}$	1144	23	21.4 %
6	$2^{31.6}$	1140	21	23.2 %
8	$2^{31.6}$	1136	23	24.7 %

The decoding success probability is about 50 %.

5.4 Applying the New Variant of Cube Attack to the On-Line Phase

In this section, we will show that the error-tolerant rate can be further improved by choosing static public variables. The key recovery problem is now converted to decoding a $[(L - \gamma \cdot n') \cdot E, n - n']$ linear code, where $1 \leq E \leq 2^{m-\bar{d}}$ represents the size of the redundant maxterm equations.

For PRESENT-80, $m = 64$ and $\bar{d} \approx 2$, then $1 \leq E \leq 2^{62}$. All the complexity remain unaltered except for the data complexity. Deriving more redundant maxterm equations will lead to higher measurements. Suppose $n' = 4$, the simulation results under various E are summarized in Table 7.

Table 7. Simulations by choosing static variables

E	Data (measurements)	Total size of L	Error probability
1	$2^{10.2}$	304	21.4 %
2^6	$2^{16.2}$	19456	29.5 %
2^8	$2^{18.2}$	77824	34.1 %
2^{10}	$2^{20.2}$	311296	38.6 %
2^{11}	$2^{21.2}$	622592	40.5 %

The decoding success probability is about 50 %.
The time complexity is $2^{27.6}$ encryptions (verification phase).
$R = \{k_{18}, k_{66}, k_{70}, k_{78}\}$.

From Table 7, it is shown that the error-tolerant rate increased with the growth of the E. It can be increased closely to 50 % on the condition that we can obtain more measurements, which means that our model can still work even if the measurement contains heavy noise.

6   Comparison and Discussions

6.1   ET-SCCA Comparisons

Compared with the original ET-SCCA [19], the error-tolerant level of the new ET-SCCA is improved significantly by utilizing the polynomial approximation and applying the new variant of cube attack. It is more flexible, since the attacker can choose appropriate size of E according to his ability (e.g., accuracy of the measurements). The comparison results are summarized in Table 8.

Table 8. Comparison between the original ET-SCCA and our model

Time complexity	Data (measurements)	Error probability	Reference
$2^{34.6}$	$2^{10.2}$	19.1 %	[19]
$2^{31.6}$	$2^{10.1}$	24.7 %	this paper
$2^{27.6}$	$2^{21.2}$	40.5 %	this paper

The success probability is about 50 % for both models.

6.2   Motivation of the New Variant of Cube Attack

The motivation of the new variant of cube attack comes from the dynamic cube attack [7]. The difference is that dynamic cube attacks transform some of the static public variables to dynamic variables and each one of these dynamic variables is assigned a function that depends on some of the cube variables and some expressions of secret variables. These functions are carefully chosen usually to zero some state bits to simplify the expression and amplify the bias of the cube tester. It requires a more complex analysis of the internal structure of the cipher. Moveover, the main purpose of dynamic cube attack is to improve the standard cube testers and construct a more efficient distinguisher, then filtering right key using this distinguisher. While the new variant (also mentioned in [15]) applied in this paper is to derive more redundant maxterm equations to facilitate the decoding process, which requires no knowledge of the round function.

6.3   About the Definition of Maxterm Equation

Recall the formal definition of maxterm equation in [8]. The maxterm equation $p_{S(I)}$ of a maxterm t_I satisfies $\deg(p_{S(I)}) \equiv 1$, which holds whenever static variables are 0s or 1s. In most applications, e.g., Trivium, all the maxterm equations are derived when they are set to 0s. However, some researchers [3] verified all the maxterm equations derived from Trivium [8] by chosen the static variables. Among 1000 runs, each of which a random IV was chosen, almost all the maxterm equations pass through a linear test with probability of about 50 %, i.e., $\deg(p_{S(I)}) \equiv 1$ cannot hold for half of the runs. All these will have a negligible influence to our simulation on PRESENT, since the number of variant maxterm equations (i.e., E) is low and we can always get enough maxterm equations by choosing the static variables.

6.4   Attacking Implementations with Masking

Masking is a widely used countermeasure against side-channel attacks. The principle is to randomly split every sensitive variable (e.g., variables involving secret keys) occurring in the computation into $d+1$ shares, where d is called the masking order and plays the role of a security parameter. Suppose a state byte S is split into $d+1$ random shares $S_0, S_1, ..., S_d$, satisfying $S_0 \oplus S_1 \oplus ... \oplus S_d = S$ and the computations are on the masked data. Suppose the attacker observe the value of each share containing noise as $S_0 \oplus e_0$, $S_1 \oplus e_1, ..., S_d \oplus e_d$, where e_i is the observation noise. By summing all these values up, $S \oplus \sum_{i=0}^{d} e_i$ can be derived. Compared with an implementation without masking, the only influence to the ET-SCCA is that the observation noise for a masking implementation is relatively higher, which is exponentially increased with the growth of the masking order d (according to piling-up lemma). However, in reality, d is small since almost all the current masking schemes suffered from the efficiency problems when d becomes bigger. Therefore, we believe that our model can still be available to a implementation with masking.

7   Conclusion and Open Problems

In this paper, we have revisited the error-tolerant side-channel cube attack and proposed a more robust model. By appropriately utilizing the polynomial approximation technique, the error-tolerant rate can be improved compared to the original ET-SCCA. We also presented an efficient way of finding the key variables that should be removed, by defining the average approximation rate. Moreover, a new variant of cube attack was proposed inspired by the idea of dynamic cube attack. The error-tolerant rate has been refined. Both theoretical analysis and simulation results indicated that the improved model is more flexible, exploiting measurements with heavy noise interference, which solves one of the open problems listed in [19]. The simulation results on PRESENT show that given about $2^{21.2}$ measurements, each with an error probability of 40.5 %, it achieves 50 % success probability of the key recovery. The error-tolerant level can be enhanced further if the attacker can obtain more measurements. Hence, we believe these results have both a theoretical and practical relevance.

A   Proof of Theorem 1

Proof. The fact that we can derive at most 2^{m-d} variant maxterm equations is obvious, since the number of static public variables is $m - d$. The master multivariate polynomial p can be represented as $p(k_1, ..., k_n, v_1, ..., v_m) = t_I \cdot p_{S(I)} + q(k_1, ..., k_n, v_1, ..., v_m)$. Since t_I is a maxterm, then the degree of $p_{S(I)}$ in secret variables is $\deg(p_{S(I)}) \equiv 1$. Then $p_{S(I)}$ can be represented as $p_{S(I)} = C + C_0 + C_1 k_1 + C_2 k_2 + ... + C_n k_n$, where $C \in \{0, 1\}$ is a constant and C_i, $0 \le i \le n$ contains only static public variables.

Since $\deg(p_{S(I)}) \equiv 1$, then $\bigvee_{i=1}^{n} C_i \neq 0$, which means that there is at least one key variables k_i with $C_i = 1$. Suppose the set of key variables with all their coefficients equal to 1 is $K = \{k_{i_1}, k_{i_2}, ..., k_{i_u}\}$ and each $C_{i_t} = 1, 1 \leq t \leq u$. By setting all the static public variables to 0, then $p_{S(I)}^{*} = C + \sum_{j=i_1}^{i_u} k_j$. The variant maxterm equations by choosing those static public variables can thus be represented as

$$p_{S(I)}' = p_{S(I)}^{*} + C_0 + \sum_{j \notin \{i_1, i_2, ..., i_u\}} C_j k_j \tag{4}$$

Then, if the chosen static public variables make all the $C_j = 0$, where $1 \leq j \leq n$ and $j \notin \{i_1, i_2, ..., i_u\}$, then equation (4) is the Type I variant. Inversely, if the chosen static public variables make that there is at least one $C_j \neq 0$, where $1 \leq j \leq n$ and $j \notin \{i_1, i_2, ..., i_u\}$, the equation (4) is the Type II variant. □

B   Proof of Corollary 1

Proof. The key recovery problem is now converted to decoding a $[(L - \gamma \cdot n') \cdot E, n - n']$ linear code, where $1 \leq E \leq 2^{m-\bar{d}}$. Suppose $L^{*} = (L - \gamma \cdot n') \cdot E$ and $n^{*} = n - n'$. Recall that the error probability p for each evaluation of the maxterm equation is $p = 1/2 - \varepsilon$. The capacity of BSC can be approximated as $C(p) \approx \varepsilon^2 \cdot 2/(ln(2))$. Simulations [16] show that the critical length $L^{*} \geq 0.35 \cdot n^{*} \cdot \varepsilon^{-2}$ provides the probability of successful decoding close to $1/2$. Thus we get $\varepsilon \geq \sqrt{\frac{0.35 \cdot n^{*}}{L^{*}}}$. Since $\varepsilon = 2^{t-1} \mu^t$ holds, then we can derive $\mu \geq (\frac{0.35 \cdot n^{*}}{L^{*}})^{\frac{1}{2 \cdot t}} \cdot 2^{\frac{1}{t}-1}$. From $q = 1/2 - \mu$, we have $q \leq \frac{1}{2} \cdot (1 - (\frac{0.35 \cdot n^{*}}{L^{*}})^{\frac{1}{2 \cdot t}} \cdot 2^{\frac{1}{t}})$. □

C   ML-decoding

Siegenthaler [16] firstly proposed the use of ML-decoding in cryptanalysis of a stream cipher by exhaustively searching through all the codewords of $[L, n]$ code. The complexity is about $O(2^n \cdot n/C(p))$. Let $A = (a_i^j)_{L \times n}$ $(1 \leq i \leq L, 1 \leq j \leq n)$ be the generator matrix of (1) and A_i denote the i-th row vector of A. The aim of the decoding is to find the closet codeword $(b_1, b_2, ..., b_L)$ to the received vector $(z_1, z_2, ..., z_L)$, and decode the key variables $\mathbf{k} = (k_1, k_2, ..., k_n)$ such that $b_i = \mathbf{k} \cdot A_i^T$, where T denotes the matrix transpose, i.e., find such $\mathbf{k}$ that minimizes $D(\mathbf{k}) = \sum_{i=1}^{L} (z_i \oplus b_i)$. It is known that ML-decoding is optimal since it has the smallest error probability among all decoding algorithms.

D   Maxterm Equations for All the 8 Leakage Functions

Table 9. Maxterms and maxterm equations

Table 10. Maxterms and maxterm equations

Leakage function of $byte_1$

Cube Indexes	Maxterm equations	Cube indexes	Maxterm equations
{2}	k_{19}	{3}	$1 + k_{18}$
{6}	k_{23}	{7}	$1 + k_{22}$
{10}	k_{27}	{11}	$1 + k_{26}$
{14}	k_{31}	{15}	$1 + k_{30}$
{18}	k_{35}	{19}	$1 + k_{34}$
{22}	k_{39}	{23}	$1 + k_{38}$
{26}	k_{43}	{27}	$1 + k_{42}$
{30}	k_{47}	{31}	$1 + k_{46}$

Leakage function of $byte_2$

Cube Indexes	Maxterm equations	Cube indexes	Maxterm equations
{34}	k_{51}	{35}	$1 + k_{50}$
{38}	k_{55}	{39}	$1 + k_{54}$
{42}	k_{59}	{43}	$1 + k_{58}$
{46}	k_{63}	{47}	$1 + k_{62}$
{50}	k_{67}	{51}	$1 + k_{66}$
{54}	k_{71}	{55}	$1 + k_{70}$
{58}	k_{75}	{59}	$1 + k_{74}$
{62}	k_{79}	{63}	$1 + k_{78}$

Leakage function of $byte_3$

Cube Indexes	Maxterm equations	Cube indexes	Maxterm equations
{1,2}	$k_{19} + k_{20}$	{1,3}	$k_{18} + k_{20}$
{1,4}	$k_{18} + k_{19}$	{2,3}	k_{17}
{2,4}	$1 + k_{17}$	{3,4}	$1 + k_{17}$
{5,6}	$k_{23} + k_{24}$	{5,7}	$k_{22} + k_{24}$
{5,8}	$k_{22} + k_{23}$	{6,7}	k_{21}
{6,8}	$1 + k_{21}$	{7,8}	$1 + k_{21}$
{9,10}	$k_{27} + k_{28}$	{9,11}	$k_{26} + k_{28}$
{9,12}	$k_{26} + k_{27}$	{10,11}	k_{25}
{10,12}	$1 + k_{25}$	{11,12}	$1 + k_{25}$
{13,14}	$k_{31} + k_{32}$	{13,15}	$k_{30} + k_{32}$
{13,16}	$k_{30} + k_{31}$	{14,15}	k_{29}
{14,16}	$1 + k_{29}$	{15,16}	$1 + k_{29}$
{17,18}	$k_{35} + k_{36}$	{17,19}	$k_{34} + k_{36}$
{17,20}	$k_{34} + k_{35}$	{18,19}	k_{33}
{18,20}	$1 + k_{33}$	{19,20}	$1 + k_{33}$
{21,22}	$k_{39} + k_{40}$	{21,23}	$k_{38} + k_{40}$
{21,24}	$k_{38} + k_{39}$	{22,23}	k_{37}
{22,24}	$1 + k_{37}$	{23,24}	$1 + k_{37}$
{25,26}	$k_{43} + k_{44}$	{25,27}	$k_{42} + k_{44}$
{25,28}	$k_{42} + k_{43}$	{26,27}	k_{41}
{26,28}	$1 + k_{41}$	{27,28}	$1 + k_{41}$
{29,30}	$k_{47} + k_{48}$	{29,31}	$k_{46} + k_{48}$
{29,32}	$k_{46} + k_{47}$	{30,31}	k_{45}
{30,32}	$1 + k_{45}$	{31,32}	$1 + k_{45}$

Leakage function of $byte_4$

Cube Indexes	Maxterm equations	Cube indexes	Maxterm equations
{33,34}	$k_{51} + k_{52}$	{33,35}	$k_{50} + k_{52}$
{33,36}	$k_{50} + k_{51}$	{34,35}	k_{49}
{34,36}	$1 + k_{49}$	{35,36}	$1 + k_{49}$
{37,38}	$k_{55} + k_{56}$	{37,39}	$k_{54} + k_{56}$
{37,40}	$k_{54} + k_{55}$	{38,39}	k_{53}
{38,40}	$1 + k_{53}$	{39,40}	$1 + k_{53}$
{41,42}	$k_{59} + k_{60}$	{41,43}	$k_{58} + k_{60}$
{41,44}	$k_{58} + k_{59}$	{42,43}	k_{57}
{42,44}	$1 + k_{57}$	{43,44}	$1 + k_{57}$
{45,46}	$k_{63} + k_{64}$	{45,47}	$k_{62} + k_{64}$
{45,48}	$k_{62} + k_{63}$	{46,47}	k_{61}
{46,48}	$1 + k_{61}$	{47,48}	$1 + k_{61}$
{49,50}	$k_{67} + k_{68}$	{49,51}	$k_{66} + k_{68}$
{49,52}	$k_{66} + k_{67}$	{50,51}	k_{65}
{50,52}	$1 + k_{65}$	{51,52}	$1 + k_{65}$
{53,54}	$k_{71} + k_{72}$	{53,55}	$k_{70} + k_{72}$
{53,56}	$k_{70} + k_{71}$	{54,55}	k_{69}
{54,56}	$1 + k_{69}$	{55,56}	$1 + k_{69}$
{57,58}	$k_{75} + k_{76}$	{57,59}	$k_{74} + k_{76}$
{57,60}	$k_{74} + k_{75}$	{58,59}	k_{73}
{58,60}	$1 + k_{73}$	{59,60}	$1 + k_{73}$
{61,62}	$k_{79} + k_{80}$	{61,63}	$k_{78} + k_{80}$
{61,64}	$k_{78} + k_{79}$	{62,63}	k_{77}
{62,64}	$1 + k_{77}$	{63,64}	$1 + k_{77}$

Leakage function of $byte_5$

Cube Indexes	Maxterm equations	Cube indexes	Maxterm equations
{1,2}	$1 + k_{20}$	{1,3}	k_{20}
{1,4}	$1 + k_{18} + k_{19}$	{2,4}	$1 + k_{17}$
{3,4}	k_{17}	{5,6}	$1 + k_{24}$
{5,7}	k_{24}	{5,8}	$1 + k_{22} + k_{23}$
{6,8}	$1 + k_{21}$	{7,8}	k_{21}
{9,10}	$1 + k_{28}$	{9,11}	k_{28}
{9,12}	$1 + k_{26} + k_{27}$	{10,12}	$1 + k_{25}$
{11,12}	k_{25}	{13,14}	$1 + k_{32}$
{13,15}	k_{32}	{13,16}	$1 + k_{30} + k_{31}$
{14,16}	$1 + k_{29}$	{15,16}	k_{29}
{17,18}	$1 + k_{36}$	{17,19}	k_{36}
{17,20}	$1 + k_{34} + k_{35}$	{18,20}	$1 + k_{33}$
{19,20}	k_{33}	{21,22}	$1 + k_{40}$
{21,23}	k_{40}	{21,24}	$1 + k_{38} + k_{39}$
{22,24}	$1 + k_{37}$	{23,24}	k_{37}
{25,26}	$1 + k_{44}$	{25,27}	k_{44}
{25,28}	$1 + k_{42} + k_{43}$	{26,28}	$1 + k_{41}$
{27,28}	k_{41}	{29,30}	$1 + k_{48}$
{29,31}	k_{48}	{29,32}	$1 + k_{46} + k_{47}$
{30,32}	$1 + k_{45}$	{31,32}	k_{45}

Leakage function of $byte_6$

Cube Indexes	Maxterm equations	Cube indexes	Maxterm equations
{33,34}	$1 + k_{52}$	{33,35}	k_{52}
{33,36}	$1 + k_{50} + k_{51}$	{34,36}	$1 + k_{49}$
{35,36}	k_{49}	{37,38}	$1 + k_{56}$
{37,39}	k_{56}	{37,40}	$1 + k_{54} + k_{55}$
{38,40}	$1 + k_{53}$	{39,40}	k_{53}
{41,42}	$1 + k_{60}$	{41,43}	k_{60}
{41,44}	$1 + k_{58} + k_{59}$	{42,44}	$1 + k_{57}$
{43,44}	k_{57}	{45,46}	$1 + k_{64}$
{45,47}	k_{64}	{45,48}	$1 + k_{62} + k_{63}$
{46,48}	$1 + k_{61}$	{47,48}	k_{61}
{49,50}	$1 + k_{68}$	{49,51}	k_{68}
{49,52}	$1 + k_{66} + k_{67}$	{50,52}	$1 + k_{65}$
{51,52}	k_{65}	{53,54}	$1 + k_{72}$
{53,55}	k_{72}	{53,56}	$1 + k_{70} + k_{71}$
{54,56}	$1 + k_{69}$	{55,56}	k_{69}
{57,58}	$1 + k_{76}$	{57,59}	k_{76}
{57,60}	$1 + k_{74} + k_{75}$	{58,60}	$1 + k_{73}$
{59,60}	k_{73}	{61,62}	$1 + k_{80}$
{61,63}	k_{80}	{61,64}	$1 + k_{78} + k_{79}$
{62,64}	$1 + k_{77}$	{63,64}	k_{77}

Leakage function of $byte_7$

Cube Indexes	Maxterm equations	Cube indexes	Maxterm equations
{1,2}	$k_{19} + k_{20}$	{1,3}	$k_{18} + k_{20}$
{1,4}	$k_{18} + k_{19}$	{2,3}	$1 + k_{17}$
{2,4}	k_{17}	{3,4}	k_{17}
{5,6}	$k_{23} + k_{24}$	{5,7}	$k_{22} + k_{24}$
{5,8}	$k_{22} + k_{23}$	{6,7}	$1 + k_{21}$
{6,8}	k_{21}	{7,8}	k_{21}
{9,10}	$k_{27} + k_{28}$	{9,11}	$k_{26} + k_{28}$
{9,12}	$k_{26} + k_{27}$	{10,11}	$1 + k_{25}$
{10,12}	k_{25}	{11,12}	k_{25}
{13,14}	$k_{31} + k_{32}$	{13,15}	$k_{30} + k_{32}$
{13,16}	$k_{30} + k_{31}$	{14,15}	$1 + k_{29}$
{14,16}	k_{29}	{15,16}	k_{29}
{17,18}	$k_{35} + k_{36}$	{17,19}	$k_{34} + k_{36}$
{17,20}	$k_{34} + k_{35}$	{18,19}	$1 + k_{33}$
{18,20}	k_{33}	{19,20}	k_{33}
{21,22}	$k_{39} + k_{40}$	{21,23}	$k_{38} + k_{40}$
{21,24}	$k_{38} + k_{39}$	{22,23}	$1 + k_{37}$
{22,24}	k_{37}	{23,24}	k_{37}
{25,26}	$k_{43} + k_{44}$	{25,27}	$k_{42} + k_{44}$
{25,28}	$k_{42} + k_{43}$	{26,27}	$1 + k_{41}$
{26,28}	k_{41}	{27,28}	k_{41}
{29,30}	$k_{47} + k_{48}$	{29,31}	$k_{46} + k_{48}$
{29,32}	$k_{46} + k_{47}$	{30,31}	$1 + k_{45}$
{30,32}	k_{45}	{31,32}	k_{45}

Leakage function of $byte_8$

Cube Indexes	Maxterm equations	Cube indexes	Maxterm equations
{33,34}	$k_{51} + k_{52}$	{33,35}	$k_{50} + k_{52}$
{33,36}	$k_{50} + k_{51}$	{34,35}	$1 + k_{49}$
{34,36}	k_{49}	{35,36}	k_{49}
{37,38}	$k_{55} + k_{56}$	{37,39}	$k_{54} + k_{56}$
{37,40}	$k_{54} + k_{55}$	{38,39}	$1 + k_{53}$
{38,40}	k_{53}	{39,40}	k_{53}
{41,42}	$k_{59} + k_{60}$	{41,43}	$k_{58} + k_{60}$
{41,44}	$k_{58} + k_{59}$	{42,43}	$1 + k_{57}$
{42,44}	k_{57}	{43,44}	k_{57}
{45,46}	$k_{63} + k_{64}$	{45,47}	$k_{62} + k_{64}$
{45,48}	$k_{62} + k_{63}$	{46,47}	$1 + k_{61}$
{46,48}	k_{61}	{47,48}	k_{61}
{49,50}	$k_{67} + k_{68}$	{49,51}	$k_{66} + k_{68}$
{49,52}	$k_{66} + k_{67}$	{50,51}	$1 + k_{65}$
{50,52}	k_{65}	{51,52}	k_{65}
{53,54}	$k_{71} + k_{72}$	{53,55}	$k_{70} + k_{72}$
{53,56}	$k_{70} + k_{71}$	{54,55}	$1 + k_{69}$
{54,56}	k_{69}	{55,56}	k_{69}
{57,58}	$k_{75} + k_{76}$	{57,59}	$k_{74} + k_{76}$
{57,60}	$k_{74} + k_{75}$	{58,59}	$1 + k_{73}$
{58,60}	k_{73}	{59,60}	k_{73}
{61,62}	$k_{79} + k_{80}$	{61,63}	$k_{78} + k_{80}$
{61,64}	$k_{78} + k_{79}$	{62,63}	$1 + k_{77}$
{62,64}	k_{77}	{63,64}	k_{77}

References

1. Aumasson, J.-P., Dinur, I., Henzen, L., Meier, W. and Shamir, A.: Efficient FPGA implementations of high-dimensional cube testers on the stream cipher Grain-128. In: Special Purpose Hardware for Attacking Cryptographic Systems-SHARCS'09' (2009)
2. Aumasson, J.-P., Dinur, I., Meier, W., Shamir, A.: Cube testers and key recovery attacks on reduced-round MD6 and trivium. In: Dunkelman, O. (ed.) FSE 2009. LNCS, vol. 5665, pp. 1–22. Springer, Heidelberg (2009)
3. Bedi, S.S., Rajesh Pillai, N.: Cube attacks on Trivium. Cryptology ePrint Archive. Report 2009/015 (2009)
4. Blum, M., Luby, M., Rubinfeld, R.: Self-testing/correcting with applications to numerical problems. J. Comput. Syst. Sci. **47**, 549–595 (1993)
5. Bogdanov, A.A., Knudsen, L.R., Leander, G., Paar, C., Poschmann, A., Robshaw, M., Seurin, Y., Vikkelsoe, C.: PRESENT: an altra-lightweight block cipher. In: Paillier, P., Verbauwhede, I. (eds.) CHES 2007. LNCS, vol. 4727, pp. 450–466. Springer, Heidelberg (2007)
6. Dinur, I., Shamir, A.: Applying cube attacks to stream ciphers in realistic scenarios. Crypt. Commun. **4**, 217–232 (2012)
7. Dinur, I., Shamir, A.: Breaking grain-128 with dynamic cube attacks. In: Joux, A. (ed.) FSE 2011. LNCS, vol. 6733, pp. 167–187. Springer, Heidelberg (2011)
8. Dinur, I., Shamir, A.: Cube attacks on tweakable black box polynomials. In: Joux, A. (ed.) EUROCRYPT 2009. LNCS, vol. 5479, pp. 278–299. Springer, Heidelberg (2009)
9. Dinur, I., Shamir, A.: Generic analysis of small cryptographic leaks. In: 2010 Workshop on Fault Diagnosis and Tolerance in Cryptography. pp. 39–48 (2010)
10. Dinur, I., Shamir, A.: Side channel cube attacks on block ciphers. Cryptology ePrint Archive. Report 2009/127 (2009)
11. Fouque, P.-A., Vannet, T.: Improving key recovery to 784 and 799 rounds of trivium using optimized cube attacks. In: Moriai, S. (ed.) FSE 2013. LNCS, vol. 8424, pp. 502–517. Springer, Heidelberg (2014)
12. Bard, G.V., Courtois, N.T., Nakahara Jr., J., Sepehrdad, P., Zhang, B.: Algebraic, AIDA/Cube and side channel analysis of KATAN family of block ciphers. In: Gong, G., Gupta, K.C. (eds.) INDOCRYPT 2010. LNCS, vol. 6498, pp. 176–196. Springer, Heidelberg (2010)
13. Lai, X.: Higher order derivatives and differential cryptanalysis. In: Blahut, R.E., Costello Jr., D.J., Maurer, U., Mittelholzer, T. (eds.) Communications and Cryptography: Two Sides of One Tapestry. The Springer International Series in Engineering and Computer Science, vol. 276, pp. 227–233. Springer, New York (1994)
14. Matsui, M.: Linear cryptanalysis method for DES cipher. In: Helleseth, T. (ed.) EUROCRYPT 1993. LNCS, vol. 765, pp. 386–397. Springer, Heidelberg (1994)
15. Quedenfeld, F.-M., Wolf, C.: Algebraic Properties of the Cube Attack. Cryptology ePrint Archive. Report 2013/800 (2013)
16. Siegenthaler, T.: Decrypting a class of stream ciphers using ciphertext only. IEEE Trans. Comput. **34**(1), 81–85 (1985)
17. Vielhaber, M.: AIDA Breaks (BIVIUM A and B) in 1 Minute Dual Core CPU Time. IACR Cryptology ePrint Archive, 402 (2009)
18. Vielhaber, M.: Breaking ONE.FIVIUM by AIDA an Algebraic IV Differential Attack. IACR Cryptology ePrint Archive, 413 (2007)

19. Li, Z., Zhang, B., Fan, J., Verbauwhede, I.: A new model for error-tolerant side-channel cube attacks. In: Bertoni, G., Coron, J.-S. (eds.) CHES 2013. LNCS, vol. 8086, pp. 453–470. Springer, Heidelberg (2013)
20. Li, Z., Zhang, B., Yao, Y., Lin, D.: Cube cryptanalysis of LBlock with noisy leakage. In: Kwon, T., Lee, M.-K., Kwon, D. (eds.) ICISC 2012. LNCS, vol. 7839, pp. 141–155. Springer, Heidelberg (2013)

A Generic Algorithm for Small Weight Discrete Logarithms in Composite Groups

Alexander May and Ilya Ozerov[✉]

Faculty of Mathematics, Horst Görtz Institute for IT-Security,
Ruhr-University Bochum, Bochum, Germany
{alex.may,ilya.ozerov}@rub.de

Abstract. Let $(\mathbb{G}, \cdot)$ be an arbitrary cyclic group of composite order N with $\mathbb{G} \simeq \mathbb{G}_1 \times \mathbb{G}_2$. We present a generic algorithm for solving the discrete logarithm problem in $\mathbb{G}$ with Hamming weight $\delta \log N$, $\delta \in (0,1)$, in time $\widetilde{O}(\sqrt{p} + \sqrt{|\mathbb{G}_2|}^{H(\delta)})$, where p is the largest prime divisor in $\mathbb{G}_1$ and $H(\cdot)$ is the binary entropy function.

Our algorithm improves on the running time of Silver-Pohlig-Hellman's algorithm whenever $\delta \neq \frac{1}{2}$. Moreover, it improves on the Meet-in-the-Middle type algorithms of Heiman, Odlyzko and Coppersmith with running time $\widetilde{O}(\sqrt{|\mathbb{G}|}^{H(\delta)})$ whenever $p < |\mathbb{G}|^{H(\delta)}$.

Keywords: Cryptanalysis · Generic discrete logarithm · Small hamming weight · Representations

1 Introduction

The hardness of the discrete logarithm problem on classical computers is one of the most central sources for constructing public key cryptography. In order to achieve minimal key-size and maximal performance, crypto designers usually choose cyclic groups $\mathbb{G}$ for which no group-specific algorithm, e.g. of index calculus type [1], is known. In these groups, the security analysis is based on the performance of generic algorithms.

However, very few generic algorithms for cyclic groups are known. Among them are Shanks' Baby-Step Giant-Step algorithm [10] and its low-memory variant, Pollard's Rho Method [9]. Both algorithms achieve a running time of $\sqrt{|\mathbb{G}|}$. Moreover, it is known by a result of Shoup [11] that generic algorithms in *prime* order groups cannot compute discrete logarithms faster than $\sqrt{|\mathbb{G}|}$.

The generic algorithm of Silver, Pohlig and Hellman [8] can be seen as a generalization of Shanks' algorithm to non-prime order groups. Let $\mathbb{G}$ be a cyclic group with $|\mathbb{G}| = N$ and prime factorization $N = \prod_{i=1}^{k} p_i^{e_i}$. Thus we have $\mathbb{G} \simeq \mathbb{G}_1 \times \ldots \times \mathbb{G}_k$ with cyclic groups $|\mathbb{G}_i| = p_i^{e_i}$. In the Silver-Pohlig-Hellman algorithm, the discrete logarithm is first computed in $\mathbb{G}_i$ modulo p_i, then lifted

---

A. May: Supported by DFG as part of GRK 1817 Ubicrypt and SPP 1736 Big Data.

I. Ozerov: Supported by DFG as part of SPP 1307 Algorithm Engineering.

© Springer International Publishing Switzerland 2014
A. Joux and A. Youssef (Eds.): SAC 2014, LNCS 8781, pp. 278–289, 2014.
DOI: 10.1007/978-3-319-13051-4_17

modulo $p_i^{e_i}$ and afterwards composed by Chinese Remaindering to the full group order N. Since this process is dominated by the running time of an individual discrete logarithm computation in $\mathbb{G}_i$ modulo p_i, the total running time is dominated by $\max_i\{\sqrt{p_i}\}$.

If we do not further restrict the discrete logarithm problem, the generic algorithms of Pollard and Silver, Pohlig and Hellman are all that we have. If we limit our discrete logarithm to a certain interval $[a, b]$ then the discrete logarithm can be computed by Pollard's kangaroo method [9] in time $\widetilde{O}(\sqrt{b-a})$, which can be seen as another variant of Shanks' algorithm.

More generic algorithms are known when we limit our discrete logarithm to a small Hamming weight. Let α be a generator of $\mathbb{G}$ with an order of bit-size n. Let $\beta = \alpha^x$ with an n-bit integer x having Hamming weight δn, $\delta \in (0, 1)$, where we call δ the relative Hamming weight of x.

A brute-force enumeration of an n-bit number x with Hamming weight δn takes time $\binom{n}{\delta n} \approx 2^{H(\delta)n}$. The algorithms of Heiman-Odlyzko [5], Coppersmith [3] and Stinson [13] split x in two parts of length $\frac{n}{2}$ and Hamming weight $\delta\frac{n}{2}$ each. This is a classical Meet-in-the-Middle approach that achieves a square-root complexity of roughly $2^{\frac{H(\delta)}{2}n} = \widetilde{O}(\sqrt{|\mathbb{G}|}^{H(\delta)})$.

1.1   Our Contribution and Related Work

We present a new algorithm that can be seen as a generalization of the Meet-in-the-Middle algorithms of Heiman-Odlyzko and Coppersmith and the Silver-Pohlig-Hellman algorithm. In spirit, our approach is similar to an algorithm of van Oorschot and Wiener [14] for the discrete logarithm problem with *small x*, as opposed to small Hamming weight x in our case. The van Oorschot-Wiener algorithm computes the CRT-representation of x modulo a factor N_1 of the group order via Silver-Pohlig-Hellman. Thus, x can be expressed as $x = x_1 N_1 + x_0$ for some known x_0. Then x_1 is easily computed via Pollard's kangaroo algorithm in time $\widetilde{O}(\sqrt{x_1})$. Thus, van Oorschot and Wiener proceed in a divide and conquer manner, where they split the computation of x in two parts.

Our algorithm also makes use of the Silver-Pohlig-Hellman algorithm as a subroutine. However, our computation of the second part is way more challenging than in the algorithm of van Oorschot and Wiener. Notice that the property of a small Hamming weight discrete logarithm does not transfer to its Chinese Remainder representation and vice versa. Nevertheless, we are able to show that parts of the Chinese Remainder representation automatically reduce the search space for small weight discrete logarithms.

In general, finding algorithms for small Hamming weight appears to be a harder problem than finding algorithms for small size, e.g. for polynomial equations there is an efficient algorithm that finds all small size integer roots due to Coppersmith [4], but there is no analogue known for small Hamming weight roots.

Let $\mathbb{G} \simeq \mathbb{G}'_1 \times \ldots \mathbb{G}'_k$ be our composite group. We write this in the form $\mathbb{G} \simeq \mathbb{G}_1 \times \mathbb{G}_2$, where we suitably combine groups. Our runtime will be dependent on the size of $|\mathbb{G}_1|$, its prime factorization, and the relative Hamming weight δ of our discrete logarithm problem. So if $k > 2$, then for a given δ we have to form $\mathbb{G}_1$ in such a way that minimizes the running time. Let $\beta = \alpha^x$ be our discrete logarithm problem in $\mathbb{G}$.

Let us first describe a simple enumeration version of our algorithm. Assume $|\mathbb{G}| = N, |\mathbb{G}_1| = N_1, |\mathbb{G}_2| = N_2$ and let n, n_1, n_2 denote the bit-sizes of N, N_1, N_2, respectively. Notice that $N = N_1 N_2$ and thus (roughly) $n = n_1 + n_2$. Let us first compute $x \bmod N_1$, that is we compute the discrete logarithm in the smaller subgroup $\mathbb{G}_1 \times \{1\} \subset \mathbb{G}$. With the Silver-Pohlig-Hellman algorithm this can be done in time $\sqrt{p}$, where p is the largest prime factor of $|\mathbb{G}_1|$. Now we enumerate all natural numbers x' which have weight δ in the upper $n - n_1 = n_2$ most significant bits and which are consistent with the computed discrete logarithm in $\mathbb{G}_1$. We are able to show that the second restriction basically determines the remaining n_1 least significant bits uniquely. Since we do this enumeration as a Meet-in-the-Middle approach, we achieve complexity

$$\sqrt{\binom{n_2}{\delta n_2}} \approx \sqrt{2^{H(\delta)n_2}} \approx \sqrt{|\mathbb{G}_2|}^{H(\delta)}.$$

We want to stress that our algorithm is not designed to attack practical cryptographic schemes. Our main goal was to combine ideas of [2,6,7,12] to obtain one of very few known generic algorithms for discrete logs. Our method is inspired by a recent subset sum algorithm of Howgrave-Graham and Joux [6] and our intention was to understand the full generality of their method in arbitrary groups. In the Howgrave-Graham-Joux algorithm the target vector $x \in \{0,1\}^n$ is represented as a sum of vectors $x_1, x_2 \in \{0,1\}^n$. But their sum is a vector sum in $\mathbb{Z}^n$, whereas our vectors represent integers and the addition $x_1 + x_2$ is in $\mathbb{Z}$, i.e. we allow carry bits that allow for new kinds of representations.

Our algorithm is also in the spirit of Stern's Information Set Decoding technique [12] for decoding random linear codes, where one part of the unknown error vector is obtained combinatorially, whereas the remaining bits are computed efficiently through simple linear algebra. Notice that like in [2,7] it is possible to combine our technique with the classical technique of [6]. We leave as an open problem whether this leads to even better results.

2   Known Generic Algorithms

In this section, we quickly repeat some standard algorithms for discrete logarithms, since we will use variations of these as subroutines in our algorithm. We start by explaining the SORT-AND-MATCH algorithm that is the basis for all Meet-in-the-Middle approaches.

Let $\alpha^x = \beta$ be a discrete logarithm instance in some group $\mathbb{G}$ generated by α. Let us write $x = x_1 + x_2$, where $x_1 \in \mathcal{S}_1, x_2 \in \mathcal{S}_2$ for some sets $\mathcal{S}_1, \mathcal{S}_2 \subset \mathbb{Z}$. Then we obtain the identity

$$\alpha^{x_1} = \beta \cdot \alpha^{-x_2}.$$

We compute a list $\mathcal{L}$ that contains the elements (α^{x_1}, x_1) for all $x_1 \in \mathcal{S}_1$. Then, we compute $(\beta \cdot \alpha^{-x_2}, x_2)$ for all $x_2 \in \mathcal{S}_2$. Any element $(\beta \cdot \alpha^{-x_2}, x_2)$ that matches an element $(\alpha^{x_1}, x_1) \in \mathcal{L}$ in its first component yields a solution $x = x_1 + x_2$ to the discrete logarithm problem. This strategy leads to the algorithm SORT-AND-MATCH.

Algorithm 1

1: **procedure** SORT-AND-MATCH$(\mathcal{S}_1, \mathcal{S}_2, \alpha, \beta)$ $\triangleright \mathcal{S}_1, \mathcal{S}_2 \subseteq \mathbb{Z}_N$
2:      Create a list $\mathcal{L}$ with entries (α^{x_1}, x_1) for all $x_1 \in \mathcal{S}_1$, sort by its first component
3:      **for all** $x_2 \in \mathcal{S}_2$ **do**
4:          Binary search for a $(\alpha^{x_1}, x_1) \in \mathcal{L}$ such that $\alpha^{x_1} = \beta/\alpha^{x_2}$
5:          **return** $x_1 + x_2$ if there is a match
6:      **end for**
7:      **return** no match
8: **end procedure**

It is not hard to see that both the time and space complexity of SORT-AND-MATCH are $\widetilde{O}(|\mathcal{S}_1| + |\mathcal{S}_2|)$, where the $\widetilde{O}(\cdot)$-notation suppresses logarithmic terms. So if $x \in \mathcal{S}$ and $x_1 \in \mathcal{S}_1$, $x_2 \in \mathcal{S}_2$ with $|\mathcal{S}_1| \approx |\mathcal{S}_2| \approx \sqrt{|\mathcal{S}|}$ then SORT-AND-MATCH achieves the square root of the time complexity that is required for simply enumerating all $x \in \mathcal{S}$. Hence our goal is to define $\mathcal{S}_1, \mathcal{S}_2$ in such a way that the maximum of their cardinalities roughly equals $\sqrt{|\mathcal{S}|}$, and that with high probability there always exist $(x_1, x_2) \in \mathcal{S}_1 \times \mathcal{S}_2$ with $x = x_1 + x_2$.

In the following, we illustrate how the selection of $\mathcal{S}_1, \mathcal{S}_2$ is done for Shanks' algorithm and for its variations due to Heiman-Odlyzko, Coppersmith and Stinson. Let $(x_{n-1}, \ldots x_0) \in \{0, 1\}^n$ be the binary representation of x, i.e. $x = \sum_{i=0}^{n-1} x_i 2^i$. Then we write $x = x_1 + x_2 = v_1 \cdot 2^{n/2} + v_2$ with $0 \le v_1, v_2 < 2^{n/2}$. Figure 1 illustrates this splitting in form of the binary representations of x_1, x_2. Notice that for ease of writing throughout this work we ignore any complications that arise from rounding terms like $\frac{n}{2}$, since this is always easy to solve.

It is obvious that the search spaces $\mathcal{S}_1, \mathcal{S}_2$ both have cardinality $2^{\frac{n}{2}} = \sqrt{|\mathcal{S}|} = \sqrt{|\{0, 1\}^n|}$ and that there always exists a pair $(x_1, x_2) \in \mathcal{S}_1 \times \mathcal{S}_2$ with $x = x_1 + x_2$. This leads to a generic discrete logarithm algorithm in $\mathbb{G}$ with time and space complexity $\widetilde{O}(2^{\frac{n}{2}}) = \widetilde{O}(\sqrt{|\mathbb{G}|})$.

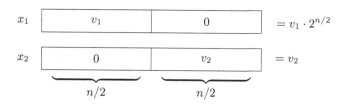

Fig. 1. Splitting

In a nutshell, when we move to discrete logarithms x whose binary representation $(x_{n-1}, \ldots, x_0)$ have Hamming weight δn, we can easily adapt the splitting of Fig. 1. Namely, we enumerate over all $v_1, v_2 \in \{0,1\}^{\frac{n}{2}}$ with Hamming weight $\delta \frac{n}{2}$. Hence, $\mathcal{S}$ consists of all numbers that can be represented as n-bit vectors with relative Hamming weight δ, whereas the numbers in $\mathcal{S}_1$, $\mathcal{S}_2$ can be represented as $\frac{n}{2}$-bit vectors with relative Hamming weight δ, where we append a $0^{\frac{n}{2}}$-string accordingly.

Let us first compare the cardinalities of $\mathcal{S}$ and $\mathcal{S}_1$, $\mathcal{S}_2$. Here, we use the well-known approximation $\binom{n}{\delta n} \approx 2^{H(\delta)n}$, that stems from Stirling's formula. This implies that $|\mathcal{S}_1| = |\mathcal{S}_2| \approx 2^{H(\delta)\frac{n}{2}} = \tilde{O}(\sqrt{|\mathbb{G}|}^{H(\delta)})$ which is equal to the square root of $|\mathcal{S}|$. Thus, we obtain a Meet-in-the-Middle algorithm with square root time and space complexity.

Notice however, that as opposed to Shanks' algorithm not every x with Hamming weight δn admits a splitting in x_1, x_2 as above, where both x_i have Hamming weight $\delta \frac{n}{2}$. The probability that a random x splits in this way is $\binom{n/2}{\delta n/2}^2 / \binom{n}{\delta n} = \Theta(\frac{1}{\sqrt{n}})$. In the algorithms of Heiman, Odlyzko and Coppersmith this problem is solved by a combinatorial structure which is called a *splitting set*, which basically allows to re-randomize the coordinates for the splitting in x_1, x_2.

A different deterministic approach due to Coppersmith is a simple application of the intermediate value theorem. Assume wlog that the weight in the v_1-part is too high, and the weight in the v_2-part is too low. Let us rotate both parts cyclically until they change places. Since a rotation by one position changes the weight in each part by at most 1, there must exist one out of the $n/2$ rotations where both parts share the same weight.

In the following, we propose a slightly different solution for guaranteeing the existence of a valid splitting, which we use in our algorithm. Namely, we choose the non-zero parts of the binary representation of the element in $\mathcal{S}_1, \mathcal{S}_2$ from $\{0,1\}^{\frac{n}{2}}$, where their relative Hamming weight lies in the interval $(\delta - \varepsilon, \delta + \varepsilon)$ for some small $\varepsilon > 0$. In this way, we can ensure that a randomly chosen x splits in $x_1 + x_2$ with appropriate Hamming weights in this interval with a probability that is exponentially close to 1, while only slightly increasing the running time. The main reason for choosing such a weight interval is that it simplifies the description and analysis of our algorithm significantly.

3   Our New Generic Discrete Log Algorithm

Let α generate a composite order group $\mathbb{G} \simeq \mathbb{G}_1 \times \mathbb{G}_2$ with $|\mathbb{G}| = N$, $|\mathbb{G}_1| = N^\tau$ and $|\mathbb{G}_2| = N^{1-\tau}$ for some $\tau \in (0,1)$. In general, there might be several ways to decompose $\mathbb{G}$ as $\mathbb{G}_1 \times \mathbb{G}_2$. We will first describe our algorithm for a fixed decomposition. Afterwards, we will minimize the running time by adjusting the decomposition accordingly.

Our new algorithm combines the Silver-Pohlig-Hellman idea with a subsequent Meet-in-the-Middle approach for enumerating small weight vectors. As described in Sect. 2, we have to define the sets $\mathcal{S}_1, \mathcal{S}_2$ that describe how we split $x = x_1 + x_2$ with $(x_1, x_2) \in \mathcal{S}_1 \times \mathcal{S}_2$. We illustrate the binary representation for our candidates (x_1, x_2) in Fig. 2. Here, the v_i have relative weight δ, whereas the w_i may have arbitrary weight.

In the following, we always assume wlog that we know the factorization of the group order N. Notice that this does not limit the applicability of our generic algorithm, since our algorithm's running time is exponential in the bit-length of N anyway, whereas the factorization of N can be computed in sub-exponential time.

The main idea of our new algorithm DLOG is as follows. We first apply the algorithm of Silver, Pohlig and Hellman to the smaller subgroup $\mathbb{G}_1$ to obtain the discrete logarithm x' modulo $M := |\mathbb{G}_1|$. Afterwards, we apply a Meet-in-the-Middle technique on the bigger subgroup $\mathbb{G}_2$ where we cut down the search space by the amount of information that is provided by x'. More precisely, knowing only $n - t$ bits of the discrete logarithm x (i.e. v_1 and v_2 in Fig. 2), it is possible to compute the remaining t consecutive bits w in polynomial time with the help of $x' = x \bmod M$.

Let us fix some useful notation. We denote $x' = [x']_M = [x]_M$, where $[\cdot]_M$ describes the smallest non-negative representative of some number modulo M, i.e. in $[0, M)$. Let us choose t such that $2^{t-1} < M \leq 2^t$. Then w is either $[w]_M := [x' - v_1 \cdot 2^{(n+t)/2} - v_2 \cdot 2^t]_M$ or $[w]_M + M$.

For any $x \in \mathbb{N}$ we denote by $\mathrm{wt}(x)$ the Hamming weight of the binary representation of x.

In lines 6 through 9 a list $\mathcal{S}_1$ is computed by enumerating all values of v_1 (see Fig. 2). We show that the remaining t-bit value w_1 can be uniquely obtained from v_1. Similarly, in lines 10 through 15 we obtain only three possible values for w_2 for each value of v_2. In total, it is sufficient to perform a Meet-in-the-Middle attack on only $n - t$ bits instead of the full n bits of the binary representation of x. Eventually, the subroutine SORT-AND-MATCH finds the discrete logarithm x,

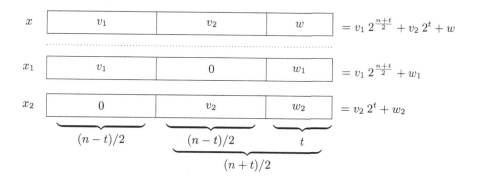

Fig. 2. New splitting

Algorithm 2

1: **procedure** DLOG($|\mathbb{G}_1|, |\mathbb{G}_2|, \alpha, \beta, \delta, \varepsilon$)
2:     $n \leftarrow \lceil \log_2(|\mathbb{G}_1| \cdot |\mathbb{G}_2|) \rceil$
3:     $t \leftarrow \lceil \log_2(|\mathbb{G}_1|) \rceil$
4:     $x' \leftarrow$ SPH($|\mathbb{G}_1|, \alpha^{|\mathbb{G}_2|}, \beta^{|\mathbb{G}_2|}$)   ▷ Use Silver-Pohlig-Hellman to get x mod $|\mathbb{G}_1|$.
5:     $M \leftarrow |\mathbb{G}_1|$
6:     $\mathcal{S}_1 \leftarrow \{\}$
7:     **for all** $0 \leq v_1 < 2^{(n-t)/2}$ with $(\delta - \varepsilon)\frac{n-t}{2} \leq \mathrm{wt}(v_1) \leq (\delta + \varepsilon)\frac{n-t}{2}$ **do**
8:         $\mathcal{S}_1 \leftarrow \mathcal{S}_1 \cup \{v_1 \cdot 2^{(n+t)/2} + [-v_1 \cdot 2^{(n+t)/2}]_M\}$
9:     **end for**
10:     $\mathcal{S}_2 \leftarrow \{\}$
11:     **for all** $0 \leq v_2 < 2^{(n-t)/2}$ with $(\delta - \varepsilon)\frac{n-t}{2} \leq \mathrm{wt}(v_2) \leq (\delta + \varepsilon)\frac{n-t}{2}$ **do**
12:         $\mathcal{S}_2 \leftarrow \mathcal{S}_2 \cup \{v_2 \cdot 2^t + [x' - v_2 \cdot 2^t]_M\}$
13:         $\mathcal{S}_2 \leftarrow \mathcal{S}_2 \cup \{v_2 \cdot 2^t + [x' - v_2 \cdot 2^t]_M - M\}$
14:         $\mathcal{S}_2 \leftarrow \mathcal{S}_2 \cup \{v_2 \cdot 2^t + [x' - v_2 \cdot 2^t]_M + M\}$
15:     **end for**
16:     **return** SORT-AND-MATCH($\mathcal{S}_1, \mathcal{S}_2, \alpha, \beta$)
17: **end procedure**

---

since we show that with overwhelming probability there is always some $x_1 \in \mathcal{S}_1$ and $x_2 \in \mathcal{S}_2$ that sum to x.

Theorem 1. *Let α be a generator of an cyclic group $\mathbb{G} \simeq \mathbb{G}_1 \times \mathbb{G}_2$ of known order N, where N has bit-size n. Let $\delta \in (0, \frac{1}{2})$ and let x be sampled uniformly at random from all elements of $\mathbb{Z}_N$ with Hamming weight δn. Let $\beta := \alpha^x$ and p be the largest prime factor of $|\mathbb{G}_1|$. Then for any $\varepsilon > 0$ with $\delta + \varepsilon \leq \frac{1}{2}$ on input $(|\mathbb{G}_1|, |\mathbb{G}_2|, \alpha, \beta, \delta)$ algorithm DLOG outputs x with probability at least $1 - \frac{4(n+1)}{|\mathbb{G}_2|\varepsilon^2}$ in time $\tilde{O}\left(\sqrt{p} + \sqrt{|\mathbb{G}_2|}^{H(\delta+\varepsilon)}\right)$ and space $\tilde{O}\left(\sqrt{|\mathbb{G}_2|}^{H(\delta+\varepsilon)}\right)$.*

Proof. Let us first define $M := |\mathbb{G}_1|$, $n := \lceil \log_2(N) \rceil$, $t := \lceil \log_2(M) \rceil$ and $x' := [x]_M$ as in DLOG. Recall that $[\cdot]_M$ denotes the least non-negative representative modulo M, and $\mathrm{wt}(\cdot)$ denotes the Hamming weight of the binary representation. For simplicity, we ignore rounding problems like with $(n - t)/2$, since they can easily be resolved without affecting the asymptotic running time.

Similar to the standard Meet-in-the-Middle approach from Sect. 2, in DLOG we decompose the unknown $x = v_1 \cdot 2^{(n+t)/2} + v_2 \cdot 2^t + w$ for some $0 \leq v_1, v_2 < 2^{(n-t)/2}$ and $0 \leq w < 2^t$, as illustrated in Fig. 2. Moreover, we require that both v_1, v_2 have some Hamming weight in the interval $[(\delta - \varepsilon)(n-t)/2, (\delta + \varepsilon)(n-t)/2]$. The proof is organized as follows. Firstly, we show that any random x possesses the correct weights for v_1, v_2 with overwhelming probability. Secondly, we show that for the correct weights, DLOG always outputs x. Thus, DLOG is of Las Vegas type. Its output is always correct, but DLOG fails on an exponentially small fraction of all input instances.

Let $x \in \mathbb{Z}_N$ be chosen uniformly at random with Hamming weight $\mathrm{wt}(x) = \delta n$. We show that $(\delta - \varepsilon)(n-t)/2 \leq \mathrm{wt}(v_1), \mathrm{wt}(v_2) \leq (\delta + \varepsilon)(n-t)/2$ holds with a probability that is at least $1 - 4(n+1)/|\mathbb{G}_2|^{\varepsilon^2}$. Let $(x_{n-1}, \ldots, x_0)$ denote the binary representation of x, and let X_i be a random variable for x_i. For simplifying our proof, we assume that x was sampled by n independent Bernoulli trials with $\mathbb{P}[X_i = 1] = \delta$ for all bits $i = 0, \ldots, n-1$. Notice that sampling x in this manner and rejecting all x that have an incorrect Hamming weight gives the same distribution as sampling x uniformly at random from all x with Hamming weight δn.

Let $I \subseteq \{0, \ldots, n-1\}$ with $|I| = (n-t)/2$ be some index set. Let $X = \sum_{i=0}^{n-1} X_i$ be the Hamming weight of x, and let $Y = \sum_{i \in I} X_i$ be the Hamming weight of coordinates I. In order to estimate DLOG's failure probability, we compute

$$\mathbb{P}\left[|Y - \delta(n-t)/2| > \varepsilon(n-t)/2 \mid X = \delta n\right],$$

which is the probability that the Hamming weight on the I-bits of x is *not* in the range between $(\delta - \varepsilon) \cdot (n-t)/2$ and $(\delta + \varepsilon) \cdot (n-t)/2$, under the condition that x has the correct Hamming weight. Notice that $\mathbb{P}[X = \delta n] \geq \mathbb{P}[X = i]$ for any $i \neq \delta n$. Since $0 \leq X \leq n$, we get $1 = \sum_{i=0}^{n} \mathbb{P}[X = i] \leq (n+1) \cdot \mathbb{P}[X = \delta n]$ and thus $\mathbb{P}[X = \delta n] \geq \frac{1}{n+1}$. This implies

$$\mathbb{P}\left[|Y - \delta(n-t)/2| > \varepsilon(n-t)/2 \mid X = \delta n\right]$$

$$\leq (n+1) \cdot \mathbb{P}\left[|Y - \delta(n-t)/2| > \varepsilon(n-t)/2\right].$$

An application of Hoeffding's inequality yields

$$(n+1) \cdot \mathbb{P}\left[|Y - \delta(n-t)/2| > \varepsilon(n-t)/2\right] \leq 2(n+1)2^{-\varepsilon^2(n-t)} \leq \frac{2(n+1)}{|\mathbb{G}_2|^{\varepsilon^2}}.$$

Hence, we obtain a probability of at most $2(n+1)/|\mathbb{G}_2|^{\varepsilon^2}$ that the relative Hamming weight for *one* of v_1, v_2 is incorrect. By the union bound, the probability that v_1 or v_2 have incorrect weight is bounded by $4(n+1)/|\mathbb{G}_2|^{\varepsilon^2}$.

It remains to show that for correct Hamming weight of v_1, v_2 DLOG always succeeds in computing x. By the correctness of our SORT-AND-MATCH routine, it suffices to show the existence of $(x_1, x_2) \in \mathcal{S}_1 \times \mathcal{S}_2$ with $x_1 + x_2 = x$.

We split x in three parts v_1, v_2, w with $x = v_1 \cdot 2^{(n+t)/2} + v_2 \cdot 2^t + w$ (see Fig. 2). Denote

$$x_1 := v_1 \cdot 2^{(n+t)/2} + w_1, x_2 := v_2 \cdot 2^t + w_2 \text{ with } 0 \leq v_1, v_2 < 2^{(n-t)/2} \text{ and}$$

$$0 \leq w_1, w_2 < 2^t.$$

In DLOG we enumerate a first list $\mathcal{S}_1$ of all possible v_1 and compute for each v_1 a corresponding w_1. We proceed with v_2 and their corresponding w_2 analogously.

In $\mathcal{S}_1$ we choose to fix $x_1 = 0 \bmod M$ — the value 0 could be any constant in $\mathbb{Z}_M$. Therefore, we compute $w_1 = -v_1 \cdot 2^{(n+t)/2} \bmod M$ and store the corresponding integer

$$x_1 := v_1 \cdot 2^{(n+t)/2} + [-v_1 \cdot 2^{(n+t)/2}]_M.$$

Notice that there always exists a $0 \le w_1 < 2^t$ with $w_1 = -v_1 \cdot 2^{(n+t)/2} \bmod M$, since $M \le 2^t$ by the choice of t.

Since we have to ensure $x_1 + x_2 = x$, we require $x_1 + x_2 = x' \bmod M$ and thus $x_2 = x' \bmod M$ by our choice of x_1. This in turn implies $w_2 = x' - v_2 \cdot 2^t \bmod M$. By construction, we obtain

$$x_1 + x_2 = v_1 \cdot 2^{(n+t)/2} + [-v_1 \cdot 2^{(n+t)/2}]_M + v_2 \cdot 2^t + [x' - v_2 \cdot 2^t]_M = x \bmod M.$$

However, this does not necessarily imply $x = x_1 + x_2$ over $\mathbb{Z}$. Especially, we have to guarantee $w_1 + w_2 = w$. Notice that by definition $w < 2^t$ and $M \le 2^t < 2M$. Since $0 \le [w_1]_M, [w_2]_M < M$ we have $0 \le [w_1]_M + [w_2]_M < 2M$.

If either $0 \le [w_1]_M + [w_2]_M < M$ and $w < M$ (case I+I in Fig. 3) or $M \le [w_1]_M + [w_2]_M < 2M$ and $M \le w$ (case II+II in Fig. 3), we are done. If $0 \le [w_1]_M + [w_2]_M < M$ and $M \le w$ (case I+II in Fig. 3), we have to add M to $[w_1]_M + [w_2]_M$. In the remaining case II+I, we have to substract M from $[w_1]_M + [w_2]_M$.

Thus, $[w_1]_M + [w_2]_M + kM = w$ holds for some $k \in \{-1, 0, 1\}$. In DLOG we choose $x_1 = v_1 \cdot 2^{(n+t)/2} + [w_1]_M \in \mathcal{S}_1$ and $x_2 = v_2 \cdot 2^t + [w_2]_M + kM \in \mathcal{S}_2$ for all $k \in \{-1, 0, 1\}$. For the correct k, we obtain $x_1 + x_2 = x$, as desired. Thus, SORT-AND-MATCH succeeds, and DLOG outputs the discrete logarithm x.

It remains to show the time and space complexities. SPH takes time $\widetilde{O}(\sqrt{p})$ with only polynomial memory consumption, when using Pollard's Rho Method as a subroutine. Notice that the complexity of the for-loops in step 7 and 11 of DLOG are dominated by the time to enumerate and store those v_i with largest weight $(\delta + \varepsilon)\frac{n-t}{2}$. Thus, our Meet-in-the-Middle attack has time and space complexity $\widetilde{O}\left(\sqrt{|\mathbb{G}_2|}^{H(\delta+\varepsilon)}\right)$. $\square$

Remark 1 (Large weight). DLOG is by definition restricted to small Hamming weight $0 < \delta \le \frac{1}{2}$. Symmetrically, DLOG can be applied to large Hamming weight $\frac{1}{2} \le \delta < 1$ by transforming the discrete logarithm instance to $\widetilde{\beta} := \alpha^{2^n - 1}/\beta$. This transforms x to $\widetilde{x} = (2^n - 1) - x$ with Hamming weight $(1 - \delta) \cdot n$.

Remark 2 (Representations). Our algorithm can be interpreted in terms of the representation technique introduced by Howgrave-Graham and Joux [6] for solving the subset sum problem. Notice that we split $w = w_1 + w_2$ with $w_1, w_2 \in \mathbb{Z}_M$. Thus, we obtain exactly M representations (w_1, w_2) of w as a sum. In our case, we

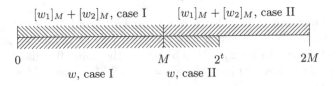

Fig. 3. $\pm M$

use the fact that exactly one representation (w_1, w_2) ensures that $x_1 = 0 \bmod M$ and $x_2 = x \bmod M$, simultaneously. We can directly compute this representation in polynomial time – once x' is known – without any further assumption on the problem instance. This differs from [6], where the authors spent exponential time to compute a representation of the solution and only receive one representation on expectation, assuming a uniform distribution of the subset sum elements.

Remark 3 (Getting rid of ε). Recall that we introduced ε to ensure that the Hamming weight of v_1 and v_2 lies within some ε-strip around its expectation with overwhelming probability. If we set $\varepsilon = 0$, then DLOG finds the discrete logarithm x only for a polynomial fraction of all x that exactly match the expected Hamming weight on v_1, v_2.

One might be tempted to use cyclic rotations of the binary representation of x, just as described at the end of Sect. 2. However, some problems arise here. If we fully rotate, we obtain a bit vector for which the Hamming weight of v_1 and v_2 matches its expectation. In this case, it might however happen that w gets split into two parts by the cyclic rotations. In this case, we were not able to bound the number of w's by a polynomial. We could also consider the case where we do not fully rotate x, but restrict to $n - t$ left rotations only such that the w-part does not split. We conjecture that the number of pathological instances where DLOG does not succeed for at least one of the $n - t$ rotations is exponentially small in this case, but we were not able to prove that.

3.1 How to Optimally Split $\mathbb{G}$ into Subgroups

It remains to show how to optimally choose the subgroups $\mathbb{G}_1$ and $\mathbb{G}_2$ dependent on the factorization of $|\mathbb{G}|$ and on the Hamming weight $\delta \cdot \log |\mathbb{G}|$ of x. Since we apply Silver-Pohlig-Hellman on $\mathbb{G}_1$, the group $\mathbb{G}_1$ should contain all prime subgroups of $\mathbb{G}$ that are smaller or as large as the largest prime subgroup of $\mathbb{G}_1$. In other words, if $N = \prod_{i=1}^{k} p_i$ is the factorization of the order of $\mathbb{G}$ and $p_1 \leq \ldots \leq p_k$, the only useful choices are $|\mathbb{G}_1| = \prod_{i=1}^{\ell} p_i$ and $|\mathbb{G}_2| = \prod_{i=\ell+1}^{k} p_i$ for $1 \leq \ell \leq k - 1$. This is because we have to spend time $\sqrt{p}$ for the maximal prime divisor p of $|\mathbb{G}_1|$ anyway. Thus, our ordering of the p_i minimizes $|\mathbb{G}_2|$ and thus the overall running time.

Among the remaining $k - 1$ choices, we have to find the best choice for ℓ. Fix an ℓ, $1 \leq \ell \leq k - 1$, and define $\tau_i := \log_N p_i$ for each $1 \leq i \leq k$. From Theorem 1, the time complexity of DLOG is

$$\widetilde{O}\left(\sqrt{p_\ell} + \sqrt{p_{\ell+1} \cdots p_k}^{H(\delta + \varepsilon)}\right) = \widetilde{O}\left((2^{n/2})^{\tau_\ell} + (2^{n/2})^{(\tau_{\ell+1} + \ldots + \tau_k) \cdot H(\delta + \varepsilon)}\right).$$

Let us define $p_0 := 1$, and thus $\tau_0 = 0$. In the case $\ell = 0$ we obtain the time complexity of the standard small weight Meet-in-the-Middle algorithm without using SPH. In the case $\ell = k$ we obtain the standard SPH without any Meet-in-the-Middle approach. Thus, our algorithm perfectly interpolates between both cases and there exist τ_i such that our algorithm improves upon SPH and standard Meet-in-the-Middle for any $0 < \ell < k$ with $\delta < \frac{1}{2}$ (cf. Fig. 4).

Theorem 2. *Let $\tau_0 := 0$. Given δ, ε with $\delta + \varepsilon \in (0, \frac{1}{2}]$ and $\tau_1, \ldots, \tau_k$ as defined above, an optimal choice for* DLOG *is to pick $0 \le \ell \le k - 1$ such that*

$$\frac{\tau_\ell}{\tau_\ell + \ldots + \tau_k} < H(\delta + \varepsilon) \le \frac{\tau_{\ell+1}}{\tau_{\ell+1} + \ldots + \tau_k}$$

and to choose $|\mathbb{G}_1| = \prod_{i=1}^{\ell} p_i$ and $|\mathbb{G}_2| = \prod_{i=\ell+1}^{k} p_i$.

Proof. First notice that

$$\bigcup_{\ell=0}^{k-1} \left(\frac{\tau_\ell}{\tau_\ell + \ldots + \tau_k}, \frac{\tau_{\ell+1}}{\tau_{\ell+1} + \ldots + \tau_k} \right]$$

defines a disjoint partition of $(0, 1]$, due to the fact that $\tau_1 + \ldots + \tau_k = 1$. Thus each $\delta + \varepsilon$ with $0 < H(\delta + \varepsilon) \le 1$ leads to a unique choice of ℓ.

Fix δ, ε and choose ℓ as defined above. We want to show that for each choice of $\ell' \ne \ell$, DLOG's complexity does not improve. Notice that there may be other choices for ℓ that achieve the same complexity.

If $\ell' < \ell$, then it is easy to see that $(\tau_{\ell+1} + \ldots + \tau_k) \cdot H(\delta + \varepsilon) \le (\tau_{\ell'+1} + \ldots + \tau_k) \cdot H(\delta + \varepsilon)$. Since $\frac{\tau_\ell}{\tau_\ell + \ldots + \tau_k} < H(\delta + \varepsilon)$, we also have $\tau_\ell < (\tau_\ell + \ldots + \tau_k) \cdot H(\delta + \varepsilon) \le (\tau_{\ell'+1} + \ldots + \tau_k) \cdot H(\delta + \varepsilon)$. Thus the complexity does not improve for $\ell' < \ell$.

If $\ell' > \ell$, obviously we have $\tau_\ell \le \tau_{\ell'}$. Additionally, we have that $(\tau_{\ell+1} + \ldots + \tau_k) \cdot H(\delta + \varepsilon) \le \tau_{\ell+1} \le \tau_{\ell'}$, since $H(\delta + \varepsilon) \le \frac{\tau_{\ell+1}}{\tau_{\ell+1} + \ldots + \tau_k}$. Therefore, the complexity also does not improve for any $\ell' > \ell$. $\qquad \square$

Figure 4 shows the time complexity for a fixed group with a size of N that is a product of 5 primes with sizes $N^{0.1}, N^{0.15}, N^{0.2}, N^{0.25}$ and $N^{0.3}$. In this case, Silver-Pohlig-Hellman (dashed line in Fig. 4, corresponding to $\ell = k$) has a time complexity of $N^{0.15}$. As we can see, our algorithm's improvement (straight line) is defined piecewise for $1 \le \ell \le k - 1$. For small values of δ, a standard small weight Meet-in-the-Middle approach (dotted line, corresponding to $\ell = 0$) yields the optimal complexity.

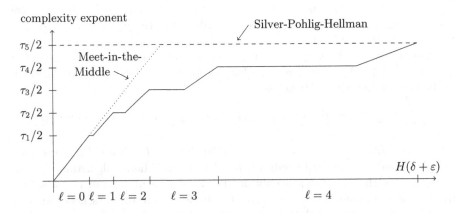

Fig. 4. Time complexity for $k = 5, \tau_1 = 0.1, \tau_2 = 0.15, \tau_3 = 0.2, \tau_4 = 0.25, \tau_5 = 0.3$

References

1. Adleman, L.M.: A subexponential algorithm for the discrete logarithm problem with applications to cryptography (abstract). In: FOCS, pp. 55–60. IEEE Computer Society (1979)
2. Becker, A., Joux, A., May, A., Meurer, A.: Decoding random binary linear codes in 2n/20: how $1 + 1 = 0$ improves information set decoding. In: Pointcheval, D., Johansson, T. (eds.) EUROCRYPT 2012. LNCS, vol. 7237, pp. 520–536. Springer, Heidelberg (2012)
3. Coppersmith, D.: Private communication to Scott Vanstone (1979)
4. Coppersmith, D.: Small solutions to polynomial equations, and low exponent rsa vulnerabilities. J. Cryptology **10**(4), 233–260 (1997)
5. Heiman, R.: A note on discrete logarithms with special structure. In: Rueppel, R.A. (ed.) EUROCRYPT 1992. LNCS, vol. 658, pp. 454–457. Springer, Heidelberg (1993)
6. Howgrave-Graham, N., Joux, A.: New generic algorithms for hard knapsacks. In: Gilbert, H. (ed.) EUROCRYPT 2010. LNCS, vol. 6110, pp. 235–256. Springer, Heidelberg (2010)
7. May, A., Meurer, A., Thomae, E.: Decoding random linear codes in 20.054n. In: Lee, D.H., Wang, X. (eds.) ASIACRYPT 2011. LNCS, vol. 7073, pp. 107–124. Springer, Heidelberg (2011)
8. Pohlig, S., Hellman, M.: An improved algorithm for computing logarithms over gf(p) and its cryptographic significance. IEEE Trans. Inf. Theory **24**, 106–110 (1978)
9. Pollard, J.M.: Monte carlo methods for index computation (mod p). Math. Comput. **32**(143), 918–924 (1978)
10. Shanks, D.: Class number, a theory of factorization and genera. In: Proceedings of Symposia in Pure Mathematics, vol. 20, AMS, pp. 415–440 (1971)
11. Shoup, V.: Lower bounds for discrete logarithms and related problems. In: Fumy, W. (ed.) EUROCRYPT 1997. LNCS, vol. 1233, pp. 256–266. Springer, Heidelberg (1997)
12. Stern, J.: A method for finding codewords of small weight. In: Cohen, G., Wolfmann, J. (eds.) Coding Theory and Applications. LNCS, vol. 388, pp. 106–113. Springer, Heidelberg (1989)
13. Stinson, D.R.: Some baby-step giant-step algorithms for the low hamming weight discrete logarithm problem. Math. Comput. **71**(237), 379–391 (2002)
14. van Oorschot, P.C., Wiener, M.: On Diffie-Hellman key agreement with short exponents. In: Maurer, U.M. (ed.) EUROCRYPT 1996. LNCS, vol. 1070, pp. 332–343. Springer, Heidelberg (1996)

Linear Biases in AEGIS Keystream

Brice Minaud[✉]

ANSSI, 51, Boulevard de la Tour-Maubourg, 75700 Paris 07 SP, France
brice.minaud@gmail.com

Abstract. AEGIS is an authenticated cipher introduced at SAC 2013, which takes advantage of AES-NI instructions to reach outstanding speed in software. Like LEX, Fides, as well as many sponge-based designs, AEGIS leaks part of its inner state each round to form a keystream. In this paper, we investigate the existence of linear biases in this keystream. Our main result is a linear mask with bias 2^{-89} on the AEGIS-256 keystream. The resulting distinguisher can be exploited to recover bits of a partially known message encrypted 2^{188} times, regardless of the keys used. We also consider AEGIS-128, and find a surprising correlation between ciphertexts at rounds i and $i+2$, although the biases would require 2^{140} data to be detected. Due to their data requirements, neither attack threatens the practical security of the cipher.

Keywords: Cryptanalysis · AEGIS · CAESAR

1 Introduction

Traditional block cipher-based encryption ensures the confidentiality of encrypted data: it is infeasible for anyone to decipher a message without knowledge of the secret encryption key. However there is a compelling need for ciphers achieving at once confidentiality and authenticity; that is, ciphers integrating a form of integrity check guaranteeing that the encrypted message does originate from its purported sender. Any tampering of the data will result in its rejection by the deciphering algorithm. The CAESAR [1] authenticated cipher competition, sponsored by the National Institute of Standards and Technology, crystallizes the community's growing interest in this type of cipher. In March 2014, first round submissions were finalized and all entries were published online, awaiting analysis.

AEGIS [9] is a particularly notable candidate in this competition. Indeed, it takes full advantage of the new AES-NI instruction set in recent Intel and AMD processors to achieve unprecedented encryption speed in software, at around half a cycle per byte. Although AEGIS was first introduced only a year ago at SAC 2013, it has already inspired other encryption designs, including PAES [10] and Tiaoxin [8]. The state update function at the core of AEGIS is simply the parallel application of a single AES round to a large state, followed by a shift and XOR. This exploits the pipeline implementation of AES-NI, which allows for the parallel computation of several AES rounds.

© Springer International Publishing Switzerland 2014
A. Joux and A. Youssef (Eds.): SAC 2014, LNCS 8781, pp. 290–305, 2014.
DOI: 10.1007/978-3-319-13051-4_18

AEGIS, like many entries in the CAESAR competition, follows a model where a large inner state leaks essentially a portion of itself every round, which is then XOR-ed with the plaintext to form the ciphertext. Moreover, like most ciphers in this family, including all duplex-like constructions [3], it delays the insertion of a plaintext block into the inner state until after the corresponding ciphertext block has been output, in order for decryption to proceed in the same direction as encryption. As such, these ciphers are not proper stream ciphers, but form an interesting hybrid, where a single round behaves like a stream cipher.

In particular, assume we have a linear distinguisher on the ciphertext with known plaintext, which we call a keystream bias by analogy with stream ciphers. That is, we know that the sum of some specific bits of the ciphertext is biased towards 0 or 1, provided the corresponding plaintext has a known value. Then, because of the stream cipher-like behavior pointed out above, if only the last block of plaintext involved varies, and the rest remains fixed as before, the sum of ciphertext bits is biased towards 0 or 1 depending on the same sum on the plaintext.

Thus, a linear distinguisher on the keystream yields an attack on the scheme, where plaintext bits of a partially known message can be recovered, provided the message is encrypted enough times. Observe that this does not require the same key be used. Indeed, this plaintext could be encrypted in entirely different sessions with different keys, as long as it is encrypted a sufficient number of times in total. This is very reminiscent of classic stream cipher attacks such as linear masking [4], as well as recent attacks on RC4 [2]. However, in the security analysis of AEGIS by its authors, as well as many CAESAR submissions displaying similar stream cipher-like behavior, this type of attacks does not seem to be taken into account. This leaves open the question of how effective they might be, which we investigate for AEGIS.

Our Contribution. In this paper, we describe linear biases in the keystream of AEGIS-128 and AEGIS-256. As far as we know, this is the first cryptanalysis of AEGIS. These biases result from the surprising property that, although the inner state of AEGIS-128 (resp. AEGIS-256) is 5 (resp. 6) times the size of its output per round, the outputs of only 3 consecutive rounds are related. This is particularly striking in the case of AEGIS-128, where we show that the outputs of rounds i and $i + 2$ are correlated.

However, the biases we find are quite small. In the case of AEGIS-256, we exhibit biases of 2^{-89} for a few linear masks, which would require 2^{188} data to be detected with good probability. This bias only requires a known plaintext to be encrypted repeatedly, with no assumption about the keys or nonces: in fact, the inner state before encryption is considered uniformly random. This distinguisher can also be exploited to recover information on a partially known plaintext encrypted 2^{188} times. Due to the data requirements involved, our attack does not threaten the practical security of the cipher. For instance, restricting the attacker to not use more than 2^{128} data in total even for AEGIS-256, independently of the keys involved, would most likely prevent this type of attack entirely.

We also investigate linear biases of AEGIS-128, and find a bias of 2^{-77} between outputs of the cipher at rounds i and $i + 2$. While this would require more than 2^{128} data, it is still worth noting, as this bias is vastly superior to any generic attack, considering the inner state is 640-bit long. In Appendix B, we investigate to what extent linear hull effects as well as multilinear techniques can be expected to reduce the data requirements. We find that around 2^{140} data would likely still be required, showing that AEGIS-128 should be safe from our attack.

The first section provides a brief description of AEGIS-128 and AEGIS-256 encryption. In the second section, we define linear biases, and study some linear biases linking substates of AEGIS. From there, we deduce biases in the keystream of AEGIS-128 and AEGIS-256. Finally, we show how these biases can be exploited to mount an attack.

1.1   Notations

For: n an integer
        X a n-bit vector
        Y a n-bit vector
        α a n-bit vector
Define: $X \oplus Y$ the bitwise XOR of X and Y
        $X \& Y$ the bitwise AND of X and Y
        $\alpha \cdot X$ the scalar product of α and X
        $|\alpha|$ the Hamming weight of α

2   Description of AEGIS-128 and AEGIS-256

When AEGIS was first introduced at SAC 2013, it came in two variants, AEGIS-128 and AEGIS-256, providing a security level of 128 and 256 bits respectively. In the CAESAR proposal, a new variant is introduced, AEGIS-128L, which fully leverages the 8-stage AES pipeline provided by Intel Sandy Bridge processors. In this paper, we focus on AEGIS-128 and AEGIS-256 in their most recent version, namely the CAESAR submission [1].

2.1   AEGIS-128

AEGIS-128 takes as parameters a 128-bit key, a 128-bit nonce, and a tag length less than or equal to 128. It proceeds in several stages: initialization, where the 640-bit inner state is initialized using the key and nonce; processing of the authenticated data, where optional associated data is integrated into the state; encryption proper, where a variable-length plaintext is encrypted into a ciphertext of the same length; and finalization, which produces an authentication tag from the inner state. Hereafter we are only interested in the encryption step. A complete description of AEGIS can be found in [9].

The inner state of AEGIS-128 consists of five 128-bit substates $S_0, \ldots, S_4$. The plaintext is divided into 128-bit blocks m_i, $i \geq 0$, and processed in successive rounds. Let us denote by $S_{i,0}, \ldots, S_{i,4}$ the values of the substates at round i. For simplicity, we set $i = 0$ when encryption begins, setting aside the initialization step as well as the processing of authenticated data.

Then we have:

$$S_{i+1,0} = S_{i,0} \oplus \mathrm{R}(S_{i,4}) \oplus m_i$$
$$S_{i+1,1} = S_{i,1} \oplus \mathrm{R}(S_{i,0})$$
$$S_{i+1,2} = S_{i,2} \oplus \mathrm{R}(S_{i,1})$$
$$S_{i+1,3} = S_{i,3} \oplus \mathrm{R}(S_{i,2})$$
$$S_{i+1,4} = S_{i,4} \oplus \mathrm{R}(S_{i,3})$$

where R denotes a single round of AES-128 [5], with no key addition. The state update function is depicted in Fig. 1.

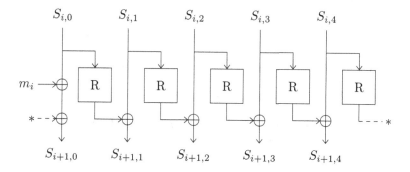

Fig. 1. State update function of AEGIS-128.

Each round, the ciphertext is output as:

$$C_i = S_{i,1} \oplus S_{i,4} \oplus (S_{i,2} \,\&\, S_{i,3}) \oplus m_i$$

2.2   AEGIS-256

AEGIS-256 takes as parameters a 256-bit key, a 256-bit nonce, and a tag length less than or equal to 128. The encryption step is very much the same as that of AEGIS-128, except the inner state consists of six rather than five 128-bit substates $S_{i,0}, \ldots, S_{i,5}$. The state update function may be written as:

$$S_{i+1,0} = S_{i,0} \oplus \mathrm{R}(S_{i,5}) \oplus m_i$$
$$S_{i+1,1} = S_{i,1} \oplus \mathrm{R}(S_{i,0})$$
$$S_{i+1,2} = S_{i,2} \oplus \mathrm{R}(S_{i,1})$$
$$S_{i+1,3} = S_{i,3} \oplus \mathrm{R}(S_{i,2})$$
$$S_{i+1,4} = S_{i,4} \oplus \mathrm{R}(S_{i,3})$$
$$S_{i+1,5} = S_{i,5} \oplus \mathrm{R}(S_{i,4})$$

Each round, the ciphertext is output as:

$$C_i = S_{i,1} \oplus S_{i,4} \oplus S_{i,5} \oplus (S_{i,2} \,\&\, S_{i,3}) \oplus m_i$$

2.3 Security Claims

AEGIS-128 and AEGIS-256 claim a security level of respectively 128 and 256 bits for plaintext confidentiality (provided the attacker did not first break the integrity of the scheme, for which a security level of 128 bits is claimed in both cases–cf. [9], Sect. 3). There is no explicit bound on data requirements.

3 Preliminaries

3.1 Linear Biases and Weights

Since we will typically deal with probabilities very close to $1/2$, it is convenient to define the *bias* of an event (as a shortcut for the bias of its probability):

Definition 1 (Bias). *The* bias *of a an event E is defined as:*

$$\mathrm{Bias}(E) = 2 \cdot \mathrm{Prob}(E) - 1$$

Definition 2 (Linear Bias). *Consider a function $F : \{0,1\}^n \to \{0,1\}^n$ from n bits to n bits. Given an* input mask *$\alpha \in \{0,1\}^n$ and* output mask *$\beta \in \{0,1\}^n$, the* linear bias *of F with masks α, β, is defined as:*

$$\mathrm{Bias}(\alpha \cdot X \oplus \beta \cdot F(X) = 0)$$

with X uniformly random in $\{0,1\}^n$.

Matsui's classic piling-up lemma [7] is commonly used to combine linear biases together.

Lemma 1 (Piling-up Lemma). *Let $X_1, \ldots, X_n$ be independent random binary variables. Then:*

$$\mathrm{Bias}(X_1 \oplus \cdots \oplus X_n = 0) = \mathrm{Bias}(X_1 = 0) \times \cdots \times \mathrm{Bias}(X_n = 0)$$

In the rest of this article, biases will often be of the form $\pm 2^{-i}$, with i an integer. This leads to the following definitions:

Definition 3 (Weight of an Event). *Let E be an event. The* weight *of E is the positive real:*

$$\mathrm{weight}(E) = -\log_2\big(|\mathrm{Bias}(E)|\big)$$

If the bias is zero, we define the weight as ∞.

Definition 4 (Weight of a Linear Bias). *The* weight *of a linear bias is the weight of its bias. That is, with the previous notations:*

$$\text{weight}(F, \alpha, \beta) = -\log_2\big(|\text{Bias}(\alpha \cdot X \oplus \beta \cdot F(X) = 0)|\big)$$

While the notion of weight is more prevalent in differential than in linear cryptanalysis, we have defined it so that it behaves exactly in the same way: due to Lemma 1, when combining linear characteristics, we simply add their weights together, making computations more readable. Since we will combine linear characteristics repeatedly, the benefits are substantial. Note that finding strong biases means we always want to minimize weights.

3.2   Linear Approximations of Bitwise AND

For x, y two independent uniformly random binary variables, it can be easily checked that their product $x\&y$ can be linearly approximated in four different ways: 0, x, y and $x \oplus y \oplus 1$, each with probability $3/4$. In particular, this implies the following lemma, which will be quite useful:

Lemma 2. *Let X, Y be two independent uniformly random variables in $\{0, 1\}^n$, and α be a linear mask in $\{0, 1\}^n$. Then:*

$$\text{weight}\big(\alpha \cdot (X\&Y) = 0\big)$$
$$= \text{weight}\big(\alpha \cdot (X\&Y \oplus X) = 0\big)$$
$$= \text{weight}\big(\alpha \cdot (X\&Y \oplus Y) = 0\big)$$
$$= \text{weight}\big(\alpha \cdot (X\&Y \oplus X \oplus Y \oplus 1) = 0\big)$$
$$= |\alpha|$$

where $|\alpha|$ denotes the Hamming weight of α. The biases are all positive.

4   Linear Biases for AEGIS-128 and AEGIS-256

4.1   Linear Biases Between Substates

The output of AEGIS-128 at round i is $C_i = S_{i,1} \oplus S_{i,4} \oplus (S_{i,2} \& S_{i,3}) \oplus m_i$. Using linear approximations of $\&$ in the previous section, this can naturally be approximated as a sum of some substates $S_{i,j}$'s. As a preliminary step towards exhibiting biases in the AEGIS-128 keystream, we point out some useful linear relations between substates $S_{i,j}$'s over three rounds.

Assume that at some round i, three consecutive plaintext blocks m_i, m_{i+1}, m_{i+2} are all-zeros. Denote $S_0 = S_{i,0}, \ldots, S_4 = S_{i,4}$. Then we can compute the value of substate 0 over the three rounds i, $i+1$, $i+2$ as:

$$S_{i,0} = S_0$$
$$S_{i+1,0} = S_0 \oplus R(S_4)$$
$$S_{i+2,0} = S_0 \oplus R(S_4) \oplus R(S_4 \oplus R(S_3))$$

We are interested in the two differences $S_{i,0} \oplus S_{i+1,0}$ and $S_{i,0} \oplus S_{i+2,0}$. Let us begin with the first:

$$S_0 \oplus S_{i+1,0} = R(S_4)$$

If we choose any linear mask α, β with $w = \text{weight}(R, \alpha, \beta)$, then by definition we have:

$$\beta \cdot (S_0 \oplus S_{i+1,0}) = \alpha \cdot S_4 \qquad \text{with weight } w \quad (1)$$

Now consider the second difference:

$$S_{i+2,0} \oplus S_{i,0} = R(S_4) \oplus R(S_4 \oplus R(S_3))$$

This is the derivative of R at point S_4 with difference $R(S_3)$. Choose two linear masks β, γ with $w' = \text{weight}(R, \beta, \gamma)$. By the piling-up lemma, we get:

$$\gamma \cdot (S_{i+2,0} \oplus S_{i,0}) = \beta \cdot S_4 \oplus \beta \cdot (S_4 \oplus R(S_3)) \qquad \text{with weight } 2w' \quad (2)$$
$$= \beta \cdot R(S_3)$$

Thus, the contribution of S_4 cancels itself out.

Finally, we can combine the previous linear approximation of R along α, β with (2) to get:

$$\gamma \cdot (S_{i+2,0} \oplus S_{i,0}) = \alpha \cdot S_3 \qquad \text{with weight } w + 2w' \quad (3)$$

Note that the above approximations also hold for $S_{i,1}, \ldots, S_{i,4}$ by shifting all $S_{i,j}$'s involved along j modulo 5. Furthermore the same equalities hold for AEGIS-256 as well, except S_3 and S_4 in all three Eqs. (1), (2), (3) become S_4 and S_5. The main takeaway in all cases is that $S_{i+1,j} \oplus S_{i,j}$ is correlated to $S_{i,j-1}$, while $S_{i+2,j} \oplus S_{i,j}$ is correlated to $S_{i,j-2}$.

In the end, we will want to choose α, γ so as to minimize $w + 2w'$ in (3). This involves considering linear propagation over two rounds of AES. Due to the branching number of 5 of the AES construction [6], we will have at least 5 active S-boxes over these two rounds. Moreover, since we want to minimize $w + 2w'$, the second round incurs twice the cost, so the optimal configuration would be to have 4 active S-boxes in the first round, and only one in the second round. This is easily achieved: choosing any linear masks at the input and output of a single S-box in the second round, then propagating the masks linearly will have the desired effect (cf. Fig. 2). In fact, there are enough degrees of freedom to pass all S-boxes with the optimal linear weight of 3. As a result, we get $w = 4 \cdot 3 = 12$ and $w' = 3$, so $w + 2w' = 18$. Appendix A gives specific values for α, β, γ.

4.2   Biases for AEGIS-128

In this section, we will exhibit a linear bias between the output of AEGIS-128 at rounds i and $i + 2$, assuming that the messages m_i, m_{i+1}, m_{i+2} are all-zeros. Let us define $S_0 = S_{i,0}, \ldots, S_4 = S_{i,4}$.

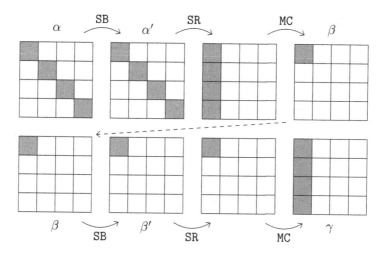

Fig. 2. Linear masks over two rounds of AES. Grey boxes denote active bytes.

Choose α, β, γ as in the previous section. Recall that $C_i = S_1 \oplus S_4 \oplus S_2 \,\&\, S_3$. Using Sect. 3.2, we can approximate C_i and C_{i+2} as:

$$\gamma \cdot C_i = \gamma \cdot (S_1 \,\oplus\, S_4 \,\oplus\, S_3) \qquad \text{with weight } |\gamma|$$
$$\gamma \cdot C_{i+2} = \gamma \cdot (S_{i+2,1} \,\oplus\, S_{i+2,4} \,\oplus\, S_{i+2,3}) \qquad \text{with weight } |\gamma|$$

It follows from Eq. (3) in the previous section that we have:

$$\gamma \cdot (C_i \oplus C_{i+2}) = \alpha \cdot (S_4 \,\oplus\, S_2 \,\oplus\, S_1) \quad \text{with weight } 3w + 6w' + 2|\gamma|$$

Now, observe that C_i may also be approximated as:

$$\alpha \cdot C_i = \alpha \cdot (S_1 \,\oplus\, S_4 \,\oplus\, S_2) \qquad \text{with weight } |\alpha|$$

We are now approximating C_i bitwise in two different ways. However, as long as α and γ have disjoint support, the two events are independent. In the remainder, we assume this is the case.

If we combine the last two equations together, we get:

$$(\alpha \oplus \gamma) \cdot C_i \oplus \gamma \cdot C_{i+2} = 0 \qquad \text{with weight } 3w + 6w' + |\alpha| + 2|\gamma|$$

This is an absolute bias on the AEGIS-128 keystream. Note that in order to simplify the presentation, we did not keep track of whether the bias is positive or negative; however, this is fixed and known.

The question now becomes how to choose α, γ so as to minimize the weight above. Details of this computation are provided in Appendix A. In the end, we obtain $|\alpha| = 5$, $|\gamma| = 9$, with all S-boxes having optimal linear bias, hence $w = 12$, $w' = 3$ as in the previous section. This yields $3w + 6w' + |\alpha| + 2|\gamma| = 77$.

4.3   Biases for AEGIS-256

Biases on the AEGIS-256 keystream are built essentially in the same way as for
AEGIS-128, except the outputs of all three rounds i, $i+1$ and $i+2$ are necessary.
Again, we assume $m_i = m_{i+1} = m_{i+2} = 0$. Recall that $C_i = S_1 \oplus S_4 \oplus S_5 \oplus S_2 \& S_3$.

We use the following approximations:

$$\alpha \cdot C_i = \alpha \cdot (S_1 \oplus S_4 \oplus S_5) \qquad \text{with weight } |\alpha|$$

$$\beta \cdot C_i = \beta \cdot (S_1 \oplus S_4 \oplus S_5 \oplus S_2 \oplus S_3) \qquad \text{with weight } |\beta|$$

$$\gamma \cdot C_i = \gamma \cdot (S_1 \oplus S_4 \oplus S_5 \oplus S_2) \qquad \text{with weight } |\gamma|$$

$$\beta \cdot C_{i+1} = \beta \cdot (S_{i+1,1} \oplus S_{i+1,4} \oplus S_{i+1,5} \oplus S_{i+1,2} \oplus S_{i+1,3}) \quad \text{with weight } |\beta|$$

$$\gamma \cdot C_{i+2} = \gamma \cdot (S_{i+2,1} \oplus S_{i+2,4} \oplus S_{i+2,5} \oplus S_{i+2,2}) \qquad \text{with weight } |\gamma|$$

Using Eq. (2) from Sect. 4.1, we have:

$$\gamma \cdot (C_i \oplus C_{i+2}) = \beta \cdot (R(S_5) \oplus R(S_2) \oplus R(S_3) \oplus R(S_0))$$
$$\text{with weight } 8w' + 2|\gamma|$$

On the other hand:

$$\beta \cdot (C_i \oplus C_{i+1}) = \beta \cdot (R(S_0) \oplus R(S_3) \oplus R(S_4) \oplus R(S_1) \oplus R(S_2))$$
$$\text{with weight } 2|\beta|$$

Summing the last two equalities yields:

$$\beta \cdot (C_i \oplus C_{i+1}) \oplus \gamma \cdot (C_i \oplus C_{i+2}) = \beta \cdot (R(S_1) \oplus R(S_4) \oplus R(S_5))$$
$$\text{with weight } 8w' + 2|\beta| + 2|\gamma|$$

Now it remains to use Eq. (1) to pass through R and get:

$$\beta \cdot (C_i \oplus C_{i+1}) \oplus \gamma \cdot (C_i \oplus C_{i+2}) = \alpha \cdot (S_1 \oplus S_4 \oplus S_5)$$
$$\text{with weight } 3w + 8w' + 2|\beta| + 2|\gamma|$$

Finally:

$$\alpha \cdot C_i \oplus \beta \cdot (C_i \oplus C_{i+1}) \oplus \gamma \cdot (C_i \oplus C_{i+2}) = 0$$
$$\text{with weight } 3w + 8w' + |\alpha| + 2|\beta| + 2|\gamma|$$

Now the question is to find α, β, γ minimizing this weight. In fact, we can
choose precisely the same α, γ as for AEGIS-128. Indeed, these masks were
chosen so as to (1) pass all S-boxes with optimal probability; (2) minimize w
as compared to w'; (3) minimize $|\alpha| + 2|\gamma|$ within the previous constraints. The
same criterions are very fitting once again; the only difference is the new β term,
but with the previous choices $|\beta| = 3$, so it is nearly optimal as well. As a result,
we have a weight of $3 \cdot 12 + 8 \cdot 3 + 5 + 2 \cdot 3 + 2 \cdot 9 = 89$.

Intuition. How the previous linear approximations were chosen so as to cancel each other out, and perhaps more importantly what made such a choice possible, may not be immediately apparent from the description of the linear characteristic itself. As a result, it may be useful to provide some intuition.

We know that $S_j \oplus S_{i+1,j} = \mathrm{R}(S_{j-1})$. From Eq. (2), $S_j \oplus S_{i+2,j} = \mathrm{R}(S_{j-1}) \oplus \mathrm{R}(S_{j-1} \oplus \mathrm{R}(S_{j-2}))$ is linearly correlated to $\mathrm{R}(S_{j-2})$, with the contribution of S_{j-1} cancelling itself out; so we may write $S_j \oplus S_{i+2,j} = \mathrm{D}(\mathrm{R}(S_{j-2}))$, where D is a purely formal notation to indicate an expression that is linearly correlated to the input of D.

On the other hand, if we approximate the & operation in C_i and C_{i+1} linearly along the same mask, and add them together, we can ensure that every $S_{i+1,j}$ is matched with the corresponding S_j, so as a result we can roughly write:

$$C_{i+1} \oplus C_i \approx \quad \mathrm{R}(S_0) \oplus \quad [\mathrm{R}(S_1)] \oplus \quad [\mathrm{R}(S_2)] \oplus \quad \mathrm{R}(S_3) \oplus \quad \mathrm{R}(S_4)$$

where the brackets denote a term that comes from a & operation and thus may be omitted at will by Sect. 3.2.

The same reasoning holds for C_{i+2}; and in the end we have:

$C_i \approx$		$S_1 \oplus$	$[S_2] \oplus$	$[S_3] \oplus$	$S_4 \oplus$	S_5
$C_{i+1} \oplus C_i \approx$	$\mathrm{R}(S_0) \oplus$	$[\mathrm{R}(S_1)] \oplus$	$[\mathrm{R}(S_2)] \oplus$	$\mathrm{R}(S_3) \oplus$	$\mathrm{R}(S_4)$	
$C_{i+2} \oplus C_i \approx$	$[\mathrm{D}(\mathrm{R}(S_0))] \oplus$	$[\mathrm{D}(\mathrm{R}(S_1))] \oplus$	$\mathrm{D}(\mathrm{R}(S_2)) \oplus$	$\mathrm{D}(\mathrm{R}(S_3)) \oplus$		$\mathrm{D}(\mathrm{R}(S_5))$

Now take the characteristic $\alpha \xrightarrow{\mathrm{R}} \beta \xrightarrow{\mathrm{R}} \gamma$ from the previous section, which is also a characteristic $\alpha \xrightarrow{\mathrm{R}} \beta \xrightarrow{\mathrm{D}} \gamma$ as can be seen in Sect. 4.1, Eq. (2). If the characteristics hold, this tells us that $\alpha \cdot S_k = \beta \cdot \mathrm{R}(S_k) = \gamma \cdot \mathrm{D}(\mathrm{R}(S_k))$. Hence, if we approximate the first line along α, the second along β, and the third along γ, if the linear characteristics hold and we add up everything, any two terms in the same column will cancel each other out. So the question becomes simply how to make an appropriate choice for each bracket in the equations above so that there is 0 or 2 terms in each column. This is exactly what we do in order to construct our linear characteristic, namely:

$C_i \approx$		$S_1 \oplus$		$S_4 \oplus$	S_5
$C_{i+1} \oplus C_i \approx$	$\mathrm{R}(S_0) \oplus$	$\mathrm{R}(S_1) \oplus$	$\mathrm{R}(S_2) \oplus$	$\mathrm{R}(S_3) \oplus$	$\mathrm{R}(S_4)$	
$C_{i+2} \oplus C_i \approx$	$\mathrm{D}(\mathrm{R}(S_0)) \oplus$		$\mathrm{D}(\mathrm{R}(S_2)) \oplus$	$\mathrm{D}(\mathrm{R}(S_3)) \oplus$		$\mathrm{D}(\mathrm{R}(S_5))$

If we look at AEGIS-128 from the same perspective, we can write:

$C_i \approx$		$S_1 \oplus$	$[S_2] \oplus$	$[S_3] \oplus$	S_4
$C_{i+1} \oplus C_i \approx$	$\mathrm{R}(S_0) \oplus$	$[\mathrm{R}(S_1)] \oplus$	$[\mathrm{R}(S_2)] \oplus$	$\mathrm{R}(S_3)$	
$C_{i+2} \oplus C_i \approx$	$[\mathrm{D}(\mathrm{R}(S_0))] \oplus$	$[\mathrm{D}(\mathrm{R}(S_1))] \oplus$	$\mathrm{D}(\mathrm{R}(S_2)) \oplus$		$\mathrm{D}(\mathrm{R}(S_4))$

After removing the second line entirely, the approximation we made is:

$C_i \approx$	$S_1 \oplus$	$S_2 \oplus$	S_4
$C_{i+2} \oplus C_i \approx$	$\mathrm{D}(\mathrm{R}(S_1)) \oplus$	$\mathrm{D}(\mathrm{R}(S_2)) \oplus$	$\mathrm{D}(\mathrm{R}(S_4))$

4.4 Exploiting the Keystream Biases

In the previous two sections, we have assumed that at some round i, three consecutive plaintexts m_i, m_{i+1}, m_{i+2} are all-zeros. From there, we have shown the existence of an absolute bias of the form:

$$\text{Bias}(\alpha \cdot C_i \oplus \beta \cdot C_{i+1} \oplus \gamma \cdot C_{i+2} \oplus b = 0) = 2^{-w}$$

In other words, we have built a distinguisher on the AEGIS keystream. However, if we no longer assume $m_{i+2} = 0$, then at round $i + 2$, the only difference in the output of the cipher is that m_{i+2} is XOR-ed into the ciphertext C_{i+2}. As a result, we have:

$$\text{Bias}(\alpha \cdot C_i \oplus \beta \cdot C_{i+1} \oplus \gamma \cdot C_{i+2} \oplus \gamma \cdot m_{i+2} \oplus b = 0) = 2^{-w}$$

Thus, the observable value $\alpha \cdot C_i \oplus \beta \cdot C_{i+1} \oplus \gamma \cdot C_{i+2}$ directly leaks information about $\gamma \cdot m_{i+2}$.

This leads to the following attack scenario. Assume the same three consecutive plaintext blocks 0, 0, m are encrypted 2^{2w} times in total, independently of the keys and nonces used. Then an attacker having access to that data would deduce the value of $\gamma \cdot m$ with good probability, just by counting the occurences of the event $\alpha \cdot C_i \oplus \beta \cdot C_{i+1} \oplus \gamma \cdot C_{i+2} = 0$ on the fly. However, the data requirements make this attack impractical, since 2^{154} and 2^{188} encryptions would be required respectively for AEGIS-128 and AEGIS-256 in order to exploit a single bias (as opposed to a multilinear approach). In Appendix B, we try to capture linear propagation in AEGIS-128 more accurately, in order to evaluate to what extent data requirements could be lowered; we conclude that AEGIS-128 seems to resist straightforward improvements of our attack, as 2^{140} data is still required.

5 Conclusion

In this article, we have constructed linear biases in the keystream of AEGIS-128 and AEGIS-256. These biases stem from dependencies between surprisingly few consecutive rounds: for AEGIS-128, linear biases exist between the outputs of rounds i and $i + 2$; while for AEGIS-256, three consecutive rounds are enough. Our main result is the construction of a linear mask with bias 2^{-89} on the keystream of AEGIS-256. This bias can be exploited to recover bits of information on a partially known plaintext encrypted 2^{188} times, regardless of the keys involved. While the biases remain too low to be a threat in practice, they are vastly superior to any generic attack, and point out an unexpected property in the keystream of AEGIS.

Acknowledgments. The author would like to thank all members of the ANSSI cryptography laboratory, especially Thomas Fuhr and Henri Gilbert, for their valuable comments and insights on this article.

A   Values of α, β, γ

Consider the situation depicted on Fig. 3, where the linear characteristic $\alpha \rightarrow \beta \rightarrow \gamma$ spans two rounds of AES (without key addition). We are trying to minimize $|\alpha| + 2|\gamma|$, while passing all S-boxes with optimal linear probability.

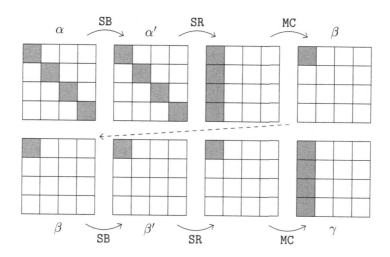

Fig. 3. Linear masks over two rounds of AES. Grey boxes denote active bytes.

First, we look for β' minimizing $|\gamma|$. With little-endian hexadecimal notations, it turns out only one value $\beta' = $ 0e reaches the minimum with $|\gamma| = 9$ (γ equals f, 8, c, 5 along one column). For this value, five choices of β allow us to pass the S-box with weight 3: $\beta = $ 09, 31, 38, c8, or f9. For each of these values, we compute α', then look for the minimal size of α such that all four S-boxes are passed with optimal probability. We find $|\alpha| = 5$, for $\beta = $ 38 (α equals 4, 3, 2, 2 along the diagonal). Observe that $|\alpha|$ is at least 4 since it has to span 4 S-boxes. Since we followed the only way to have $|\gamma| = 9$, which is optimal; $|\alpha|$ is within 1 of being optimal; and we are trying to minimize $|\alpha| + 2|\gamma|$, we have found the unique optimal choice.

More accurately, it is the unique optimal choice once we have fixed our choice for the active S-box in the second round. In fact, we could choose any one of the other 15 S-boxes, and use the exact same values of linear masks β, β' at the entrance of that S-box (namely, 38 and 0e). Indeed, the circulant nature of the AES MixColums matrix means that a given mask at the input (resp. output) of one S-box will propagate to a permutation of the same masks at the output (resp. input) of the previous (resp. next) layer of S-boxes, and hence in our case will yield the same sizes for α and γ. Thus there are 16 choices for α, γ with identical properties for our purpose; one per choice of active S-box in the middle round.

B   Refined Linear Model of AEGIS-128

In Sect. 4.2, we have found a linear mask with bias 2^{-77} on the AEGIS-128 keystream. Detecting this bias would require 2^{154} data, which is significantly more than is sensible with a security parameter of 128 bits. However, the bias on an actual AEGIS-128 keystream may be slightly different, due to the independence assumptions required by our analysis. Additionaly, there may be linear hull effects strengthening or weakening the bias. In any case, other biases of comparable strength undoubtedly exist, and could be exploited in a multilinear attack. With all this in mind, it may be worth wondering whether the 2^{154} data requirement has some chance of being brought down below 2^{128}.

It so happens that for AEGIS-128, there is a fairly elegant way of simultaneously taking into account many of the effects listed above. In our previous analysis, we used standard linear cryptanalysis techniques to follow the propagation of a bias along a few AES-based transformations. This amounts to modelling the transformations in a certain way, materialized by independence assumptions. However in the case of AEGIS-128, large parts of the transformations can be computed with complete accuracy by looking at byte distributions, without the need to model anything.

If we recap the previous analysis in Sect. 4.2, we approximate C_i and $C_i \oplus C_{i+2}$ linearly, and from there we obtain the following two sums:

$$S_1 \oplus S_2 \oplus S_4$$
$$and: \quad R(S_2) \oplus R(S_2 \oplus R(S_1)) \oplus R(S_3) \oplus R(S_3 \oplus R(S_2)) \oplus R(S_0) \oplus R(S_0 \oplus R(S_4))$$
$$= D(R(S_1)) \oplus D(R(S_2)) \oplus D(R(S_4)).$$

where D is a purely formal notation denoting the fact that its input and output are linearly correlated (cf. Sect. 4.1); then we use the fact that X and $D(R(X))$ are correlated. More precisely, we first relate X to $R(X)$, then $R(X)$ to $D(R(X))$. Thus the propagation is decomposed in two steps, which we can picture as:

$$S_1 \oplus S_2 \oplus S_4 \quad \rightarrow \quad R(S_1) \oplus R(S_2) \oplus R(S_4) \quad \rightarrow \quad D(R(S_1)) \oplus D(R(S_2)) \oplus D(R(S_4))$$

So our propagation "factors" through the value $R(S_1) \oplus R(S_2) \oplus R(S_4)$: that is to say, all information we had on $S_1 \oplus S_2 \oplus S_4$ is first translated as information on $R(S_1) \oplus R(S_2) \oplus R(S_4)$; after which only information on $R(S_1) \oplus R(S_2) \oplus R(S_4)$ is used to deduce information on $D(R(S_1)) \oplus D(R(S_2)) \oplus D(R(S_4))$. Moreover, with our linear masks, only a single S-box is active in $R(S_1) \oplus R(S_2) \oplus R(S_4)$, so actually the whole propagation factors through the value of $R(S_1) \oplus R(S_2) \oplus R(S_4)$ on a single byte.

The idea for our new model is that we are going to compute the actual distribution of $R(S_1) \oplus R(S_2) \oplus R(S_4)$ on one byte from the knowledge of $C_i = S_1 \oplus S_2 \ \& \ S_3 \oplus S_4$. Then we are going to use this full distribution, rather than a single linear mask, to compute a distribution of $S_{i+2,3} \oplus S_3 \ \oplus \ S_{i+2,4} \oplus S_4 \ \oplus \ S_{i+2,1} \oplus S_1$, which is our linear approximation of $C_i \oplus C_{i+2}$ along a linear mask γ. Thus we hope to compute the bias of $\gamma \cdot (C_i \oplus C_{i+2})$ more accurately. A good motivation for this model is that the two steps above: linking knowledge of C_i to the distribution of $R(S_1) \oplus R(S_2) \oplus R(S_4)$ on one byte; and then the distribution of $R(S_1) \oplus R(S_2) \oplus R(S_4)$ on one byte to the bias of the linear approximation of $\gamma \cdot (C_i \oplus C_{i+2})$, can be computed with perfect precision within complexity at most 2^{32}, as we show below. Hence the only loss of precision results from "factoring" through $R(S_1) \oplus R(S_2) \oplus R(S_4)$; but as we saw in the previous paragraph, we were already making this approximation when we used standard linear characteristic techniques.

Thus, assume we know some specific value for $C_i = S_1 \oplus S_2 \,\&\, S_3 \oplus S_4$. Then we can actually compute the distribution of a single byte of $R(S_1) \oplus R(S_2) \oplus R(S_4)$ with full precision. Indeed, if we denote by SB the SubBytes layer of AES, from $S_1 \oplus S_2 \,\&\, S_3 \oplus S_4$ we can compute the distribution of $SB(S_1) \oplus SB(S_2) \oplus SB(S_4)$ in an exact manner, since each byte depends only on the value S_1, S_2, S_3, S_4 on the same byte; so we need only guess 4 bytes simultaneously.

From there, we can also compute the distribution of $R(S_1) \oplus R(S_2) \oplus R(S_4)$ on a single byte exactly, since it is simply the independent sum of the previous byte distributions through the MixColumns matrix. Moreover, in the end, this distribution depends on only 16 bytes in total, which is 128 bits, so we can simply count how many choices lead to a specific value using a 128-bit integer, and the resulting distribution is prefectly precise. Thus, from knowledge of $C_i = S_1 \oplus S_2 \,\&\, S_3 \oplus S_4$, it is possible to compute the distribution of $R(S_1) \oplus R(S_2) \oplus R(S_4)$ on one byte with full precision.

Now for the second step of the propagation, we want to compute the distribution of the following value (i.e. the linear approximation of $C_i \oplus C_{i+2}$ along the mask γ):

$$S_{i+2,3} \oplus S_3 \ \oplus \ S_{i+2,4} \oplus S_4 \ \oplus \ S_{i+2,1} \oplus S_1$$
$$= R(S_2) \oplus R(S_2 \oplus R(S_1)) \ \oplus \ R(S_3) \oplus R(S_3 \oplus R(S_2)) \ \oplus \ R(S_0) \oplus (S_0 \oplus R(S_4))$$

from the known distribution of:

$$R(S_1) \oplus R(S_2) \oplus R(S_4)$$

More precisely, we are interested in the distribution of the first value before MixColumns (which is the last operation applied to each component), on a single byte, so R reduces to one S-box layer (and a permutation of the bytes).

At first sight, it suffices to guess the values of all $R(S_i)$'s on this one byte. This only involves guessing 5 bytes, requiring 2^{40} operations, which is reasonable. However a better algorithm is possible by observing that the contribution of S_0 and S_4 is independent from the rest on both sides. As a result, it suffices to compute these two distributions separately, then add them together:

$$R(S_1) \oplus R(S_2) \rightarrow R(S_2) \oplus R(S_2 \oplus R(S_1)) \ \oplus \ R(S_3) \oplus R(S_3 \oplus R(S_2))$$
$$R(S_4) \rightarrow R(S_0) \oplus (S_0 \oplus R(S_4))$$

Thus the complexity drops down to 2^{32} operations.

The end result is that for a fixed value of $C_i = S_1 \oplus S_2 \,\&\, S_3 \oplus S_4$, we can compute the distribution of $R(S_1) \oplus R(S_2) \oplus R(S_4)$ on one byte without any approximation. Then from this distribution, we can compute the distribution of one byte of the linear approximation of $C_i \oplus C_{i+2}$ before `MixColumns`, which is what we measure from $C_i \oplus C_{i+2}$ using our linear masks, modulo the cost of the linear approximation along γ, which is $2|\gamma|$.

We implemented this model, and results correlate fairly well with our previous analysis. In particular, we recover the fact that the values of β' we chose yields the strongest bias, although one other value seems as strong (namely 12), which is not too surprising since it is one of the two second-best candidates as far as minimizing $|\gamma|$. The main difference is that we find a bias close to 2^{-72} when we fix C_i to some random value and measure $\gamma \cdot (C_i \oplus C_{i+2})$ according to our model, rather than 2^{-77} when we used pure linear masking. We surmise this is mostly due to more information being taken into account at the input, resulting overall in more information at the output; although both steps of the new model behave slightly better than expected.

On the other hand, few output masks yield biases in this vicinity. If we exploit the best bias at 2^{-72}, across all 16 possible second-round S-boxes (cf. Appendix A), $2^{144-4} = 2^{140}$ data would still be required to mount an attack. With additional improvements, one could hope to further reduce data requirements, but at this point it seems very unlikely that data requirements could fall below 2^{128}. Thus, the main conclusion of our model seems to be that AEGIS-128 remains resistant to straightforward improvements of our attack.

References

1. CAESAR- Competition for Authenticated Encryption: Security, Applicability, and Robustness. General secretary D.J. Bernstein (2013). http://competitions.cr.yp.to/caesar.html
2. AlFardan, N., Bernstein, D.J., Paterson, K.G., Poettering, B., Schuldt, J.: On the security of RC4 in TLS. In: USENIX Security Symposium (2013), Presented at FSE 2013 as an invited talk by D.J. Bernstein (2013). http://www.isg.rhul.ac.uk/tls/
3. Bertoni, G., Daemen, J., Peeters, M., Van Assche, G.: Duplexing the sponge: single-pass authenticated encryption and other applications. In: Miri, A., Vaudenay, S. (eds.) SAC 2011. LNCS, vol. 7118, pp. 320–337. Springer, Heidelberg (2012)
4. Coppersmith, D., Halevi, S., Jutla, C.S.: Cryptanalysis of stream ciphers with linear masking. In: Yung, M. (ed.) CRYPTO 2002. LNCS, vol. 2442, pp. 515–532. Springer, Heidelberg (2002)
5. Daemen, J., Rijmen, V.: AES proposal: RIJNDAEL. Advanced Encryption Standard submission (1999). http://jda.noekeon.org/
6. Daemen, J., Rijmen, V.: The wide trail design strategy. In: Honary, B. (ed.) Cryptography and Coding 2001. LNCS, vol. 2260, pp. 222–238. Springer, Heidelberg (2001)
7. Matsui, M.: Linear cryptanalysis method for DES cipher. In: Helleseth, T. (ed.) EUROCRYPT 1993. LNCS, vol. 765, pp. 386–397. Springer, Heidelberg (1994)
8. Nikolić, I.: Tiaoxin-346. CAESAR submission (2014). http://competitions.cr.yp.to/round1/tiaoxinv1.pdf

9. Wu, H., Preneel, B.: AEGIS: a fast authenticated encryption algorithm. CAESAR submission, updated from Cryptology ePrint Archive Report 2013/695, updated from SAC 2013 version (2014). http://competitions.cr.yp.to/round1/aegisv1.pdf
10. Ye, D., Wang, P., Hu, L., Wang, L., Xie, Y., Sun, S., Wang, P.: PAES v1: paralleliz- able authenticated encryption schemes based on AES round function. CAESAR submission (2014). http://competitions.cr.yp.to/round1/paesv1.pdf

Chaskey: An Efficient MAC Algorithm
for 32-bit Microcontrollers

Nicky Mouha[1]([⊠]), Bart Mennink[1], Anthony Van Herrewege[1], Dai Watanabe[2],
Bart Preneel[1], and Ingrid Verbauwhede[1]

[1]Dept. of Electrical Engineering, ESAT/COSIC, KU Leuven and iMinds,
Ghent, Belgium
{nicky.mouha,bart.mennink,anthony.vanherrewege,bart.preneel,
ingrid.verbauwhede}@esat.kuleuven.be
[2]Yokohama Research Laboratory, Hitachi, Yokohama, Japan
dai.watanabe.td@hitachi.com

Abstract. We propose Chaskey: a very efficient Message Authentication Code (MAC) algorithm for 32-bit microcontrollers. It is intended for applications that require 128-bit security, yet cannot implement standard MAC algorithms because of stringent requirements on speed, energy consumption, or code size. Chaskey is a permutation-based MAC algorithm that uses the Addition-Rotation-XOR (ARX) design methodology. We prove that Chaskey is secure in the standard model, based on the security of an underlying Even-Mansour block cipher. Chaskey is designed to perform well on a wide range of 32-bit microcontrollers. Our benchmarks show that on the ARM Cortex-M3/M4, our Chaskey implementation reaches a speed of 7.0 cycles/byte, compared to 89.4 cycles/byte for AES-128-CMAC. For the ARM Cortex-M0, our benchmark results give 16.9 cycles/byte and 136.5 cycles/byte for Chaskey and AES-128-CMAC respectively.

Keywords: Microcontroller · Message authentication code · Standard model security · Permutation-based · ARX

1   Introduction

Message Authentication Code (MAC) algorithms are one of the basic building blocks for cryptographic systems. A MAC algorithm processes a message m and a secret key K to generate a tag τ. It should be hard for an attacker to construct a forgery: that is, to generate a valid combination of (m, τ) without knowledge of the secret key K. Thereby, the MAC algorithm ensures the authenticity of the message m.

Over the years, a large variety of MAC algorithms have been proposed. Some of the most commonly used algorithms today are CMAC [30,38], HMAC [6,60],

This work was supported in part by the Research Council KU Leuven: GOA TENSE (GOA/11/007) and OT/13/071. Nicky Mouha and Bart Mennink are Postdoctoral Fellows of the Research Foundation – Flanders (FWO).

A. Joux and A. Youssef (Eds.): SAC 2014, LNCS 8781, pp. 306–323, 2014.
DOI: 10.1007/978-3-319-13051-4_19

and UMAC [14]. CMAC is based on a block cipher, usually AES or Triple-DES, whereas HMAC uses a hash function such as MD5, SHA-1, or SHA-2, and UMAC is based on a universal hash function combined with a standard cryptographic primitive such as a block cipher or a hash function.

Unlike most other MAC algorithms, a nonce input is required for MAC algorithms based on universal hash functions [19,61]. This includes MAC algorithms such as UMAC [14], Poly1305-AES [9], and GMAC [31]. The nonce should not be reused, or this would lead to a forgery attack. Furthermore, Poly1305-AES and GMAC become insecure when tags are truncated [34]. We note that currently used MAC algorithms based on universal hash functions typically make use of multiplications. On several microcontrollers, the number of cycles required to execute an integer multiplication instruction is data-dependent, which makes the implementations potentially vulnerable to timing attacks [45].

For MAC algorithms that are based on hash functions, the block size is typically very large: for MD5, SHA-1, SHA-2, and the upcoming SHA-3 [10], messages are processed in blocks of at least 512 bits. For very short messages, this will result in a large overhead. But also for longer messages, it is generally undesirable for typical microcontrollers to process such large blocks. This is because many load and store operations are required to move data back and forth between the limited number of registers and the RAM, which significantly increases the time, energy, and code size of the MAC algorithm implementation.

A similar issue appears for block-cipher-based MAC algorithms, which typically use AES or Triple-DES. On typical microcontrollers, the key schedule of these block ciphers increases the register pressure: round keys must be either precomputed and stored in RAM, or computed on the fly. Furthermore, on 32-bit platforms, the S-box operations of AES and Triple-DES require extensive use of bit masking operations to implement the S-box operations, which again negatively impacts the speed of the implementation. Finally, we note that MAC algorithms based on reduced-round block ciphers such as ALPHA-MAC [22] and Pelican MAC [23] have been proposed, yet their performance gain is small for very short messages because a full-round block cipher is used for both initialization and finalization.

Chaskey

We present *Chaskey*, a permutation-based MAC algorithm that overcomes these issues. Chaskey takes a 128-bit key K and processes a message m in 128-bit blocks using a 128-bit permutation π. This permutation is based on the Addition-Rotation-XOR (ARX) design methodology. Its design is inspired by the permutation of SipHash [3], however with 32-bit instead of 64-bit words.

Chaskey has the following features:

- **Dedicated Design.** Chaskey is a dedicated design for 32-bit microcontroller architectures. The addition and XOR operations are performed on 32-bit words, and each of these operations requires only one instruction on these architectures.

- **Cross-Platform Versatility.** We took into account that certain microcontrollers do not support variable-length bit rotations and bit shifts. By choosing some rotation constants to be multiples of 8, these bit rotations are efficiently implemented by swapping 8-bit or 16-bit registers.
- **Efficient Implementation.** Benchmarks on an ARM Cortex-M4 show that Chaskey requires only 7.0 cycles/byte for long (≥ 128 byte) messages, and 10.6 cycles/byte for short (16 byte) messages. It has been implemented in only 402 bytes of ROM. Results for the Cortex-M0 are very good as well: 16.9 cycles/byte for long messages, 21.3 cycles/byte for short ones, and 414 bytes of ROM for the implementation. There is, roughly speaking, a linear relation between the number of cycles and energy consumption [21]. We therefore expect Chaskey to be very energy efficient as well.
- **Resistance Against Timing Attacks.** On all microcontroller architectures that we are aware of, every instruction of Chaskey takes a constant time to execute. The total number of cycles depends only on the message length. Therefore, Chaskey is inherently secure against timing attacks.
- **Key Agility.** Chaskey does not have a key schedule, as keys are simply XORed into the state. Updating the key in Chaskey requires generating a new uniformly random 128-bit key, and only two shifts and two conditional XORs on 128-bit words to generate two subkeys.
- **Tag Truncation.** Chaskey is robust under tag truncation. Unlike for example GMAC [34], the best attack on Chaskey with short tags is tag guessing. We recommend $|\tau| \geq 64$ for typical applications. Shorter tags may be used after careful analysis of the probability of occasionally accepting an inauthentic message.
- **Nonces are Optional.** Several MAC algorithms (including GMAC [31], VMAC [46], and Poly1305-AES [9]) require a nonce, and become completely insecure if this nonce is reused (see e.g. [40]). Chaskey does not require a nonce, and therefore avoids these issues altogether.
- **Provably Secure.** We prove that Chaskey is secure, based on the security of an Even-Mansour [32,33] block cipher based on π, up to about $D = 2^{64}$ blocks of chosen plaintexts and $T = 2^{128}/D$ off-line block cipher evaluations.
- **Patent-Free.** We are unaware of any patents or patent applications related to Chaskey.

The name Chaskey is derived from Chasqui, also written as Chaski. Chasquis were fast runners that delivered messages in the Inca empire. They were of short stature, and could cover large distances through mountainous areas with little nutrition available to them [52].

2   Preliminaries

Table 1 summarizes the notation used in this paper. Throughout, n is both the key size and the block size. While the Chaskey algorithm is introduced for $n = 128$, we remark that our statements on the Chaskey mode of operation are independent of this specific choice of n.

We interchangeably consider an element a of $GF(2^n)$ as an n-bit string $a[n-1]a[n-2]\ldots a[0]$ and as the polynomial $a(x) = a[n-1]x^{n-1}+a[n-2]x^{n-2}+\ldots+a[0]$ with binary coefficients. Let $f(x)$ be an irreducible polynomial of degree n with binary coefficients. For $n = 128$, we choose $f(x) = x^{128} + x^7 + x^2 + x + 1$. Then to multiply two elements a and b, we represent them as two polynomials $a(x)$ and $b(x)$, and calculate $a(x)b(x) \bmod f(x)$. For example, we show how to multiply an element by x in Algorithm 1. Note that x corresponds to bit string $0^{126}10$, which is 2 in decimal notation.

When converting between bit strings and arrays of 32-bit words, we always use little endian byte ordering. Inside every byte, bit numbering starts with the least significant bit.

Table 1. Notation.

Notation	Description
$x \| y$	concatenation of bit strings x and y
$|x|$	length of bit string x
$x + y$	addition of x and y modulo 2^{32} (in text)
$x \boxplus y$	addition of x and y modulo 2^{32} (in figures)
$x \lll s$	rotation of x to the left by s positions
$x \ll s$	shift of x to the left by s positions
$x \oplus y$	bitwise exclusive OR (XOR) of x and y
$\Delta^{\oplus}x$	XOR difference of x and x': $\Delta x = x \oplus x'$
0^a	bit string consisting of a times 0
$\mathrm{right}_t(a)$	select the t least significant bits of a
$x[i]$	bit selection: bit at position i of word x, where $i = 0$ is the least significant bit

3   Specification of Chaskey

3.1   Mode of Operation

Chaskey uses an n-bit key K to process a message m of arbitrary size into a tag τ of $t \leq n$ bits. For every key K, two subkeys K_1, K_2 are generated as shown in Algorithm 2.

The message m is split into ℓ blocks $m_1, m_2, \ldots, m_\ell$ of n bits each, except for the last block m_ℓ which may be incomplete. We define that an empty message $m = \varnothing$ consists of one empty block: $|m_1| = 0$. An n-bit permutation π then iterates over the message, as specified in Algorithm 3 and illustrated in Fig. 1.

An alternative description of Chaskey based on an Even-Mansour [32,33] block cipher E with $2n$-bit key and n-bit block size is given in Algorithm 4. This block-cipher-based description is equivalent to Chaskey once we define E using π as $E_{X\|Y}(m) = \pi(m \oplus X) \oplus Y$. The purpose of this block-cipher-based alternative is to reduce the security of Chaskey to the security of the underlying block cipher E. A security proof will be given in Sect. 5. This security proof views Chaskey-B as a variant of FCBC by Black and Rogaway [15,16], shown in Algorithm 5.

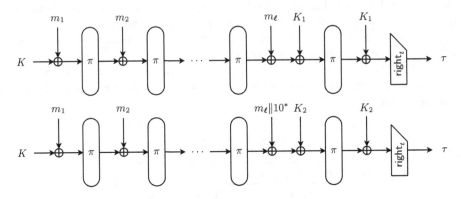

Fig. 1. The Chaskey mode of operation when $|m_\ell| = n$ (top), and when $0 \le |m_\ell| < n$ (bottom). The round function of permutation π is shown in Fig. 2, the subkeys K_1 and K_2 are generated according to Algorithm 2, and $m_\ell \| 10^*$ is shorthand for $m_\ell \| 10^{n-|m_\ell|-1}$.

From this block-cipher-based description, it can be seen that Chaskey is similar to the three-key MAC constructions proposed by Black and Rogaway [15, 16]. Their constructions are variants of CBC-MAC [1,37] that are secure for variable-length messages and avoid padding for messages of an integer number of blocks. As in CMAC [30,38], our algorithm requires only one n-bit key, from which two n-bit subkeys are generated. However, unlike CMAC, Chaskey does not require any block cipher calls to generate these two subkeys, only two shifts and two conditional XORs on 128-bit words.

Chaskey also differs from the CBC-MAC variants in literature because its underlying block cipher uses an Even-Mansour construction and as it uses the same subkey twice in the last two subkey XORs: before and after the last permutation call. Therefore, it is possible that this subkey (or part thereof) can remain inside the registers of the microcontroller. This reduces the number of load and store operations, which are very expensive on typical microcontrollers.

Every key K must be chosen independently and uniformly at random from the entire key space. To avoid attacks with a practical complexity of off-line permutation evaluations, as will be explained in Sect. 6.1, we restrict the total number of blocks to be authenticated under the same key K to at most 2^{48}. This corresponds to refreshing the key after at most 4 petabytes of data. To avoid tag guessing attacks, we recommend that the tag size $|\tau| \ge 64$. Changing $|\tau|$ always requires selecting a new key K uniformly at random.

Algorithm 1. TimesTwo

1: **proc** TimesTwo(a)
2:    **if** $a[127] = 0$ **then**
3:        **return** $(a \ll 1) \oplus 0^{128}$
4:    **else**
5:        **return** $(a \ll 1) \oplus 0^{120}10000111$

Algorithm 2. SubKeys

1: **proc** SubKeys(K)
2:    $K_1 \leftarrow$ TimesTwo(K)
3:    $K_2 \leftarrow$ TimesTwo(K_1)
4:    **return** (K_1, K_2)

Algorithm 3. Chaskey

1: **proc** Chaskey$^\pi$(K, m)
2:    $(K_1, K_2) \leftarrow$ SubKeys(K)
3:    $m_1 \| \ldots \| m_\ell \leftarrow m$
4:    $h_1 \leftarrow K$
5:    **for** $i = 1, \ldots, \ell - 1$ **do**
6:        $h_{i+1} \leftarrow \pi(h_i \oplus m_i)$
7:    **if** $|m_\ell| = n$ **then**
8:        $L \leftarrow K_1$
9:    **else**
10:        $m_\ell \leftarrow m_\ell \| 10^{n-|m_\ell|-1}$
11:        $L \leftarrow K_2$
12:    $h_{\ell+1} \leftarrow \pi(h_\ell \oplus m_\ell \oplus L) \oplus L$
13:    **return** $\tau \leftarrow \mathsf{right}_t(h_{\ell+1})$

Algorithm 4. Chaskey-B

1: **proc** Chaskey-B^E(K, m)
2:    $(K_1, K_2) \leftarrow$ SubKeys(K)
3:    $m_1 \| \ldots \| m_\ell \leftarrow m$
4:    $h_1 \leftarrow 0^n$
5:    **for** $i = 1, \ldots, \ell - 1$ **do**
6:        $h_{i+1} \leftarrow E_{K\|K}(h_i \oplus m_i)$
7:    **if** $|m_\ell| = n$ **then**
8:        $L \leftarrow K_1$
9:    **else**
10:        $m_\ell \leftarrow m_\ell \| 10^{n-|m_\ell|-1}$
11:        $L \leftarrow K_2$
12:    $h_{\ell+1} \leftarrow E_{K\oplus L\|L}(h_\ell \oplus m_\ell)$
13:    **return** $\tau \leftarrow \mathsf{right}_t(h_{\ell+1})$

Algorithm 5. FCBC [15, 16]

1: **proc** FCBC($(p_1, p_2, p_3), m$)
2:
3:    $m_1 \| \ldots \| m_\ell \leftarrow m$
4:    $h_1 \leftarrow 0^n$
5:    **for** $i = 1, \ldots, \ell - 1$ **do**
6:        $h_{i+1} \leftarrow p_1(h_i \oplus m_i)$
7:    **if** $|m_\ell| = n$ **then**
8:        $q \leftarrow p_2$
9:    **else**
10:        $m_\ell \leftarrow m_\ell \| 10^{n-|m_\ell|-1}$
11:        $q \leftarrow p_3$
12:    $h_{\ell+1} \leftarrow q(h_\ell \oplus m_\ell)$
13:    **return** $\tau \leftarrow h_{\ell+1}$

3.2   Permutation π

The permutation π is built using three operations: addition modulo 2^{32}, bit rotations, and XOR (ARX). The structure is the same as that of SipHash [3], but with 32-bit instead of 64-bit words and different rotation constants. Although SipHash has been proposed only very recently, it has found its way into several widely used software packages. For example, SipHash is used inside the hash table implementations of FreeBSD, Python, Perl, and Ruby. Both Chaskey and SipHash use the 2-input MIX operation of Skein [35], one of the finalists of the SHA-3 competition [53].

In Chaskey, the permutation π consists of eight applications of a round function. This round function is specified in Fig. 2.

Although we are confident that 8 rounds is enough for a secure construction, we recommend that implementers include the 16-round variant Chaskey-LTS (*long term security*) as a fallback in case of cryptanalytical breakthroughs. Chaskey-LTS consumes roughly twice the number of cycles and thus twice the amount of energy as Chaskey, but is still much faster than AES-CMAC. As only the number of rounds is different, it is possible to implement both Chaskey and Chaskey-LTS with negligible overhead in code size.

Note that half of the rotation constants of π are chosen to be multiples of eight. This is because a variety of microcontrollers do not support rotations and shifts over arbitrary amounts, e.g. the Renesas H8/300 CPU supports only one-bit rotations and shifts, the Renesas H8/2000 supports one-bit and two-bit rotations and shifts, and Microchip's 8-bit microcontrollers (PIC10/12/16/18) support one-bit rotations. Due to our choice of constants, implementation on

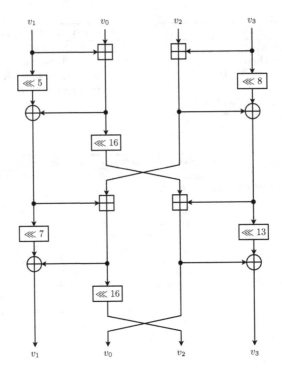

Fig. 2. A round of the Chaskey permutation π, defined as: $v_0\|v_1\|v_2\|v_3 \leftarrow \pi(v_0\|v_1\|v_2\|v_3)$. We intentionally swapped v_0 and v_1, as this reduces the number of crossing lines in the figure.

8- and 16-bit microcontrollers will be more efficient than had these constants been chosen at random. They furthermore allow us to implement Chaskey efficiently on a wide range of 32-bit microcontrollers, yet we have found that they do not seem to make π weaker against cryptanalytical attacks.

4   Implementation Results

We implemented Chaskey on several microcontroller platforms. We provide implementation results on ARM Cortex-M0 and -M4 platforms, and compare these to AES-128-CMAC on the same platforms. All our implementations have been compiled with GNU Tools for ARM Embedded Processors version 4.7.3 20121207. The Cortex-M0 benchmarks are executed on an STM32F030R8 microcontroller of STMicroelectronics, the Cortex-M4 ones on an STM32F401RE.

We compare the results for our Chaskey implementation with what is, to the best of our knowledge, the fastest available AES implementation for the ARM Cortex-M series: SharkSSL [55,56]. Since no AES-128-CMAC benchmarks are available for this implementation, we instead compare with AES-128-ECB, which is guaranteed to be at least as fast and small as AES-128-CMAC. Note

Table 2. Benchmark results for Chaskey and AES-128-CMAC on Cortex-M0/M4. AES-128-CMAC is implemented using AES code from the MAGEEC framework. AES-128-ECB on Cortex-M0/M3 is based on figures from SharkSSL [55,56]. Note that compiling with speed optimization flags does not always result in the fastest implementation.

Microcontroller	Algorithm	Data [byte]	gcc flags	ROM size [byte]	Speed [cycles/byte]
		Speed optimized			
Cortex-M0	AES-128-ECB (SharkSSL)	n/a	n/a	8 380	124.4
	AES-128-CMAC	128	-O2	13 492	136.5
	Chaskey	16	-O2	1 308	21.3
		128	-O2	1 308	18.3
Cortex-M3/M4	AES-128-ECB (SharkSSL)	n/a	n/a	4 854	66.7
	AES-128-CMAC	128	-O2	28 524	105.0
	Chaskey	16	-O2	908	10.6
		128	-O2	908	7.0
		Size optimized			
Cortex-M0	AES-128-ECB (SharkSSL)	n/a	n/a	4 398	112.7
	AES-128-CMAC	128	-Os	11 664	140.0
	Chaskey	16	-Os	414	21.8
		128	-Os	414	16.9
Cortex-M3/M4	AES-128-ECB (SharkSSL)	n/a	n/a	3 922	86.1
	AES-128-CMAC	128	-Os	10 952	89.4
	Chaskey	16	-Os	402	16.1
		128	-Os	402	11.2

that we list SharkSSL results for the Cortex-M3, since Cortex-M4 results are not available. However, the architecture of both microcontrollers is extremely similar, and thus results are expected to be the same.

Results for the various implementations are shown in Table 2. In all of our own benchmarks, round keys are precomputed, and time required to do so is not included in the listed numbers.

5 Proof of Security

We focus on the security of the Chaskey mode of operation. For this, we consider $n, t \in \mathbb{N}$ to be arbitrary values. Denote by $\mathsf{block}(k, n)$ the set of all block ciphers with k-bit key and n-bit block size, and let $\mathsf{perm}(n)$ denote the set of all permutations on n bits. Note that for $E \in \mathsf{block}(k, n)$, we have $E_K \in \mathsf{perm}(n)$ for all $K \in \{0, 1\}^k$. The definitions below follow Bellare et al. [7] and Iwata and Kurosawa [38,39].

MAC Security. Let $\mathcal{H} : \mathcal{K} \times \{0, 1\}^* \to \{0, 1\}^t$ be a MAC function.

$$\mathbf{Adv}_{\mathcal{H}}^{\mathsf{mac}}(q, D, r) = \max_{\mathcal{A}} \Pr \left(\begin{array}{c} K \xleftarrow{\$} \mathcal{K}, (m, \tau) \xleftarrow{\$} \mathcal{A}^{\mathcal{H}_K} ; \\ \mathcal{H}_K(m) = \tau \text{ and } m \text{ never queried} \end{array} \right),$$

where the maximum is taken over all adversaries making at most q queries of total length at most D blocks and running in time r.

3PRP Security. The strength of a block cipher E is conventionally expressed as the PRP (pseudorandom permutation) security. In Chaskey-B (see Algorithm 4) we use a block cipher $E \in \mathrm{block}(2k, n)$ on input of three different keys: $E_{K\|K}$, $E_{K\oplus K_1\|K_1}$, and $E_{K\oplus K_2\|K_2}$, where K_1, K_2 are generated as shown in Algorithm 2. As the keys (K, K_1, K_2) are dependent, so are the three different usages of E. As such, a slightly more involved security notion is needed, which we call 3PRP. For ease of presentation, the definition is adapted to the specific key generation and block cipher use mode of Chaskey.

$$\mathbf{Adv}_E^{\mathrm{3prp}}(D, r) = \max_{\mathcal{A}} \left| \Pr\left(\begin{array}{c} K \xleftarrow{\$} \{0,1\}^k,\ (K_1, K_2) \leftarrow \mathrm{SubKeys}(K) \ ; \\ \mathcal{A}^{E_{K\|K}, E_{K\oplus K_1\|K_1}, E_{K\oplus K_2\|K_2}} = 1 \end{array} \right) - \Pr\left(p_1, p_2, p_3 \xleftarrow{\$} \mathrm{perm}(n)\ ;\ \mathcal{A}^{p_1, p_2, p_3} = 1 \right) \right|,$$

where the maximum is taken over all adversaries making at most D queries and running in time r.

The proof consists of two phases. Theorem 1 states the security of Chaskey-B in the standard model, based on any E with $2n$-bit key and n-bit block size. This result is generalized in the ideal permutation model to Chaskey in Theorem 2, once we use $E_{X\|Y}(m) = \pi(m \oplus X) \oplus Y$ for $\pi \in \{0,1\}^n$.

Theorem 1. *Let $K \xleftarrow{\$} \{0,1\}^n$ and consider* Chaskey-B$_K^E : \{0,1\}^* \to \{0,1\}^t$. *Then,*

$$\mathbf{Adv}_{\text{Chaskey-B}}^{\mathrm{mac}}(q, D, r) \leq \frac{2D^2}{2^n} + \frac{1}{2^t} + \mathbf{Adv}_E^{\mathrm{3prp}}(D, r).$$

Theorem 2. *Let $K \xleftarrow{\$} \{0,1\}^n$, assume that $\pi \xleftarrow{\$} \mathrm{perm}(n)$, and let us consider* Chaskey$_K^\pi : \{0,1\}^* \to \{0,1\}^t$. *Then,*

$$\mathbf{Adv}_{\text{Chaskey}}^{\mathrm{mac}}(q, D, r) \leq \frac{2D^2}{2^n} + \frac{1}{2^t} + \frac{D^2 + 2DT}{2^n},$$

where T is defined as r/r_π for r_π denoting the running time of one evaluation of π.

The proofs of Theorems 1 and 2 can be found in the full version of the paper.[1]

6   Cryptanalysis

6.1   Attack Setting

In this section, we give an overview of the cryptographic properties of the Chaskey permutation π, and the two-key Even-Mansour block cipher $E_{X\|Y}(m) =$

---

[1] http://eprint.iacr.org/2014/386

$\pi(m \oplus X) \oplus Y$. Note even if π has structural weaknesses, Theorem 1 guarantees that Chaskey remains secure as long as $E_{K\|K}$, $E_{K\oplus K_1\|K_1}$, and $E_{K\oplus K_2\|K_2}$ are secure Even-Mansour block ciphers that are indistinguishable from each other. In particular, attackers are restricted to the following setting:

Uniformly Random Key K. Every implementation of Chaskey should ensure that the n-bit key K is chosen uniformly at random from the entire key space. In this way, Chaskey completely avoids all attacks on $E_{K\|K}$ using weak keys [24], known keys [43] or related keys [8,11,12]. In a weak-key attack, the attacker knows that the key K is chosen from a smaller subset of the key space. The attacker controls the value of K in a known-key attack, which in the case of the Even-Mansour block cipher corresponds to an attack on the underlying permutation π. In a related-key attack, the attacker obtains encryptions under different keys, and will know (or even control) the relationship among these keys.

Data Complexity D Below $2^{n/2}$ Chosen Plaintexts. No encryption device is allowed to perform close to $2^{n/2}$ block cipher calls under the same key. This is because after about $2^{n/2}$ block cipher calls, an internal collision attack [54] becomes likely. The same restriction applies to all iterated MAC constructions with an n-bit state. We will now explain that the data complexity under the same key should be restricted further to avoid attacks with a practical time complexity.

Time Complexity T Below $2^n/D$ Block Cipher Evaluations. Even and Mansour [32,33] proved that any attack on their construction requires about T block cipher evaluations and $2^n/D$ known plaintexts. Dunkelman et al. [29] described a key recovery attack on the Even-Mansour construction to show that this bound is tight. As they clarify, this tight bound holds for both single-key and two-key Even-Mansour. To avoid attacks with a practical time complexity, the specification restricts the total number of blocks under the same key K to at most 2^{48}. This limit assumes that performing about 2^{80} off-line permutation evaluations is impractical for the attacker. Implementations that require a higher security level should rekey more frequently. We note that the amortized cost of rekeying is usually negligible, and rekeying does not require additional cryptographic components if Chaskey is also used as a key derivation function (KDF) [20].

No Chosen Ciphertext Attacks. The attacker cannot make any decryption queries $E_{K\|K}^{-1}$, $E_{K\oplus K_1\|K_1}^{-1}$, or $E_{K\oplus K_2\|K_2}^{-1}$, for the simple reason that Chaskey implementations do not contain the decryption function, and the corresponding keys are secret.

Tag Guessing Has Probability $2^{-|\tau|}$. The probability of constructing a forgery by guessing the tag is $2^{-|\tau|}$. Guessing a tag correctly for Chaskey does not make additional forgeries easier. The specification recommends that $|\tau| \geq 64$, which ensures that the probability of guessing τ correctly after 2^{32} trials is less than one in a billion. If it is acceptable to occasionally accept an inauthentic message as authentic (e.g. in certain voice communication applications [34]), the use of shorter tags may be carefully considered.

Implementation Attacks. Chaskey is inherently secure against timing attacks, as its execution time depends only on the message length $|m|$, and not on the secret key K. However, a straightforward implementation of Chaskey provides no resistance against hardware side channel attacks, nor to fault attacks. Furthermore, note that if the internal state of Chaskey is recovered and $|\tau| = n$, it is easy to recover the secret key K from any (m, τ)-pair.

6.2 Cryptanalysis of the Block Cipher

We now proceed with our cryptanalysis results for the block ciphers $E_{K\|K}$, $E_{K\oplus K_1\|K_1}$, and $E_{K\oplus K_2\|K_2}$ using π as the underlying permutation.

Standard Differential Cryptanalysis. We searched for differential characteristics of $E_{K\|K}$ that are linear in $GF(2)$, which means the output difference of every addition is the XOR of the two input differences. This was done by formulating this problem as the search for low-weight codewords in a linear code [58].

The best found characteristics for $1, 2, \ldots, 8$ rounds are shown in Table 3. We show only the input and output differences; the linearity property can be used to find the internal differences. We calculated the characteristic probability in two ways: by determining the probability of every addition using the Lipmaa-Moriai formula [49] and multiplying these probabilities, and by using Leurent's ARX Toolkit [47,48] to obtain a more accurate estimate that takes certain dependencies between operations into account.

In Table 4, we give the differences after every round of the best found differential characteristic for eight rounds, which corresponds to the last characteristic in Table 3. It is interesting to note that this characteristic has what can be described as an *hourglass* structure: the differences are sparse in the middle of the characteristics (located only in the most significant bits), and gradually become denser towards the outer rounds. The same observation also holds for all other characteristics of Table 3.

In Table 3, probabilities below 2^{-128} indicate that a characteristic exists only with some probability. Although such characteristics are not usable in an attack, it is important to explore them from a design point of view. Table 3 shows that Even-Mansour block ciphers based on π have a very large security margin against even very advanced variants of differential cryptanalysis attacks, especially as the data complexity in any attack on Chaskey is limited to 2^{64}.

Note that it is possible that better (possibly non-linear) characteristics exist, or that the probability of a given characteristic is lower than the probability of the corresponding differential. However, we expect that these effects will not be significant enough to invalidate our security claim against differential cryptanalysis.

Table 3. Best found differential characteristics for $1, 2, \ldots, 8$ rounds of the permutation π. Only the input and output differences are shown. Each of these characteristics is linear, this property can be used to determine the internal differences. We calculate the characteristic probability in two ways: assuming independence of every operation and using the Lipmaa-Moriai formula, as well as by Leurent's ARX toolkit for a more refined estimate.

# Rounds	$\Delta_{\mathrm{in}}^{\oplus}(v_0, v_1, v_2, v_3) \to \Delta_{\mathrm{out}}^{\oplus}(v_0, v_1, v_2, v_3)$	Lipmaa-Moriai	Leurent
1	$(00000000, 00000000, 80000000, 00000000)$ $\to (80000000, 80000000, 80000000, 80001000)$	1	1
2	$(00008400, 00000400, 00000000, 00000000)$ $\to (80008080, 00000040, 00000000, 80109080)$	2^{-4}	2^{-4}
3	$(00000008, 00000008, 00008181, 00000081)$ $\to (80109080, 80009810, 80009010, 92008082)$	2^{-16}	2^{-16}
4	$(C0240100, 44202100, 0C200008, 0C200000)$ $\to (10409000, 00547800, 00101840, 12408210)$	2^{-37}	2^{-37}
5	$(C8226120, 4C224101, 084C6908, 0C046900)$ $\to (E8001014, 08912214, 00802210, EA120916)$	2^{-73}	$2^{-73.1}$
6	$(1AC8DA46, 73C0D20A, 9282B2A3, 02947AA1)$ $\to (6A00109B, 50B7698C, 12866000, 68037999)$	2^{-133}	$2^{-132.8}$
7	$(8C74CC70, 7F3690AE, 5403A321, D1852232)$ $\to (DBCD9AC0, 293EC4DB, 08036B1F, B195C08B)$	2^{-208}	$2^{-205.6}$
8	$(90EA132B, 88490EDB, 45854D95, E6A41996)$ $\to (726DC8C0, 097D6D14, 24592382, 2C2329AF)$	2^{-293}	$2^{-289.9}$

Truncated Differential Cryptanalysis. We used the same techniques that were applied to Salsa20 [4] to find truncated differentials for $E_{K\|K}$. More specifically, we introduced differences in the most significant bits of the inputs, and searched for statistical biases in the output bits. We found such biases for up to four rounds of the block cipher. For example, if in the plaintext $\Delta^{\oplus}v_1[31]$ and $\Delta^{\oplus}v_2[31]$ are both 1, then we found experimentally that $\Delta^{\oplus}v_2[16]$ after four rounds has a bias of about $2^{-12.48}$ towards 0. We tried out all combinations of input differences in the most significant bits of the four input words, but did not find biases in any of the output bit differences after five rounds or more, when experimenting with sets of 2^{30} samples.

Meet-in-the-Middle Attacks. The idea behind a meet-in-the-middle attack is to separate the mathematical equations that describe a block cipher into two or more groups, in such a way that some variables do not appear in at least one of the groups of equations. After three rounds of π, full diffusion occurs: every input bit affects every output bit. Similarly, π^{-1} also reaches full diffusion after three rounds. As eight rounds of π consist of almost three full diffusions, meet-

Table 4. Best found linear differential characteristic for 8 rounds of π. This is the characteristic given in the last row of Table 3. If we assume independence of every operation and use the Lipmaa-Moriai formula for every addition, we find a probability of 2^{-293}. Leurent's ARX toolkit can be used to refine this probability to $2^{-289.9}$. Note the *hourglass* structure: differences are sparse in the middle, and gradually become denser towards the outer rounds.

Round$_i$	$\Delta^{\oplus}v_0$	$\Delta^{\oplus}v_1$	$\Delta^{\oplus}v_2$	$\Delta^{\oplus}v_3$	$\Pr[\text{Round}_{i-1} \to \text{Round}_i]$
0	90EA132B	88490EDB	45854D95	E6A41996	
1	1AC8DA46	73C0D20A	B2A39282	02947AA1	2^{-76}
2	0C200008	08200008	81048100	81000085	2^{-55}
3	00000000	00000000	00008080	00800000	2^{-15}
4	00000000	80000000	80000000	00000000	2^{-1}
5	00000000	80008850	80008010	10000000	2^{-4}
6	18400010	18C02200	10010240	08421212	2^{-19}
7	6A00109B	50B7698C	12866000	68037999	2^{-39}
8	726DC8C0	097D6D14	24592382	2C2329AF	2^{-84}

in-the-middle attacks should not be applicable to Even-Mansour block ciphers based on π.

Note that the attacker is not allowed to perform chosen-ciphertext attacks, which limits the power of advanced meet-in-the-middle attacks, using the splice-and-cut technique that was introduced for hash function cryptanalysis [2, 59] and subsequently applied to block ciphers as well [18, 62].

A further extension of splice-and-cut meet-in-the-middle attacks are biclique attacks [17, 42]. Most applications of bicliques offer only slight improvements over brute force attacks [57]. Although brute-force-like attacks provide insight into the security of ciphers in the absence of other shortcut attacks, they do not affect the practical security of the cipher.

Rotational Cryptanalysis. A randomly chosen key K ensures that the input of the permutation π when used in an Even-Mansour block cipher will (with very high probability) have an asymmetrical state, thereby preventing rotational attacks [41].

Slide Attacks. Because every round of π is identical, slide attacks [13] are applicable to π. However, in a slide attack, about $2^{n/2}$ plaintext-ciphertext pairs are required before a slid pair is found. Therefore, slide attacks have a data complexity that goes beyond our security bound, and do not pose a threat to π, nor to Even-Mansour block ciphers based on π.

Fixed Points. Because π contains only the modular addition, XOR, and bitwise rotation operations, the permutation has the following fixed point: $\pi(0^n) = 0^n$.

Fixed-points are a type of differentiability attack [50]. When π is used inside the $E_{K\|K}$ block cipher, this fixed point corresponds to $E_{K\|K}(K) = K$. If K is chosen uniformly at random, this relationship only holds with probability 2^{-n} for any plaintext chosen by the attacker. Similar observations hold for $E_{K \oplus K_1\|K_1}$ and $E_{K \oplus K_2\|K_2}$. Although it may seem to be a bold move from a design point of view to allow that $E_{0^{2n}}(0^n) = 0^n$, we note that this property also holds for the stream cipher Trivium [25,27] and the block cipher KATAN [26]. However, no attacks have been found that break the full version of these ciphers.

Dependency Between Key and Subkeys. As shown by Algorithm 1, the subkeys K_1 and K_2 are generated from the key K as $K_1 = xK$ and $K_2 = x^2 K_1$. Theorem 1 requires that an attacker cannot distinguish $E_{K\|K}$, $E_{(x+1)K\|xK}$, and $E_{(x^2+1)K\|x^2K}$ from each other. As shown by Theorem 2, this assumption holds if the underlying permutation π is an ideal permutation. We now argue that even if the permutation π of Sect. 3.2 is used instead, an attacker cannot distinguish these three block ciphers. Because of the rotational relations between the key K and the subkeys K_1 and K_2, rotational cryptanalysis [41] seems to be a promising technique. However, the fact that $(x + 1)K$ and xK, as well as $(x^2 + 1)K$ and x^2K both differ by K, seems to effectively preclude rotational cryptanalysis to distinguish $E_{(x+1)K\|xK}$ or $E_{(x^2+1)K\|x^2K}$ from $E_{K\|K}$, or from each other. Furthermore, the security proof assumes that individual queries to the three aforementioned block ciphers are permitted, whereas an attacker can in practice only observe τ.

Other Attacks. We do not consider zero-sum attacks [5] and cube attacks [28] to be a threat for ARX ciphers, because the addition operation ensures that for every output bit, the polynomial expression in $GF(2)$ representing this bit in terms of its inputs will be of sufficiently high degree. Moreover, rebound attacks [51] are not known to be relevant to secret-key algorithms.

7 Conclusion

Chaskey is a permutation-based MAC algorithm, with at its core an ARX-based permutation π based on SipHash. Alternatively, Chaskey can also be interpreted as a block-cipher-based MAC algorithm based on an underlying Even-Mansour block cipher.

Inspired by the block-cipher-based CMAC, Chaskey avoids padding for messages of an integer number of blocks. Its subkey generation is even more efficient than CMAC, as it does not require any block cipher calls.

We proved that Chaskey is secure, based on the 3PRP-indistinguishability of three underlying Even-Mansour block ciphers. Assuming that the permutation π used in these Even-Mansour block ciphers is ideal, we proved that Chaskey is secure up to about $D = 2^{n/2}$ chosen plaintexts and about $T = 2^n/D$ queries to π or π^{-1}.

We remark, however, that the efficient permutation π designed for Chaskey shows properties that allow it to be distinguished from an ideal permutation. For example, it is easy to find a fixed point: $\pi(0^n) = 0^n$. Fortunately, this observation does not extend to an attack when this permutation is used inside an Even-Mansour block cipher, as finding this fixed point implies knowledge of the secret key.

Therefore, we explored the distinguishability of the three Even-Mansour block ciphers from a cryptanalysis point of view. After investigating a wide variety of currently known cryptanalysis attacks, we found no shortcut attacks resulting from using our proposed eight-round permutation π instead of an ideal permutation. We recommend, however, that implementers also support a 16-round Chaskey-LTS as a fallback in case of cryptanalytical breakthroughs.

Our benchmarks showed that Chaskey performs very well on ARM Cortex-M microcontrollers. We measured that our straightforward Chaskey implementations are between 7 to 15 times faster than AES-128-CMAC in speed-optimized implementations, and at about 10 times smaller in area optimized implementations. Because of the roughly linear relation between cycle count and energy consumption, Chaskey is therefore much more energy efficient as well. Although 32-bit microcontrollers were our main target platform, Chaskey is also expected to perform well on 8-bit and 16-bit platforms.

References

1. Ambler, E.: Computer Data Authentication. FIPS PUB 113, National Institute of Standards and Technology (NIST), May 1985. http://csrc.nist.gov/publications/fips/fips113/fips113.html
2. Aoki, K., Sasaki, Y.: Preimage attacks on one-block MD4, 63-step MD5 and more. In: Avanzi, R.M., Keliher, L., Sica, F. (eds.) SAC 2008. LNCS, vol. 5381, pp. 103–119. Springer, Heidelberg (2009)
3. Aumasson, J.-P., Bernstein, D.J.: SipHash: a fast short-input PRF. In: Galbraith, S., Nandi, M. (eds.) INDOCRYPT 2012. LNCS, vol. 7668, pp. 489–508. Springer, Heidelberg (2012)
4. Aumasson, J.-P., Fischer, S., Khazaei, S., Meier, W., Rechberger, C.: New features of Latin dances: analysis of Salsa, ChaCha, and Rumba. In: Nyberg, K. (ed.) FSE 2008. LNCS, vol. 5086, pp. 470–488. Springer, Heidelberg (2008)
5. Aumasson, J.-P., Meier, W.: Zero-sum distinguishers for reduced Keccak-f and for the core functions of Luffa and Hamsi. Presented at the Rump Session of Cryptographic Hardware and Embedded Systems - CHES 2009 (2009)
6. Bellare, M., Canetti, R., Krawczyk, H.: Keying hash functions for message authentication. In: Koblitz [44], pp. 1–15
7. Bellare, M., Kilian, J., Rogaway, P.: The security of cipher block chaining. In: Desmedt, Y.G. (ed.) CRYPTO 1994. LNCS, vol. 839, pp. 341–358. Springer, Heidelberg (1994)
8. Bellare, M., Kohno, T.: A theoretical treatment of related-key attacks: RKA-PRPs, RKA-PRFs, and applications. In: Biham, E. (ed.) EUROCRYPT 2003. LNCS, vol. 2656, pp. 491–506. Springer, Heidelberg (2003)
9. Bernstein, D.J.: The Poly1305-AES message-authentication code. In: Gilbert, H., Handschuh, H., [36], pp. 32–49

10. Bertoni, G., Daemen, J., Peeters, M., Van Assche, G.: The Keccak SHA-3 submission. Submission to the NIST SHA-3 Competition (Round 3) (2011)
11. Biham, E.: New types of cryptanalytic attacks using related keys (Extended abstract). In: Helleseth, T. (ed.) EUROCRYPT 1993. LNCS, vol. 765, pp. 398–409. Springer, Heidelberg (1994)
12. Biham, E.: New types of cryptanalytic attacks using related keys. J. Cryptology **7**(4), 229–246 (1994)
13. Biryukov, A., Wagner, D.: Slide attacks. In: Knudsen, L.R. (ed.) FSE 1999. LNCS, vol. 1636, pp. 245–259. Springer, Heidelberg (1999)
14. Black, J., Halevi, S., Krawczyk, H., Krovetz, T., Rogaway, P.: UMAC: fast and secure message authentication. In: Wiener, M. (ed.) CRYPTO 1999. LNCS, vol. 1666, pp. 216–233. Springer, Heidelberg (1999)
15. Black, J., Rogaway, P.: CBC MACs for arbitrary-length messages: the three-key constructions. In: Bellare, M. (ed.) CRYPTO 2000. LNCS, vol. 1880, pp. 197–215. Springer, Heidelberg (2000)
16. Black, J., Rogaway, P.: CBC MACs for arbitrary-length messages: the three-key constructions. J. Cryptology **18**(2), 111–131 (2005)
17. Bogdanov, A., Khovratovich, D., Rechberger, C.: Biclique cryptanalysis of the full AES. In: Lee, D.H., Wang, X. (eds.) ASIACRYPT 2011. LNCS, vol. 7073, pp. 344–371. Springer, Heidelberg (2011)
18. Bogdanov, A., Rechberger, C.: A 3-subset meet-in-the-middle attack: cryptanalysis of the lightweight block cipher KTANTAN. In: Biryukov, A., Gong, G., Stinson, D.R. (eds.) SAC 2010. LNCS, vol. 6544, pp. 229–240. Springer, Heidelberg (2011)
19. Carter, J.L., Wegman, M.N.: Universal classes of hash functions. J. Comput. Syst. Sci. **18**(2), 143–154 (1979)
20. Chen, L.: Recommendation for key derivation using pseudorandom functions (Revised). NIST Special Publication 800–108, National Institute of Standards and Technology (NIST), October 2009. http://csrc.nist.gov/publications/nistpubs/800-108/sp800-108.pdf
21. de Clercq, R., Uhsadel, L., Van Herrewege, A., Verbauwhede, I.: Ultra low-power implementation of ECC on the ARM Cortex-M0+. In: 51th Design Automation Conference (DAC 2014), pp. 1–6. IEEE, San Francisco (2014)
22. Daemen, J., Rijmen, V.: A new MAC construction ALRED and a specific instance ALPHA-MAC. In: Gilbert, H., Handschuh, H., [36], pp. 1–17
23. Daemen, J., Rijmen, V.: The Pelican MAC function. IACR Cryptology ePrint Archive, Report 2005/88 (2005)
24. Davies, D.W.: Some regular properties of the 'Data Encryption Standard' algorithm. In: Chaum, D., Rivest, R.L., Sherman, A.T. (eds.) CRYPTO '82, pp. 89–96. Plenum Press, New York (1982)
25. De Cannière, C.: TRIVIUM: a stream cipher construction inspired by block cipher design principles. In: Katsikas, S.K., López, J., Backes, M., Gritzalis, S., Preneel, B. (eds.) ISC 2006. LNCS, vol. 4176, pp. 171–186. Springer, Heidelberg (2006)
26. De Cannière, C., Dunkelman, O., Knežević, M.: KATAN and KTANTAN — a family of small and efficient hardware-oriented block ciphers. In: Clavier, C., Gaj, K. (eds.) CHES 2009. LNCS, vol. 5747, pp. 272–288. Springer, Heidelberg (2009)
27. De Cannière, C., Preneel, B.: TRIVIUM. In: Robshaw, M., Billet, O. (eds.) New Stream Cipher Designs. LNCS, vol. 4986, pp. 244–266. Springer, Heidelberg (2008)
28. Dinur, I., Shamir, A.: Cube attacks on tweakable black box polynomials. In: Joux, A. (ed.) EUROCRYPT 2009. LNCS, vol. 5479, pp. 278–299. Springer, Heidelberg (2009)

29. Dunkelman, O., Keller, N., Shamir, A.: Minimalism in cryptography: the Even-Mansour scheme revisited. In: Pointcheval, D., Johansson, T. (eds.) EUROCRYPT 2012. LNCS, vol. 7237, pp. 336–354. Springer, Heidelberg (2012)

30. Dworkin, M.: Recommendation for Block Cipher Modes of Operation: The CMAC Mode for Authentication. NIST special publication 800-38B, National Institute of Standards and Technology (NIST), May 2005. http://csrc.nist.gov/publications/nistpubs/800-38B/SP_800-38B.pdf

31. Dworkin, M.: Recommendation for Block Cipher Modes of Operation: Galois/Counter Mode (GCM) and GMAC. NIST special publication 800-38D, National Institute of Standards and Technology (NIST), November 2007. http://csrc.nist.gov/publications/nistpubs/800-38D/SP-800-38D.pdf

32. Even, S., Mansour, Y.: A construction of a cipher from a single pseudorandom permutation. In: Matsumoto, T., Imai, H., Rivest, R.L. (eds.) ASIACRYPT 1991. LNCS, vol. 739, pp. 210–224. Springer, Heidelberg (1993)

33. Even, S., Mansour, Y.: A construction of a cipher from a single pseudorandom permutation. J. Cryptology 10(3), 151–162 (1997)

34. Ferguson, N.: Authentication weaknesses in GCM. Comments submitted to NIST Modes of Operation Process, May 2005

35. Ferguson, N., Lucks, S., Schneier, B., Whiting, D., Bellare, M., Kohno, T., Callas, J., Walker, J.: The Skein Hash Function Family. Submission to the NIST SHA-3 Competition (Round 3) (2010). http://www.skein-hash.info/sites/default/files/skein1.3.pdf

36. Gilbert, H., Handschuh, H. (eds.): FSE 2005. LNCS, vol. 3557. Springer, Heidelberg (2005)

37. ISO/IEC: Information Technology: Information Technology - Security Techniques - Message Authentication Codes (MACs) - Part 1: Mechanisms Using a Block Cipher. ISO/IEC 9797–1:2011 (2011)

38. Iwata, T., Kurosawa, K.: OMAC: one-key CBC MAC. In: Johansson, T. (ed.) FSE 2003. LNCS, vol. 2887, pp. 129–153. Springer, Heidelberg (2003)

39. Iwata, T., Kurosawa, K.: Stronger security bounds for OMAC, TMAC, and XCBC. In: Johansson, T., Maitra, S. (eds.) INDOCRYPT 2003. LNCS, vol. 2904, pp. 402–415. Springer, Heidelberg (2003)

40. Joux, A.: Authentication Failures in NIST version of GCM. Comments submitted to NIST Modes of Operation Process, June 2006

41. Khovratovich, D., Nikolić, I.: Rotational cryptanalysis of ARX. In: Hong, S., Iwata, T. (eds.) FSE 2010. LNCS, vol. 6147, pp. 333–346. Springer, Heidelberg (2010)

42. Khovratovich, D., Rechberger, C., Savelieva, A.: Bicliques for preimages: attacks on Skein-512 and the SHA-2 family. In: Canteaut, A. (ed.) FSE 2012. LNCS, vol. 7549, pp. 244–263. Springer, Heidelberg (2012)

43. Knudsen, L.R., Rijmen, V.: Known-key distinguishers for some block ciphers. In: Kurosawa, K. (ed.) ASIACRYPT 2007. LNCS, vol. 4833, pp. 315–324. Springer, Heidelberg (2007)

44. Koblitz, N. (ed.): CRYPTO 1996. LNCS, vol. 1109. Springer, Heidelberg (1996)

45. Kocher, P.C.: Timing attacks on implementations of Diffie-Hellman, RSA, DSS, and other systems. In: Koblitz [44], pp. 104–113

46. Krovetz, T.: Message authentication on 64-bit architectures. In: Biham, E., Youssef, A.M. (eds.) SAC 2006. LNCS, vol. 4356, pp. 327–341. Springer, Heidelberg (2007)

47. Leurent, G.: Analysis of differential attacks in ARX constructions. In: Wang, X., Sako, K. (eds.) ASIACRYPT 2012. LNCS, vol. 7658, pp. 226–243. Springer, Heidelberg (2012)

48. Leurent, G.: Construction of differential characteristics in ARX designs application to Skein. In: Canetti, R., Garay, J.A. (eds.) CRYPTO 2013, Part I. LNCS, vol. 8042, pp. 241–258. Springer, Heidelberg (2013)

49. Lipmaa, H., Moriai, S.: Efficient algorithms for computing differential properties of addition. In: Matsui, M. (ed.) FSE 2001. LNCS, vol. 2355, pp. 336–350. Springer, Heidelberg (2002)

50. Maurer, U.M., Renner, R.S., Holenstein, C.: Indifferentiability, impossibility results on reductions, and applications to the random oracle methodology. In: Naor, M. (ed.) TCC 2004. LNCS, vol. 2951, pp. 21–39. Springer, Heidelberg (2004)

51. Mendel, F., Rechberger, C., Schläffer, M., Thomsen, S.S.: The rebound attack: cryptanalysis of reduced Whirlpool and Grøstl. In: Dunkelman, O. (ed.) FSE 2009. LNCS, vol. 5665, pp. 260–276. Springer, Heidelberg (2009)

52. Mozans, H.J.: Along the Andes and down the Amazon, vol. 2. D. Appleton and Company, New York (1911)

53. National Institute of Standards and Technology: Announcing Request for Candidate Algorithm Nominations for a New Cryptographic Hash Algorithm (SHA-3) Family. Federal Register 27(212), 62212–62220, November 2007. http://csrc.nist.gov/groups/ST/hash/documents/FR_Notice_Nov07.pdf

54. Preneel, B., van Oorschot, P.C.: MDx-MAC and building fast MACs from hash functions. In: Coppersmith, D. (ed.) CRYPTO 1995. LNCS, vol. 963, pp. 1–14. Springer, Heidelberg (1995)

55. RealTimeLogic: SHARKSSL v2.3.3 Crypto Library Benchmarks with ARM Cortex-M0@24MHz + ARM GCC 4.5.1 (2014). http://realtimelogic.com/products/sharkssl/Cortex-M0/

56. RealTimeLogic: SHARKSSL/RAYCRYPTO v2.4 Crypto Library Benchmarks with ARM Cortex-M3@50MHz + IAR EWARM 6.40 (2014). http://realtimelogic.com/products/sharkssl/Cortex-M3/

57. Rechberger, C.: On bruteforce-like cryptanalysis: new meet-in-the-middle attacks in symmetric cryptanalysis. In: Kwon, T., Lee, M.-K., Kwon, D. (eds.) ICISC 2012. LNCS, vol. 7839, pp. 33–36. Springer, Heidelberg (2013)

58. Rijmen, V., Oswald, E.: Update on SHA-1. In: Menezes, A. (ed.) CT-RSA 2005. LNCS, vol. 3376, pp. 58–71. Springer, Heidelberg (2005)

59. Sasaki, Y., Aoki, K.: Preimage attacks on step-reduced MD5. In: Mu, Y., Susilo, W., Seberry, J. (eds.) ACISP 2008. LNCS, vol. 5107, pp. 282–296. Springer, Heidelberg (2008)

60. Turner, J.M.: The Keyed-Hash Message Authentication Code (HMAC). FIPS PUB 198-1, National Institute of Standards and Technology (NIST), July 2008. http://csrc.nist.gov/publications/fips/fips198-1/FIPS-198-1_final.pdf

61. Wegman, M.N., Carter, J.L.: New hash functions and their use in authentication and set equality. J. Comput. Syst. Sci. **22**(3), 265–279 (1981)

62. Wei, L., Rechberger, C., Guo, J., Wu, H., Wang, H., Ling, S.: Improved meet-in-the-middle cryptanalysis of KTANTAN (Poster). In: Parampalli, U., Hawkes, P. (eds.) ACISP 2011. LNCS, vol. 6812, pp. 433–438. Springer, Heidelberg (2011)

Fast Point Multiplication Algorithms
for Binary Elliptic Curves
with and without Precomputation

Thomaz Oliveira[1]($\boxtimes$), Diego F. Aranha[2], Julio López[2],
and Francisco Rodríguez-Henríquez[1]

[1] Computer Science Department, CINVESTAV-IPN, Mexico City, Mexico
thomaz.figueiredo@gmail.com
[2] Institute of Computing, University of Campinas, Campinas, Brazil

Abstract. In this paper we introduce new methods for computing
constant-time variable-base point multiplications over the Galbraith-Lin-
Scott (GLS) and the Koblitz families of elliptic curves. Using a left-to-right
double-and-add and a right-to-left halve-and-add Montgomery ladder over
a GLS curve, we present some of the fastest timings yet reported in the
literature for point multiplication. In addition, we combine these two procedures to compute a multi-core protected scalar multiplication. Furthermore, we designed a novel regular τ-adic scalar expansion for Koblitz
curves. As a result, using the regular recoding approach, we set the speed
record for a single-core constant-time point multiplication on standardized
binary elliptic curves at the 128-bit security level.

Keywords: Binary elliptic curves · Scalar multiplication · Software
implementation

1 Introduction

From a cryptographic perspective, one of the most interesting consequences of
the Snowden revelations is the increased awareness about the importance of
implementing security protocols that offer the Perfect Forward Secrecy (PFS)
property. The PFS property guarantees that in a given protocol, none of its past
short term session keys can be derived from the long term server's private key.
One tangible example of this situation is the recent announcement by the Internet Engineering Task Force that the Transport Layer Security (TLS) protocol
version 1.3, will no longer include cipher suites based on RSA key transport primitives [34]. Instead, the client-server secret key establishment will be performed
via either the Ephemeral Diffie-Hellman or the Elliptic Curve Ephemeral Diffie-
Hellman (ECDHE) methods. Because of the significant performance advantage
of the latter over the former, it is anticipated that in the years to come, ECDHE
will be the favorite choice for establishing a TLS shared secret.

J. López — The author was supported in part by the Intel Labs University Research
Office.

© Springer International Publishing Switzerland 2014
A. Joux and A. Youssef (Eds.): SAC 2014, LNCS 8781, pp. 324–344, 2014.
DOI: 10.1007/978-3-319-13051-4_20

The specifications of all the TLS protocol versions [8–10] include support for prime and binary field elliptic curve cryptographic primitives. In the case of binary elliptic curves, the TLS protocol supports a selection of several standardized random curves as well as Koblitz curves [23] at the 80-, 128-, 192- and 256-bit security levels. Koblitz curves allow performance improvements, due to the availability of the Frobenius automorphism τ. Also, their generation is inherently *rigid* (in the SafeCurves sense [2]), where the only degree of freedom in the curve generation process consists in choosing a suitable prime degree extension m that produces a curve with almost-prime order. This severely limits the possibility of "1-in-a-million attacks" [35] aiming to reach a weak curve after testing many random seeds.

Point multiplication is the single most important operation of (hyper) elliptic curve cryptography, for that reason, considerable effort has been directed towards achieving fast and compact software/hardware implementations of it. A major result that has influenced the latest implementations was found in 2009, when Galbraith, Lin and Scott (GLS), building on a previous technique introduced by Gallant, Lambert and Vanstone (GLV) [14], constructed efficient endomorphisms for a class of elliptic curves defined over the quadratic field $\mathbb{F}_{q^2}$, where q is a prime number [13]. Taking advantage of this result, the authors of [13] performed a 128-bit security level point multiplication that took 326,000 clock cycles on a 64-bit processor. Since then, a steady stream of algorithmic and technological advances has translated into a significant reduction in the number of clock cycles required to compute a (hyper) elliptic curve constant-time variable-base-point multiplication at the 128-bit security level [1,4,5,11,16,24,38].

The authors of [11,24] targeted a twisted Edwards GLV-GLS curve defined over $\mathbb{F}_{p^2}$, with $p = 2^{127} - 5997$. That curve is equipped with a degree-4 endomorphism allowing a fast point multiplication computation that required just 92,000 clock cycles on an Ivy Bridge processor [11]. Bos *et al.* [5] and Bernstein *et al.* [1], presented an efficient point multiplication on the Kummer surface associated with the Jacobian of a genus 2 curve defined over a field generated by the prime $p = 2^{127} - 1$. Each iteration of the Montgomery ladder presented in [1] costs roughly 25 field multiplications, which implemented on a Haswell processor permits to compute a point multiplication in 72,000 clock cycles.

In 2014, Oliveira *et al.* introduced the λ-projective coordinate system that leads to faster binary field elliptic curve arithmetic [31,32]. The authors applied that coordinate system into a binary GLS curve that admits a degree-2 endomorphism and a fast field arithmetic associated with the quadratic field extension of the binary field $\mathbb{F}_{2^{127}}$. When implemented on a Haswell processor, this approach permits to perform one constant-time point multiplication computation in just 60,000 clock cycles.

Contributions of This Paper. This work presents new methods aimed to perform fast constant-time variable-base-point multiplication computation for both random and Koblitz binary elliptic curves of the form $y^2 + xy = x^3 + ax^2 + b$. In the case of random binary elliptic curves, we introduce a novel right-to-left variant of the classical Montgomery-López-Dahab ladder algorithm presented in [25], which efficiently adapted the original ladder idea introduced by Peter Montgomery in his 1987 landmark paper [26]. The new variant presented in this work does not require point doublings, but instead, it uses the efficient point halving operation available on binary elliptic curves. In contrast with the algorithm presented in [25] that does not admit the benefit of precomputed tables, our proposed variant *can* take advantage of this technique, a feature that could be proved valuable for the fixed-base-point multiplication scenario. Moreover, we show that our new right-to-left Montgomery ladder formulation can be nicely combined with the classical ladder to attain a high parallel acceleration factor for a constant-time multi-core implementation of the point multiplication operation. As a second contribution, we present a procedure that adapts the regular scalar recoding of [21] to the task of producing a regular τ-NAF scalar recoding for Koblitz curves. This approach has faster precomputation than related recodings [30] and allows us to achieve a speed record for single-core constant-time point multiplication on standardized binary elliptic curves at the 128-bit security level.

The remainder of this paper is organized as follows. In Sect. 2 we give a short description of the GLS and Koblitz curves, their arithmetic and their security. In Sect. 3 we present new variants of the Montgomery ladder for binary elliptic curves. Then, in Sect. 4, we introduce a regular τ-NAF recoding amenable for producing protected point multiplication implementations on Koblitz curves. In Sect. 5, we present our experimental implementation results and finally, we draw our conclusions in Sect. 6.

2 Mathematical Background

2.1 Quadratic Field Arithmetic

A binary extension field $\mathbb{F}_q$, $q = 2^m$, can be constructed by taking an degree-m polynomial $f(x) \in \mathbb{F}_2[x]$ irreducible over $\mathbb{F}_2$, where the field elements in $\mathbb{F}_q$ are the set of binary polynomials of degree less than m. Quadratic extensions of a binary extension field can be built using a degree two monic polynomial $g(u) \in \mathbb{F}_q[u]$ irreducible over $\mathbb{F}_q$. In this case, the field $\mathbb{F}_{q^2}$ is isomorphic to $\mathbb{F}_q[u]/(g(u))$ and its elements can be represented as $a_0 + a_1 u$, with $a_0, a_1 \in \mathbb{F}_q$. Operations in the quadratic extension are performed coefficient-wise. For instance, the multiplication of two elements $a, b \in \mathbb{F}_{q^2}$ is computed at the cost of three multiplications in the base field using the customary Karatsuba formulation,

$$a \cdot b = (a_0 + a_1 u) \cdot (b_0 + b_1 u) \qquad (1)$$
$$= (a_0 b_0 + a_1 b_1) + (a_0 b_0 + (a_0 + a_1) \cdot (b_0 + b_1))u,$$

with $a_0, a_1, b_0, b_1 \in \mathbb{F}_q$.

In [31,32], the authors developed an efficient software library for the field $\mathbb{F}_{2^m}$ and its quadratic extension $\mathbb{F}_{2^{2m}}$, with $m = 127$, generated by means of the irreducible trinomials $f(x) = x^{127} + x^{63} + 1$ and $g(u) = u^2 + u + 1$, respectively.

The computational cost of the field arithmetic in the quadratic extension field gets significantly reduced by using that towering approach. To be more concrete, let M and m denote the cost of one field multiplication over $\mathbb{F}_{q^2}$ and $\mathbb{F}_q$, respectively. The execution of the arithmetic library of [32] on the Sandy Bridge and Haswell microprocessors yields a ratio M/m of just 2.23 and 1.51, respectively. These experimental results are considerably better than the theoretical ratio $M/m = 3$ that one could expect from the Karatsuba formulation of Eq. (1). The aforementioned performance speedup can be explained from the fact that the towering field approach permits a much better usage of the processor's pipelined execution unit, which potentially can improve the speed of one 64-bit carry-less multiplication[1] from 7 clock cycles to the maximum achievable throughput of just 2 clock cycles [12].

2.2   GLS Binary Elliptic Curves

Let $E_{a,b}(\mathbb{F}_{q^2})$ denote the additive abelian group formed by the point at infinity $\mathcal{O}$ and the set of affine points $P = (x,y)$ with $x,y \in \mathbb{F}_{q^2}$ that satisfy the ordinary binary elliptic curve equation given as,

$$E : y^2 + xy = x^3 + ax^2 + b, \qquad (2)$$

defined over $\mathbb{F}_{q^2 = 2^{2m}}$, with $a \in \mathbb{F}_{q^2}$ and $b \in \mathbb{F}_{q^2}^*$. Let $\#E_{a,b}(\mathbb{F}_{q^2})$ denote the size of the group $E_{a,b}(\mathbb{F}_{q^2})$, and let us assume that $E_{a,b}(\mathbb{F}_{q^2})$ includes a subgroup $\langle P \rangle$ of prime order r.

The point multiplication operation, denoted by $Q = kP$, corresponds to adding P to itself $k - 1$ times, with $k \in [0, r - 1]$. The average cost of computing kP by a random n-bit scalar k using the traditional double-and-add method is about $nD + \frac{n}{2}A$, where D and A are the cost of doubling and adding a point, respectively. If the elliptic curve E of Eq. (2) is equipped with a non-trivial efficiently computable endomorphism ψ such that $\psi(P) = \delta P \in \langle P \rangle$, for some $\delta \in [2, r - 2]$. Then the point multiplication can be computed à la GLV as,

$$Q = kP = k_1 P + k_2 \psi(P) = k_1 P + k_2 \cdot \delta P,$$

where the subscalars $|k_1|, |k_2| \approx n/2$, can be found by solving a closest vector problem in a lattice [13]. Having split the scalar k into two parts, the computation of $kP = k_1 P + k_2 \psi(P)$ can be performed by applying simultaneous multiple point multiplication techniques [18] that translates into a saving of half of the doublings required by the execution of a single point multiplication kP.

Inspired by the GLS technique of [13], Hankerson, Karabina and Menezes presented in [17] a family of binary GLS curves defined over the field $\mathbb{F}_{q^2}$, with $q = 2^m$, which admits a two-dimensional endomorphism. This endomorphism can be computed at the inexpensive cost of just three additions in $\mathbb{F}_q$. Furthermore, by carefully choosing the elliptic curve parameters a, b of Eq. (2), the authors of [17] showed that it is possible to find members of that family of GLS curves with an almost-prime group order of the form $\#E_{a,b}(\mathbb{F}_{q^2}) = hr$, with $h = 2$ and where r is a $(2m - 1)$-bit prime number.

---

[1] Corresponding to the Intel's PCLMULQDQ instruction.

Security of GLS Curves. Given a point $Q \in \langle P \rangle$, the Elliptic Curve Discrete Logarithm Problem (ECDLP) consists of finding the unique integer $k \in [0, r-1]$ such that $Q = kP$. To the best of our knowledge, the most powerful attack for solving the ECDLP on binary elliptic curves was presented in [33] (see also [20,36]), with an associated computational complexity of $O(2^{c \cdot m^{2/3} \log m})$, where $c < 2$, and where m is a prime number. This is worse than generic algorithms with time complexity $O(2^{m/2})$ for all prime field extensions m less than $N = 2000$, a bound that is well above the range used for performing elliptic curve cryptography [33]. On the other hand, since the elliptic curve of Eq. (2) is defined over a quadratic extension of the field $\mathbb{F}_q$, the generalized Gaudry-Hess-Smart (gGHS) attack [15,19] to solve the ECDLP on the curve E, applies. To prevent this attack, it suffices to verify that the constant b of $E_{a,b}(\mathbb{F}_{q^2})$ is not weak. Nevertheless, the probability that a randomly selected $b \in \mathbb{F}_q^*$ is a weak parameter, is negligibly small [17].

2.3   Koblitz Curves

A Koblitz curve, also known as an anomalous binary curve or subfield curve, is defined as the set of affine points $P = (x, y) \in \mathbb{F}_q \times \mathbb{F}_q$, $q = 2^m$, that satisfy the Weierstraß equation $E_a : y^2 + xy = x^3 + ax^2 + 1$, $a \in \{0, 1\}$, together with a point at infinity denoted by $\mathcal{O}$. In λ-affine coordinates, where the points are represented as $P = (x, \lambda = x + \frac{y}{x})$, $x \neq 0$, the λ-affine form of the above equation becomes [32], $(\lambda^2 + \lambda + a)x^2 = x^4 + 1$. A Koblitz curve forms an abelian group denoted as $E_a(\mathbb{F}_{2^m})$ of order $2(2-a)r$, for an odd prime r, where its group law is defined by the point addition operation.

Frobenius Map. Since their introduction in [23], Koblitz curves were extensively studied for their additional structure that allows, in principle, a performance speedup in the point multiplication computation. The Frobenius map $\tau : E_a(\mathbb{F}_q) \to E_a(\mathbb{F}_q)$ defined by $\tau(\mathcal{O}) = \mathcal{O}$, $\tau(x, y) = (x^2, y^2)$, is a curve automorphism satisfying $(\tau^2 + 2)P = \mu\tau(P)$ for $\mu = (-1)^{1-a}$ and all $P \in E_a(\mathbb{F}_q)$. By solving the equation $\tau^2 + 2 = \mu\tau$, the Frobenius map can be seen as the complex number $\tau = \frac{\mu + \sqrt{-7}}{2}$. Notice that in λ-coordinates the Frobenius map action remains the same, because, $\tau(x, \lambda) = (x^2, \lambda^2) = (x^2, x^2 + \frac{y^2}{x^2})$, which corresponds to the λ-representation of $\tau(x, y)$. Let $\mathbb{Z}[\tau]$ be the ring of polynomials in τ with coefficients in $\mathbb{Z}$. Since the Frobenius map is highly efficient, as long as it is possible to convert an integer scalar k to its τ-representation $k = \sum_{i=0}^{l-1} u_i \tau^i$, its action can be exploited in a point multiplication computation by adding multiples $u_i \tau^i(P)$, with $u_i \tau^i \in \mathbb{Z}[\tau]$. Solinas [37] proposed exactly that, namely, a τ-adic scalar recoding analogous to the signed digit scalar *Non-Adjacent Form* representation.

Security of Koblitz Curves. From the security point of view, it has been argued that the availability of additional structure in the form of endomorphisms can be a potential threat to the hardness of elliptic curve discrete logarithms [3], but limitations observed in approaches based on isogeny walks is evidence contrariwise [22]. Furthermore, the generation of Koblitz curves satisfy by definition the *rigidity* property. Constant-time compact implementations for Koblitz curves are also easily obtained by specializing the Montgomery-López-Dahab ladder algorithm [25] for $b = 1$, although we show below that this is not the most efficient constant-time implementation strategy possible. Another practical advantage is the adoption of Koblitz curves by several standards bodies [27], which guarantee interoperability and availability of implementations in many hardware and software platforms.

3   New Montgomery Ladder Variants

This Section presents algorithms for computing the scalar multiplication through the Montgomery ladder method. Here, we let P be a point in a binary elliptic curve of prime order $r > 2$ and k a scalar of bit length n. Our objective is to compute $Q = kP$.

Algorithm 1. Left-to-right Montgomery ladder [26]

Input: $P = (x, y)$, $k = (1, k_{n-2}, \ldots, k_1, k_0)$
Output: $Q = kP$
 1: $R_0 \leftarrow P$; $R_1 \leftarrow 2P$;
 2: **for** $i = n - 2$ **downto** 0 **do**
 3:     **if** $k_i = 1$ **then**
 4:         $R_0 \leftarrow R_0 + R_1$; $R_1 \leftarrow 2R_1$
 5:     **else**
 6:         $R_1 \leftarrow R_0 + R_1$; $R_0 \leftarrow 2R_0$
 7:     **end if**
 8: **end for**
 9: **return** $Q = R_0$

Algorithm 1 describes the classical left-to-right Montgomery ladder approach for point multiplication [26], whose key algorithmic idea is based on the following observation. Given a base point P and two input points R_0 and R_1, such that their difference, $R_0 - R_1 = P$, is known, the x-coordinates of the points, $2R_0$, $2R_1$ and $R_0 + R_1$, are fully determined by the x-coordinates of P, R_0 and R_1.

More than one decade after its original proposal in [26], López and Dahab presented in [25] an optimized version of the Montgomery ladder, which was specifically crafted for the efficient computation of point multiplication on ordinary binary elliptic curves. In this scenario, compact formulae for the point addition and point doubling operations of Algorithm 1 can be derived from the following result.

Lemma 1 [25]. *Let* $P = (x, y)$, $R_1 = (x_1, y_1)$, *and* $R_0 = (x_0, y_0)$ *be elliptic curve points, and assume that* $R_1 - R_0 = P$, *and* $x_0 \neq 0$. *Then, the x-coordinate of the point* $(R_0 + R_1)$, x_3, *can be computed in terms of* x_0, x_1, *and* x *as follows,*

$$x_3 = \begin{cases} x + \frac{x_0 \cdot x_1}{(x_0 + x_1)^2} & R_0 \neq \pm R_1 \\ x_0^2 + \frac{b}{x_0^2} & R_0 = R_1 \end{cases} \tag{3}$$

Moreover, the y-coordinate of R_0 *can be expressed in terms of* P, *and the x-coordinates of* R_0, R_1 *as,*

$$y_0 = x^{-1}(x_0 + x) \left[(x_0 + x)(x_1 + x) + x^2 + y \right] + y \tag{4}$$

Let us denote the projective representation of the points R_0, R_1 and $R_0 + R_1$, without considering their y-coordinates as, $R_0 = (X_0, -, Z_0)$, $R_1 = (X_1, -, Z_1)$ and $R_0 + R_1 = (X_3, -, Z_3)$. Then, for the case $R_0 = R_1$, Lemma 1 implies,

$$\begin{cases} X_3 = X_0^4 + b \cdot Z_0^4 \\ Z_3 = X_0^2 \cdot Z_0^2 \end{cases} \tag{5}$$

Furthermore, for the case $R_0 \neq \pm R_1$, one has that,

$$\begin{cases} Z_3 = (X_0 \cdot Z_1 + X_1 \cdot Z_0)^2 \\ X_3 = x \cdot Z_3 + (X_0 \cdot Z_1) \cdot (X_1 \cdot Z_0) \end{cases} \tag{6}$$

From Eqs. (5) and (6) it follows that the computational cost of each ladder step in Algorithm 1 is of 5 multiplications, 1 multiplication by the curve b-constant, 4 or 5 squarings[2] and 3 additions over the binary extension field where the elliptic curve has been defined.

In the rest of this Section, we will present a novel right-to-left formulation of the classical Montgomery ladder.

3.1 Right-to-Left Double-and-Add Montgomery-LD Ladder

Algorithm 2 presents a right-to-left version of the classical Montgomery ladder procedure. At the end of the i-th iteration, the points in the variables R_0, R_1 are, $R_0 = 2^{i+1}P$, and $R_1 = \ell P + \frac{P}{2}$, where ℓ is the integer represented by the i rightmost bits of the scalar k. The variable R_2 maintains the relationship, $R_2 = R_0 - R_1$ from the initialization (step 1), until the execution of the last iteration of the main loop (steps 2–9). This comes from the fact that at each iteration, if $k_i = 1$, then the difference $R_0 - R_1$ remains unchanged. If otherwise, $k_i = 0$, then both R_2 and R_0 are updated with their respective original values plus R_0, which ensures that $R_2 = R_0 - R_1$, still holds. Notice however that, although the difference $R_2 = R_0 - R_1$, is known, it may vary throughout the iterations.

---

[2] Either $b = 1$ or $\sqrt{b}$ is precomputed. Formula (5) can also be computed as $Z_3 = (X_0 \cdot Z_0)^2$ and $X_3 = (X_0^2 + \sqrt{b} \cdot Z_0^2)^2$.

Algorithm 2. Montgomery-LD double-and-add scalar multiplication (right-to-left)

Input: $P = (x, y)$, $k = (k_{n-1}, k_{n-2}, \ldots, k_1, k_0)$
Output: $Q = kP$
 1: $R_0 \leftarrow P$; $R_1 \leftarrow \frac{P}{2}$; $R_2 \leftarrow \frac{P}{2} = (R_0 - R_1)$;
 2: **for** $i = 0$ **to** $n - 1$ **do**
 3:    **if** $k_i = 1$ **then**
 4:       $R_1 \leftarrow R_1 + R_0$;
 5:    **else**
 6:       $R_2 \leftarrow R_2 + R_0$;
 7:    **end if**
 8:    $R_0 \leftarrow 2R_0$;
 9: **end for**
10: **return** $Q = R_1 - \frac{P}{2}$

As stated in Lemma 1, the point additions of steps 4 and 6 in Algorithm 2 can be computed using the x-coordinates of the points R_0, R_1 and R_2, according to the following analysis. If $k_i = 1$, then the x-coordinate of $R_0 + R_1$ is a function of the x-coordinates of R_0, R_1 and R_2, because $R_2 = R_0 - R_1$. If $k_i = 0$, the x-coordinate of $R_2 + R_0$ is a function of the x-coordinates of the points R_0, R_1 and R_2, because $R_0 - R_2 = R_0 - (R_0 - R_1) = R_1$. Hence, considering the projective representation of the points $R_0 = (X_0, -, Z_0)$, $R_1 = (X_1, -, Z_1)$, $R_2 = (X_2, -, Z_2)$ and $R_0 + R_1 = (X_3, -, Z_3)$, where all the y-coordinates are ignored, and assuming $R_0 \neq \pm R_1$, we have,

$$\begin{cases} T = (X_0 \cdot Z_1 + X_1 \cdot Z_0)^2 \\ Z_3 = Z_2 \cdot T \\ X_3 = X_2 \cdot T + Z_2 \cdot (X_0 \cdot Z_1) \cdot (X_1 \cdot Z_0) \end{cases} \tag{7}$$

From Eqs. (5) and (7), it follows that the computational cost of each ladder step in Algorithm 2 is of 7 multiplications, 1 multiplication by the curve b-constant, 4 or 5 squarings and 3 additions over the binary field where the elliptic curve lies.

Although conceptually simple, the above method has several algorithmic and practical shortcomings. The most important one is the difficulty to recover, at the end of the algorithm, the y-coordinate of R_1, as in none of the available points $(R_0, R_1$ and $R_2)$ the corresponding y-coordinate is known. This may force the decision to use complete projective formulae for the point addition and doubling operations of steps 4, 6 and 8, which would be costly. Finally, we stress that to guarantee that the case $R_0 = R_2$ will never occur, it is sufficient to initialize R_1 with $\frac{P}{2}$, and perform an affine subtraction at the end of the main loop (step 10).

In the following subsection we present a halve-and-add right-to-left Montgomery ladder algorithm that alleviates the above shortcomings and still achieves a competitive performance.

3.2   Right-to-Left Halve-and-Add Montgomery-LD Ladder

Algorithm 3 presents a right-to-left Montgomery ladder procedure similar to Algorithm 2, but in this case, all the point doubling operations are substituted with point halvings. A left-to-right approach using halve-and-add with Montgomery ladder was published in [29], however, this method requires one inversion per iteration, which degrades its efficiency due to the cost of this operation.

Algorithm 3. Montgomery-López-Dahab halve-and-add (right-to-left)

Input: $P = (x, y)$, $k' = (k'_{n-1}, k'_{n-2}, \ldots, k'_1, k'_0)$
Output: $Q = kP$
1: **Precomputation:** $x(P_i)$, where $P_i = \frac{P}{2^i}$, for $i = 0, \ldots, n$
2: $R_1 \leftarrow P_n$; $R_2 \leftarrow P_n$;
3: **for** $i = 0$ **to** $n - 1$ **do**
4:     $R_0 \leftarrow P_{n-1-i}$;
5:     **if** $k'_i = 1$ **then**
6:         $R_1 \leftarrow R_0 + R_1$;
7:     **else**
8:         $R_2 \leftarrow R_0 + R_2$;
9:     **end if**
10: **end for**
11: $R_1 \leftarrow R_1 - P_n$
12: **return** R_1

As in any halve-and-add procedure, an initial step before performing the actual computation consists of processing the scalar k such that it can be equivalently represented with negative powers of two. To this end, one first computes $k' \equiv 2^{n-1}k \bmod r$, with $n = |r|$. This implies that, $k \equiv \sum_{i=1}^{n} k'_{n-i}/2^{i-1} \bmod r$ and therefore, $kP = \sum_{i=1}^{n} k'_{n-i}(\frac{1}{2^{i-1}}P)$. Then, in the first step of Algorithm 3, n halvings of the base point P are computed. We stress that all the precomputed points $P_i = \frac{P}{2^i}$, for $i = 0, \ldots, n$ can be stored in affine coordinates. In fact, just the x-coordinate of each one of the above n points must be stored (with the sole exception of the point P_n, whose y-coordinate is also computed and stored).

As in the preceding algorithm notice that at the end of the i-th iteration, the points in the variables R_0, R_1 are, $R_0 = \frac{P}{2^{n-i-1}}$, and $R_1 = \ell P + P_n$, where in this case ℓ is the integer represented as, $\ell = \sum_{j=0}^{i} \frac{k'_j}{2^{n-j}} \bmod r$. Notice also that the variable R_2 maintains the relationship, $R_2 = R_0 - R_1$, until the execution of the last iteration of the main loop (steps 3–10). This comes from the fact that at each iteration, if $k_i = 1$, then the difference $R_0 - R_1$ remains unchanged. If otherwise, $k_i = 0$, then both R_2 and R_0 are updated with their respective original values plus R_0, which ensures that $R_2 = R_0 - R_1$, still holds.

Since at every iteration, the values of the points R_0, R_1 and $R_0 - R_1$, are all known, the compact point addition formula (7) can be used. In practice, this is also possible because the y-coordinate of the output point kP can be readily recovered using Eq. 4, along with the point $2P$. Moreover, since the points in the precomputed table were generated using affine coordinates, it turns out that the z-coordinate of the point R_0 is always 1 for all the iterations of the main loop. This simplifies (7) as,

$$
\begin{cases}
T = (X_0 \cdot Z_1 + X_1)^2 \\
Z_3 = Z_2 \cdot T \\
X_3 = X_2 \cdot T + Z_2 \cdot (X_0 \cdot Z_1) \cdot (X_1)
\end{cases}
\tag{8}
$$

Hence, the computational cost per iteration of Algorithm 3 is of 5 multiplications, 1 squaring, 2 additions and one point halving over the binary field where the elliptic curve lies.

GLS Endomorphism. The efficient computable endomorphism provided by the GLS curves can be used to implement the 2-GLV method on the Algorithm 3. As a result, only $n/2$ point halving operations must be computed. Besides the speed improvement, the 2-GLV method reduces to a half the number of precomputed points that must be stored.

3.3 Multi-core Montgomery Ladder

As proposed in [38], by properly recoding the scalar, one can efficiently compute the scalar multiplication in a multi-core environment. Specifically, given a scalar k of size n, we fix a constant t which establishes how many scalar bits will be processed by the double-and-add, and by the halve-and-add procedures. This is accomplished by computing, $k' = 2^t k \bmod r$, which yields,

$$
k = \underbrace{\frac{k'_0}{2^t} + \frac{k'_1}{2^{t-1}} + \cdots + \frac{k'_{t-1}}{2^1}}_{halve\text{-}and\text{-}add} + \underbrace{\frac{k'_t}{2^0} + 2^1 k'_{t+1} + 2^2 k'_{t+2} + \cdots + 2^{(n-1)-t} k'_{n-1}}_{double\text{-}and\text{-}add}
$$

In a two-core setting, it is straightforward to combine the left-to-right and right-to-left Montgomery ladder procedures of Algorithms 1 and 3, and distribute them to both cores. In this scenario, the number of necessary pre-computed halved points reduces to $\sim \frac{n}{4}$. In a four-core platform, we can apply the GLS endomorphism to the left-to-right Montgomery ladder (Algorithm 1). Even though the GLV technique is ineffective for the classical Montgomery algorithm (due to the fact that we cannot share the point doublings between the base point and its endomorphism), the method permits an efficient splitting of the algorithm workload into two cores. In this way, one can use the first two cores for computing t-digits of the GLV subscalars k_1 and k_2 by means of Algorithm 3, while we allocate the other two cores to compute the rest of the scalar's bits using Algorithm 1, as shown in Algorithm 6 (see Appendix A).

Table 1. Montgomery-LD algorithms cost comparison. In this table, M, M_a, M_b, S, I denote the following field operations: multiplication, multiplication by the curve a-constant, multiplication by the curve b-constant, squaring and inversion. The point halving operation is denoted by H.

	Method			Cost
1-core	Alg. 1: Montgomery-LD (double-and-add, left-to-right)		pre/post	$10M + 1S + 1I$
			sc. mult.	$n(5M + 1M_b + 4S)$
	Alg. 3: Montgomery-LD-2-GLV (halve-and-add, right-to-left)		pre/post	$48M + 1M_a + 13S + 3I$
			sc. mult.	$(\frac{n}{2} + 1)H + n(5M + 1S)$
2-core	Montgomery-LD-2-GLV (double-and-add, left-to-right)	core I	pre/post	$25M + 1M_a + 5S + 2I$
			sc. mult.	$(n - t_2)(5M + 1M_b + 4S)$
	Montgomery-LD-2-GLV (halve-and-add, right-to-left)	core II	pre/post	$46M + 2M_a + 12S + 2I$
			sc. mult.	$(\frac{t_2}{2} + 1)H + t_2(5M + 1S)$
	Overhead			$15M + 5S + 1I$
4-core	Montgomery-LD-2-GLV (double-and-add, left-to-right)	cores I & II	pre/post	$10M + 1S + 1I$
			sc. mult.	$(\frac{n}{2} - t_4)(5M + 1M_b + 4S)$
	Montgomery-LD-2-GLV (halve-and-add, right-to-left)	cores III & IV	pre/post	$16M + 1M_a + 4S + 1I$
			sc. mult.	$(\frac{t_4}{2} + 1)H + t_4(5M + 1S)$
	Overhead			$34M + 1M_a + 12S + 1I$

3.4   Cost Comparison of Montgomery Ladder Variants

Table 1 shows the computational costs associated to the Montgomery ladder variants described in this Section. The constants t_2 and t_4 represent the values of the parameter t chosen for the two- and four-core implementations, respectively.[3] All Montgomery ladder algorithms require a basic post-computation cost to retrieve the y-coordinate, which demands ten multiplications, one squaring and one inversion. Due to the application of the GLV technique, the Montgomery-LD-2-GLV halve-and-add version (corresponding to Algorithm 3), requires some few extra operations, namely, the subtraction of a point and the addition of two accumulators, which is performed using the López-Dahab (LD) projective coordinate formulae. In the end, one extra inversion is needed to convert the point representation from LD-projective coordinates to affine coordinates.

In the case of the parallel versions, the overhead is given by the post-computation done in one single core. The exact costs are mainly determined by the accumulator additions that are performed via full and mixed LD-projective formulae. In all of the timings reported in Sect. 5, we consider the LD-projective to affine coordinate transformation cost.

4   A Novel Regular τ-Adic Approach

4.1   Recoding in τ-Adic Form

The recoding approach proposed by Solinas finds an element $\rho \in \mathbb{Z}[\tau]$, of as small norm as possible, such that $\rho \equiv k \pmod{\frac{\tau^m - 1}{\tau - 1}}$. A τ-adic expansion with

---

[3] In our implementations (see Subsect. 5.3 below), the values used for the parameters t_2 and t_4 ranged from 53 to 55.

Algorithm 4. Regular width-w τ-recoding for m-bit scalar

Input: $w, t_w, \alpha_u = \beta_u + \gamma_u \tau$ for $u = \{\pm 1, \pm 3, \pm 5, \ldots, \pm 2^{w-1} - 1\}, \rho = r_0 + r_1 \tau \in \mathbb{Z}[\tau]$
with odd r_0, r_1

Output: $\rho = \displaystyle\sum_{i=0}^{\lceil \frac{m+2}{w-1} \rceil} u_i \tau^{i(w-1)}$

```
 1: for i ← 0 to ⌈ (m+2)/(w-1) ⌉ - 1 do        14: if r₀ ≠ 0 and r₁ ≠ 1 then
 2:    if w = 2 then                           15:    uᵢ ← r₀ + r₁τ
 3:       uᵢ ← ((r₀ - 2r₁) mod 4) - 2          16: else
 4:       r₀ ← r₀ - uᵢ                         17:    if r₁ ≠ 0 then
 5:    else                                    18:       uᵢ ← r₁
 6:       u ← (r₀ + r₁t_w mod 2ʷ) - 2^(w-1)    19:    else
 7:       if u > 0 then s ← 1 else s ← -1      20:       uᵢ ← r₀
 8:       r₀ ← r₀ - sβ_u, r₁ ← r₁ - sγ_u, uᵢ ← sα_u  21: end if
 9:    end if                                  22: end if
10:    for j ← 0 to (w - 2) do
11:       t ← r₀, r₀ ← r₁ + μr₀/2, r₁ ← -t/2
12:    end for
13: end for
```

average non-zero density $\frac{1}{3}$ can be obtained by repeatedly dividing ρ by τ and assigning the remainders to the digits u_i to obtain $k = \sum_{i=0}^{i=l-1} u_i \tau^i$. An alternative approach that does not involve multi-precision divisions, is to compute an element $\rho' = k$ partmod$\left(\frac{\tau^m - 1}{\tau - 1}\right)$ by performing a partial reduction procedure [37]. A width-w τ-NAF expansion with non-zero density $\frac{1}{w+1}$, where at most one of any w consecutive coefficients is non-zero, can also be obtained by repeatedly dividing ρ' by τ^w and assigning the remainders to the digit set $\{0, \pm\alpha_1, \pm\alpha_3, \ldots, \pm\alpha_{2^{w-1}-1}\}$, for $\alpha_i = i \bmod \tau^w$. Under reasonable assumptions, this window-based recoding has length $l \leq m + 1$ [37].

In this section, a regular recoding version of the (width-w) τ-NAF expansion is derived. The security advantages of such recoding are the predictable length and locations of non-zero digits in the expansion. This eliminates any side-channel information that an attacker could possibly collect regarding the operation executed at any iteration of the scalar multiplication algorithm (point doubling/Frobenius map or point addition). As long as querying a precomputed table of points to select the second operand of a point addition takes constant time, the resulting algorithm should be resistant against any timing-based side-channel attacks.

Let us first consider the integer recoding proposed by Joye and Tunstall [21]. They observed that any odd integer i in the interval $[0, 2^w)$ can be written as $i = 2^{w-1} + (-(2^{w-1} - i))$. Repeatedly dividing an odd n-bit integer $k - ((k \bmod 2^w) - 2^{w-1})$ by 2^{w-1} maintains the parity and assigns the remainders to the digit set $\{\pm 1, \ldots, \pm(2^{w-1} - 1)\}$, producing an expansion of length $\lceil 1 + \frac{n}{w-1} \rceil$ with non-zero density $\frac{1}{w-1}$. Our solution for the problem of finding a regular τ-adic expansion employs the same intuition, as explained next.

Let $\phi_w : \mathbb{Z}[\tau] \to \mathbb{Z}_{2^w}$ be a surjective ring homomorphism induced by $\tau \mapsto t_w$, for $t_w^2 + 2 \equiv \mu t_w \pmod{2^w}$, with kernel $\{\alpha \in \mathbb{Z}[\tau] : \tau^w \text{ divides } \alpha\}$. An element $i = i_0 + i_1\tau$ from $\mathbb{Z}[\tau]$ with odd integers $i_0, i_1 \in [0, 2^w)$ satisfies the analogous property $\phi_w(i) = 2^{w-1} + (-(2^{w-1} - \phi_w(i)))$. Repeated division of $(r_0 + r_1\tau) - (((r_0 + r_1\tau) \bmod \tau^w) - \tau^{w-1})$ by τ^{w-1}, correspondingly of $\phi_w(\rho') = (r_0 + r_1 t_w) - ((r_0 + r_1 t_w \bmod 2^w) - 2^{w-1})$ by 2^{w-1}, obtains remainders that belong to the set $\{0, \pm\alpha_1, \pm\alpha_3, \ldots, \pm\alpha_{2^{w-1}-1}\}$. The resulting expansion always has length $\lceil 1 + \frac{m+2}{w-1} \rceil$ and non-zero density $\frac{1}{w-1}$. Algorithm 4 presents the recoding process for any $w \geq 2$. The resulting recoding can also be seen as an adaption of the SPA-resistant recoding of [30], mapping to the digit set $\{0, \pm\alpha_1, \pm\alpha_3, \ldots, \pm\alpha_{2^{w-1}-1}\}$ instead of integers. While the non-zero densities are very similar, our scheme provides a performance benefit in the precomputation step, since the Frobenius map is usually faster than point doubling and preserves affine coordinates and consequently faster point additions.

4.2 Left-to-Right Regular Approach

Algorithm 5 presents a complete description of a regular scalar multiplication approach that uses as a building block the regular width-w τ-recoding procedure just described.

For benchmarking purposes, we also included a baseline implementation of the customary Montgomery López-Dahab ladder. This allows easier comparisons with related work and permits to evaluate the impact of incomplete reduction in the field arithmetic performance (cf. Subsect. 5.2).

Algorithm 5. Protected scalar multiplication

Input: $P = (x, \lambda)$, $k \in \mathbb{Z}$, width w
Output: $Q = kP$
1: Compute $\rho' = r_0 + r_1\tau = k \text{ partmod} \left(\frac{\tau^m - 1}{\tau - 1} \right)$
2: if $2|r_0$ then $r_0' = r_0 + 1$
3: if $2|r_1$ then $r_1' = r_1 + 1$

4: Compute width-w length-l regular τ-adic of $r_0' + r_1'\tau$ as $\sum_{i=0}^{\lceil 1 + \frac{m+2}{w-1} \rceil} u_i \tau^{i(w-1)}$ (Alg. 4)
5: **for** $i \in \{1, \ldots, 2^{w-1} - 1\}$ **do**
6:     Compute $P_u = \alpha_u P$
7:
8: $Q \leftarrow \mathcal{O}$
9: **for** $i = l - 1$ **downto** 0 **do**
10:     $Q \leftarrow \tau^{w-1}(Q)$
11:     Perform a linear pass to recover P_{u_i}
12:     $Q \leftarrow Q + P_{u_i}$
13: **end for**
14: **return** $Q = Q - (r_0' - r_0)P - (r_1' - r_1)\tau(P)$.

---

5   Implementation Issues and Results

In this Section, we discuss several implementation issues. We also present our experimental results and we compare them against state-of-the-art protected point multiplication implementations at the 128-bit security level.

5.1   Mechanisms to Achieve a Constant-Time GLS-Montgomery Ladder Implementation

To protect the previously described algorithms against timing attacks, we observed the following precautions:

Branchless Code. The main loop, the pre- and post-computation phases are implemented by a completely branch-free code.

Data Veiling. To guarantee a constant memory access pattern in the main loop of the Montgomery ladder algorithms, we proposed an efficient data veiling method, as described in Algorithm 7 of Appendix B. Algorithm 7 evaluates the actual and the previous scalar bits to decide whether the variables containing the Montgomery-LD accumulators values should or should not be masked. This strategy saves a considerable portion of the computational effort associated to Algorithm 1 of [4].

Field Arithmetic. Two of the base field arithmetic operations over $\mathbb{F}_q$ were implemented through look-up tables, namely, the half-trace and the multiplicative inverse operations. The half-trace is used to perform the point halving primitive, which is required in the pre-computation phase of the Montgomery-LD halve-and-add algorithm. The multiplicative inverse is one of the operations in the y-coordinate retrieval procedure, at the end of the Montgomery ladder algorithms. Also, whenever post-computational additions are necessary, inverses must be performed to convert a point from LD-projective to affine coordinates.

Although we are aware of the existence of protocols that consider the base point as a secret information [6], in which case one *could not* consider that our software provides protection against timing attacks, in the vast majority of protocols, the base point is public. Consequently, any attacks aimed at the two field operations mentioned above would be pointless.

5.2   Mechanisms to Achieve a Constant-Time Koblitz Implementation

Implementing Algorithm 5 in constant time needs some care, since all of its building blocks must be implemented in constant time.

Finite Field Arithmetic. Modern implementations of finite field arithmetic can make extensive use of vector registers, removing timing variances due to the cache hierarchy. For our illustrative implementation of curve NIST-K283, we closely follow the arithmetic described in Bluhm-Gueron [4], adopting the incomplete reduction improvement proposed by Negre-Robert [28].

Integer Recoding. All the branches in Algorithm 4 need to be eliminated by conditional execution statements to protect leakage of the scalar k. Moreover, to remove the remaining sign-related branches, multiple precision integer arithmetic must be implemented in complement of two. If two constants, say β_u, γ_u, are stored in a precomputed table, then they need to be recovered by a linear pass across the table in constant time. Finally, the partial reduction step producing ρ' must also be implemented in constant time by removing all of its branches. Notice that the requirement for r_0, r_1 to be odd is not a problem, since partial reduction can be modified to always result in odd integers, with a possible correction at the end of the scalar multiplication by performing a (protected) conditional subtraction of points (line 14 of Algorithm 5).

5.3 Results

Our implementation was mainly designed for the Intel Haswell processor family, which supports vectorial sets such as SSE and AVX, a carry-less multiplication and some bit manipulation instructions. The programming was done in C with the support of assembly inline code. The compilation was performed via GCC version 4.7.3 with the flags `-m64 -march=core-avx2 -mtune=core-avx2 -O3 -fomit-frame-pointer -funroll-loops`. Finally, the timings were collected on an Intel Core i7-4700MQ, with the Turbo Boost and Hyperthreading features disabled[4].

Table 2 presents the experimental timings obtained for the most prominent building blocks required for computing the point multiplication operation on the GLS and Koblitz binary elliptic curves.

We present in Table 3 a comparison of our timings against a selection of state-of-the-art implementations of the point multiplication operation on binary and prime elliptic curves. Due to the Montgomery-LD point doubling efficiency, which costs 49 % less than a point halving, the GLS-Montgomery-LD-double-and-add achieved the fastest timing in the one-core setting, with 70,800 clock cycles. This is 13 % faster than the performance obtained by the GLS-Montgomery-LD-halve-and-add algorithm. In the known-base point setting, we can ignore the GLS-Montgomery-LD-halve-and-add pre-computation expenses associated with its table of halved points. In that case, we can compute the scalar multiplication in an estimated time of 44,600 clock cycles using a table of just 4128 bytes.

---

[4] We intend to submit our software to the ECRYPT Benchmarking of Cryptographic Systems (eBACS) SUPERCOP toolkit in the near future.

Table 2. Timings (in clock cycles) for the elliptic curve operations in the Intel Haswell platform.

Elliptic curve operation	GLS $E/\mathbb{F}_{2^{254}}$	
	cycles	op/M^1
Halving	184	4.181
Montgomery-LD D&A (left-to-right) Addition (Eq. (6))	161	3.659
Montgomery-LD H&A (right-to-left) Addition (Eq. (8))	199	4.522
Montgomery-LD Doubling^a (Eq. (5))	95	2.159

Elliptic curve operation	Koblitz $E/\mathbb{F}_{2^{283}}$	
	cycles	op/M^1
Frobenius	70	1.235
Integer τ-adic recoding (Alg. 4) ($w = 5$)	8,900	156.863
Point addition	602	10.588

^{1}Ratio to multiplication.

^{a}The flexibility for finding a curve b-constant, provided by the GLS curves, allow us to have a small $\sqrt{b}$ (see Appendix C). As a consequence, we used the Eq. (5) alternative formula.

Furthermore, the GLS-Montgomery-LD-halve-and-add is crucial for implementing the multi-core versions of the Montgomery ladder. When compared with our one-core double-and-add implementation, Table 3 reports a speedup of 1.36 and 2.03 in our two- and four-core Montgomery ladder versions, respectively. Here, besides the overhead costs commented in Sect. 3, we can clearly perceive the usual multicore management penalty. Finally, we observe that our GLS-Montgomery-LD-double-and-add surpasses by 48 %, 40 % and 2 % the Montgomery ladder implementations of [4] (Random), [4] (Koblitz) and [1], respectively.

As for our Koblitz implementations, the fast τ endomorphism allows us to have a regular-recoding implementation that outperforms a standard Montgomery ladder for Koblitz curves by 18 %. In addition, our fastest Koblitz code surpasses by 16 % the recent implementation reported in [4][5]. Finally, note that, in spite of the fact that the τ endomorphism is 26 % faster than the Montgomery-LD point doubling, the superior efficiency of the GLS quadratic field arithmetic produces faster results for the GLS Montgomery ladder algorithms.

---

[5] We could not reproduce the timing of 118,000 cycles with the code available from [4], which indicates that TurboBoost could be possibly turned on their benchmarks. Considering this, our implementation of Koblitz-Montgomery-LD becomes 9 % faster than [4], reflecting the savings from partial reduction, and the speedup achieved by the Koblitz-regular implementation increases to 26 %.

Table 3. Timings (in clock cycles) for 128-bit level scalar multiplication with timing-attack resistance in the Intel Ivy Bridge (I) and Haswell (H) architectures.

	Method	Cycles	Arch
State-of-the-art implementations	Montgomery-DJB-chain (prime) [7]	148,000	I
	Random-Montgomery-LD ladder (binary) [4]	135,000	H
	Genus-2-Kummer (prime) [5]	122,000	I
	Koblitz-Montgomery-LD ladder (binary) [4]	118,000	H
	Twisted-Edwards-4-GLV (prime) [11]	92,000	I
	Genus-2-Kummer Montgomery ladder (prime) [1]	72,200	H
	GLS-2-GLV double-and-add (binary, λ) [32]	60,000	H
Our Work	Koblitz-Montgomery-LD double-and-add (left-to-right)	122,000	H
	Koblitz-regular τ-and-add (left-to-right, $w = 5$)	99,000	H
	GLS-Montgomery-LD-2-GLV halve-and-add (**Algorithm 3**)	80,800	H
	GLS-Montgomery-LD double-and-add (**Algorithm 1**)	70,800	H
	2-core GLS-Montgomery-LD-2-GLV halve-and-add/double-and-add	52,000	H
	4-core GLS-Montgomery-LD-2-GLV halve-and-add/double-and-add (**Algorithm 6**)	34,800	H

6   Conclusion

We presented several algorithms that permit to compute a constant-time high-security point multiplication operation over two families of binary elliptic curves, namely, the GLS and the Koblitz curves. Although this work was completely focused on a high-end desk computation of the variable-base point multiplication, the possibility of applying Algorithm 3 to the fixed-base point multiplication setting is highly appealing since that procedure requires a comparatively small pre-computed table of roughly $2n \cdot (n + 1)$ bits for computing a point multiplication at the n-bit security level. The above combined with the Montgomery ladder unique feature of performing all the computations using only two point coordinates, should be attractive for deployments of public key cryptography on constrained computing environments.

A   Multi-core Montgomery Ladder

Here we present the four-core GLS-Montgomery-LD ladder algorithm. Given t_4 the integer constant that establishes the workload of each algorithm, $P \in E(\mathbb{F}_{q^2})$, and the scalar k represented as $k_1 + k_2 \cdot \delta$ using the GLS-GLV method, cores I and II are both responsible for computing $\lfloor \frac{n}{2} \rfloor - t_4$ bits of the subscalars k_1 and k_2 using the Montgomery-LD double-and-add method. In turn, the cores III and IV, both compute t_4 bits of k_1 and k_2 with the Montgomery-LD halve-and-add algorithm. In the end, on a single core, it is necessary to add all the accumulators Q_i, for $i = 0 \ldots 3$.

Algorithm 6. Parallel Montgomery ladder scalar multiplication (four-core)

Input: $P \in E(\mathbb{F}_{q^2})$ of order r, scalar k of bit length n, integer constant t_4
Output: $Q = kP$
 $k' \leftarrow 2^{t_4}k \bmod r$
 Represent $k' = k'_1 + k'_2\lambda$, where $\psi(P) = \lambda P$

{Initialization}
 $R_0 \leftarrow \mathcal{O}, R_1 \leftarrow P$
 for $i = \lceil \frac{n}{2} \rceil$ **downto** t_4 **do**
 $\quad b \leftarrow k'_{1,i} \in \{0,1\}$
 $\quad R_{1-b} \leftarrow R_{1-b} + R_b$
 $\quad R_b \leftarrow 2R_b$
 end for
 $Q_0 \leftarrow R_0$
 {Barrier}   **Core I**
	{Initialization}
	 $R_0 \leftarrow \mathcal{O}, R_1 \leftarrow P$
	 for $i = \lceil \frac{n}{2} \rceil$ **downto** t_4 **do**
	 $\quad b \leftarrow k'_{2,i} \in \{0,1\}$
	 $\quad R_{1-b} \leftarrow R_{1-b} + R_b$
	 $\quad R_b \leftarrow 2R_b$
	 end for
	 $Q_1 \leftarrow R_0$
	 {Barrier} **Core II**

{Precomputation}
 for $i = 1$ **to** $t_4 + 1$ **do**
 $\quad P_i \leftarrow \frac{P}{2^i}$
 end for
 {Initialization}
 $R_1 \leftarrow P_{t_4+1}, R_2 \leftarrow P_{t_4+1}$
 for $i = 0$ **to** $t_4 - 1$ **do**
 $\quad R_0 \leftarrow P_{t_4-i}$
 $\quad b \leftarrow k'_{1,i} \in \{0,1\}$
 $\quad R_{2-b} \leftarrow R_{2-b} + R_0$
 end for
 $Q_2 \leftarrow R_1 - P_{t_4+1}$
 {Barrier}   **Core III**
	{Precomputation}
	 for $i = 1$ **to** $t_4 + 1$ **do**
	 $\quad P_i \leftarrow \frac{P}{2^i}$
	 end for
	 {Initialization}
	 $R_1 \leftarrow P_{t_4+1}, R_2 \leftarrow P_{t_4+1}$
	 for $i = 0$ **to** $t_4 - 1$ **do**
	 $\quad R_0 \leftarrow P_{t_4-i}$
	 $\quad b \leftarrow k'_{2,i} \in \{0,1\}$
	 $\quad R_{2-b} \leftarrow R_{2-b} + R_0$
	 end for
	 $Q_3 \leftarrow R_1 - P_{t_4+1}$
	 {Barrier} **Core IV**

$\quad$ **return** $Q = Q_0 + Q_2 + \psi(Q_1 + Q_3)$

B   Memory Access Pattern

The following data veiling algorithm ensures a fixed memory access pattern for all Montgomery-LD ladder algorithms. Given the two Montgomery-LD ladder accumulators A and B, and the scalar $k = (k_{n-1}, k_{n-2}, \ldots k_0)$, this method allows us, in the beginning of the i-th main loop iteration, to use the bits k_{i-1} and k_i to decide if A and B will or will not be swapped. As a result, it is not necessary to reapply the procedure at the end of the i-th iteration.

Algorithm 7. Data veiling algorithm

Input: Scalar digits k_i and k_{i-1}, Montgomery-LD ladder accumulators A and B
Output: Montgomery-LD ladder accumulators A and B

$mask \leftarrow 0 - (k_{i-1} \oplus k_i)$
$tmp \leftarrow A \oplus B$
$tmp \leftarrow tmp \wedge mask$
$A \leftarrow A \oplus tmp$
$B \leftarrow B \oplus tmp$
return A, B

C GLS Elliptic Curve Parameters

For achieving a greater benefit from the multiplication by the b-constant in the Montgomery-LD doubling formula $X_3 = X_0{}^4 + bZ_0{}^4 = (X_0{}^2 + \sqrt{b}Z_0{}^2)^2$ we carefully selected a GLS curve with a 64-bit b-parameter square-root. As a result, we saved two carry-less multiplication and a dozen of SSE instructions per field multiplication. Next, we describe the parameters, as polynomials represented in hexadecimal, for our GLS curve $E_{a,b}/\mathbb{F}_{q^2} : y^2 + xy = x^3 + ax^2 + b$.

- $a = u$
- $b = $ 0x540451444104015441015405405151011
- $\sqrt{b} = $ 0xE2DA921E91E38DD1

The 253-bit prime order r of the main subgroup of $E_{a,b}/\mathbb{F}_{q^2}$ is,

$r = $ 0x1FFFFFFFFFFFFFFFFFFFFFFFFFFFFFFFFA6B89E49D3FECD828CA8D66BF4B88ED5.

Also, the integer δ such that $\psi(P) = \delta P$ for all $P \in E_{a,b}$ is,

$\delta = $ 0x74AEFB81EE8A42E9E9D0085E156A8EFBA3D302F9C74D737FA00360F9395C788.

The base point $P = (x, y)$ of order r used in this work is,

$x = $ 0x4A21A3666CF9CAEBD812FA19DF9A3380 + 0x358D7917D6E9B5A7550B1B083BC299F3 · u

$y = $ 0x6690CB7B914B7C4018E7475D9C2B1C13 + 0x2AD4E15A695FD54011BA179D5F4B44FC · u.

Finally, the towering of our field $\mathbb{F}_q \cong \mathbb{F}_2[x]/(f(x))$ and its quadratic extension $\mathbb{F}_{q^2} \cong \mathbb{F}_q[u]/(g(x))$ is constructed by means of the irreducible trinomials $f(x) = x^{127} + x^{63} + 1$ and $g(u) = u^2 + u + 1$.

References

1. Bernstein, D.J., Chuengsatiansup, C., Lange, T., Schwabe, P.: Kummer strikes back: new DH speed records. Cryptology ePrint Archive, Report 2014/134 (2014). http://eprint.iacr.org/
2. Bernstein, D.J., Lange, T.: SafeCurves: choosing safe curves for elliptic-curve cryptography. http://safecurves.cr.yp.to

3. Bernstein, D.J., Lange, T.: Security dangers of the NIST curves. Invited talk, International State of the Art Cryptography Workshop, Athens, Greece (2013)
4. Bluhm, M., Gueron, S.: Fast software implementation of binary elliptic curve cryptography. Cryptology ePrint Archive, Report 2013/741 (2013). http://eprint.iacr.org/
5. Bos, J.W., Costello, C., Hisil, H., Lauter, K.: Fast cryptography in genus 2. In: Johansson, T., Nguyen, P.Q. (eds.) EUROCRYPT 2013. LNCS, vol. 7881, pp. 194–210. Springer, Heidelberg (2013)
6. Chatterjee, S., Karabina, K., Menezes, A.: A new protocol for the nearby friend problem. In: Parker, M.G. (ed.) Cryptography and Coding 2009. LNCS, vol. 5921, pp. 236–251. Springer, Heidelberg (2009)
7. Costello, C., Hisil, H., Smith, B.: Faster compact Diffie–Hellman: endomorphisms on the x-line. In: Nguyen, P.Q., Oswald, E. (eds.) EUROCRYPT 2014. LNCS, vol. 8441, pp. 183–200. Springer, Heidelberg (2014)
8. Dierks, T., Allen, C.: The TLS Protocol Version 1.0. RFC 2246 (Proposed Standard), January 1999. Obsoleted by RFC 4346, updated by RFCs 3546, 5746, 6176
9. Dierks, T., Rescorla, E.: The Transport Layer Security (TLS) Protocol Version 1.1. RFC 4346 (Proposed Standard), April 2006. Obsoleted by RFC 5246, updated by RFCs 4366, 4680, 4681, 5746, 6176
10. Dierks, T., Rescorla, E.: The Transport Layer Security (TLS) Protocol Version 1.2. RFC 5246 (Proposed Standard), August 2008
11. Faz-Hernández, A., Longa, P., Sánchez, A.H.: Efficient and secure algorithms for GLV-based scalar multiplication and their implementation on GLV-GLS curves. In: Benaloh, J. (ed.) CT-RSA 2014. LNCS, vol. 8366, pp. 1–27. Springer, Heidelberg (2014)
12. Fog, A.: Instruction Tables: List of Instruction Latencies, Throughputs and Micro-operation Breakdowns for Intel, AMD and VIA CPUs. http://www.agner.org/optimize/instruction_tables.pdf. Accessed 14 May 2014
13. Galbraith, S.D., Lin, X., Scott, M.: Endomorphisms for faster elliptic curve cryptography on a large class of curves. In: Joux, A. (ed.) EUROCRYPT 2009. LNCS, vol. 5479, pp. 518–535. Springer, Heidelberg (2009)
14. Gallant, R.P., Lambert, R.J., Vanstone, S.A.: Faster point multiplication on elliptic curves with efficient endomorphisms. In: Kilian, J. (ed.) CRYPTO 2001. LNCS, vol. 2139, pp. 190–200. Springer, Heidelberg (2001)
15. Gaudry, P., Hess, F., Smart, N.P.: Constructive and destructive facets of Weil descent on elliptic curves. J. Cryptol. 15, 19–46 (2002)
16. Gueron, S., Krasnov, V.: Fast prime field elliptic curve cryptography with 256 bit primes. Cryptology ePrint Archive, Report 2013/816 (2013). http://eprint.iacr.org/
17. Hankerson, D., Karabina, K., Menezes, A.: Analyzing the Galbraith-Lin-Scott point multiplication method for elliptic curves over binary fields. IEEE Trans. Comput. 58(10), 1411–1420 (2009)
18. Hankerson, D., Menezes, A., Vanstone, S.: Guide to Elliptic Curve Cryptography. Springer, Secaucus (2003)
19. Hess, F.: Generalising the GHS attack on the elliptic curve discrete logarithm problem. LMS J. Comput. Math. 7, 167–192 (2004)
20. Huang, Y.-J., Petit, C., Shinohara, N., Takagi, T.: Improvement of Faugère et al.'s method to solve ECDLP. In: Sakiyama, K., Terada, M. (eds.) IWSEC 2013. LNCS, vol. 8231, pp. 115–132. Springer, Heidelberg (2013)

21. Joye, M., Tunstall, M.: Exponent recoding and regular exponentiation algorithms. In: Preneel, B. (ed.) AFRICACRYPT 2009. LNCS, vol. 5580, pp. 334–349. Springer, Heidelberg (2009)

22. Koblitz, A.H., Koblitz, N., Menezes, A.: Elliptic curve cryptography: the serpentine course of a paradigm shift. J. Number Theory **131**(5), 781–814 (2011)

23. Koblitz, N.: CM-curves with good cryptographic properties. In: Feigenbaum, J. (ed.) CRYPTO 1991. LNCS, vol. 576, pp. 279–287. Springer, Heidelberg (1992)

24. Longa, P., Sica, F.: Four-dimensional Gallant-Lambert-Vanstone scalar multiplication. J. Cryptol. **27**(2), 248–283 (2014)

25. López, J., Dahab, R.: Fast multiplication on elliptic curves over $GF(2^m)$ without precomputation. In: Koç, Ç.K., Paar, C. (eds.) CHES 1999. LNCS, vol. 1717, pp. 316–327. Springer, Heidelberg (1999)

26. Montgomery, P.L.: Speeding the Pollard and elliptic curve methods of factorization. Math. Comput. **48**, 243–264 (1987)

27. National Institute of Standards and Technology. Recommended Elliptic Curves for Federal Government Use. NIST Special Publication (1999). http://csrc.nist.gov/csrc/fedstandards.html

28. Négre, C., Robert, J.-M.: Impact of optimized field operations AB, AC and $AB + CD$ in scalar multiplication over binary elliptic curve. In: Youssef, A., Nitaj, A., Hassanien, A.E. (eds.) AFRICACRYPT 2013. LNCS, vol. 7918, pp. 279–296. Springer, Heidelberg (2013)

29. Nègre, C., Robert, J.-M.:. New parallel approaches for scalar multiplication in elliptic curve over fields of small characteristic (2013). http://hal.archives-ouvertes.fr/docs/00/90/84/63/PDF/parallelization-ecsm8.pdf

30. Okeya, K., Takagi, T., Vuillaume, C.: Efficient representations on Koblitz curves with resistance to side channel attacks. In: Boyd, C., González Nieto, J.M. (eds.) ACISP 2005. LNCS, vol. 3574, pp. 218–229. Springer, Heidelberg (2005)

31. Oliveira, T., López, J., Aranha, D.F., Rodríguez-Henríquez, F.: Lambda coordinates for binary elliptic curves. In: Bertoni, G., Coron, J.-S. (eds.) CHES 2013. LNCS, vol. 8086, pp. 311–330. Springer, Heidelberg (2013)

32. Oliveira, T., López, J., Aranha, D.F., Rodríguez-Henríquez, F.: Two is the fastest prime: lambda coordinates for binary elliptic curves. J. Cryptogr. Eng. **4**(1), 3–17 (2014)

33. Petit, C., Quisquater, J.-J.: On polynomial systems arising from a Weil descent. In: Wang, X., Sako, K. (eds.) ASIACRYPT 2012. LNCS, vol. 7658, pp. 451–466. Springer, Heidelberg (2012)

34. Salowey, J.: Confirming consensus on removing RSA key transport from TLS 1.3. Transport Layer Security Working Group of the IETF Mailing List, 3 May 2014

35. Scott, M.: Re: NIST announces set of elliptic curves (1999). https://groups.google.com/forum/message/raw?msg=sci.crypt/mFMukSsORmI/FpbHDQ6hM_MJ

36. Shantz, M., Teske, E.: Solving the elliptic curve discrete logarithm problem using Semaev polynomials, Weil descent and Gröbner basis methods - an experimental study. Cryptology ePrint Archive, Report 2013/596 (2013). http://eprint.iacr.org/

37. Solinas, J.A.: Efficient arithmetic on Koblitz curves. Des. Codes Crypt. **19**(2–3), 195–249 (2000)

38. Taverne, J., Faz-Hernández, A., Aranha, D.F., Rodríguez-Henríquez, F., Hankerson, D., López, J.: Speeding scalar multiplication over binary elliptic curves using the new carry-less multiplication instruction. J. Cryptogr. Eng. **1**, 187–199 (2011)

Partial Key Exposure Attacks on RSA: Achieving the Boneh-Durfee Bound

Atsushi Takayasu$^{(\boxtimes)}$ and Noboru Kunihiro

The University of Tokyo, Tokyo, Japan
a-takayasu@it.k.u-tokyo.ac.jp, kunihiro@k.u-tokyo.ac.jp

Abstract. Several algorithms have been proposed for factoring RSA modulus N when attackers know the most or the least significant $(\beta - \delta) \log N$ bits of secret exponents $d < N^\beta$. The attacks are expected to work when $\beta < 1 - 1/\sqrt{2}$ with full size public exponent e considering Boneh and Durfee's result for small secret exponent attacks on RSA. However, previous attacks do not always work in this condition when attackers know only a small amount of information on secret exponent, that is, δ is close to β. In this paper, we propose the improved algorithms for partial key exposure attacks which cover Boneh and Durfee's bound when $\delta = \beta$. Our algorithms are the best among all known results when attackers know the most significant bits of $d \le N^{9/16}$ or the least significant bits of $d \le N^{(9-\sqrt{21})/12}$. In our algorithm constructions, we construct basis matrices for lattices which are not triangular and analyze the determinant by using unravelled linearization. The analysis enables us to make better use of the algebraic structures of modular polynomials, that is, we can select appropriate lattice bases or construct appropriate lattice bases.

Keywords: RSA · Cryptanalysis · Partial key exposure · Coppersmith's method · Lattices

1 Introduction

1.1 Background

Small Secret Exponent RSA. When small secret exponent $d < N^\beta$ is used, RSA cryptosystem becomes efficient for the decryption cost or the signature generation cost. However, Wiener [32] revealed the vulnerability. They claimed that public modulus N can be factored in polynomial time when $\beta < 1/4$.

Boneh and Durfee [5] revisited the attack and further improved the result. They used lattice-based Coppersmith's method to solve modular equations [8, 20]. At first, they constructed lattices which provide Wiener's bound $\beta < 1/4$. Next, they added some extra polynomials in the lattice bases and improved the bound to $\beta < (7-2\sqrt{7})/6 = 0.28474\cdots$. Finally, they achieved a stronger bound $\beta < 1 - 1/\sqrt{2} = 0.29289\cdots$ by extracting sublattices. To achieve the stronger bound, they used lattices which are not full-rank. Since the determinant of such lattices are difficult to compute, the analysis of the bound is involved.

© Springer International Publishing Switzerland 2014
A. Joux and A. Youssef (Eds.): SAC 2014, LNCS 8781, pp. 345–362, 2014.
DOI: 10.1007/978-3-319-13051-4_21

Partial Key Exposure Attacks on RSA. Boneh, Durfee and Frankel [6] introduced several attacks on RSA with small public exponent e. Their attacks make good use of the knowledge of the most significant bits (MSBs) or the least significant bits (LSBs) of secret exponent d. After that, such partial key exposure situations have been practically reported using side channel attacks, cold boot attacks [14]. Therefore, estimating the security of RSA with partial knowledge of the secret key has become increasingly important problem. See also [15,16,22,23,28].

Blömer and May [4] improved the attacks using Coppersmith's method to solve modular equations [8,20]. Blömer and May's work revealed that partial key exposure RSA is vulnerable for larger public exponent e. Ernst et al. [13] improved the attack to full size encryption exponent e using Coppersmith's method to find small roots of polynomials over the integers [9,12]. In addition, they proposed analogous attacks with full size public exponent e and small secret exponent d. In this paper, we study the situation:

- the prime factors p, q are the same bit size, $q < p < 2q$,
- the public exponent e is full size, the bit length of e is $\log N$,
- the secret exponent $d < N^\beta$ is small, $0 < \beta \leq 1$,
- in addition to public keys (N, e), attackers know d_0 which is $(\beta - \delta) \log N$ the most or the least significant bits of the secret exponent d with $0 \leq \delta \leq \beta$.

Partial key exposure situation with $\delta = \beta$, when attackers know no information of secret exponent d, is the same situation as Boneh and Durfee's work [5]. Therefore, the attack should always work when $\beta < 1 - 1/\sqrt{2}$. However, Ernst et al.'s results [13] only achieved the Boneh and Durfee's weaker bound $\beta < (7 - 2\sqrt{7})/6$ when $\delta = \beta$.

At PKC 2009, Aono [1] improved the algorithm for the LSBs partial key exposure attacks using Coppersmith's method to solve modular equations [8,20]. Aono used lattices which are not full rank and the basis matrices are not triangular. The result covers Boneh and Durfee's stronger bound $\beta < 1 - 1/\sqrt{2}$ when $\delta = \beta$. However, the attack is not applicable to the MSBs partial key exposure case. Sarkar, Gupta and Maitra [29] analyzed the MSBs partial key exposure attacks using Coppersmith's method to solve modular equations [8,20]. Though their attack partially covers Ernst et al.'s bound, they cannot improve it. To construct algorithms for the MSBs partial key exposure attacks that cover Boneh and Durfee's stronger bound remains an open problem.

Unravelled Linearization. Herrmann and May [18] introduced a new technique for lattice constructions, *unravelled linearization*. To solve nonlinear modular equations, consider the linear modular polynomials using *linearization*. In addition, unravelled linearization makes use of the lost algebraic structure using *unravelling*, which partially unravel the linearized variables in basis matrices. This operation transform basis matrices which are not triangular to be triangular and enables us to analyze the lattices which are not full rank easily. At PKC 2010, Herrmann and May [19] gave an elementary proof for Boneh and

Durfee's attack to achieve the stronger bound $\beta < 1 - 1/\sqrt{2}$. They used unravelled linearization and transformed Boneh and Durfee's lattices to be full rank with triangular basis matrices. Compared with original Boneh and Durfee's proof [5], the elegant technique enables us to extract appropriate sublattices easily. In addition, several results [2,18,21,31] have been reported to improve the previous results with the technique.

Collecting Helpful Polynomials. To maximize solvable root bounds, it is crucial to select appropriate polynomials in lattice bases. To examine which polynomials to be selected, May introduced the notion of *helpful* in his survey [26]. They called the polynomials whose sizes of diagonals in the basis matrices are smaller than the size of the modulus helpful polynomials. Helpful polynomials reduce the determinant of the lattices and enable us to obtain better bounds.

At a glance, the notion is completely trivial. However, Takayasu and Kunihiro [30] made use of the notion and provided the improved lattice constructions. They claimed that as many helpful polynomials as possible should be selected in lattice bases as long as the basis matrices are triangular. Based on the strategy, they improved the algorithms to solve two forms of modular multivariate linear equations [7,17]. The two algorithms were improved with full rank lattices with triangular basis matrices. That means though the analyses of triangular basis matrices are easy, that do not mean selections of appropriate lattice bases are trivial. Takayasu and Kunihiro's results [30] imply that the notion of helpful enables us to determine the appropriate polynomial selections.

1.2   Our Contributions

In this paper, we use Coppersmith's method to solve modular equations [8,20] and propose a improved algorithms for partial key exposure attacks on RSA for both the MSBs and the LSBs cases. Both our algorithms achieve Boneh and Durfee's stronger bound $\beta < 1 - 1/\sqrt{2}$ when $\delta = \beta$. Since we consider bivariate equations, our algorithms work under the assumption that polynomials obtained by the LLL reduced bases are algebraically independent as in previous works [1,4,5,13,29]. The assumption may be valid since few negative cases have been reported.

For the MSBs partial key exposure attacks, this is the first result to cover Boneh and Durfee's stronger bound when $\delta = \beta$.

Theorem 1. *When we know the most significant* $(\beta - \delta) \log N$ *bits of secret exponent* d, *public moduli* N *can be factored in polynomial time in* $\log N$ *and* $1/\epsilon$ *provided that*

(i) $\delta \leq \dfrac{1 + \beta - \sqrt{-1 + 6\beta - 3\beta^2}}{2} - \epsilon, \beta \leq \dfrac{1}{2}$,

(ii) $\delta \leq \dfrac{\tau}{2} - \dfrac{\tau^2}{3} + \dfrac{1}{6\tau} \cdot \dfrac{(\tau - 2(\beta - \delta))^3}{2 + 2\delta - 4\beta} - \epsilon, \tau = 1 - \dfrac{2\beta - 1}{1 - 2\sqrt{1 + \delta - 2\beta}}$,

$\dfrac{1}{2} < \beta \leq \dfrac{9}{16}$.

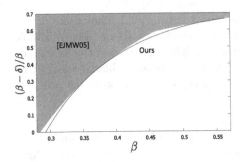
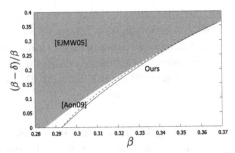

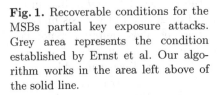

Fig. 1. Recoverable conditions for the MSBs partial key exposure attacks. Grey area represents the condition established by Ernst et al. Our algorithm works in the area left above of the solid line.

Fig. 2. Recoverable condition for the LSBs partial key exposure attacks. Grey area represents the condition established by Ernst et al. Aono's algorithm works in the area left above of the broken line. Our algorithm works in the area left above of the solid line.

We solve the same modular equations as Sarkar et al. [29], though Sarkar et al.'s algorithm does not even cover Boneh and Durfee's weaker bound. To the best of our knowledge, this is the first result to analyze the basis matrices which are not triangular for the MSBs partial key exposure attacks. Unravelled linearization enables us to analyze algebraic structures of modular polynomials in detail. Though we use the same polynomials as Sarkar et al. in lattice bases, we change the selections. Figure 1 compares the solvable root bounds of our algorithm and Ernst et al.'s algorithms [13]. When $\beta \leq 9/16 = 0.5625$, our algorithm is superior to the previous ones.

For the LSBs partial key exposure attacks, our algorithms cover Boneh and Durfee's stronger bound when $\delta = \beta$ and are superior to Aono's algorithm [1].

Theorem 2. *When we know the least significant* $(\beta - \delta) \log N$ *bits of secret exponent d, public moduli N can be factored in polynomial time in* $\log N$ *and* $1/\epsilon$ *provided that*

$$\delta \leq \frac{1 + \beta - \sqrt{-1 + 6\beta - 3\beta^2}}{2} - \epsilon, \beta \leq \frac{9 - \sqrt{21}}{12}.$$

We solve the same modular equations as Aono. First, we use unravelled linearization and transform Aono's basis matrices to be triangular. This transformation reveals a bottleneck of Aono's lattice constructions. We change polynomials in lattice bases in order to make full use of algebraic structures of modular polynomials. Figure 2 compares the solvable root bounds of our algorithm, Ernst et al.'s algorithm [13], and Aono's algorithm [1]. When $\beta < (9 - \sqrt{21})/12 = 0.36811\cdots$, our algorithms are superior to the previous ones.

1.3   Roadmap

The organization of this paper is as follows. In Sect. 2, we recall the RSA key generation and formulate the MSBs and the LSBs partial key exposure attacks. In Sect. 3, we introduce Coppersmith's method to solve modular equations, the technique and the strategy for the lattice constructions. In Sect. 4, we analyze the MSBs partial key exposure attack and prove Theorem 1. In Sect. 5, we analyze the LSBs partial key exposure attack and prove Theorem 2.

2   Formulations of Partial Key Exposure Attacks

We recall that the RSA key generation is described as

$$ed = 1 + \ell\phi(N), \text{where } \phi(N) = (p-1)(q-1) = N - (p+q-1).$$

In the MSBs partial key exposure case, we know d_0 which is the most significant bits of secret exponent d. We rewrite $d = d_0 M + d_1$ with an integer $M := 2^{\lfloor \delta \log N \rfloor}$, d_1 is the unknown part of d. In this case, we can easily calculate an approximation to ℓ, $\ell_0 = \lfloor (ed_0 - 1)/N \rfloor$. We rewrite $\ell = \ell_0 + \ell_1$. The size of the unknown ℓ_1 is bounded by N^γ with $\gamma = \max\{\delta, \beta - 1/2\}$. This analysis is written in [4] in detail. Again, looking at the RSA key generation,

$$e(d_0 M + d_1) = 1 + (\ell_0 + \ell_1)(N - (p+q-1)).$$

We consider the modular polynomial

$$f_{MSBs}(x,y) = 1 + (\ell_0 + x)(N + y) \pmod{e}.$$

This polynomial has the roots $(x,y) = (\ell_1, -(p+q-1))$. Sizes of the roots are bounded by X, Y where $X := N^\gamma, Y := 3N^{1/2}$. We can factor the RSA modulus N, if we can find the roots of $f_{MSBs}(x,y)$.

In the LSBs partial key exposure case, we know d_0 which is the least significant bits of secret exponent d. We rewrite $d = d_1 M + d_0$ with an integer $M := 2^{\lfloor (\beta - \delta) \log N \rfloor}$, d_1 is the unknown part of d. In this case, we cannot calculate an approximation to ℓ. Again, look at the RSA key generation,

$$e(d_1 M + d_0) = 1 + \ell(N - (p+q-1)).$$

We consider the modular polynomial

$$f_{LSBs}(x,y) = 1 - ed_0 + x(N + y) \pmod{eM}.$$

This polynomial has the roots $(x,y) = (\ell, -(p+q-1))$. Sizes of the roots are bounded by X, Y where $X := N^\beta, Y := 3N^{1/2}$. We can factor the RSA modulus N, if we can find the roots of $f_{LSBs}(x,y)$.

3   Coppersmith's Method to Solve Modular Equations

The Overview of the Method. At EUROCRYPT 1996, Coppersmith introduced the lattice based method to solve modular univariate equations in polynomial time. The method reveals several vulnerabilities of RSA cryptosystems. See [10,11,25–27] for more information. This method can be heuristically extended to bivariate cases with reasonable assumption. In this paper, we explain the reformulation by Howgrave-Graham [20]. For bivariate polynomials $h(x,y) = \sum h_{i_X,i_Y} x^{i_X} y^{i_Y}$, define a norm of the polynomials as $\|h(x,y)\| := \sqrt{\sum h_{i_X,i_Y}^2}$. The following Howgrave-Graham's lemma enables us to solve modular equations by finding roots of polynomials over the integers.

Lemma 1 (Howgrave-Graham's lemma [20]). *Let $h(x,y)$ be a bivariate integer polynomial which consists of at most n monomials. Let W, m, X, Y be positive integers. When the polynomial $h(x,y)$ satisfies*

1. *$h(\bar{x}, \bar{y}) = 0 \pmod{W^m}$, where $|\bar{x}| < X, |\bar{y}| < Y$,*
2. *$\|h(xX, yY)\| < W^m / \sqrt{n}$.*

Then $h(\bar{x}, \bar{y}) = 0$ holds over the integers.

To solve bivariate equations, we should find two polynomials that satisfy Howgrave-Graham's lemma. We use lattices and the LLL algorithm to find such low norm polynomials. Let $\mathbf{b}_1, \ldots, \mathbf{b}_n$ be linearly independent k-dimensional vectors. The lattice $L(\mathbf{b}_1, \ldots, \mathbf{b}_n)$ spanned by the basis vectors $\mathbf{b}_1, \ldots, \mathbf{b}_n$ is defined as $L(\mathbf{b}_1, \ldots, \mathbf{b}_n) = \{\sum_{j=1}^{n} c_j \mathbf{b}_j : c_j \in \mathbb{Z}\}$. We call n the rank of the lattice, and k the dimension of the lattice. When $n = k$, lattices are described as full rank. The basis matrix of the lattice B is defined as the $n \times k$ matrix that has basis vectors $\mathbf{b}_1, \ldots, \mathbf{b}_n$ in each row. In this paper, we use only full rank lattices. The determinant of a full rank lattice is computed by $\mathrm{vol}(L(B)) = |\det(B)|$.

In 1982, Lenstra, Lenstra and Lovász proposed the LLL algorithm [24], which find short lattice vectors in polynomial time.

Proposition 1 (LLL algorithm [24]). *Given k-dimensional basis vectors $\mathbf{b}_1, \ldots, \mathbf{b}_n$, the LLL algorithm finds short lattice vectors $\mathbf{v}_1, \mathbf{v}_2$ that satisfy*

$$\|\mathbf{v}_1\| \leq 2^{(n-1)/4}(\mathrm{vol}(L))^{1/n}, \|\mathbf{v}_2\| \leq 2^{n/2}(\mathrm{vol}(L))^{1/(n-1)}.$$

These norms are all Euclidean norms. The running time of the LLL algorithm is $O(n^5 k (\log B)^3)$ where $\log B$ represents the maximum input length.

To solve the modular equations $h(x,y) = 0 \pmod{W}$, we create n polynomials $h_1(x,y), \ldots, h_n(x,y)$ that have the same roots as the original solutions modulo W^m with a positive integer m. We generate basis vectors $\mathbf{b}_1, \ldots, \mathbf{b}_n$ whose elements are the coefficients of the polynomials $h_1(xX, yY), \ldots, h_n(xX, yY)$, respectively. The polynomials modulo W^m whose coefficients correspond to any lattice vectors spanned by $\mathbf{b}_1, \ldots, \mathbf{b}_n$ have the same roots as the original solutions. If two polynomials $p_1(x,y)$ and $p_2(x,y)$ whose coefficients correspond to short lattice vectors $\mathbf{v}_1, \mathbf{v}_2$ satisfy Howgrave-Graham's lemma, we can find the roots over the integers. This operation can easily be done by computing the Gröbner bases or resultant of $p_1(x, y), p_2(x, y)$.

We should note that the polynomials $p_1(x, y)$ and $p_2(x, y)$ have no assurance of algebraic independency. We assume that these polynomials are algebraic independent, and the resultant will not vanish. This assumption might be valid, since few negative cases have been reported.[1]

Unravelled Linearization. Boneh and Durfee [5] solved modular equations

$$f_{BD}(x, y) := 1 + N(x + y) = 0 \pmod{e}$$

for small secret exponent attacks on RSA. They selected shift-polynomials

$$g_{[u,i]}^{BD1}(x, y) := x^{u-i} f_{BD}(x, y)^i e^{m-i}, \text{ for } u = 0, 1, \ldots, m, i = 0, 1, \ldots, u,$$
$$g_{[u,j]}^{BD2}(x, y) := y^j f_{BD}(x, y)^u e^{m-u}, \text{ for } u = 0, 1, \ldots, m, j = 0, 1, \ldots, \lfloor (1 - 2\beta)u \rfloor,$$

in the lattice bases. The selection generates the basis matrix which is not triangular. That means there are some shift-polynomials which have several new monomials when added in the basis matrix. To avoid the situation, Herrmann and May [19] use the linearization $z := 1 + xy$. The linearization reduces the number of monomials of $f_{BD}(x, y)$. We partially apply the linearization to some monomials and the basis matrix becomes triangular. This operation enables us to compute the determinant of the lattice easily. See [19] for detailed analysis.

Collecting Helpful Polynomials. May [26] defined the notion of helpful compared with sizes of diagonals and a size of a modulus. Helpful polynomials contribute to the conditions for modular equations to be solved. Since each polynomial may affect not only the diagonal but also several other diagonals in our analyses, we cannot examine which polynomials to be selected with the previous definition of helpful polynomials. Therefore, we redefine the notion which covers the previous definition.

Definition 1 (Helpful Polynomials). *To solve equations with a modulus W, consider a basis matrix B. We add a new shift-polynomial $h_{[i',j']}(x, y)$ and construct a new basis matrix B^+. We call $h_{[i',j']}(x, y)$ a helpful polynomial, provided that*

$$\frac{\det(B^+)}{\det(B)} \leq W^m.$$

Conversely, if the inequality does not hold, we call $h_{[i',j']}(x, y)$ an unhelpful polynomial.

---

[1] We should note that in Bernstein et al.'s [3] millions of experiments with very small lattice dimension, the heuristic assumption fails in many cases. However, they propose the method to recover small solutions in such cases. See the paper for detailed information.

4   Partial Key Exposure Attack: The Most Significant Bits Case

4.1   Previous Works

In the MSBs partial key exposure case, Ernst et al. [13] found the small roots of polynomials over the integers

$$g^{EJMW1}(x, y, z) = 1 - ed_0 M + ex + y(N + z),$$

$$\text{or } g^{EJMW2}(x, y, z) = 1 - ed_0 M + ex + (\ell_0 + y)(N + z),$$

to factor N. The polynomial $g^{EJMW1}(x, y, z)$ has the roots $(x, y) = (-d_1, \ell, -(p + q - 1))$, and the polynomial $g^{EJMW2}(x, y, z)$ has the roots $(x, y) = (-d_1, \ell_1, -(p + q - 1))$. Their algorithms work provided that

(1) $\gamma \leq \frac{5}{6} - \frac{1}{3}\sqrt{1 + 6\beta} - \epsilon$,
(2) $\gamma \leq \frac{3}{16} - \epsilon$ and $\beta \leq \frac{11}{16}$,
(3) $\gamma \leq \frac{1}{3} + \frac{1}{3}\beta - \frac{1}{3}\sqrt{4\beta^2 + 2\beta - 2} - \epsilon$ and $\beta \geq \frac{11}{16}$.

The condition (1) can be obtained by finding the roots of the polynomial $g^{EJMW1}(x, y, z)$. The conditions (2), (3) can be obtained by finding the roots of the polynomial $g^{EJMW2}(x, y, z)$ with $\gamma = \delta$ and $\gamma = \beta - 1/2$, respectively. The condition yields Boneh and Durfee's weaker bound $\beta < (7 - 2\sqrt{7})/6$ when $\delta = \beta$.

Sarkar et al. [29] solved the modular equation $f_{MSBs}(x, y) = 0$ to factor N. To solve the modular equation, they used shift-polynomials

$$g^{MSBs1}_{[u,i]}(x, y) := x^{u-i} f_{MSBs}(x, y)^i e^{m-i},$$

$$g^{MSBs2}_{[u,j]}(x, y) := y^j f_{MSBs}(x, y)^u e^{m-u}.$$

Both shift-polynomials modulo e^m have the same roots as the original solutions, that is, $g^{MSBs1}_{[u,i]}(\ell_1, -(p + q - 1)) = 0 \pmod{e^m}$ and $g^{MSBs2}_{[u,j]}(\ell_1, -(p + q - 1)) = 0 \pmod{e^m}$. They selected shift-polynomials

$$g^{MSBs1}_{[u,i]}(x, y) \quad \text{for } u = 0, 1, \ldots, \lfloor m/4\gamma \rfloor, i = 0, 1, \ldots, \max\{m, u\},$$

$$g^{MSBs2}_{[u,j]}(x, y) \quad \text{for } u = 0, 1, \ldots, m, i = 1, 2, \ldots, u,$$

in the lattice bases. This selection generates triangular basis matrices with diagonals $X^u Y^i e^{m-i}$ for $g^{MSBs1}_{[u,i]}(x, y)$, and $X^u Y^{u+j} e^{m-u}$ for $g^{MSBs2}_{[u,j]}(x, y)$. The condition for the algorithm to work is the same as (2) of Ernst et al.'s condition.

4.2   Our Lattice Constructions

In this section, we explain our improved lattice constructions. At first, we consider the case for $\beta \leq 1/2$.

For Smaller d. To solve the modular equation $f_{MSBs}(x,y) = 0$, we use the same shift-polynomials $g_{[u,i]}^{MSBs1}(x,y), g_{[u,j]}^{MSBs2}(x,y)$ as Sarkar et al. However, we change the selections. To construct the basis matrix, we use shift-polynomials

$$g_{[u,i]}^{MSBs1}(x,y) \text{ for } u = 0,1,\ldots,m, i = 0,1,\ldots,u,$$

$$g_{[u,j]}^{MSBs2}(x,y) \text{ for } u = 0,1,\ldots,m, j = 1,2,\ldots,\lfloor 2(\beta - \gamma)m + (1 + 2\gamma - 4\beta)u \rfloor,$$

in the lattice bases. The selections of shift-polynomials generate basis matrices which are not triangular. However, we partially apply the linearization $z = 1 + (\ell_0 + x)y$ and the basis matrices can be transformed into triangular. The size of the root for the linearized variable z is bounded by $Z := 3N^{1/2+\beta}$. In general, we reveal the following property.

Lemma 2. *We define the polynomial order $\prec$ as*

$$g_{[u,i]}^{MSBs1}(x,y), g_{[u,j]}^{MSBs2}(x,y) \prec g_{[u',i']}^{MSBs1}(x,y), g_{[u',j']}^{MSBs2}(x,y), \text{ if } u < u',$$

$$g_{[u,i]}^{MSBs1}(x,y) \prec g_{[u',j']}^{MSBs2}(x,y), \text{ if } u = u'$$

$$g_{[u,i]}^{MSBs1}(x,y) \prec g_{[u',i']}^{MSBs1}(x,y), \text{ if } u = u', i < i',$$

$$g_{[u,j]}^{MSBs2}(x,y) \prec g_{[u',j']}^{MSBs2}(x,y), \text{ if } u = u', j < j'.$$

Ordered in this way, the basis matrices become triangular with diagonals $X^{u-\lceil l^{MSBs}(i)\rceil} Y^{i-\lceil l^{MSBs}(i)\rceil} Z^{\lceil l^{MSBs}(i)\rceil} e^{m-i}$ for $g_{[u,i]}^{MSBs1}(x,y)$, and $X^{u-\lceil l^{MSBs}(u+j)\rceil} Y^{u+j-\lceil l^{MSBs}(u+j)\rceil} Z^{\lceil l^{MSBs}(u+j)\rceil} e^{m-u}$ for $g_{[u,j]}^{MSBs2}(x,y)$, where

$$l^{MSBs}(k) := \max\left\{0, \frac{k - 2(\beta - \gamma)m}{2 + 2\gamma - 4\beta}\right\}.$$

The proof is written in the full version.

The linearization technique enables us to select shift-polynomials more flexibly with the constraint for basis matrices to be triangular. Therefore, we can eliminate some unhelpful polynomials and add helpful polynomials compared with Sarkar et al.'s basis matrices. To maximize the solvable root bounds, our collections of shift-polynomials are determined by the following lemma.

Lemma 3. *When $\beta \leq 1/2$, assume there are shift-polynomials $g_{[u,i]}^{MSBs1}(x,y)$ for $u = u' + j', \ldots, m, i = u' + j'$ and $g_{[u,j]}^{MSBs2}(x,y)$ for $u = u' + 1, \ldots, u' + j' - 1, j = u' + j' - u$ in lattice bases. In this case, shift-polynomials $g_{[u',j']}^{MSBs2}(x,y)$ are helpful polynomials when $u' = 0, 1, \ldots, m, j' = 1, \ldots, \lfloor 2(\beta - \gamma) + (1 + 2\gamma - 4\beta)u \rfloor$. Shift-polynomials $g_{[u',j']}^{MSBs2}(x,y)$ are unhelpful polynomials when $u' = 0, 1, \ldots, m, j' > 2(\beta - \gamma) + (1 + 2\gamma - 4\beta)u'$.*

Proof. Consider the basis matrix B. We add a new shift-polynomial $g_{[u',j']}^{MSBs2}(x,y)$ and construct the basis matrix B^+. The value $\det(B^+)/\det(B)$ can be computed as

$$\frac{\det(B^+)}{\det(B)} = Y^{j'} Z^{u'} e^{m-u'} \times \left(\frac{XY}{Z}\right)^{m-u'}$$

$$\approx N^{\frac{1}{2}j' + (\frac{1}{2} + \beta)u' + m - u' - (\beta - \gamma)(m - u')}.$$

This value is smaller than the size of the modulus $e^m \approx N^m$, when

$$j' \leq 2(\beta - \delta)m + (1 + 2\delta - 4\beta)u'.$$

That is, Lemma 3 is proved.   □

We prove the bound (i) of Theorem 1. We can rewrite the diagonals as $X^{u' - \lceil l^{MSBs}(j')\rceil} Y^{j' - \lceil l^{MSBs}(j')\rceil} Z^{\lceil l^{MSBs}(j')\rceil} e^{m - \min\{u', j'\}}$ for $j' = 0, 1, \ldots, 2(1 - \beta)m$, $u' = \lceil l^{MSBs}(j')\rceil, \ldots, m$. Ignoring low order terms of m, we compute the dimension

$$n = \sum_{j'=0}^{\lfloor 2(1-\beta)m \rfloor} \sum_{u'=\lceil l^{MSBs}(j)\rceil}^{m} 1 = \left(\frac{1}{2} + 2(\beta - \gamma) + \frac{1 + 2\gamma - 4\beta}{2} \right) m^2,$$

and the determinant of the lattices $\det(B) = X^{s_X} Y^{s_Y} Z^{s_Z} e^{s_e}$ where

$$s_X = \sum_{j'=0}^{\lfloor 2(1-\beta)m \rfloor} \sum_{u'=\lceil l^{MSBs}(j)\rceil}^{m} (u' - \lceil l^{MSBs}(j')\rceil)$$
$$= \left(\frac{1}{6} + (\beta - \gamma) + \frac{1 + 2\gamma - 4\beta}{6} \right) m^3,$$

$$s_Y = \sum_{j'=0}^{\lfloor 2(1-\beta)m \rfloor} \sum_{u'=\lceil l^{MSBs}(j)\rceil}^{m} (j' - \lceil l^{MSBs}(j')\rceil)$$
$$= ((\beta - \gamma) + 2(\beta - \gamma)^2 + (\beta - \gamma)(1 + 2\gamma - 4\beta) + \frac{1 + 2\gamma - 4\beta}{6}$$
$$+ \frac{(1 + 2\gamma - 4\beta)^2}{6})m^3,$$

$$s_Z = \sum_{j'=0}^{\lfloor 2(1-\beta)m \rfloor} \sum_{u'=\lceil l^{MSBs}(j')\rceil}^{m} \lceil l^{MSBs}(j')\rceil = \left(\frac{1}{6} + \frac{1 + 2\gamma - 4\beta}{6} \right) m^3,$$

$$s_e = \sum_{j'=0}^{\lfloor 2(1-\beta)m \rfloor} \sum_{u'=\lceil l^{MSBs}(j)\rceil}^{m} (m - \min\{u', j'\})$$
$$= \left(\frac{1}{3} + (\beta - \gamma) + \frac{1 + 2\gamma - 4\beta}{6} \right) m^3,$$

We can find solutions of $f_{MSBs}(x, y) = 0$ provided that $(\det(B))^{1/n} < e^m$, that is,

$$2\gamma^2 - 2(1 + \beta)\gamma + 2\beta^2 - 2\beta + 1 > 0.$$

This condition yields the bound

$$\gamma < \frac{1 + \beta - \sqrt{-1 + 6\beta - 3\beta^2}}{2}.$$

It is clear that $\gamma = \max\{\delta, \beta - 1/2\} = \delta$ when $\beta \le 1/2$. Therefore, the bound (i) of Theorem 1 is proved.

For Larger d. In the following, we briefly summarize the case for $1/2 < \beta \le 9/16$. The detailed analysis is written in the full version.

When $\beta > 1/2$, $2(\beta - \gamma)m + (1 + 2\gamma - 4\beta)u < 0$ for larger $u > -2(\beta - \gamma)m/(1 + 2\gamma - 4\beta)$. Since we select shift-polynomials $g_{[u,i]}^{MSBs1}(x, y)$ for $-2(\beta - \gamma)m/(1 + 2\gamma - 4\beta) < u \le m$ which are unhelpful polynomials and do not contribute for basis matrices to be triangular, we should redefine collections of shift-polynomials. We use shift-polynomials

$$g_{[u,i]}^{MSBs1}(x, y) \text{ with } u = 0, 1, \ldots, m, i = 0, 1, \ldots, \min\{u, t\},$$

$$g_{[u,j]}^{MSBs2}(x, y) \text{ with } u = 0, 1, \ldots, m,$$
$$j = 1, 2, \ldots, \min\{\lfloor 2(\beta - \gamma)m + (1 + 2\gamma - 4\beta)u \rfloor, t - u\},$$

in the lattice bases. The parameter $\tau = t/m$ should be optimized later[2]. The selections of shift-polynomials generate basis matrices which are not triangular. However, we partially apply the linearization $z = 1 + (l_0 + x)y$ and the basis matrices can be transformed into triangular. That means Lemma 2 holds.

We prove the bound (ii) of Theorem 1. We can rewrite the diagonals as $X^{u' - \lceil l^{MSBs}(j') \rceil} Y^{j' - \lceil l^{MSBs}(j') \rceil} Z^{\lceil l^{MSBs}(j') \rceil} e^{m - \min\{u', j'\}}$ for $j' = 0, 1, \ldots, t, u' = \lceil l^{MSBs}(j) \rceil, \lceil l^{MSBs}(j) \rceil + 1, \ldots, m$. Ignoring low order term of m, we compute the dimension

$$n = \sum_{j'=0}^{t} \sum_{u' = \lceil l^{MSBs}(j) \rceil}^{m} 1 = mt - \frac{1}{2} \cdot \frac{(t - 2(\beta - \gamma)m)^2}{2 + 2\gamma - 4\beta},$$

and the determinant of the lattices $\det(B) = X^{s_X} Y^{s_Y} Z^{s_Z} e^{s_e}$ where

$$s_X = \sum_{j'=0}^{t} \sum_{u' = \lceil l^{MSBs}(j') \rceil}^{m} (u' - \lceil l^{MSBs}(j') \rceil)$$

$$= \frac{m^2 t}{2} - \frac{1}{6} \cdot \frac{(t - 2(\beta - \gamma)m)^3}{(2 + 2\gamma - 4\beta)^2} - s_Z,$$

$$s_Y + s_Z = \sum_{j'=0}^{t} \sum_{u' = \lceil l^{MSBs}(j') \rceil}^{m} j'$$

---

[2] These collections and optimization of the parameter τ are based on the notion of *consecutive helpful polynomials* defined in [30]. See the paper in detail.

$$= \frac{1}{6} \cdot \frac{t^3 - 8(\beta - \gamma)^3 m^3}{2 + 2\gamma - 4\beta} + \frac{t^2}{2} \left(m - \frac{t - 2(\beta - \gamma)m}{2 + 2\gamma - 4\beta} \right)$$

$$s_Z = \sum_{j'=0}^{t} \sum_{u'=\lceil l^{MSBs}(j') \rceil}^{m} \lceil l^{MSBs}(j') \rceil$$

$$= \frac{m}{2} \cdot \frac{(t - 2(\beta - \gamma)m)^2}{2 + 2\gamma - 4\beta} - \frac{1}{3} \cdot \frac{(t - 2(\beta - \gamma)m)^3}{(2 + 2\gamma - 4\beta)^2}$$

$$s_e = \sum_{j'=0}^{t} \sum_{u'=\lceil l^{MSBs}(j') \rceil}^{m} (m - \min\{u', j'\})$$

$$= -\frac{m}{2} \cdot \frac{(t - 2(\beta - \gamma)m)^2}{2 + 2\gamma - 4\beta} + \frac{1}{6} \cdot \frac{(t - 2(\beta - \gamma)m)^3}{(2 + 2\gamma - 4\beta)^2} + m^2 t - \frac{mt^2}{2} + \frac{1}{6}t^3$$

We can find solutions $f_{MSBs}(x, y) = 0$ provided that $(\det(B))^{1/n} < e^m$, that is,

$$\gamma\tau - \frac{\tau^2}{2} + \frac{\tau^3}{3} < \frac{1}{6} \cdot \frac{(\tau - 2(\beta - \gamma))^3}{2 + 2\gamma - 4\beta}.$$

Note that Sarkar et al.'s condition can be written as $\gamma\tau - \tau^2/2 + \tau^3/3 < 0$ with $\tau = (1/4 - \gamma)/\gamma$. We can improve the result since $\tau - 2(\beta - \gamma) > 0, 2 + 2\gamma - 4\beta > 0$ and the right hand side of the inequality is positive. To maximize the solvable root bound, we set the parameter

$$\tau = 1 - \frac{2\beta - 1}{1 - 2\sqrt{1 + \gamma - 2\beta}}$$

and obtain the bound (ii) of Theorem 1.

5   Partial Key Exposure Attack: The Least Significant Bits Case

5.1   Previous Works

In the LSBs partial key exposure case, Ernst et al. [13] found the small roots of polynomials over the integers

$$g^{EMJW3}(x, y, z) = 1 - ed_0 + eMx + y(N + z),$$

to factor N. The polynomial $g^{EJMW3}(x, y, z)$ has the roots $(x, y) = (-d_1, \ell, -(p + q - 1))$. Their algorithm works provided that

$$\delta \leq \frac{5}{6} - \frac{1}{3}\sqrt{1 + 6\beta} - \epsilon.$$

When $\delta = \beta$, the condition yields Boneh and Durfee's weaker bound [5] $\beta < (7 - 2\sqrt{7})/6$.

Blömer and May [4] consider LSBs key exposure attacks with small public exponents e and full size secret exponents d. Though the situation is slightly different from the one considered in this paper, their lattice construction provides the same bound as Ernst et al.'s algorithm. Blömer and May solve the modular equation $f_{LSBs}(x, y) = 0$ to factor N. To solve the modular equation, they used shift-polynomials

$$g_{[u,i]}^{LSBs1}(x,y) := x^{u-i} f_{LSBs}(x,y)^i (eM)^{m-i},$$
$$g_{[u,j]}^{LSBs2}(x,y) := y^j f_{LSBs}(x,y)^u (eM)^{m-u}.$$

Both shift-polynomials modulo $(eM)^m$ have the same roots as the original solutions, that is, $g_{[u,i]}^{LSBs1}(\ell, -(p+q-1)) = 0 \pmod{(eM)^m}$, $g_{[u,j]}^{LSBs2}(\ell, -(p+q-1)) = 0 \pmod{(eM)^m}$. They selected shift-polynomials

$g_{[u,i]}^{LSBs1}(x,y)$ with $u = 0, 1, \ldots, m, i = 0, 1, \ldots, u$,

$g_{[u,j]}^{LSBs2}(x,y)$ with $u = 0, 1, \ldots, m, j = 1, 2, \ldots, \lfloor (1 - 2\delta)m/2 \rfloor$,

in the lattice bases. This selections generate triangular basis matrices with diagonals $X^u Y^i e^{m-i}$ for $g_{[u,i]}^{LSBs1}(x,y)$, and $X^u Y^{u+j} e^{m-u}$ for $g_{[u,j]}^{LSBs2}(x,y)$. Their algorithm works provided that $\delta \le \frac{5}{6} - \frac{1}{3}\sqrt{1 + 6\beta} - \epsilon$. The bound corresponds to Ernst et al.'s bound.

Aono [1] improved the attack and firstly achieved Boneh and Durfee's stronger bound [5]. To improve the bound, Aono considered the other modular polynomial,

$$f_{LSBs1}(x,y) = 1 + x(N + y) \pmod{e}.$$

The roots of the polynomials are $(x, y) = (\ell, -(p + q - 1))$ which are the same as $f_{LSBs}(x, y)$. To construct the basis matrix, Aono used shift-polynomials $g_{[u,i]}^{LSBs1}, g_{[u,j]}^{LSBs2}(x, y)$, and

$$g_{[u,k]}^{LSBs3}(x,y) := y^k f_{LSBs1}(x,y)^u e^{m-u} M^m.$$

Shift-polynomials $g_{[u,k]}^{LSBs3}(x, y)$ modulo $(eM)^m$ have the same roots as the original solutions, that is, $g_{[u,k]}^{LSBs3}(\ell, -(p + q - 1)) = 0 \pmod{(eM)^m}$. Aono selected shift-polynomials

$g_{[u,i]}^{LSBs1}(x,y)$ with $u = 0, 1, \ldots, m, i = 0, 1, \ldots, u$,

$g_{[u,j]}^{LSBs2}(x,y)$ with $u = 0, 1, \ldots, m, j = 1, 2, \ldots, t$,

$g_{[u,k]}^{LSBs3}(x,y)$ with $u = \lceil t/(1 - 2\beta) \rceil, \lceil t/(1 - 2\beta) \rceil + 1, \ldots, m$,

$$k = t + 1, t + 2, \ldots, \lfloor (1 - 2\beta)u \rfloor,$$

with $t = \sqrt{2(1 - 2\beta)(\beta - \delta)}m$ in the lattice bases. This selections of shift-polynomials generate basis matrices which are not triangular. Aono bounded the determinant of the lattice by computing Gram-Schmidt orthogonal bases. The algorithm works provided that

$$2\beta^2 - 3\beta + 2\tau(\beta - \delta) - \delta + 1 > 0,$$

when $1 + 2\delta - 4\beta > 0$. When $\delta = \beta$, this yields Boneh and Durfee's stronger bound $\beta < 1 - 1/\sqrt{2}$. When $1 + 2\delta - 4\beta \leq 0$, Aono's lattice construction becomes the same as Blömer and May [4].

5.2   Our Observation of Aono's Lattice Using Unravelled Linearization

As we showed, the basis matrix constructed by Aono [1] is not triangular. However, we reveal that the basis matrix can be transformed into triangular with linearization $z = 1 + xy$. The size of the root for the linearized variable z is bounded by $Z := 3N^{1/2+\beta}$.

Lemma 4. *We define the polynomial order $\prec$ as*

$$g_{[u,i]}^{LSBs1}(x,y) \prec g_{[u,j]}^{LSBs2}(x,y) \prec g_{[u,k]}^{LSBs3}(x,y),$$

$$g_{[u,i]}^{LSBs1}(x,y) \prec g_{[u',i']}^{LSBs1}(x,y), \ if \ u < u' \ or \ u = u', i < i',$$

$$g_{[u,j]}^{LSBs2}(x,y) \prec g_{[u',j']}^{LSBs2}(x,y), \ if \ u < u' \ or \ u = u', j < j',$$

$$g_{[u,k]}^{LSBs3}(x,y) \prec g_{[u',k']}^{LSBs3}(x,y), \ if \ u < u' \ or \ u = u', k < k'.$$

Ordered in this way, the basis matrix becomes triangular with diagonals $X^u Y^i \times (eM)^{m-i}$ for $g_{[u,i]}^{LSBs1}(x,y)$, $X^u Y^{u+j}(eM)^{m-u}$ for $g_{[u,j]}^{LSBs2}(x,y)$, and $Y^k Z^u \times e^{m-u} M^m$ for $g_{[u,k]}^{LSBs3}(x,y)$.

The proof is written in the full version.

5.3   Our Lattice Constructions

In this section, we propose the improved algorithm for LSBs partial key exposure attacks when $1 + 2\delta - 4\beta > 0$. We change the shift-polynomials used in the lattice bases. We use the shift-polynomial $g_{[u,i]}^{LSBs1}(x,y)$, and

$$g_{[u,k]}^{LSBs4}(x,y) := y^k f_{LSBs}(x,y)^{u - \lceil l^{LSBs}(k)\rceil} f_{LSBs1}(x,y)^{\lceil l^{LSBs}(k)\rceil}$$

$$\times e^{m-u} M^{m-(u-\lceil l^{LSBs}(k)\rceil)},$$

where

$$l^{LSBs}(k) = \max\left\{0, \frac{k - 2(\beta-\delta)m}{1 + 2\delta - 4\beta}\right\}.$$

Shift-polynomials $g_{[u,k]}^{LSBs4}(x,y)$ modulo $(eM)^m$ have the same roots as the original solutions, that is, $g_{[u,k]}^{LSBs4}(\ell, -(p+q-1)) = 0 \pmod{(eM)^m}$. We selected shift-polynomials

$g_{[u,i]}^{LSBs1}(x,y)$ with $u = 0, 1, \ldots, m, i = 0, 1, \ldots, u$,

$g_{[u,k]}^{LSBs4}(x,y)$ with $u = 0, 1, \ldots, m, k = 1, 2, \ldots, \lfloor 2(\beta-\delta)m + (1+2\delta-4\beta)u\rfloor$, in the lattice bases. Though the selections generate basis matrices which are not triangular, we partially apply the linearization $z = 1 + xy$ and the basis matrix can be transformed into triangular. In general, we reveal the following property.

Lemma 5. *We define the polynomial order $\prec$ as*

$$g_{[u,i]}^{LSBs1}(x,y) \prec g_{[u,k]}^{LSBs4}(x,y),$$

$$g_{[u,i]}^{LSBs1}(x,y) \prec g_{[u',i']}^{LSBs1}(x,y), \; if \; u < u' \; or \; u = u', i < i',$$

$$g_{[u,k]}^{LSBs4}(x,y) \prec g_{[u',k']}^{LSBs4}(x,y), \; if \; u < u' \; or \; u = u', k < k'.$$

Ordered in this way, the basis matrix becomes triangular with diagonals $X^u Y^i \times (eM)^{m-i}$ for $g_{[u,i]}^{LSBs1}(x,y)$, and $X^{u-\lceil l^{LSBs}(k)\rceil} Y^{u-\lceil l^{LSBs}(k)\rceil+k} Z^{\lceil l^{LSBs}(k)\rceil} \times e^{m-u} M^{m-(u-\lceil l^{LSBs}(k)\rceil)}$ for $g_{[u,k]}^{LSBs4}(x,y)$.

The proof is written in the full version.

Lemmas 4, 5 clarify the point of our improvements. When $l^{LSBs}(k) = 0$, $g_{[u,k]}^{LSBs4}(x,y) = g_{[u,k]}^{LSBs2}(x,y)$. When $l^{LSBs}(k) > 0$, the diagonals of $g_{[u,k]}^{LSBs4}(x,y)$ in our basis matrices are smaller than that of $g_{[u,k]}^{LSBs3}(x,y)$ in Aono's basis matrices with respect to powers of M. Therefore, we can improve the bound when shift-polynomials $g_{[u,k]}^{LSBs4}(x,y)$ with $l^{LSBs}(k) > 0$ are used.

To maximize the solvable root bounds, our collection of shift-polynomials is determined by the following lemma.

Lemma 6. *Assume that there are shift-polynomials $g_{[u,k]}^{LSBs4}(x,y)$ for $(u,k) = (u'+1, k'+1), (u'+2, k'+2), \ldots, (m, m-u'+k')$ in B. Shift-polynomials $g_{[u',k']}^{LSBs4}(x,y)$ are helpful polynomials when $u' = 0, 1, \ldots, m, k' = 1, 2, \ldots, \lfloor 2(\beta - \delta)m + (1+2\delta - 4\beta)u' \rfloor$. Shift-polynomials $g_{[u',k']}^{LSBs4}(x,y)$ are unhelpful polynomials when $u' = 0, 1, \ldots, m, k' > 2(\beta - \delta)m + (1 + 2\delta - 4\beta)u'$.*

Proof. Consider the basis matrix B. We add a new shift-polynomial $g_{[u',k']}^{LSBs4}(x,y)$ and construct the basis matrix B^+. The value $\det(B^+)/\det(B)$ can be computed as

$$\frac{\det(B^+)}{\det(B)} = Y^{k'} Z^{u'} e^{m-u'} M^m \times \left(\frac{1}{M}\right)^{u'}$$

$$\approx N^{\frac{1}{2}k' + \left(\frac{1}{2}+\beta\right)u' + m - u' + (\beta-\delta)u'}.$$

This value is smaller than the size of the modulus $(eM)^m \approx N^{(1+\beta-\delta)m}$, when

$$j' \leq 2(\beta - \delta)m + (1 + 2\delta - 4\beta)u'.$$

That is, Lemma 6 is proved. $\square$

Note that Lemma 6 does not hold when $1 + 2\delta - 4\beta < 0$. Since our assumption that there are shift-polynomials $g_{[u,k]}^{LSBs4}(x,y)$ for $(u,k) = (u'+1, k'+1), (u'+2, k'+2), \ldots, (m, m-u'+k')$ in B does not hold.

We prove the bound of Theorem 2. Ignoring low order terms of m, we compute the dimension

$$n = \sum_{u=0}^{m}\sum_{i=0}^{u}1 + \sum_{u=0}^{m}\sum_{k=1}^{\lfloor 2(\beta-\delta)m+(1+2\delta-4\beta)u\rfloor}1 = \left(\frac{1}{2}+2(\beta-\delta)+\frac{1+2\delta-4\beta}{2}\right)m^2,$$

and the determinant of the lattice $\det(B) = X^{s_X}Y^{s_Y}Z^{s_Z}e^{s_e}M^{s_M}$ where

$$s_X = \sum_{u=0}^{m}\sum_{i=0}^{u}(u-i) = \frac{1}{3}m^3,$$

$$s_Y = \sum_{u=0}^{m}\sum_{i=0}^{u}i + \sum_{u=0}^{m}\sum_{k=1}^{\lfloor 2(\beta-\delta)m+(1+2\delta-4\beta)u\rfloor}k$$
$$= \left(\frac{1}{6}+2(\beta-\delta)^2+(\beta-\delta)(1+2\delta-4\beta)+\frac{(1+2\delta-4\beta)^2}{6}\right)m^3,$$

$$s_Z = \sum_{u=0}^{m}\sum_{k=1}^{\lfloor 2(\beta-\delta)m+(1+2\delta-4\beta)u\rfloor}u = \left((\beta-\delta)+\frac{1+2\delta-4\beta}{3}\right)m^3,$$

$$s_e = \sum_{u=0}^{m}\sum_{i=0}^{u}(m-i) + \sum_{u=0}^{m}\sum_{k=1}^{\lfloor 2(\beta-\delta)m+(1+2\delta-4\beta)u\rfloor}(m-u)$$
$$= \left(\frac{1}{3}+(\beta-\delta)+\frac{1+2\delta-4\beta}{6}\right)m^3,$$

$$s_M = \sum_{u=0}^{m}\sum_{i=0}^{u}(m-i) + \sum_{u=0}^{m}\sum_{k=1}^{\lfloor 2(\beta-\delta)m+(1+2\delta-4\beta)u\rfloor}(m-(u-\lceil l^{LSBs}(k)\rceil))$$
$$= \left(\frac{1}{3}+(\beta-\delta)+\frac{1+2\delta-4\beta}{3}\right)m^3.$$

We can find solutions of $f_{LSBs}(x,y) = 0, f_{LSBs1}(x,y) = 0$ provided that $(\det(B))^{1/n} < (eM)^m$, that is,

$$2\delta^2 - 2(1+\beta)\delta + 2\beta^2 - 2\beta + 1 > 0.$$

This condition yields the bound

$$\delta < \frac{1+\beta-\sqrt{-1+6\beta-3\beta^2}}{2}.$$

Therefore, the bound of Theorem 2 is proved.

References

1. Aono, Y.: A new lattice construction for partial key exposure attack for RSA. In: Jarecki, S., Tsudik, G. (eds.) PKC 2009. LNCS, vol. 5443, pp. 34–53. Springer, Heidelberg (2009)

2. Bauer, A., Vergnaud, D., Zapalowicz, J.-C.: Inferring sequences produced by nonlinear pseudorandom number generators using Coppersmith's methods. In: Fischlin, M., Buchmann, J., Manulis, M. (eds.) PKC 2012. LNCS, vol. 7293, pp. 609–626. Springer, Heidelberg (2012)

3. Bernstein, D.J., Chang, Y.-A., Cheng, C.-M., Chou, L.-P., Heninger, N., Lange, T., van Someren, N.: Factoring RSA keys from certified smart cards: Coppersmith in the wild. In: Sako, K., Sarkar, P. (eds.) ASIACRYPT 2013, Part II. LNCS, vol. 8270, pp. 341–360. Springer, Heidelberg (2013)

4. Blömer, J., May, A.: New partial key exposure attacks on RSA. In: Boneh, D. (ed.) CRYPTO 2003. LNCS, vol. 2729, pp. 27–43. Springer, Heidelberg (2003)

5. Boneh, D., Durfee, G.: Cryptanalysis of RSA with private key d less than $n^{0.292}$. IEEE Trans. Inf. Theory **46**(4), 1339–1349 (2000)

6. Boneh, D., Durfee, G., Frankel, Y.: An attack on RSA given a small fraction of the private key bits. In: Ohta, K., Pei, D. (eds.) ASIACRYPT 1998. LNCS, vol. 1514, pp. 25–34. Springer, Heidelberg (1998)

7. Cohn, H., Heninger, N.: Approximate common divisors via lattices. In: ANTS-X, 2012. IACR Cryptology ePrint Archive, Report 2011/437 (2011). http://eprint.iacr.org/2011/437

8. Coppersmith, D.: Finding a small root of a univariate modular equation. In: Maurer, U.M. (ed.) EUROCRYPT 1996. LNCS, vol. 1070, pp. 155–165. Springer, Heidelberg (1996)

9. Coppersmith, D.: Finding a small root of a bivariate integer equation; factoring with high bits known. In: Maurer, U.M. (ed.) EUROCRYPT 1996. LNCS, vol. 1070, pp. 178–189. Springer, Heidelberg (1996)

10. Coppersmith, D.: Small solutions to polynomial equations, and low exponent RSA vulnerabilities. J. Cryptol. **10**(4), 233–260 (1997)

11. Coppersmith, D.: Finding small solutions to small degree polynomials. In: Silverman, J.H. (ed.) CaLC 2001. LNCS, vol. 2146, pp. 20–31. Springer, Heidelberg (2001)

12. Coron, J.-S.: Finding small roots of bivariate integer polynomial equations revisited. In: Cachin, C., Camenisch, J.L. (eds.) EUROCRYPT 2004. LNCS, vol. 3027, pp. 492–505. Springer, Heidelberg (2004)

13. Ernst, M., Jochemsz, E., May, A., de Weger, B.: Partial key exposure attacks on RSA up to full size exponents. In: Cramer, R. (ed.) EUROCRYPT 2005. LNCS, vol. 3494, pp. 371–386. Springer, Heidelberg (2005)

14. Halderman, J.A., Schoen, S.D., Heninger, N., Clarkson, W., Paul, W., Calandrino, J.A., Feldman, A.J., Appelbaum, J., Felten, E.W.: Lest we remember: cold boot attacks on encryption keys. In: Proceedings of the USENIX Security Symposium 2008, pp. 45–60 (2008)

15. Henecka, W., May, A., Meurer, A.: Correcting errors in RSA private keys. In: Rabin, T. (ed.) CRYPTO 2010. LNCS, vol. 6223, pp. 351–369. Springer, Heidelberg (2010)

16. Heninger, N., Shacham, H.: Reconstructing RSA private keys from random key bits. In: Halevi, S. (ed.) CRYPTO 2009. LNCS, vol. 5677, pp. 1–17. Springer, Heidelberg (2009)

17. Herrmann, M., May, A.: Solving linear equations modulo divisors: on factoring given any bits. In: Pieprzyk, J. (ed.) ASIACRYPT 2008. LNCS, vol. 5350, pp. 406–424. Springer, Heidelberg (2008)

18. Herrmann, M., May, A.: Attacking power generators using unravelled linearization: when do we output too much? In: Matsui, M. (ed.) ASIACRYPT 2009. LNCS, vol. 5912, pp. 487–504. Springer, Heidelberg (2009)

19. Herrmann, M., May, A.: Maximizing small root bounds by linearization and applications to small secret exponent RSA. In: Nguyen, P.Q., Pointcheval, D. (eds.) PKC 2010. LNCS, vol. 6056, pp. 53–69. Springer, Heidelberg (2010)

20. Howgrave-Graham, N.: Finding small roots of univariate modular equations revisited. In: Darnell, Michael J. (ed.) Cryptography and Coding 1997. LNCS, vol. 1355, pp. 131–142. Springer, Heidelberg (1997)

21. Kunihiro, N.: On optimal bounds of small inverse problems and approximate GCD problems with higher degree. In: Gollmann, D., Freiling, F.C. (eds.) ISC 2012. LNCS, vol. 7483, pp. 55–69. Springer, Heidelberg (2012)

22. Kunihiro, N., Honda, J.: RSA meets DPA: recovering RSA secret keys from noisy analog data. IACR Cryptology ePrint Archive, Report 2014/513 (2014). http://eprint.iacr.org/2014/513

23. Kunihiro, N., Shinohara, N., Izu, T.: Recovering RSA secret keys from noisy key bits with erasures and errors. In: Kurosawa, K., Hanaoka, G. (eds.) PKC 2013. LNCS, vol. 7778, pp. 180–197. Springer, Heidelberg (2013)

24. Lenstra, A.K., Lenstra Jr., H.W., Lovász, L.: Factoring polynomials with rational coefficients. Math. Ann. **261**, 515–534 (1982)

25. May, A.: New RSA vulnerabilities using lattice reduction methods. Ph.D. thesis, University of Paderborn (2003)

26. May, A.: Using LLL-reduction for solving RSA and factorization problems: a survey (2010). http://www.cits.rub.de/permonen/may.html

27. Nguyên, P.Q., Stern, J.: The two faces of lattices in cryptology. In: Silverman, J.H. (ed.) CaLC 2001. LNCS, vol. 2146, pp. 146–180. Springer, Heidelberg (2001)

28. Paterson, K.G., Polychroniadou, A., Sibborn, D.L.: A coding-theoretic approach to recovering noisy RSA keys. In: Wang, X., Sako, K. (eds.) ASIACRYPT 2012. LNCS, vol. 7658, pp. 386–403. Springer, Heidelberg (2012)

29. Sarkar, S., Sen Gupta, S., Maitra, S.: Partial key exposure attack on RSA – improvements for limited lattice dimensions. In: Gong, G., Gupta, K.C. (eds.) INDOCRYPT 2010. LNCS, vol. 6498, pp. 2–16. Springer, Heidelberg (2010)

30. Takayasu, A., Kunihiro, N.: Better lattice constructions for solving multivariate linear equations modulo unknown divisors. In: Boyd, C., Simpson, L. (eds.) ACISP. LNCS, vol. 7959, pp. 118–135. Springer, Heidelberg (2013)

31. Takayasu, A., Kunihiro, N.: Cryptanalysis of RSA with multiple small secret exponents. In: Susilo, W., Mu, Y. (eds.) ACISP 2014. LNCS, vol. 8544, pp. 176–191. Springer, Heidelberg (2014)

32. Wiener, M.J.: Cryptanalysis of short RSA secret exponents. IEEE Trans. Inf. theory **36**(3), 553–558 (1990)

Solving the Discrete Logarithm of a 113-bit Koblitz Curve with an FPGA Cluster

Erich Wenger$^{(\boxtimes)}$ and Paul Wolfger

Institute for Applied Information Processing and Communications,
Graz University of Technology, Inffeldgasse 16a, 8010 Graz, Austria
Erich.Wenger@iaik.tugraz.at, Paul.Wolfger@student.tugraz.at

Abstract. Using FPGAs to compute the discrete logarithms of elliptic curves is a well-known method. However, until to date only CPU clusters succeeded in computing new elliptic curve discrete logarithm records. This work presents a high-speed FPGA implementation that was used to compute the discrete logarithm of a 113-bit Koblitz curve. The core of the design is a fully unrolled, highly pipelined, self-sufficient Pollard's rho iteration function. An 18-core Virtex-6 FPGA cluster computed the discrete logarithm of a 113-bit Koblitz curve in extrapolated 24 days. Until to date, no attack on such a large Koblitz curve succeeded using as little resources or in such a short time frame.

Keywords: Elliptic curve cryptography · Discrete logarithm problem · Koblitz curve · Hardware design · FPGA · Discrete logarithm record

1 Introduction

It is possible to repeatedly fold a standard letter-sized sheet of paper at the midway point about six to seven times. In 2012, some MIT students [28] were able to fold an 1.2 Km long toilet paper 13 times. And every time the paper was folded, the number of layers on top of each other doubled. Therefore, the MIT students ended up with $2^{13} = 8192$ layers of paper on top of each other. And poor Eve's job was to manually count all layers one by one.

Similar principles apply in cryptography, although bigger numbers are involved. In Elliptic Curve Cryptography (ECC), where $\lceil \log_2 n \rceil$-bit private keys are used, Eve does not have to iterate through all possible n keys. Instead, Eve would use the more efficient parallelizable Pollard's rho algorithm that finishes in approximately $\sqrt{n}$ steps. The omnipresent question is how big n has to be such that even the most powerful adversaries are not able to reconstruct a private key. Especially in embedded, cost-sensitive applications, it is important to use keys that are only as large as necessary.

Discrete logarithms over elliptic curves were computed in the past [9,18] and several experimental baselines were established. Since then, committees [3,6] steadily increased the minimal n by simply applying Moore's law. However, it is necessary to practically compute discrete logarithms to check to which margin the current standards hold.

© Springer International Publishing Switzerland 2014
A. Joux and A. Youssef (Eds.): SAC 2014, LNCS 8781, pp. 363–379, 2014.
DOI: 10.1007/978-3-319-13051-4_22

The task of computing a discrete logarithm can be split into the work done by researchers and the work done by machines. This paper presents *both* a *novel hardware architecture* and the *discrete logarithm of a 113-bit Koblitz curve*. The highly pipelined, high-speed, and practically extensively tested ECC Breaker FPGA design was used to solve the discrete logarithm of a 113-bit Koblitz curve in extrapolated[1] 24 days using mere 18 FPGAs. Therefore, ECC breaker is the first FPGA design to be used to solve a large discrete logarithm. Further, based on ECC Breaker it is even possible to compute discrete logarithms of even larger binary-field elliptic curves. Substantiated by practical experimentation, this paper will make a notable contribution to the community's activity of breaking larger elliptic curves and carving new standards.

This paper is structured as follows: Sect. 2 gives an overview on related work. Section 3 revisits some mathematical foundations and Sect. 4 summarizes the experiments with different iteration functions. The most suitable iteration function was implemented in hardware, which is described in Sect. 5. As the design is flexible enough to attack larger elliptic curves, Sect. 6 gives runtime and cost approximations. Section 7 summarizes the learned lessons and Sect. 8 concludes the paper. Appendix A gives an overview of the targeted curve parameters and pseudo-randomly chosen target points.

2   Related Work

Certicom [10] introduced ECC challenges in 1997 to increase industry acceptance of cryptosystems based on the elliptic curve discrete logarithm problem (ECDLP). They published system parameters and point challenges for different security levels. Since then, the hardest solved Certicom challenges are ECCp-109 for prime-field based elliptic curves, done by Monico *et al.* using a cluster of about 10,000 computers (mostly PCs) for 549 days, and ECC2-109 for binary-field based elliptic curves, also done by Monico *et al.*, computing on a cluster of around 2,600 computers (mostly PCs) for around 510 days. Harley *et al.* [18] solved an ECDLP over a 109-bit Koblitz-curve Certicom challenge (ECC2K-108) with public participation using up to 9,500 PCs in 126 days.

As the Certicom challenges lie far apart (ECCp-131 is 2,048 times more complex than ECCp-109), also self-generated challenges have been broken. A discrete logarithm defined over a 112-bit prime-field elliptic curve was solved by Bos *et al.* [9], utilizing 200 Playstation 3s for 6 months. A single playstation reached a throughput of $42 \cdot 10^6$ iterations per second (IPS). *This work* presents the discrete logarithm of a 113-bit binary-field Koblitz curve that used 18 FPGAs for extrapolated 24 days and reached a throughput of $165 \cdot 10^6$ iterations per FPGA per second.

Further, several attempts and approximations were done in order to attack larger elliptic curves using FPGAs. Dormale *et al.* [24] targeted ECC2-113, ECC2-131, and ECC2-163 using Xilinx Spartan 3 FPGAs performing up to $20 \cdot 10^6$ IPS. Most promising is the work of Bailey *et al.* [5] who attempt to break

---

[1] The attack actually run for 47 days but not all FPGAs were active at all times.

ECC2K-130 using Nvidia GTX 295 graphics cards, Intel Core 2 Extreme CPUs, Sony PlayStation 3s, and Xilinx Spartan 3 FPGAs. Their FPGA implementation has a throughput of $33.7 \cdot 10^6$ IPS and was later improved by Fan et al. [12] to process $111 \cdot 10^6$ IPS. Other FPGA architectures were proposed by Güneysu et al. [16], Judge and Schaumont [20], and Mane et al. [21]. Güneysu et al.'s Spartan 3 architecture performs about $173 \cdot 10^3$ IPS, Judge and Schaumont's Virtex 5 architecture executes $2.87 \cdot 10^6$ IPS, and Mane et al.'s Virtex 5 architecture does $660 \cdot 10^3$ IPS.

So far, none of their FPGA implementations have been successful in solving ECDLPs. This work on the other hand presents a *practically tested architecture* which can be used to attack both *Koblitz curves* and *binary-field Weierstrass curves*.

3   Mathematical Foundations

To ensure a common vocabulary, it is important to revisit some of the basics. Hankerson et al. [17] and Cohen et al. [11] shall be consulted for further details.

3.1   Elliptic Curve Cryptography

This paper focuses on Weierstrass curves that are defined over binary extension fields $K = \mathbb{F}_{2^m}$. The curves are defined as $E/K : y^2 + xy = x^3 + ax^2 + b$, where a and b are system parameters and a tuple of x and y which fulfills the equation is called a point $P = (x, y)$. Using multiple points and the *chord-and-tangent* rule, it is possible to derive an additive group of order n, suitable for cryptography. The number of points on an elliptic curve is denoted as $\#E(K) = h \cdot n$, where n is a large prime and the cofactor h is typically in the range of 2 to 8. The core of all ECC-based cryptographic algorithms is a scalar multiplication $Q = kP$, in which the scalar $k \in [0, n-1]$ is multiplied with a point P to derive Q, where both points are of order n.

As computing $Q = kP$ can be costly, a lot of research was done on the efficient and secure computation of $Q = kP$. A subset of binary Weierstrass curves, known as Koblitz curves (also known as anomalous binary curves), have some properties which make them especially interesting for fast implementations. They may make use of a map $\sigma(x, y) = (x^2, y^2)$, $\sigma(\infty) = (\infty)$, an automorphism of order m known as a *Frobenius automorphism*. This means that there exists an integer λ^j such that $\sigma^j(P) = \lambda^j P$. Another automorphism, which is not only applicable to Koblitz curves, is the negation map. The negative of a point $P = (x, y)$ is $-P = (n-1)P = (x, x+y)$.

3.2   Elliptic Curve Discrete Logarithm Problem (ECDLP)

The security of ECC lies in the intractability of the ECDLP: Given the two points Q and P, connected by $Q = kP$, it should be practically infeasible to compute the scalar $0 \leq k < n$. As standardized elliptic curves are designed such that neither the Pohlig-Hellman attack [25], nor the Weil and Tate pairing attacks [13,23], nor the Weil descent attack [15] apply, the standard algorithm to compute a discrete logarithm is Pollard's rho algorithm [26].

The main idea behind Pollard's rho algorithm is to define an iteration function f that defines a random cyclic walk over a graph. Using Floyd's cycle-finding algorithm, it is possible to find a pair of colliding triples, each consisting of a point X_i and two scalars c_i and d_i. As $X_1 = X_2 = c_1P + d_1Q = c_2P + d_2Q$ and $(d_2 - d_1)Q = (d_2 - d_1)kP = (c_1 - c_2)P$, it is possible to compute $k = (c_1 - c_2)(d_2 - d_1)^{-1} \mod n$. Pollard's rho algorithm expects to encounter such a collision after $\sqrt{\frac{\pi n}{2}}$ steps.

In order to parallelize the attack efficiently, Van Oorschot and Wiener [30] introduced the concept of distinguished points. Distinguished points are a subset of points, which satisfy a particular condition. Such a condition can be a specific number of leading zero digits of a point's x-coordinate or a particular range of Hamming weights in normal basis. Those distinguished points are stored in a central database, but can be computed in parallel. The number of instances running in parallel is linearly proportional to the achievable speedup. Note that each instance starts with a random starting triple and uses one of the iteration functions f which are discussed in the following section.

4   Selecting the Iteration Function

As the iteration function will be massively parallelized and synthesized in hardware, it is crucial to evaluate different iteration functions and select the most suitable one. In this work, the iteration functions by Teske [29], Wiener and Zuccherato [31], Gallant et al. [14], and Bailey et al. [4] were checked for their practical requirements and achievable computation rates. Table 1 summarizes the experiments done in software on a 41-bit Koblitz curve.

What all iteration functions have in common is that they update a state, henceforth referred to as triple, consisting of two scalars $c_i, d_i \in [0..n-1]$ and a point $X_i = c_iP + d_iQ$. An iteration function f deterministically computes $X_{i+1} = f(X_i)$ and updates c_{i+1} and d_{i+1} accordingly, such that $X_{i+1} = c_{i+1}P + d_{i+1}Q$ holds. A requirement on f is that it should be easily computable and to have the characteristics of a random function.

Teske's r-adding walk [29] is a nearly optimal choice for an iteration function. It partitions the elliptic curve group into r distinct subsets $\{S_1, S_2, ..., S_r\}$ of roughly equal size. If a point X_i is assigned to S_j, the iteration function computes $f(X_i) = X_i + R[j]$, with $R[]$ being an r-sized table consisting of linear combinations of P and Q. After approximately $\sqrt{\frac{\pi n}{2}}$ steps, Teske's r-adding walk finds two colliding points for all types of elliptic curves.

The Frobenius automorphism of Koblitz curves can not only be used to speedup the scalar multiplication, but also to improve the expected runtime of a parallelized Pollard's rho by a factor of $\sqrt{m}$. Wiener and Zuccherato [31], Gallant et al. [14], and Bailey et al. [4] proposed iteration functions which should achieve this $\sqrt{m}$-speedup.

Table 1. Implementation results of all tested iteration functions.

Reference	Iteration function	Expected iterations	Measured iterations
Teske [29]	$f(X_i) = X_i + R[j]$	$929 \cdot 10^3$	$906 \cdot 10^3$
Wiener and Zuccherato [31]	$f(X_i) = \min_{0 \leq l < m} \{\sigma^l(X_i + R[j])\}$	$145 \cdot 10^3$	$147 \cdot 10^3$
Gallant et al. [14]	$f(X_i) = X_i + \sigma^l(X_i)$	$145 \cdot 10^3$	$166 \cdot 10^3$
Bailey et al. [4]	$f(X_i) = X_i + \sigma^{(l \mod 8)+3}(X_i)$	$145 \cdot 10^3$	$183 \cdot 10^3$

Wiener and Zuccherato [31] proposed to calculate $f(X_i) = \sigma^l(X_i + R[j]) \; \forall \, l \in [0, m-1]$ and choose the point, which has the smallest x-coordinate when interpreted as an integer. Gallant et al. [14] introduced an iteration function based on a labeling function $\mathcal{L}$, which maps the equivalence classes defined by the Frobenius automorphism to some set of representatives. The iteration function is then defined as $f(X_i) = X_i + \sigma^l(X_i)$, where $l = \text{hash}_m(\mathcal{L}(X_i))$. Bailey et al. [4] suggested to compute $f(X_i) = X_i + \sigma^{(l \mod 8)+3}(X_i)$ to reduce the complexity of the iteration function.

Additionally to the Frobenius automorphism, it is possible to use a negation map to improve the expected runtime by a factor of $\sqrt{2}$. The negation map compares X_i with $-X_i$ and selects the point with the smaller y-coordinate when interpreted as an integer. Although the potential speed-up seems very promising, there is an unfortunate challenge associated with the negation map; the problem of fruitless cycles which is discussed in Sect. 7.

In order to make sure that the potential iteration functions work as promised, a 41-bit Koblitz curve was used to evaluate the iteration functions with a C implementation on a PC (cf. Table 1). As labeling function $\mathcal{L}$, the Hamming weight of the x-coordinate in normal basis was used. The hash function was disregarded. Table 1 summarizes the average number of iterations (computing 100 ECDLPs) of all tested iteration functions using four parallel threads. The experiments showed that the average number of iterations of Gallant's and Bailey's iteration functions are 13-24 % higher compared to the iteration function by Wiener and Zuccherato. Additionally, with a probability of 14-20 % some of the parallel threads produced identical sequences of distinguished points. Restarting the threads regularly or on-demand would counter this problem. Not countering the problem of fruitless threads would increase the average runtime of Gallant's iteration function by another 29 %.

As Wiener and Zuccherato's iteration function achieved the best speed and does not have the problem of fruitless threads, it was chosen to be implemented in hardware. Additionally, by leaving out the automorphism, the hardware can be used to attack general binary-field Weierstrass curves as well.

5   ECC Breaker Hardware

Before the actual hardware architecture and its most critical components are discussed, it is important to review the basic design assumptions and core ideas of the ECC Breaker design.

5.1   Basic Assumptions and Decisions

ASIC Vs FPGA Design. In literature it is possible to find a lot of FPGA and ASIC designs optimized for some crucial characteristic. Some authors even dare to compare FPGA and ASIC results. However, several of the largest components in ASIC designs, e.g., registers, RAM, or integer multipliers, are for free in an FPGA design. For instance, every slice comes with several registers. Therefore, adding pipeline stages in a logic-heavy FPGA design is basically for free. For this paper, Xilinx Virtex-6 ML605 evaluation boards were chosen because of availability. Note that all following design decisions were made to maximize the performance of ECC Breaker on this particular board.

Design Goals. As Pollard's rho algorithm is perfectly parallelizeable, the design goal clearly is to maximize the throughput per (given) area. Note that the speed (iterations per seconds) of an attack is linearly proportional to the throughput and inversely proportional to the chip area (more instances per FPGA also increase the speed). Therefore, the most basic design decision was whether to go for many small or a single large FPGA design.

Core Idea. In earlier designs, many area-efficient architectures were considered, each coming with a single $\mathbb{F}_{2^m}$ multiplier, $\mathbb{F}_{2^m}$ squarer, and $\mathbb{F}_{2^m}$ adder per instance. The main problems of these designs were the costly multiplexers and the low utilization of the hardware. Therefore the design principle of ECC Breaker is a single, fully unrolled, fully pipelined iteration function. *In order to keep all pipeline stages busy, the number of pipeline stages equals the amount of triples processed within ECC Breaker.*

ECC Breaker Versus Related Work. (i) In the current setup, the interface between ECC Breaker and desktop is a simple, slow, serial interface. This might be a challenge for related implementations but not for ECC Breaker. The implemented iteration function does not require mechanisms to detect fruitless cycles or threads and the on-chip distinguished points (triple) storage assures that only distinguished triples have to be read. (ii) The proposal by Fan *et al.* [12] to perform simultaneous inversion and therefore save some finite-field multipliers was not picked because their proposal introduces many multiplexers and a complex control logic, and might not fully utilize the available hardware. (iii) Further, ECC Breaker stands on its own by coming with prime field $\mathbb{F}_n$ arithmetic which has only a minor impact on the size of the hardware ($< 3\%$). Additionally, it proved indispensable during development that the generated distinguished triples could be easily verified.

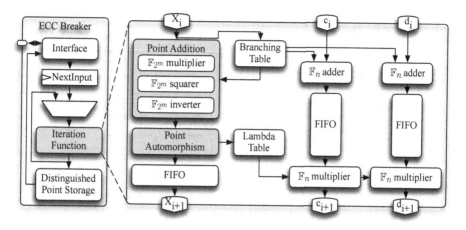

Fig. 1. Top-level view of ECC breaker on the left. Iteration Function on the right.

Generalization of ECC Breaker. Although the current version of ECC Breaker is carefully optimized for a 113-bit binary-field Koblitz curve, the underlying architecture and design approach is also suitable for larger elliptic curves, e.g., a 131-bit Koblitz curve. In Sect. 6, approximations of the expected runtimes and potential costs to attack such a larger curve are given.

5.2   The Architecture

The basic architecture of ECC Breaker is presented in Fig. 1. The core of ECC Breaker is a circular, self-sufficient, fully autonomous iteration function. A (potentially slow) interface is used to write the `NextInput` register. If the current stage of the pipeline is not active, the pipeline is fed with the triple from the `NextInput` register. This is done until all stages of the pipeline process data. If a point is distinguished, it is automatically added to the distinguished point storage (a sufficiently large block RAM). At periodic but time-insensitive intervals the host computer can read all distinguished points that accumulated within the storage.

The iteration function itself consists of several components: a point addition module, a point automorphism module, two $\mathbb{F}_n$ adders, two $\mathbb{F}_n$ multipliers, two block-RAM based tables and several block-RAM based FIFOs to care for data-dependencies.

5.3   ECC Breaker Components

Point Addition Module. No matter which iteration function is selected, an affine point addition module is always necessary. In the case of binary Weierstrass curves, the formulas for a point addition $(x_3, y_3) = (x_1, y_1) + (x_2, y_2)$ are $x_3 = \mu^2 + \mu + x_1 + x_2 + a$ and $y_3 = \mu(x_1 + x_3) + x_3 + y_1$, with $\mu = (y_1 + y_2)/(x_1 + x_2)$. Special cases of points being equivalent, inverse of each other, or the identity are not handled by the hardware as they are very unlikely to occur in practice.

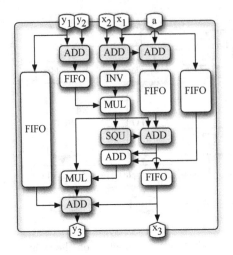

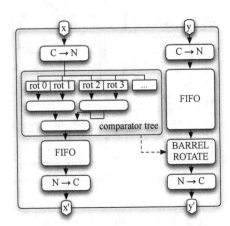

Fig. 2. Simplified point addition module. The grey shaded blocks are without registers.

Fig. 3. Point automorphism unit with m comparator units.

Figure 2 shows the implemented point addition module which directly maps the formulas from above. Two $\mathbb{F}_{2^m}$ multiplier, one $\mathbb{F}_{2^m}$ inverter, and five FIFOs are necessary to compute a point addition in 184 cycles. Note that it is not possible to get rid of the costly inversion as the result of the point addition must be available in affine coordinates (cf. [24]).

Point Automorphism Module. In order to speed-up Pollard's rho algorithm for Koblitz curves, it is necessary to uniquely map m points from the same equivalence class to a single point. As ECC Breaker follows Wiener and Zuccherato's [31] approach of interpreting the field elements as integers and comparing them, it was necessary to design a module that does m squarings and m comparisons as efficiently as possible. This module relies on normal basis representation and is depicted in Fig. 3. It converts x and y into normal basis, finds the smallest x within the normal basis, rotates y appropriately, and transforms x and y back into a canonical polynomial representation. As the m exponents of x ($x^{(2^i)}$) are computed by simple rewiring (x rotated by i steps), and the smallest x is found using a binary comparison tree, no canonical $\mathbb{F}_{2^m}$ squarer is needed.

As optimization, only the $t = 70$ most significants bits of x are compared. This means that if two numbers with t equivalent most significant bits are compared, no unique minimum is found. However, the probability for that is only 2^{-t}. For $i = \sqrt{\frac{\pi n}{2m}}$ iterations and $m \cdot i$ comparisons, the probability for not selecting the smaller value is only $1 - (1 - 2^{-t})^{m \cdot i} = 0.00081$ for $m = 113$.

In respect to the overall design, the point automorphism module requires 14 % of all slices and is about 5.6 times smaller than the point addition module (in terms of slices). The majority of the point automorphism module is the comparator tree. The basis transformations are fairly cheap and make up only 20 % of the point automorphism module.

$\mathbb{F}_{2^m}$ **inverse.** An Euclidean-based inversion algorithm is not deterministic and therefore hard to compute with a pipelined hardware module. Therefore, ECC Breaker computes the inverse using Fermat's little theorem; an inversion by exponentiation. Fortunately, an exponentiation with 2^{m-2} can be computed very efficiently using Itoh and Tsujii's [19] exponentiation trick, needing 112 squarers and 8 multipliers for $m = 113$: $a = a^{2^1-1} \rightarrow a^{2^2-1} \rightarrow a^{2^3-1} \rightarrow a^{2^6-1} \rightarrow a^{2^7-1} \rightarrow a^{2^{14}-1} \rightarrow a^{2^{28}-1} \rightarrow a^{2^{56}-1} \rightarrow a^{2^{112}-1} \rightarrow a^{2^{113}-2} = a^{-1}$.

$\mathbb{F}_{2^m}$ **normal basis.** The advantage of a normal basis is that a squaring is a simple rotation operation. The disadvantage of a normal basis is that a $\mathbb{F}_{2^m}$ multiplication is fairly complex to compute. ECC breaker uses per default a normal, canonical polynomial representation.

Only within the point automorphism module the normal basis rendered advantageous. The necessary matrix multiplication for a basis transformation can be implemented very efficiently. As the matrix is constant, on average $m/2$ of the input signals are xored per output signal. Based on the results from Table 2, 666 LUTs are needed per basis transformation.

Experiments show that the normal basis could also reduce the area of the consecutive squaring units within the $\mathbb{F}_{2^m}$ inverse. The 14, 28, and 56 squarers currently need 1 582, 3 164, and 6 328 LUTs, respectively. Doing two basis transformations and a rotation within normal basis would actually save area. Also, accumulating the two transformation matrices into a single matrix would further reduce the area. However, as all squarers together only need 11 % of all slices and 10 % of all LUTs, the potential area improvement is rather limited. Therefore, contrary to [5,12], ECC Breaker only uses a normal basis number representation within the point automorphism module.

$\mathbb{F}_{2^m}$ **multiplier.** As in total ten $\mathbb{F}_{2^m}$ multipliers are needed for the point addition module and the $\mathbb{F}_{2^m}$ inversion module, the $\mathbb{F}_{2^m}$ multipliers have the largest effect on the area footprint of the ECC Breaker design. For ECC Breaker, the following multiplier designs on a Virtex-6 FPGA were evaluated (post-synthesis): (i) A simple 113-bit parallel polynomial multiplier needs 5,497 LUTs. (ii) A Mastrovito multiplier [22] interprets the $\mathbb{F}_{2^m}$ multiplication as matrix multiplication and performs both a polynomial multiplication and the reduction step simultaneously. Unfortunately, it needs 7,104 LUTs. A polynomial multiplication and reduction with the used pentanomial can be implemented much more efficiently. (iii) Bernstein [7] combines some refined Karatsuba and Toom recursions for his batch binary Edwards multiplier. The code from [8] for a 113-bit polynomial multiplier needs 4,409 LUTs. (iv) Finally, the best results were achieved with a slightly modified binary Karatsuba multiplier, described by Rodrıguez-Henrıquez and Koç [27]. Their recursive algorithm was applied down to a 16×16-bit

Table 2. Hierarchical representation of final hardware design (post place-and-route).

Entity	Instances	Cycles	Registers	LUTs	Slices
top	1		58,784	62,655	100%
iteration function	1	210	57,332	60,826	98%
point addition	1	184	35,691	43,177	79%
$\mathbb{F}_{2^m}$ inverse	1	168	29,809	35,126	65%
$\mathbb{F}_{2^m}$ multiplier	8	7	14,958	28,273	51%
$\mathbb{F}_{2^m}$ squarer	112	1	12,543	6,325	11%
$\mathbb{F}_{2^m}$ multiplier	2	7	3,738	7,127	13%
point automorphism	1	16	15,189	14,372	14%
comparator tree	1	7	13,238	10,529	10%
basis transformation	4	1	452	2,664	3%
$\mathbb{F}_n$ multiplier	2	26	3,650	2,000	2%
$\mathbb{F}_n$ adder	2	9	1,308	1,051	1%

multiplier level, which is synthesized as standard polynomial multiplier. The formulas for the resulting multiplier structure are given in Appendix B. The design only requires 3,757 LUTs. At last the design was equipped with several pipeline stages such that it can be clocked with high frequencies.

$\mathbb{F}_{2^m}$ **multiplier.** Computing prime-field multiplications in hardware can be a troublesome and very resource-intensive task. In the case of a Virtex-6, dedicated DSP slices were used for integer multiplications. As a result, the two $\mathbb{F}_n$ multipliers are very resource efficient, requirering only 2×145 DSP slices and 2% of all slices.

6   Results and Transferability of Results

The construction of the current ECC Breaker design was an iterative process that continuously optimized the speed, the area, and the power consumption of all components. To make maximal use of the available resources, the available block RAMs and DSP slices were used whenever possible. Table 2 gives the number of registers and LUTs needed for all components of a 113-bit Koblitz-curve ECC Breaker design. The design was synthesized and mapped with Xilinx ISE 14.6.

ECC Breaker requires (post place-and-route) 47% of all available slices (17,782/37,680), 41% of all LUTs (62,657/150,720), 19% of all registers (58,788/301,440), 37% of all DSP macros (290/768), and less than 10% of all block RAMs. The biggest components are the point addition module and the $\mathbb{F}_{2^m}$ inverse module. Although already extensively optimized, the 10 $\mathbb{F}_{2^m}$ multipliers require about 64% of all slices. As the place-and-route tool performs optimizations across module borders, the slice counts of all components are just approximations.

Table 3. ECC Breaker on different FPGAs (post synthesis).

Series	Part Number	LUTs	of total	Registers	max Freq. [MHz]	Develop. Kit	Price [USD]
Point Addition w/o Automorphism							
Virtex-6	XC6VLX240T	57,294	38%	37,060	261	ML605	2,495
Spartan-6	XC6SLX150T	57,686	62%	37,715	147	LX150T	995
Point Addition w/ Automorphism							
Virtex-6	XC6VLX240T	86,409	57%	55,881	261	ML605	2,495
Artix-7	XC7A200T	86,478	64%	55,848	264	AC701	999
Virtex-7	XC7VX485T	86,391	28%	55,704	313	VC707	3,495
Kintex-7	XC7K325T	86,391	42%	55,704	313	KC705	1,695

6.1 ECC Breaker on Different FPGAs

As the VHDL code is portable, the suitability of ECC Breaker was also evaluated for other Xilinx FPGAs. Although size was a secondary optimization goal, ECC Breaker has been designed for a particular Virtex-6 FPGA. So it does not come with surprise that ECC Breaker does only fit into certain FPGAs that come with certain features. For instance, the used ML605 development board incorporates a Virtex-6 FPGA that comes with 768 DSPs (of which 290 are used).

An overview of synthesis results on different FPGAs is given in Table 3. Fortunately, the ECC Breaker design is perfectly suitable for all kind of the latest Xilinx Virtex-7 (targets high performance designs), Kintex-7 (targets best performance per cost), and Artix-7 (low cost) FPGA devices. The Virtex-7 and Kintex-7 development boards can even fit multiple ECC Breaker instances. However, the Kintex-7 KC705 development board fits the most instances per cost. The prices, taken from www.avnet.com [1], do not contain taxes and do not contain potential bulk discounts.

Also smaller FPGAs were considered. Especially the Spartan-6 LX150T-series is of special interest as they are part of SciEngines RIVYERA's S6-LX150 FPGA cluster [2]. Unfortunately, Spartan-6 FPGAs come with 180 DSPs at maximum and therefore the LX150T could be only used to attack non-Koblitz binary-field curves.

6.2 Expected Runtimes

Using the synthesis results from Table 3, several performance approximations of different elliptic-curve targets were performed (cf. Table 4). Note that the results are very optimistic as they are post-synthesis, the FPGAs are running at the maximum frequency, and a single FPGA contains multiple instances of ECC Breaker.

Computing a discrete logarithm of a 113-bit Koblitz curve (denoted as ECC2K-112) can be done in about 3 days, when a cluster of 256 Spartan-6 FPGAs would be available. With the same cluster, it would be possible to attack a 113-bit binary-field curve (denoted as ECC2-112) in 28 days.

Table 4. Approximations of best-case runtimes and costs for different targets.

Series	Target	Freq. [MHz]	Inst-ances	FPGAs	Costs [10^3 USD]	Iterations	exp. Runt. [days]
Virtex-6	ECC2K-112	261	1	17	42	$8.5 \cdot 10^{15}$	22
Spartan-6	ECC2K-112	147	1	256	255	$8.5 \cdot 10^{15}$	3
Virtex-6	ECC2-112	261	2	17	42	$90.3 \cdot 10^{15}$	118
Spartan-6	ECC2-112	147	1	256	255	$90.3 \cdot 10^{15}$	28
Kintex-7	ECC2K-130	313	2	590	1,000	$4,055.4 \cdot 10^{15}$	127
Kintex-7	ECC2-131	313	2	5,900	10,001	$46,239.1 \cdot 10^{15}$	145
Kintex-7	ECC2-163	313	1	589,971	1,000,001	$3,030.3 \cdot 10^{21}$	189,934

The Certicom challenge ECC2K-130 targeted by Bailey *et al.* [5] and Fan *et al.* [12] can be computed in 127 days, assuming a budget of one million USD. The Certicom challenge ECC2-131 can be computed in 145 days, assuming a budget of ten million USD. Targeting the smallest standardized NIST curve B-163 would take 520 years, assuming a budget of one billion USD, which would be a reasonable budget of certain agencies. However, if there were one billion USD to be spent, it would be more reasonable to go for a dedicated ASIC design.

7   Lessons Learned

After spending months of research time on continuously improving the ECC Breaker design, there are some important insights that need to be discussed.

Maximum Achievable Frequency. As ECC Breaker is a fairly complex and large design, the hardware synthesizer reached its limit when it comes to maximum frequency approximations. In most cases, it was only possible to reach a fraction of the theoretically given frequency after mapping and routing.

Limited by Power Consumption. However, it was not even possible to run the ML605 development boards at maximum post-map-and-route frequency. An average power consumption of about 12 A at the internal power supply resulted in a sporadic emergency switch-off with which the power controller protected itself. This is rather strange, considering that the internal power supply is designed to support 20 Ampere. Therefore it was necessary to reduce the clock frequency further in order to achieve a stable operation. Finally, ECC Breaker was running at 165 MHz even though the synthesizer approximated a maximum clock frequency of 275 MHz.

Multiple Instances per FPGA. It was previously mentioned that some FPGAs fit multiple ECC Breaker instances. However, it has yet to be tested whether two ECC Breaker instances per FPGA at lower clock frequency outperform a single instance, clocked with a higher frequency. Especially under consideration of the previously discussed routing and power problems, multiple ECC Breaker instances per FPGA might not be feasible.

Fruitless Cycles. An initial implementation made use of a negation map. However, the possibility of a fruitless cycle in which $X_{i+1} = f(X_i) = X_i + R[j] - R[j] = X_i$ rendered the hardware implementation with negation map useless. The probability that a fruitless cycle occurs is $p = \frac{1}{2 \cdot m \cdot r}$, r being the number of branches and m the size of the automorphism. The probability to encounter a fruitless cycle after i iterations is $1 - (1 - p)^i$. Given 1,024 branches, an automorphism of size 113, and a clock rate of 165 MHz, the iteration function was trapped in a cycle with a probability of 99 % after less than one second. It is subject to future research how to efficiently get rid of the fruitless-cycle problem in a fully pipelined hardware design.

8   Conclusion

This work presents a circular, self-sufficient, highly pipelined, fully autonomous hardware design that was used to practically compute the discrete logarithm of a 113-bit Koblitz curve within extrapolated 24 days on mere 18 Virtex-6 FPGAs. However, because of the scalability and adaptability of ECC Breaker, even more complex results can be expected. This work will bring the community one step closer to solving the ECC2K-130 challenge.

Acknowledgments. The authors are really grateful to Wolfgang Kastl and Jürgen Fuß from the University of Applied Sciences Upper Austria who provided sixteen ML605 boards and continuously supported us and the authors would like to thank the reviewers for their helpful comments.

This work has been supported in part by the Austrian Government through the research program FIT-IT under the project number 835917 (project NewP@ss), by the European Commission through the FP7 program under project number 610436 (project MATTHEW), and the Secure Information Technology Center-Austria (A-SIT).

A   Targeted Curve and Target Point Pair Selection

The selection of the curve parameters for a 113-bit Koblitz curve are quite straightforward. However, to proof that the discrete logarithm was actually computed without knowing it in advance, a point generation function was needed. The Sage code in Listing 1.1 was used to deterministically and pseudo-randomly generate two points with order n using Sage 5.12. As P and Q are generated pseudo-randomly, their discrete logarithm is unknown. The Sage script also checks the point orders and the validity of the computed result. Table 5 summarizes all parameters needed for the discrete logarithm computation.

Table 5. Curve parameters of targeted 113-bit elliptic curve.

m	113
irreducible polynomial	$x^{113} + x^5 + x^3 + x^2 + 1$
irreducible polynomial	0x20000000000000000000000000000002d
elliptic curve E	$y^2 + xy = x^3 + ax^2 + b$
curve parameter a	1
curve parameter b	1
order n	0xffffffffffffffffdbf91af6dea73
cofactor h	2
point $P.x$	0x0a27644cfced9667d2084f8be061c
point $P.y$	0x0d5acd887d5585dd75c5d07165699
point $Q.x$	0x189037f88aed8e32400b16d2b1a6e
point $Q.y$	0x00e4718fb1e9f50f845ff162ff59c
scalar k such that $Q = kP$	0x276c233740d817000b80478fde46

B   Binary Karatsuba $\mathbb{F}_{2^{113}}$ Multiplier

Algorithm 1 gives the top-level $\mathbb{F}_{2^{113}}$ multiplier formulas. KS64, KS32, and KS16 are 64-bit, 32-bit, and 16-bit binary Karatsuba multipliers, respectively.

Algorithm 1. Calculate $c = a \times b$, with a, b being m-bit binary polynomials.

Input: a, b
Output: $c = a \times b$
1: $m_{ab1} \leftarrow (a[112..64] \oplus a[63..0]) \times (b[112..64] \oplus b[63..0])$ ▷ KS64
2: $c_{l1} \leftarrow a[63..0] \times b[63..0]$ ▷ KS64
3: $c_{l2} \leftarrow a[95..64] \times b[95..64]$ ▷ KS32
4: $c_{l3} \leftarrow a[111..96] \times b[111..96]$ ▷ KS16
5: $m_{ab2} \leftarrow (a[95..64] \oplus a[111..96]) \times (b[95..64] \oplus b[111..96])$ ▷ KS32

6: $m_{a3} \leftarrow b[112] \times a[111..96]$ 12: $m_2 \leftarrow m_{ab2} \oplus c_{l2} \oplus c_3$
7: $m_{b3} \leftarrow a[112] \times b[111..96]$ 13: $c_2[62..0] \leftarrow c_{l2}$
8: $m_3 \leftarrow m_{a3} \oplus m_{b3}$ 14: $c_2[97..64] \leftarrow c_3$
9: $c_3[32] \leftarrow a[112] \times b[112]$ 15: $c_2[94..32] \leftarrow c_2[94..32] \oplus m_2$
10: $c_3[30..0] \leftarrow c_{l3}$
11: $c_3[31..16] \leftarrow c_3[31..16] \oplus m_3$ 16: $m_1 \leftarrow m_{ab1} \oplus c_{l1} \oplus c_2$
                                    17: $c[126..0] \leftarrow c_{l1}$
                                    18: $c[225..128] \leftarrow c_2$
                                    19: $c[190..64] \leftarrow c[190..64] \oplus m_1$

Listing 1.1. Sage code to verify P, Q, and $Q = kP$.

```
m=113
a=1
b=1
h=2 n=0xfffffffffffffffffdbf91af6dea73
k=0x276c233740d817000b80478fde46 FF =
sage.rings.finite_rings.finite_field_ext_pari.\
    FiniteField_ext_pari;
K = FF(2**m,'x')
x=K.gen()
E = EllipticCurve(K, [1,a,0,0,b])

def str_to_poly(str):
    I=Integer(str, base=16)
    v=K(0)
    for i in range(0,K.degree()):
        if (I >> i) & 1 > 0:
            v = v + x^i
    return v

def poly_to_str(poly):
    vec=poly._vector_()
    string =""
    for i in range(0,len(vec)):
        string = string + str(vec[len(vec) - i - 1])
    return hex(Integer(string, base=2))

import hashlib
PX = str_to_poly(hashlib.sha256(str(0)).hexdigest())
PY=PolynomialRing(K, 'PY').gen()
P_ROOTS = (PY^2+PX*PY+PX^3+a*PX^2+b).roots()
P=E([PX,P_ROOTS[0][0]]); P=P*h

QX = str_to_poly(hashlib.sha256(str(1)).hexdigest())
Q_ROOTS = (PY^2+QX*PY+QX^3+a*QX^2+b).roots()
Q=E([QX,Q_ROOTS[0][0]]); Q=Q*h

print 'P.x:', poly_to_str(P[0])
print 'P.y:', poly_to_str(P[1])
print 'Q.x:', poly_to_str(Q[0])
print 'Q.y:', poly_to_str(Q[1])
print k*P==Q, is_prime(n), (n*P).is_zero(), (n*Q).is_zero()
```

References

1. Avnet Inc, Feb 2014. http://www.avnet.com/
2. SciEngines GmbH, Feb 2014. http://www.sciengines.com/
3. Babbage, S., Catalano, D., Cid, C., de Weger, B., Dunkelman, O., Gehrmann, C., Granboulan, L., Güneysu, T., Hermans, J., Lange, T., Lenstra, A., Mitchell, C., Näslund, M., Nguyen, P., Paar, C., Paterson, K., Pelzl, J., Pornin, T., Preneel, B., Rechberger, C., Rijmen, V., Robshaw, M., Rupp, A., Schläffer, M., Vaudenay, S., Vercauteren, F., Ward. M.: ECRYPT II yearly report on algorithms and keysizes (2011–2012). http://www.ecrypt.eu.org/ Sep 2012
4. Bailey, D.V., Baldwin, B., Batina, L., Bernstein, D.J., Birkner, P., Bos, J.W., van Damme, G., de Meulenaer, G., Fan, J., Güneysu, T., Gurkaynak, F., Kleinjung, T., Lange, T., Mentens, N., Paar, C., Regazzoni, F., Schwabe, P., Uhsadel, L.: The certicom challenges ECC2-X. IACR cryptology ePrint archive, Report 2009/466 (2009)
5. Bailey, D.V., Batina, L., Bernstein, D.J., Birkner, P., Bos, J.W., Chen, H.-C., Cheng, C.-M., van Damme, G., de Meulenaer, G., Perez, L.J.D., Fan, J., Güneysu, T., Gurkaynak, F., Kleinjung, T., Lange, T., Mentens, N., Niederhagen, R., Paar, C., Regazzoni, F., Schwabe, P., Uhsadel, L., Herrewege, A.V., Yang. B.-Y.: Breaking ECC2K-130. IACR cryptology ePrint archive, Report 2009/541 (2009)
6. Barker, E., Roginsky, A.: Recommendation for cryptographic key generation. NIST Spec. Publ. **800**, 133 (2012)
7. Bernstein, D.J.: Batch binary edwards. In: Halevi, S. (ed.) CRYPTO 2009. LNCS, vol. 5677, pp. 317–336. Springer, Heidelberg (2009)
8. Bernstein, D.J.: Binary Batch Edwards 113-bit Multiplier (May 2009). http://binary.cr.yp.to/bbe251/113.gz
9. Bos, J.W., Kaihara, M.E., Kleinjung, T., Lenstra, A.K., Montgomery, P.L.: Solving a 112-bit prime elliptic curve discrete logarithm problem on game consoles using sloppy reduction. Int. J. Appl. Crypt. **2**(3), 212 (2012)
10. Certicom Research: The Certicom ECC Challenge (Nov 1997). https://www.certicom.com/index.php/the-certicom-ecc-challenge
11. Cohen, H., Frey, G., Avanzi, R., Doche, C., Lange, T., Nguyen, K., Vercauteren, F. (eds.): Handbook of Elliptic and Hyperelliptic Curve Cryptography: Discrete Mathematics and its Applications. Handbook of Elliptic and Hyperelliptic Curve Cryptography. Chapman and Hall/CRC, Boca Raton (2006)
12. Fan, J., Bailey, D.V., Batina, L., Güneysu, T., Paar, C., Verbauwhede, I.: Breaking elliptic curve cryptosystems using reconfigurable hardware. In: Field Programmable Logic and Applications (FPL), pp. 133–138. IEEE (2010)
13. Frey, G., Rück, H.-G.: A remark concerning m-divisibility and the discrete logarithm in the divisor class group of curves. Math. Comput. **62**(206), 865–874 (1994)
14. Gallant, R., Lambert, R., Vanstone, S.: Improving the parallelized Pollard lambda search on anomalous binary curves. Math. Comput. Am. Math. Soc. **69**(232), 1699–1705 (2000)
15. Gaudry, P., Hess, F., Smart, N.P.: Constructive and destructive facets of weil descent on elliptic curves. J. Cryptol. **15**(1), 19–46 (2002)
16. Güneysu, T., Paar, C., Pelzl, J.: Attacking elliptic curve cryptosystems with special-purpose hardware. In: FPGA, pp. 207. ACM Press (2007)
17. Hankerson, D., Vanstone, S., Menezes, A.J.: Guide to Elliptic Curve Cryptography. Springer, New York (2004)

18. Harley, R.: Elliptic curve discrete logarithms: ECC2K-108 (2000). http://cristal. inria.fr/harley/ecdl7/readMe.html

19. Itoh, T., Tsujii, S.: A fast algorithm for computing multiplicative inverses in $GF(2^m)$ using normal bases. Inf. Comput. **78**(3), 171–177 (1988)

20. Judge, L., Schaumont, P.: A flexible hardware ECDLP engine in bluespec. In: Special-Purpose Hardware for Attacking Cryptographic Systems (SHARCS) (2012)

21. Mane, S., Judge, L., Schaumont, P.: An integrated prime-field ECDLP hardware accelerator with high-performance modular arithmetic units. In: Reconfigurable Computing and FPGAs, pp. 198–203. IEEE, Nov. 2011

22. Mastrovito, E.D.: VLSI designs for multiplication over finite fields $GF(2^m)$, pp. 297–309. In: Applied Algebra, Algebraic Algorithms and Error-Correcting Codes (1988)

23. Menezes, A.J., Okamoto, T., Vanstone, S.A.: Reducing elliptic curve logarithms to logarithms in a finite field. Trans. Inf. Theory **39**(5), 1639–1646 (1993)

24. Meurice de Dormale, G., Bulens, P., Quisquater, J.-J.: Collision search for elliptic curve discrete logarithm over $GF(2^m)$ with FPGA. In: Paillier, P., Verbauwhede, I. (eds.) CHES 2007. LNCS, vol. 4727, pp. 378–393. Springer, Heidelberg (2007)

25. Pohlig, S.C., Hellman, M.E.: An improved algorithm for computing logarithms over $GF(p)$ and its cryptographic significance. Trans. Inf. Theory **24**(1), 106–110 (1978)

26. Pollard, J.M.: A monte carlo method for factorization. BIT Numer. Math. **15**(3), 331–334 (1975)

27. Rodrıguez-Henrıquez, F., Koç, Ç.: On fully parallel karatsuba multipliers for $GF(2^m)$, pp. 405–410. In Computer Science and Technology (2003)

28. Stier, C.: Students break record by folding toilet paper 13 times. http://www. newscientist.com/blogs/nstv/2012/01/paper-folding-limits-pushed.html. Jan 2012

29. Teske, E.: Speeding up pollard's rho method for computing discrete logarithms. In: Buhler, J.P. (ed.) ANTS 1998. LNCS, vol. 1423, pp. 541–554. Springer, Heidelberg (1998)

30. van Oorschot, P.C., Wiener, M.J.: Parallel collision search with cryptanalytic applications. J. Cryptol. **12**(1), 1–28 (1999)

31. Wiener, M., Zuccherato, R.J.: Faster attacks on elliptic curve cryptosystems. In: Tavares, S., Meijer, H. (eds.) SAC 1998. LNCS, vol. 1556, pp. 190–200. Springer, Heidelberg (1999)

Author Index

Printed in the United States
By Bookmasters